WITHDRAWN

Advances in
Physical Organic Chemistry

Advances in Physical Organic Chemistry

Volume 13

Edited by

V. Gold

Department of Chemistry
King's College
University of London

Associate Editor

D. Bethell

The Robert Robinson Laboratories
University of Liverpool

1976

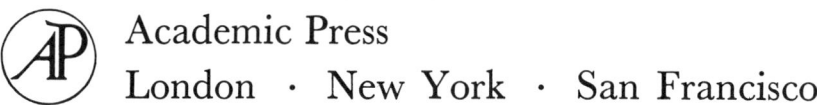

Academic Press
London · New York · San Francisco

A Subsidiary of Harcourt Brace Jovanovich, Publishers

ACADEMIC PRESS INC. (LONDON) LTD
24-28 Oval Road, London NW1

United States Edition published by
ACADEMIC PRESS INC.
111 Fifth Avenue, New York, New York 10003

Copyright © 1976 By Academic Press Inc. (London) Ltd

All Rights Reserved

No part of this book may be reproduced in any form by
photostat, microfilm, or any other means,
without written permission from the publishers

Library of Congress Catalog Card Number: 62—22125
ISBN 0-12-033513-1

PRINTED IN GREAT BRITAIN BY
WILLIAM CLOWES & SONS LIMITED
LONDON, COLCHESTER AND BECCLES

Contributors to Volume 13

N. L. Allinger, *Department of Chemistry, University of Georgia, Athens, Georgia 30602, U.S.A.*

Edward M. Arnett, *Department of Chemistry, University of Pittsburgh, Pittsburgh, Pennsylvania 15260, U.S.A.*

A. J. Bard, *Department of Chemistry, University of Texas, Austin, Texas 78712, U.S.A.*

Janos H. Fendler, *Department of Chemistry, Texas A & M University, College Station, Texas 77843, U.S.A.*

A. Ledwith, *Donnan Laboratories, University of Liverpool, Liverpool L69 3BX.*

Steven N. Rosenthal, *Department of Chemistry, Texas A & M University, College Station, Texas 77843, U.S.A.*

Gianfranco Scorrano, *Istituto di Chimica Organica, Centro CNR Meccanismi di Reazioni Organiche, Università di Padova, Italy.*

H. J. Shine, *Department of Chemistry, Texas Tech University, Lubbock, Texas 79409, U.S.A.*

Contents

Contibutors to Volume 13 v

Calculation of Molecular Structure and Energy by Force-Field Methods — *N. L. Allinger*

1. Introduction 2
2. Calculational Approaches to Molecular Structure . . 4
3. Summary of Studies Reported 26
4. Alkanes 28
5. Alkenes 47
6. Alkynes 50
7. Molecules containing Delocalized Electronic Systems . 52
8. Other Hydrocarbons 58
9. Molecules containing Heteroatoms 58
10. Reaction Rates 72
11. Conclusions 75

Protonation and Solvation in Strong Aqueous Acids — *Edward M. Arnett and Gianfranco Scorrano*

1. Introduction 84
2. Standard Free Energies of Proton Transfer and Solvation in Aqueous Acid 88
3. Measured Heats of Ionization in Various Acidic Media . 106
4. Heats of Solvation 131
5. Acidity Functions and Solvation 142
6. Summary 146

Formation, Properties and Reactions of Cation Radicals in Solution — *A. J. Bard, A. Ledwith and H. J. Shine*

1. Formation of Cation Radicals 156
2. Electrochemical Methods of Formation and Investigation of Cation Radicals 197
3. Physical Properties of Cation Radicals 210
4. Electron Transfer Reactions of Cation Radicals . . . 218
5. Reactions of Cation Radicals with Nucleophiles . . . 226
6. Cation-Radicals from Bypyridylium Salts 254
7. Concluding Remarks 264

^{13}C NMR Spectroscopy in Macromolecular Systems of Biochemical Interest — *Steven N. Rosenthal and Janos H. Fendler*

1. Introduction 280
2. Carbohydrates 287
3. Nucleic Acids and Constituents 323
4. Proteins and their Residues 350
5. Lipids 382
6. Macromolecular Model Systems 390

Calculation of Molecular Structure and Energy by Force-Field Methods

N. L. ALLINGER

Department of Chemistry, University of Georgia, Athens, Georgia 30602, U.S.A.

1. Introduction 2
   Molecular Structure 2
   Molecular Energy 3
2. Calculational Approaches to Molecular Structure 4
   The *Ab Initio* Method, and Simplifications of it 4
   The Force-Field Method 7
   The Force-Field 10
   The Mechanical Model 12
   Energy Minimization Schemes 22
3. Summary of Studies Reported 26
4. Alkanes 28
   Structural Results 28
   Heats of Formation 39
   Strain Energy 45
5. Alkenes 47
   Structural Results 47
   Heats of Formation 50
6. Alkynes 50
7. Molecules Containing Delocalized Electronic Systems . . . 52
8. Other Hydrocarbons 58
9. Molecules Containing Heteroatoms 58
   Silanes and Related Compounds 59
   Carbonyl Compounds 60
   Thiols and Thioethers 66
   Halides 68
   Other compounds 68
   Electrostatic effects 69
10. Reaction Rates 72
11. Conclusions 75
    References 76

1. INTRODUCTION

Molecular Structure

A hundred years ago, when the organic chemist spoke of the "structure of a molecule", he meant knowing which atom was bound to which; if he could write a Kekulé formula for the compound, he considered that he knew its structure. After the recognition of the tetrahedral nature of carbon (van't Hoff, 1875), a small complication was added to structure in the nature of stereochemistry, but apart from that, the organic chemist's understanding of molecular structure did not change very much up until about 1950. After the appearance of Barton's pioneering paper outlining the importance of conformational analysis (Barton, 1950), the organic chemist began to worry in more detail about the "structures" of the molecules he worked with. Just how much more he worried depended on just what his interests were.

On the other hand, during the 1930s and later, the physicist or physical chemist who was interested in vibrational or microwave spectroscopy, or in x-ray crystallography or electron diffraction, had an understanding of the word "structure" that was a good deal more detailed. Interpretation of the data obtainable by these physical methods required that one understood "structure" in the sense that one knew in more or less detail the bond lengths in the molecule, the bond angles, and the torsional angles. Since about 1950, the interests of the average organic chemist have gradually expanded to include areas formerly reserved to the spectroscopist and diffractionist. To interpret quantities in which organic chemists are now routinely interested, such as coupling constant in the nmr spectrum, for example, or optical rotatory dispersion curves, it is necessary to have an understanding of molecular structure which is a good deal deeper than would have been of interest to an organic chemist 100 years ago, or even 20 years ago.

There are a great many methods for determining molecular structure which are in widespread use today. These include x-ray crystallography, electron diffraction, microwave and vibrational spectroscopy, and potentially useful but still embryonic methods involving nmr spectra. Of these, there is no doubt but that x-ray crystallography is of the greatest practical use at the present time. The reason for this is simple. An x-ray diffraction pattern of a typical molecule gives hundreds, often thousands, of pieces of data. All of these data must be accurately interpreted to find the structure; the structure which is found adequately accounts for this large number of data. In general, no other method gives such a large volume of precise, *interpretable* numerical information, and therefore no other method gives such good structural detail. X-ray crystallography suffers from some serious drawbacks nonetheless.

The first problem, sometimes an insuperable one, is that a compound must be obtained as a crystal in order for it to be studied. If a suitable crystal is available, then the crystallographic study of the molecule proceeds rapidly in the ordinary case. Many structures are quite straightforward, others are more difficult, and occasionally impossible.

The shortcomings of the crystallographic method described above are, in a statistical sense, not very serious, but for any given molecule they may be serious indeed. It may be possible under favorable circumstances to determine a complete structure by this method in a few days, but the average time required per structure is at present probably on the order of a month.

An ultimate limitation to the use of the x-ray method is, however, that one obtains information about the molecule as it is in the crystal, not as it is in the gas phase or in solution (where most reactions occur). For the most part, information on the crystal is good enough, but occasionally one needs additional information. This is particularly true when the molecule has a number of conformations available to it. The major conformation in solution is usually the same as the one found in the crystal, but this is not always the case. Molecules which are fairly flexible are, in a statistical sense, very different in a noncrystalline phase.

Molecular Energy

The other major topic with which this review will concern itself is the internal energy of a molecule. This is a subject of wide interest to organic chemists, because equilibria are dependent upon relative internal energies (among other things), and most of chemistry is concerned in one way or another with equilibria, especially if, following Eyring, reactions are viewed as involving equilibria between ground and transition states.

Methods for determining the internal energy of an organic molecule traditionally depend ultimately on heats of combustion.

Such measurements can be carried out with a high degree of precision, so that heats of formation and derivable properties such as strain energies can be determined to within a few tenths of a kcal mole^{-1} in favorable cases (for a recent compilation of such data, see Cox and Pilcher, 1970). These include most hydrocarbons up to about C_8, but for larger molecules the accuracy attainable decreases. The information available regarding simple compounds other than hydrocarbons, such as ketones, alcohols, sulfides, amines and the like is pretty scarce, which attests to the relatively large amount of effort required to obtain it. Unfortunately, many cases are not very favorable and, even for hydrocarbons, there are some inexplicable difficulties recorded in the literature. For example, the heat of formation of adamantane is reported by one research group to be -30.65 ± 0.98 kcal mole^{-1} (Boyd *et al.*, 1971) and by a second group to be -32.96 ± 0.19 kcal mole^{-1} (Mansson *et al.*, 1970).

2. CALCULATIONAL APPROACHES TO MOLECULAR STRUCTURE

Considering the difficulties involved in the experimental determination of molecular structure, one might ask if modern theory is not good enough for us to determine structure by calculation, and bypass the experimental difficulties completely. The answer is a very qualified one. Under some circumstances structure can be quickly and accurately determined by calculation, but often this is not possible. Calculational methods are therefore tending to augment the arsenal of the structural chemist, but experimental methods are in no immediate danger of becoming obsolete.

The Ab Initio *Method, and Simplifications of it*

When one wants to determine the structure of a molecule by calculation, as opposed to by experiment, there are a variety of ways in which one may proceed. The most direct way in principle would be by solution of the Schrödinger equation for a given nuclear configuration, followed by an adjustment of nuclear configuration so as to minimize the energy of the molecule. This can be done exactly for the hydrogen molecule ion. For more complex cases, approximations are necessary. The most sophisticated method commonly used at present is called the *ab initio* method (see, for example, Pople and Beveridge, 1970). There are three serious limitations in this method that might be mentioned here. First, the wave function is an infinite series of terms. It is truncated to some conveniently manageable size (the basis set) for purposes of calculation. For molecules containing up to several first row atoms, the truncation as ordinarily carried out introduces a sizeable error in the absolute energy, but it gives a fairly good structure, and good relative energies between different isomers, or different nuclear arrangements of a given isomer. The situation will evidently be improved automatically as computers become more efficient and the use of larger basis sets becomes routine.

The second problem that is sometimes lost sight of by organic chemists, is that the *ab initio* method as in current use is in fact a *self-consistent field method*. In this method, each electron is allowed to move in the average field of the remaining electrons, instead of treating each of the electronic motions independently. If the latter were done, it would be found that the motions of the electrons are strongly "correlated". If we take two electrons moving in a p-orbital for example, for the first electron there is a 50% chance that it will be found in the top lobe and a 50% chance that it will be found in the bottom lobe. The same is true for the second electron. In the self-consistent field method, therefore, the two electrons are both found in the top lobe 25% of the time, in the bottom lobe 25% of the time, and in different lobes 50% of the time. In actuality, because of the "electron correlation", the two electrons will indeed each spend half their time in each lobe, but they will spend almost all of their time in different lobes. This correlation lowers considerably the repulsion energy between the electrons. Just how serious the effect is on the final results depends upon what it is that one is trying to calculate. If one wants to calculate the total energy of the system by the SCF method, the results will be very poor, because the repulsions between the electrons in a molecule are very large when expressed as kcal mole^{-1}. The correlation energies of two different isomers are almost exactly the same, however, so differences between isomers are rather well calculated. This electron correlation is the reason for the so-called van der Waals attraction between molecules, so obviously a self-consistent field method will fail totally to show this kind of phenomenon.

The third problem which must be considered when the results of an *ab initio* calculation are compared with experiment is that the theoretical result pertains to a motionless model, while the actual compound is undergoing vibrational motions. These vibrations have a substantial effect on the energy of the molecule, even at $0°K$ (zero-point energy). At higher temperatures, as higher vibrational levels become populated, the experimental energy is further increased. Since the molecular vibrations are in general anharmonic, bond lengths, and other structural parameters, do not correspond exactly to the *ab initio* values, especially at room temperature. The *ab initio* calculation does not *per se* take these effects into account.

The situation at present seems to be that an *ab initio* calculation with a sufficiently large basis set will give quite good results for most problems of interest with small molecules (Clark and Stewart, 1970; Schaefer, 1972; Hariharan and Pople, 1974). Occasionally the results are not very good. At present, it is not always possible to predict into which of the above categories a given calculation will fall. The major practical problem to the use of this method at present is one of computer time. Molecules which contain more than four or five first-row atoms can be treated for one, or a few, nuclear configurations. To determine such a structure by minimizing the energy with respect to all degrees of freedom of such a molecule is at present possible, but expensive. Since the amount of calculation required for a molecule goes up by approximately the fourth power of the number of orbitals, adding an extra atom or two, or adding a second row atom increases computational time greatly.

There are a number of semi-empirical quantum mechanical methods which seek to reduce the amount of computation required in the *ab initio* calculation, and reproduce the results. Among the popular methods at present are those referred to as CNDO, INDO and the like (Pople and Beveridge, 1970). These methods are faster than an equivalent *ab initio* calculation by perhaps one or two orders of magnitude for small molecules but by more for larger molecules (as with the NDO methods the time required is approximately proportional to the square of the number of orbitals). While such methods require much less computing time than does the *ab initio* method, the requirement is still impractically large if the energy must be minimized for each degree of freedom, except for rather small or symmetrical molecules. In addition, the approximations used in these methods introduce considerable uncertainties into the results which are obtained.

A semi-empirical method which takes a somewhat different track is the MINDO method of Dewar (Bingham and Dewar, 1972). This method differs from previously mentioned semi-empirical methods in that it seeks to reproduce not the *ab initio* results which are in some respects inaccurate anyway, but rather the experimental results. The formalism is similar to that used in a CNDO calculation. However, by adjusting other quantities in the calculation, an effort is made to compensate for the lack of explicit inclusion of electron correlation. There are both strengths and weaknesses to this method. When suitably parameterized, it *may* give results which are more useful and dependable than those obtained by *ab initio* calculation. At present this can be done very well in some respects, not so well in other respects. The computation time is similar to that required by the CNDO method.

The extended Hückel method (Hoffmann, 1963) might be mentioned here. This is a simple, very fast method, which can be used to gain qualitative insights into problems, but, because of the extensive approximations made, it cannot compete with the other methods mentioned in terms of general reliability and accuracy. It has the attractive feature that it enables reasonably straightforward interpretations to be made of various phenomena, but it is not generally applicable to the kind of problem being discussed here.

The Force-Field Method

Finally, we come to the *force-field method*, which is also known as the *Westheimer method* or the *molecular mechanics method*. An excellent review of this method was published earlier (Williams *et al.*, 1968) covering the literature through January 1968. This chapter will concentrate on more recent results.

This method differs fundamentally from the other calculational methods previously mentioned. The Born–Oppenheimer approximation says that one can separate the nuclear from the electronic motions in a molecule, and the previously discussed quantum mechanical methods have to do with the electronic system, after the nuclear positions have been established (or assumed). To determine structures by such methods, one must repeat the calculation for a number of different nuclear positions, and locate the energy minimum in some way. Unless the structure is known at the outset, one therefore requires not just a single calculation, but many calculations, in order to determine the actual structure.

The force-field method involves the other part of the Born–Oppenheimer approximation, that is the positioning of the nuclei. The electronic system is not considered explicitly, but its effects are of course taken into account indirectly. This method is often referred to as a "classical" approach, not because the equations and parameters are derived from classical mechanics, but rather because it is assumed that a set of equations exist which are of the form of the classical equations of motion. The problem from this point of view is one of establishing just which equations are necessary, and determining the numerical values for the constants which appear in the equations. In general there is no limit as to what functions may be chosen or what parameters are to be used, except that the force-field must duplicate the experimental data.

The fundamental idea behind the force-field method can be briefly restated in the following way. We have a great deal of experimental information regarding small molecules, such as bond lengths, angles, strain energies, and so on. A large molecule consists of the same features we already know about in small molecules, but combined and strung together in various ways. Can we, with the help of current structural theory, formulate the structure of a large molecule in terms of the elementary features of the small molecules? The answer is that we can, and the device used to bring this about is the force-field.

The force-field method which will be discussed here presents a useful method for determining the structure and energy of a molecule. Both structure and energy seem to be of primary importance. There are many other properties of a molecule that can be determined, after the structure and/or energy are known, or they can be determined simultaneously with the structure and energy. These include the Raman and infrared spectra, and the other thermodynamic functions. Since these quantities are of limited use to organic chemists, we will concentrate here on just structure and energy (gas phase) at 25°C.

The force-field method is a method which, like the other calculational procedures mentioned above, has certain shortcomings. It is an empirical method, and is based on a large volume of experimental data. These data must exist for a given class of compounds before the method can be developed and applied to any particular compound in that class.

The main strength of the force-field method lies in its speed and ease of application. To determine the structure and energy of an average molecule in which one might be interested, and which does not exceed the limits of the method and of the available data, say two or three conformations, may require one man-day or so and a few minutes of computer time. It is not necessary to have available a sample of the compound, and of course the major difficulty in experimental structural studies is often the synthesis of the compound in the first place. Whether the compound has been synthesized (or is even synthesizeable) is of no importance in this method.

At this point it is necessary to decide at just what level the problem of structure and energy is to be attacked. Real molecules at room temperature occupy a series of vibrational states, and the atoms are not at rest but rather are vibrating. The energy of the total assemblage varies with temperature, because the molecules occupy the vibrational levels according to a Boltzmann distribution. The kind of calculation usually carried out by the molecular mechanics method gives us, as far as structure goes, a potential function, and the molecule is regarded as being a rigid structure, at rest at an energy minimum. This type of structure is adequate for many purposes. If it is not adequate for a given purpose, the vibrational levels may be calculated from the classical vibrational frequency, or quantum mechanically, by standard methods of statistical mechanics. We will return to this point later.

The question is often asked "Why use the force-field method, if one can use the *ab initio* method?" The answer is, that if one has the capability for using either method, the force-field method is always much to be preferred, if applicable. The reason for this is that the force-field method generally gives results as good as can be obtained. The force-field method is faster than the *ab initio* method by perhaps a factor of 10^3 for small molecules. Since the time requirements for the *ab initio* method go up as an n^4 (where n is the number of *orbitals*) while the time requirements for the force field method go up by N^2 (where N is the number of *atoms* in this case), the advantage of the force-field method becomes much greater as molecular size increases. Therefore, as a matter of practical economics, the force-field method is always to be preferred within the limits of its applicability. The big advantage of the *ab initio* method is that, as long as one is working with nuclei and electrons, no additional parameters need to be known. It is thus quite generally applicable. With the force-field method, if one wants to include an atom or bonding arrangement not previously studied, or if one wants to include a range of bond angles, or bond lengths which have not been previously covered, then more experimental data are required, or the results of the force-field method are in doubt. So, in the long run, one is going to use the force-field method when one can, and the *ab initio* method (or suitable variant) when one must.

In certain special cases, partial force-field calculations have been used. The most important example concerns applications to proteins. Since protein molecules are so large, total energy minimization would be extremely time consuming. Since ordinarily there is very little strain in a protein due to geometric distortion other than torsion, the simplification has been made that bond lengths and bond angles have their normal values, and remain constant. The energy then needs to be minimized only with respect to torsional degrees of freedom of the peptide chain. This saves greatly on computer time, and gives results which are probably about as good as one would get from total energy minimization. This is a rather specialized case, which will not be discussed here. Interested readers are referred to existing reviews covering that application (Scheraga, 1968; Ramachandran and Sasisekharan, 1968).

The Force-Field

We begin the development of a force-field by imagining a mechanical model of a molecule. This general idea can be traced back to Andrews (1930). A molecule is considered to be a series of masses (atoms) attached by springs (bonds). Deformation of the structure will result in an energy change which we can calculate if we know the necessary force laws and constants involved. These force laws and constants constitute the force-field. The latter is developed by working backwards, i.e. by deciding what kinds of forces and constants are needed to reproduce the known structures of small molecules. Initially we want to pick quantities that correspond to the chemist's way of thinking about things, since then one can use intuition and conventional ideas and methods in understanding the results. There are two general kinds of force-fields in wide use at present (Wilson *et al.*, 1955), the "valence force-field" and the "Urey-Bradley force-field". For a given number of parameters, these seem to give results that are about equivalent (Snyder and Schachtschneider, 1965).

It is important to retain sight of the fact that force-field calculations are done on a "molecular model". This model is assigned properties which reproduce experimental facts, but this does not mean that it is in every respect a faithful reproduction of the molecule under study. It only means that the particular information which has been used to develop the model is reproduced by the model. We are now at a point where there are perhaps half a dozen or so force-fields for hydrocarbons in common use today. These force-fields have been parameterized to reproduce different properties, and they generally reproduce those particular properties very well, and also many other properties of hydrocarbons fairly well. With respect to molecular structure and energy, the results are substantially the same for all of these force-fields even though the details of the force-fields are rather different. One must therefore not confuse the molecular model with physical reality. Until now, those who have developed force-fields have been largely interested in reproducing structural and energic data. The point has now been reached where this can be done quite well, and we can now turn to the important point of trying to understand molecules better by seeing which phenomena are consistently reproduced in different force-fields. Interpretations of phenomena which change abruptly from one force-field to another are of questionable physical significance. As force-fields become further refined, they may tend to converge upon one another with respect to certain of these phenomena. If so, a quantitative understanding of these phenomena would seem attainable. If the force-fields do not tend to converge on one another, it simply means that we cannot obtain information concerning physical reality with respect to those particular phenomena from the models being used.

As will be shown in what follows, current force-fields are some distance from this point of "convergence" today. Some of them have stiff force constants, and soft atoms, and some have the reverse. Thus changes, or errors, if you will, in one group of parameters can be compensated for with adjustment of another group.

Until the last few years, most force-fields described in the literature were developed exclusively for the interpretation of vibrational spectra. Such force-fields are usually of limited use in the present context, because typically they neglect many or all of the van der Waals interactions, and were not parameterized with structure and energy (apart from the spacing of the vibrational levels) in mind. Recently Lifson (Lifson and Warshel, 1968; Warshel and Lifson, 1970) and Boyd (Boyd, 1968; Shieh et al., 1969; Boyd et al., 1971; Chang et al., 1970) have each developed force-fields aimed at fitting simultaneously to vibrational spectra, structure, and energy (including the thermodynamic functions). Bartell (Jacob et al., 1967) has started with spectroscopic force constants and developed additional parameters to yield a force-field which will fit to structural data. While not internally consistent, Bartell believes this should be a reasonable approximation, and it seems to be. Such force-fields hold considerable promise for the future and at present they are among the best ones available for structure and energy.

The Mechanical Model

To develop a classical valence force-field, we will picture the molecule as though it were a series of masses joined together by springs [1].

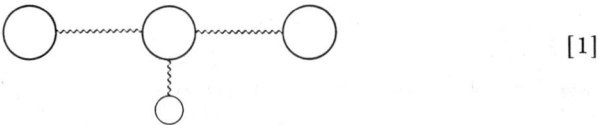   [1]

We know that bonds tend to have a certain "natural" length. If a bond is stretched, we can assume initially that Hooke's law applies, as it would to a spring. A similar relationship applies for bending angles. Thus we can define a stretching energy (E_s) and a bending energy (E_θ) for a molecule, by eqns (1) and (2) in which the summations are over all of the bond lengths (l) and bond angles (θ) in the molecule.

$$E_s = \sum \frac{k_s}{2} (l - l_0)^2 \quad (1)$$

$$E_\theta = \sum \frac{k_\theta}{2} (\theta - \theta_0)^2 \quad (2)$$

Hooke's law is a good approximation for small deformations only. For larger deformations, one may add additional terms (cubic, quartic, etc.), or substitute a Morse potential for stretching. In general, simple potential functions are used when possible, and more complicated ones when necessary. Sufficiently complicated functions will reproduce any desired properties, but with additional labor. Additionally, the more parameters that are added, the more the results become obscured and removed from an intuitive understanding.

It should be clear that in order for this scheme to work, the force constants (k) and natural values for things such as angles and lengths must be what the spectroscopists call "transferable". For example, the force constant for stretching a C–C bond must be the same in butane as it is in pentane or hexane. It need not be the same for all C–C bonds, but it needs to be the same for all C–C bonds of a given class, and the class must be precisely definable. A general survey of spectroscopic literature indicates that force constants are similar, but not identical in different molecules. This similarity was sufficient to encourage early workers in these calculations that transferability would be found to exist. In a detailed study directed at alkanes in which an effort was made to obtain an optimized set of "transferable" force constants, such were in fact found (Snyder and Schachtschneider, 1965).

In addition to the stretching and bending terms, we know that there are van der Waals interactions between all pairs of atoms which are not bonded to one another, nor to a common atom. (These latter two cases are excluded in the van der Waals calculations, because, if the atoms are bonded together, the van der Waals interaction is partly allowed for in the bond stretching, and, if they are bonded to a common atom, it is partly allowed for in the bond bending.) The sum of van der Waals interactions may be represented by E_{vdw}. The attractive part of the van der Waals curve is a result of electron correlation and is inversely proportional to the sixth power of the distance separating the atoms. The repulsive part of the potential is more steeply dependent upon distance; an inverse power in the range of 9–12 is used to express this part of the function. Alternatively, an exponential of similar slope is used. It is possible to cast the equation into a very general form (3), in which only two parameters are used to specify the interactions between any pair of atoms. This function is sometimes referred to as the "Hill function". In eqn (3), the c's are

$$E_{vdw} = \epsilon \left[-c_1 \left(\frac{r^*}{r} \right)^6 + c_2 \exp(-c_3 r/r^*) \right] \quad (3)$$

universal constants, ϵ is an energy parameter, and r is the interatomic distance. The function shows an energy minimum at a distance r^*, which is defined as the sum of the van der Waals radii of the two interacting atoms (Hill, 1948). If c_1, c_2, and c_3 are taken to be universal constants, only two parameters are necessary to describe the interaction potential between any pair of atoms (excluding those which undergo chemical reaction, of course). The parameters necessary are ϵ and the sum of the van der Waals radii. These can be looked upon as scaling factors which apply to an otherwise universal curve (Fig. 1). This function works well for rare gases (Hill, 1948), and small molecules like N_2 and CH_4. If one wishes to apply it to atoms which are parts of molecules, there are new problems which must be faced. First, the assumption is generally made, although not usually stated, that the interactions can be summed over pairs of atoms, and higher conglomerates may be neglected. It is also assumed that the interaction is the same (as long as the atoms are not bound to one

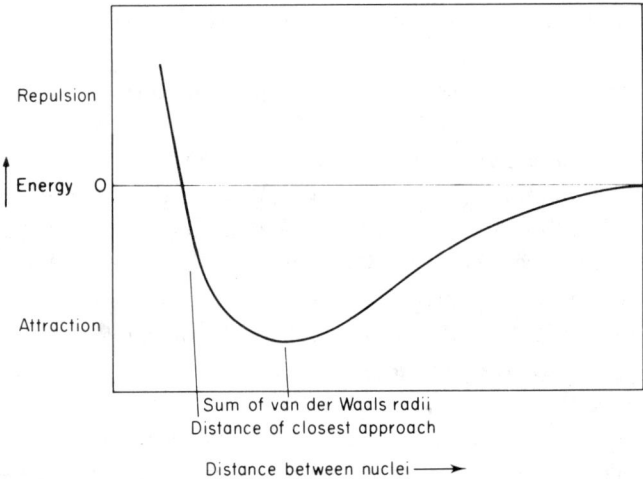

Figure 1. The interaction potential (van der Waals potential) for a pair of rare gas atoms (schematic).

another or to a common atom) regardless of whether the atoms are part of the same molecule or different molecules, and regardless of the nature of the material between them. Accepting the limitations posed by those approximations, one must establish values for r^* (Lennard-Jones, 1931) and ϵ for each pair of atoms. It is generally assumed that the r^*-values for atoms are additive. The original approximation of Hill [which comes from polarizability theory (Hirschfelder et al., 1954)] took the value for ϵ for a pair of dissimilar atoms to be the geometric mean of the value for identical pairs of the different atoms interacting. This procedure is still sometimes used, while others prefer to assign ϵ as an independent parameter.

Initially, there were difficulties with the r^*-values. Most common sources list "van der Waals radii of atoms", which are based directly or indirectly on a tabulation by Pauling (Pauling, 1960; Bondi, 1964), and which actually represent the "distance of closest approach" between these atoms in crystals. The closest distance to which a pair of atoms will come in a crystal is a good deal closer than the sum of their van der Waals radii, as the following illustration will show (Allinger et al., 1967a).

Suppose one had two rare gas atoms at the position of the energy minimum. This is clearly the sum of their van der Waals radii. If instead of two atoms, however, we have two molecules, say two hexane molecules, and these lie side by side, do the closest hydrogens on the two molecules approach to within the sum of their van der Waals radii? Suppose we align the two molecules side by side, so that the nearest pairs of hydrogens on the two chains are at the distance assigned as the sum of their van der Waals radii. Is this the distance to which they will approach in the crystal? The answer is no, for the following reason. Even though the nearest pairs of hydrogens are at the position of an energy minimum with respect to one another, there are many other atoms in the chain on the left which are all further than the sum of the van der Waals radii from many other atoms in the right chain. Between the two molecules, therefore, there is a sizeable attraction. The molecules will therefore move toward one another, and as they do so the nearest pairs of hydrogens will begin to repel one another. When the slope of the sum of the repulsions just balances the slope of the sum of the attractions between all the other atoms, the two molecules will come to their minimum energy position. This means that there will be a sizeable interpenetration of the van der Waals radii of the nearest hydrogens. In the crystal the molecule is not only attracted to its nearest neighbors, but also to a lesser extent to the next nearest neighbors, and to a still lesser extent to the next set of neighbors and so on. In general, we can conclude that the van der Waals radius of an atom will be *larger* than its distance of closest approach in a crystal.

Numerical calculations have been carried out to try to decide just what values should be used for the van der Waals constants of hydrogen and carbon. Williams (1966, 1967; see also Ferro and Hermans, 1970) used a great deal of crystal data to evaluate the necessary parameters but did not fit intramolecular interactions. The n-hexane (Warshel and Lifson, 1970; Allinger et al., 1967), and n-octane (Warshel and Lifson, 1970) crystals have been used as the model to fit to by others, along with the intramolecular data. It has been found that a van der Waals radius for hydrogen in the range of 1·5–1·8 Å is reasonable, when coupled with the proper value for ϵ (softer if a larger radius is used) and suitable carbon parameters, and will reproduce approximately the spacing in the alkane crystals, as well as their heat of sublimation*.

The van der Waals radius for carbon is less precisely determined. Since the carbon atoms in a saturated hydrocarbon are buried below the hydrogens, the value chosen for carbon has much less effect upon the quantities being calculated. (Diamond is a special case, where the interactions between the carbons can be clearly studied.) The radius of carbon is generally taken to be in the range of 1·7–2·1 Å (Williams et al., 1968; Lifson and Warshel, 1968; Warshel and Lifson, 1970; Boyd, 1968; Shieh et al., 1969; Boyd et al., 1971; Chang et al., 1970; Engler et al., 1973; Wertz and Allinger, 1974). Aromatic carbon is susceptible to crystal study, for example the interplanar distance in graphite can be fitted too. Such fitting requires a carbon radius of about 1·9 Å (assuming that the carbon is spherical) (Boyd, 1968; Shieh et al., 1969; Boyd et al., 1971; Chang et al., 1970; Allinger et al., 1968b). It is not certain that the same values apply to aliphatic carbon, since the electron distribution about the two kinds of carbon is different.

When we recognize the relationship between the van der Waals radius and the distance of closest approach, and have numerical data for hydrogen, carbon, and the rare gases, we can make some estimate of the van der Waals radii for the remaining atoms of the periodic table. As one goes from left to right across the periodic table, the van der Waals radii of the atoms decrease along a given row. The reason for this is that the attraction from increased nuclear charge pulls the electron cloud in closer, since it outweighs the electron–electron repulsion within the cloud. The rare gases, at the far right of the table, must have the smallest van der Waals radii of the atoms in their respective rows. These considerations permit one to assign a series of van der Waals radii as shown in Table 1. We have used values of ϵ for the first and second row atoms which correspond to the same amount of attraction as is found for the corresponding rare gases, as

---

* That the van der Waals radius of hydrogen (in hydrocarbons) must be about 1·5 Å, instead of Pauling's value (1·2 Å) seems to have first been recognized by Bartell (1960) in connection with early structural force-field work. The significance of this work was slow to be recognized, however, and even today the value of 1·2 Å is widely seen in the literature. See also, McCullough and Lindenmeyer (1972) for an important analysis of alkane crystal data.

this is known to be at least approximately correct (Hill, 1948). For hydrogen and carbon, we have adjusted these values to be consistent with the spacing and heat of sublimation of the normal hexane crystal, while simultaneously fitting to the other geometric and thermochemical information available for a large group of hydrocarbons. These values are also given in Table 1. In Fig. 2 are shown plots that correspond to selected data in Table 1. These are a consistent set of van der Waals data, which we use, and which were arrived at as discussed above. Other workers use somewhat different curves for carbon and hydrogen, and these various curves all work fairly well in the proper force-field. Representative curves are given in Fig. 2. No systematic table of van der Waals radii (like Table 1)

TABLE 1

Van der Waals Constants Radius (Å) — (ϵ)

				H	He
				1·50 (0·063)	1·48 (0·017)
B	C	N	O	F	Ne
1·80 (0·027)	1·85 (0·030) sp^2, sp 1·75 (0·041) sp^3	1·70 (0·038)	1·65 (0·046)	1·60 (0·056)	1·54 (0·070)
	Si	P	S	Cl	A
	2·10 (0·137)	2·05 (0·157)	2·00 (0·184)	1·95 (0·214)	1·92 (0·235)
	Ge	As	Se	Br	Kr
	2·25 (0·198)	2·20 (0·226)	2·15 (0·260)	2·10 (0·296)	2·07 (0·326)
	Sn			I	
	2·40 (0·272)			2·25 (0·400)	

seems to have appeared in the literature since the distinction between van der Waals radii and distances of closest approach was recognized.

An item about which there is a divergence in viewpoint at present concerns the interactions between non-identical atoms. If we use the Hill function as a vehicle for the discussion (not all workers use this function, but their functions can be recast in this form), there are just two variables to consider: a distance parameter (r^*) and a hardness parameter (ϵ). Most workers take the sum of the radii of the individual atoms interacting as r^*, but this is not universally done. The parameter ϵ was taken originally by Hill to be the product of the square roots of the ϵ-values for the interacting atoms, which is what simple physical considerations seem to dictate (Hirschfelder et al.,

1954). Some workers take ϵ for a non-identical pair of atoms to be an independent parameter, not related to the ϵ-values for the identical pairs. This difference in choosing ϵ seems to be the main cause of the difference in van der Waals properties of atoms used by different workers. To fit available structural information, hydrogen needs to be a good deal harder if ϵ for the carbon–hydrogen interaction is taken to be the product of the square roots of the ϵ-values of the separate atoms than it does if one can use a smaller ϵ for the carbon–hydrogen interaction. In that case, the hydrogens can

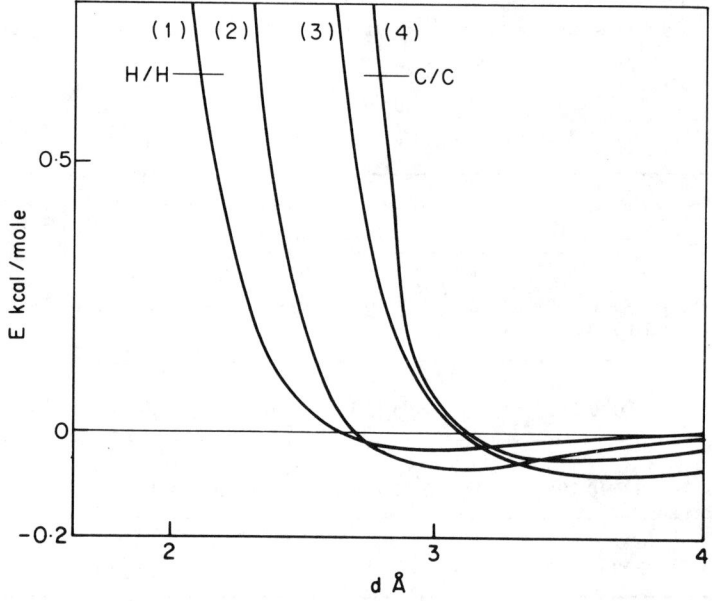

Figure 2. Representative van der Waals curves from current force-fields: (1) H/H; (4) C/C; Bartell (Jacob *et al.*, 1967); (2) H/H; (3) C/C; Allinger (Wertz and Allinger, 1974).

be considerably softer to give the same structural results. In Fig. 2 are shown the H/H and C/C interaction curves currently in use in two different force-fields.

One wonders, of course, if a hydrogen atom is really as "big" as indicated by curve 2, or as "small" as indicated by curve 1 in Fig. 2. This is not known at present, and indeed, it is not clear if this question is a meaningful one. At present we can only say that these curves give models which reproduce experimental structures about equally well, in the context of appropriate force-fields. The other popular force-fields use curves similar to these. These curves do not differ nearly as much as some of those which were in use ten years or so ago (Williams et al., 1968; Hendrickson, 1961, 1962, 1964, 1967).

If we were to calculate the energy of a molecule like ethane including only the stretching, bending, and van der Waals terms, we should find that the eclipsed form is less stable than the staggered form, but not sufficiently less stable. In order to calculate this eclipsing energy correctly, we must include a torsional term in the calculation, which for ethane has the form of eqn (4), where E_ω is

$$E_\omega = \frac{V_0}{2} (1 - \cos 3\omega) \qquad (4)$$

the torsional energy, V_0 is a force constant, and ω is the H–C–C–H dihedral angle. The torsional function in general can be written as a Fourier series (5) and summed over all non-terminal bonds.

$$E_\omega = \sum \left[\frac{V_1}{2} (1 + \cos \omega) + \frac{V_2}{2} (1 - \cos 2\omega) + \frac{V_3}{2} (1 + \cos 3\omega) + \cdots \right] \qquad (5)$$

The total energy of the molecule, then, relative to an arbitrary "natural" point, is so far given by eqn (6). This equation applies to

$$E = E_s + E_\theta + E_{vdw} + E_\omega \qquad (6)$$

hydrocarbons which are not highly strained, and works very well. The force constants needed are either picked from fitting to known structures, or spectroscopic values are used. The latter are not quite adequate in some cases, because the spectroscopists who deduced them probably used a somewhat different force-field (probably omitting van der Waals interactions) but they are approximately correct.

Of course, the organic chemist is interested in molecules other than relatively unstrained hydrocarbons. Consequently, he quickly tires of the force-field mentioned above, and wants to refine it so that it will handle properly more strained molecules and compounds which are not hydrocarbons. From here on, depending on what the person working with the force-field feels is important, there are various directions the refinement may take.

The cyclobutane molecule turns out to be an important one for force-field development. The reason for this is that it typifies a large number of small ring compounds which have C–C–C angles that are deformed very much below the tetrahedral value. The problem with cyclobutane can be briefly formulated in the following way. While the angular deformation would be minimized by having the ring planar, the ring in fact puckers considerably (20-40 degrees) from planarity, which leads to an even further compression of the bond angles. Presumably, the relief of torsional strain is sufficient to overcome the additional bending strain. In addition, the bond lengths in cyclobutane are unusually long (Wright and Salem, 1972; Pasternak and Meyer, 1972), about 1·548 Å, as opposed to 1·535 Å for cyclohexane and open chains. Parameters which work well for unstrained compounds calculate the bond length to be too short, and the molecule comes out to be planar. Furthermore, using spectroscopic force constants, the strain energy is calculated to be much too large.

Clearly, we need somehow to reduce the calculated bending energy for cyclobutane. The most direct approach is to say that a four-membered ring is different from other hydrocarbons, and that the bond angle does not have a natural value which is nearly tetrahedral, but rather has a value which is about 90 degrees. Given that cyclobutane carbons are different from other carbons, one can then choose the whole new parameter set, and thereby fit cyclobutane perfectly well. This approach has been used, and works for cyclobutane itself (Boyd, 1968; Shieh et al., 1969; Boyd et al., 1971; Chang et al., 1970). In addition, these parameters should be extendable to other cyclobutane rings which are not too different from the parent cyclobutane. However, this is not a completely satisfactory solution to the general problem. In norbornane, for example, there is a bond angle in the bridge which is about 92 degrees, similar to that found in cyclobutane. If ordinary constants are not suitable for cyclobutane, one can hardly expect them to be suitable for norbornane, which is almost equally deformed at one carbon. Looking through the literature, one can find compounds which have angles at many different values between those found in cyclobutane and the tetrahedral values. Obviously we do not want to find a new parameter set for every value of the bond angle, so a more general approach is desirable.

One obvious way to attack the problem is simply to reduce the value of the bending force constant sufficiently so that the heat of formation can be accurately reproduced. While the heat of formation can be taken care of in this way, the cyclobutane ring by such a calculation will be planar, or nearly so, rather than puckered. In order to get the pucker and the energy both right, we need a smaller value for the curve at 90 degrees than we would get from the spectroscopic force constants, and we also need a smaller slope to the curve. The first attempts to solve this problem consisted of simply drawing *ad hoc* curves, which had these characteristics (Allinger et al., 1968a; Wiberg and Lampman, 1966). This dealt adequately with cyclobutane, but the validity of such curves was in doubt where there were intermediate degrees of bending.

A more recent approach to the problem has been to extend the bending function beyond the quadratic term, by adding a cubic term (Engler et al., 1973). The absolute value of the cubic term was used, so that while the function is harmonic near the minimum, for large bendings both the energy and the slope of the curve are reduced. Such a function meets the requirements indicated above, and one might hope that it would fit intermediate degrees of bending as well as at the minimum and at 90 degrees. In fact, judging by comparison with experimental geometries for various angles, the curve seems to work quite well for angles in the 90–120° range.

Another problem concerns the long bonds in cyclobutane. When we pinch an angle down to a value as small as 90 degrees, there is obviously a considerable van der Waals repulsion between the end carbons, although with the ordinary valence force-field such a repulsion is not explicitly accounted for. Nonetheless, it is presumed to occur, and if outside constraints require this angle to be compressed, then the repulsion should force the bonds to stretch. Similarly, if an angle is opened beyond the tetrahedral value, its repulsion is reduced and the bond lengths can contract. The apparent solution to the problem is to add a stretch–bend interaction term into the energy expression, a device well known to spectroscopists. Thus we have an energy of the form represented by eqn (7). When

$$E_{s\theta} = k_{s\theta}(l - l_0)(\theta - \theta_0) \tag{7}$$

the stretch–bend force constant ($k_{s\theta}$) is properly chosen, the bond lengths in cyclobutane indeed stretch out to the desired value, while remaining at normal values for six-membered rings, and open chains. With a Urey–Bradley force field a geminal interaction (repulsion) is explicitly included, and leads to a similar bond stretching.

To get cyclobutane to pucker, instead of using a cubic term to soften the bending function, one can use a torsion–bend interaction term (8) (Wertz and Allinger, 1974).

$$E_{\omega\theta} = k_{\omega\theta}(\omega - \omega_0)(\theta - \theta_0) \tag{8}$$

where $k_{\omega\theta}$ is a torsion–bend interaction force constant.

All of these different methods of modifying the bending function to account for cyclobutane and related molecules have previously been used in the literature, and each method works for the area where it was tested. Which procedure will have the greatest range of applicability and superiority in the long run remains to be seen.

Energy Minimization Schemes

Once one is able to calculate the energy of the molecule by a suitable program utilizing equations of the sort given in the previous section, then one can worry about the details of the calculation itself. The structure of the molecule will correspond to that geometry where the energy is at a minimum. Therefore if we write the total energy of the molecule as in eqn (9), in principle, all we

$$E = E_s + E_\theta + E_{vdw} + E_\omega + \text{cross terms, etc.} \tag{9}$$

need to do to find the structure is take the derivative of this equation with respect to each of the degrees of freedom of the molecule, and find the place (or places) at which each of those derivatives is simultaneously equal to zero. There is a variety of mathematical techniques which can be used to do this, and different methods seem to have advantages in different cases, as will be discussed below. Suffice it to say that utilizing the more efficient of these methods with the aid of a computer in the class of the IBM 360/65, it is possible to minimize the energy of an average conformation containing about 10 carbons in less than 1 minute, or for 25 carbons, perhaps 3–5 minutes.

The early work on this problem was carried out by Westheimer (1956, see also Kitaigorodskii, 1960, 1961; Eliel *et al.*, 1965) using hand calculations, and later by Hendrickson (1961, 1962, 1964, 1967) using a computer. The energy minimization schemes of these workers, while well suited to their particular problems, lacked generality. The first general scheme for finding an energy minimum for a molecule was put forth by Wiberg (1965). This was a steepest descent method, which can be briefly described as follows. The energy of the molecule was calculated with the coordinates corresponding to the initial trial geometry. Atom 1 was then moved along the x coordinate by a small test increment, and the energy was recalculated. Atom 1 was then moved in turn along the y and z coordinates, each time the energy being recalculated. Atoms 2, 3,

4 ... etc. were moved similarly in turn. The atoms were then simultaneously moved in directions that would yield a reduction in energy by varying amounts (correction terms). The greater the reduction in energy had been for the test increment, the further the atoms were moved. After one application of these correction terms to the coordinates, the total energy was recalculated. It would be found either to have gone down or to have remained constant. If the energy remained constant, the molecule was already at the energy minimum. If the energy went down, the geometry was adjusted by again applying the correction terms, and the process was repeated until the energy ceased to decrease. A new set of correction terms was then calculated, and the process was repeated. This series of calculations was continued until no further reduction in energy was possible.

There are both strengths and weaknesses to the steepest descent method of minimizing the energy. An important strength is that the method is free from hang-ups; in our hands, the method was never found to leave the geometry hanging on a saddle point, which was sometimes a difficulty with more sophisticated methods. Energy minimization proceeded quite rapidly when the molecule was far from its optimum geometry, but slowed down as the energy minimum was approached, and became very slow at the end.

An improvement on the simple steepest descent method was made by Schleyer (Engler et al., 1973), and utilized the "pattern search" procedure. In this case the method proceeds as before, but information regarding the direction of motion of the atoms is saved from one iteration to the next, and the correction terms for succeeding motions are summed, and then applied. The advantage of this method is that if a particular atom is moving down a long hill with small slope, while the steepest descent method requires many iterations to move it, the pattern search method accelerates the motion, because the motion is repeatedly in the same direction, and the size of the correction term increases.

It is very difficult to compare "true" efficiencies in minimization algorithms. Depending on the sophistication of the program, the machine system used, and other variables difficult to trace, computation times vary widely, even for the same procedure. It would seem that the pattern search method should be considerably faster than the steepest descent method, other things being equal.

A more sophisticated procedure for locating the energy minima is the Newton-Raphson method, and variants thereof. In this method the function of the energy with respect to the coordinates is analytically differentiated with respect to each coordinate. The second derivatives are then taken, again with respect to each coordinate. It is then desired to move the atoms to points where all of the first partial derivatives are simultaneously equal to zero. This can be accomplished by solution of $3N$-6 simultaneous equations. Because of approximations used in deriving the equations, one does not usually reach the energy minimum on the first try, and the process is repeated with the improved coordinates as many times as required.

The strengths of this method are that analytical differentiation is much faster than numerical differentiation, and the atoms can be moved in one iteration to an approximation of the energy minimum geometry which is a much greater improvement over the trial geometry than can be obtained by the steepest descent method. The programming is much more complicated, because the derivatives with respect to the coordinates must be obtained for the general case. Variations of this method were used by Boyd (Boyd, 1968; Shieh et al., 1969; Boyd et al., 1971; Chang et al., 1970), Bartell (Jacob et al., 1967), and Lifson (Lifson and Warshel, 1968; Warshel and Lifson, 1970). A practical disadvantage of such a procedure is that one must calculate sufficient elements to fill out a $(3N$-$6) \times (3N$-$6)$ matrix, which then must be diagonalized. The computing speed for this procedure is a lot faster than for the steepest descent method, and probably is faster than for the pattern search method in a typical case.

Our earlier scheme is a modification of the Newton–Raphson method, where the atoms are moved one at a time rather than simultaneously (Allinger et al., 1971). It is therefore not necessary to calculate all of the elements to fill out a $3(N$-$6) \times 3(N$-$6)$ matrix, nor is it necessary to diagonalize such a matrix. The principles of the method are outlined as follows. It is assumed that the energy surface in the vicinity of an energy minimum can be approximated by the function (10), where x, y, and z represent the Cartesian coordinates

$$E = Ax^2 + By^2 + Cz^2 + Dxy + Eyz + Fxz + Gx + Hy + Iz + J \qquad (10)$$

of the atom in question, and A-J have numerical values that one wants to determine. A necessary condition for an energy minimum is that the partial derivatives of the energy with respect to each coordinate equal zero. If we take our initial structure, and work on one atom at a time, what we need to do is differentiate the above equation with respect to x, y, and z, and set each of the resulting equations equal to zero, and iterate over all atoms. Working only with the first derivatives, we can proceed in the following way. If we neglect the cross terms (the terms with the coefficients D, E, and F), we will have three equations containing six unknowns. To solve this system, we must obtain three additional equations. What we can do is move our atom a small amount, in each x, y, and z coordinate (using the steepest descent criterion) and recalculate the derivatives again. This will give us now six equations, containing six unknowns, and we may therefore solve for the coordinates corresponding (approximately) to the energy minimum. We can therefore apply this procedure in turn to the individual atoms, and then keep on applying it and approach the energy minimum more and more closely.

When a direct test was made of our first derivative program against Boyd's program for the n-hexane molecule, it was found that our scheme was approximately twice as fast. Since the calculation time required per iteration for the matrix method will increase at a rate which is proportional somewhere in between the square of the number of atoms (if calculation of the matrix elements is the slow step), or to the cube of the number of atoms (if diagonalization is the slow step), whereas for our method the rate is in between the first and second powers of the number of atoms, it would seem that the speed advantage of our method will increase with increasing molecular size.

An alternative and faster way to proceed has more recently been developed (Wertz, 1974). In the method described above, one was in essence calculating three of the second derivatives analytically, and neglecting the other three second derivatives. In the current method, all six second derivatives are analytically evaluated, in addition to the three first derivatives, all at the same point. This gives us, including the cross derivatives, a total of nine equations and nine unknowns, so this system may be solved directly. In practice, the second derivative scheme runs approximately three times faster than the first derivative scheme.

Still further improvements in our program were possible. A current version of the program (Wertz, 1974) is faster than the original steepest descent method by a factor of about 400, the actual energy minimization for a molecule the size of hexane requiring only about 4 seconds of CPU time on the IBM 360/65, while a steroid may require 1-2 minutes.

There are special cases when the full matrix method is expected to be faster than the one-atom-at-a-time method, however. If one is trying to carry out a pseudo-rotation of the twist form of a cyclohexane derivative, for example, it is a very slow procedure with the one-at-a-time method. The problem is that the atoms must move cooperatively in order to bring about pseudo-rotation. This cooperative motion is better accomplished with the full matrix method.

3. SUMMARY OF STUDIES REPORTED

Because calculations of the force-field variety have become so accurate and efficient, a great many studies using such calculations to attack problems in molecular structure have been and are currently being carried out. These studies give, as a minimum, the bond lengths, bond angles, and torsional angles of the molecules examined, in addition to information concerning energy. The amount of such data now accumulated is too voluminous to present here in any detail. There are several ways in which one might try to summarize this information. One way is with a table, such as Table 2. In this table we have summarized in a concise (and necessarily not very precise) way the classes or kinds of compounds that have been studied by currently available force-fields. Actually, in the last few years quite a few force-fields have been described in the literature. Most of these have been applied to only limited numbers and/or kinds of compounds, so that it is hard to know what their limits are. They usually consist of minor variations from one of the more thoroughly studied force-fields. Therefore the tables which follow are limited to only the six force-fields for which extensive usage has also been reported. Since these six force-fields are still viable, early results have been and are being continually replaced by newer and for the most part better results. When a research group which developed one of these six force-fields presents a new version of a force-field, and describes some results obtained with it, little or no mention is generally made as to the results that would be obtained with the new force-field in areas studied previously with an earlier version. Since in general the research worker develops a new force-field to fit results which the old ones did not fit satisfactorily, it is assumed that the new force-field is checked against older results, and they generally come out at least as well as previously. Since it would usually be too time consuming to check all previous work,

TABLE 2

Areas of Applicability of Current Force-Fields[a]

Authors	Simple alkanes	Congested alkanes	Cycloalkanes (size) C3	C4	C5-C7	C8-C12	Polycyclics	Alkenes	Alkynes	Conjugated alkenes	Silanes	Thianes	Halides	Ketones	Aromatic	Acid deriv.
Allinger	2	2	1	1	2	2	2	2	2	2	2	1	1	2	2	
Altona	2		1		2[b]		1									
Bartell	2	2						1								
Boyd	1			1	2		2									
Lifson	2	1			2	2		2		2				1	1	
Schleyer	2	2		1	2	2	2									1

[a] The significance of the entries 1 and 2 is explained on p. 28.
[b] For a review concerning five-membered rings, see Altona, 1971.

only selected examples are generally chosen, and it is always possible that something has gone awry without being noticed. Having pointed this out, the assumption is hereafter made that the later version of a force-field from any one research group reproduces essentially all of their previously described results which are not specifically revised.

In Table 2 a summary of the different classes of structure that have been studied with the six most widely used force-fields is given. A 2 in a given column indicates a relatively thorough study, and good fit to experimental data. A 1 indicates a study which has been either not very thorough, or which does not give as good results as one might desire. If no entry is made, insufficient information is available to permit any evaluation, usually, few or no studies were reported for that particular class. For classes of compounds not listed in the table, either they are included as part of a closely related class, or insufficient studies have been reported to warrant listing at all. A number of additional literature references concerning hydrocarbons not specifically discussed here will be given in a general listing (page 59).

A discussion of the compounds studied by compound class is given in the following sections.

4. ALKANES

Structural Results

A major objective of molecular mechanics is to fit and predict structures of molecules. To do this, we have to define what we mean by "structure". There are quite a few different ways of measuring bond length, for example, and for each of these different ways, there are several definitions as to just what is meant by bond length (for a more complete discussion of this problem, see Robiette, 1973). The nuclei in an actual molecule are undergoing vibrational motion, and these motions are generally anharmonic. The experimental structure corresponds to a thermal average over the occupied vibrational states. Usually force-fields are now calibrated to agree with structures obtained by either electron or x-ray diffraction. Electron diffraction is generally applied to gases, and x-ray diffraction to crystals, and in each case the electron cloud about the nucleus does the actual scattering, and it is the location of this cloud that is being measured.

The quantity that is usually available from electron diffraction work (given the symbol r_g) and the "bond length" from x-ray diffraction are usually the same to within the limits of our interest. These numbers are not the same as spectroscopic values obtained from infrared or microwave spectra, which measure different quantities, but they are fairly similar. None of these measurements give directly the equilibrium internuclear distance (r_e) which the theoretician is interested in.

The differences in bond length obtained by different methods are fortunately small, but for C—C bonds, microwave measurements give bond lengths that are on the average about 0·006 Å shorter than those obtained by diffraction measurements (Allinger et al., 1971; Lide, 1962; Kuchitsu et al., 1967–68). The carbonyl bond (C=O), on the other hand, is consistently longer by microwave measurements than by diffraction (Allinger et al., 1972). Since microwave data are available for a relatively small number of molecules, and diffraction data are what one commonly wants to compare with, it is best to fit the force field calculations to diffraction structures in so far as they are available, and to other structures only as a last resort. A detailed discussion of the problem is given by Robiette (1973), to which the interested reader is referred for additional information.

Apart from the above problems, one needs also to consider the special case of the hydrogen atom. With other elements the electron cloud is rather symmetric about the nucleus. However, with hydrogen the bonding electrons are the only ones present, and since bond formation pulls electron density in between the atoms bonded, the electron cloud assigned to hydrogen is no longer very symmetrical with respect to the nucleus. Electron and x-ray diffraction methods, which look at the electron cloud, therefore see hydrogens in a different place from spectroscopic (or neutron diffraction) methods which look at the nucleus. In addition, the scattering ability of hydrogen is low, especially for x-rays, and the hydrogen positions are poorly located by the latter method. While an ordinary C—H bond has a length of about 1·11 Å spectroscopically, the value is reduced to about 1·10 by electron diffraction, and may be in the range of 1·05–1·09 by x-ray diffraction. This may be partly due to experimental error in the usual sense, but is also partly due to the different definition of the bond length, depending on the method of measurement. Of course, C—H bonds are important in organic molecules, and some decision must be made as to how to handle this problem. Experimental studies of crystal packing by Williams (1966, 1967) indicated that the hydrogen can be treated as spherical, with the electron cloud moved along the C—H bond away from the hydrogen nucleus by a small amount (5–10% of the bond length). This is the kind of approach used in many molecular mechanics calculations (Allinger *et al.*, 1967; Warshel and Lifson, 1970). The effective shape of the covalently bound fluorine atom has also been briefly commented on (Nyburg and Szymanski, 1968).

It should be noted that if the electron cloud is centred away from the nucleus, then a dipole is present. To neglect the presence of this dipole is to introduce an internal inconsistency into the calculations, but this is commonly done nonetheless.

All of the force-fields include in their parameterization the data from small alkanes. Consequently, bond lengths and angles, torsional barriers, and equilibrium constants between conformations are all calculated to approximately within experimental error. Thus the structures of ethane, propane, and butane, for example, are well calculated by all force-fields. Similarly the rotational barrier in ethane is reproduced properly, as is the gauche-*anti* energy difference for butane. The latter has a value which is known with a relatively large percentage error (0.6 ± 0.2 kcal mole^{-1}) (Bartell and Kohl, 1963). This energy difference is extremely important in cyclic compounds, and in particular in methylcyclohexane. The axial-equatorial energy difference in the latter compound is due to the same "gauche interaction", but, since the experimental number is known with a smaller percentage error (1.9 ± 0.1 kcal mole^{-1}, gas phase) (Allinger *et al.*, 1968c and references therein), fitting it has been crucial. Early studies showed that this number could not be reproduced using a value of 1.2 Å for the van der Waals radius of hydrogen, and were important in establishing a better value for that quantity (Allinger *et al.*, 1967).

Just why is gauche butane less stable than *anti* butane? According to the textbooks, it is because of the repulsion between the two methyl groups in the gauche form, which is relieved in the *anti* form. According to some of the force-field calculations, this is not the whole story, and, indeed, it may not be the story at all. One can look at the energy difference between gauche and *anti* butane in terms of the interactions which actually occur, as calculated by the force-field. The results depend to a moderate extent on the force-field used. We have had access to only two force-fields, and so can only compare these two in detail at this time. The first is that of Schleyer in which the carbons are rather large compared to the other force-fields in current usage.* With Schleyer's force-field, one-third of the gauche butane interaction is due to repulsion between these carbons. Another one-third of the energy is due to repulsions between the methyl hydrogens, together with distortions which the molecule undergoes to relieve the repulsions. Finally, one-third of the effect is not due to the methyls at all, but rather due to the hydrogens on carbons 2 and 3 which are gauche to one another. In our force-field, the hydrogens are much harder than in most other force-fields, and we find that more than one-half of the interaction energy is due to the gauche hydrogens. It seems likely that other current force-fields, which have softer hydrogens than we use, and softer carbons than Schleyer uses, will give results in between the two sets quoted above. It would therefore seem that other current force-fields (which fit the methylcyclohexane data) will attribute a good fraction, say 30–50%, of the gauche butane interaction to an interaction between gauche hydrogens, not between the methyls. This is a rather different viewpoint from that assumed at the outset of the development of conformational analysis, and its correctness and importance remain to be fully explored.

The comparison of the results of Schleyer's force-field and ours quoted above points up a very interesting fact. While the two force-fields give almost indistinguishable results insofar as the total energies and the overall structures of hydrocarbons, the numerical details of the interactions in the two force-fields are substantially different. However, as will be delineated further below, there is independent evidence which we interpret as indicating that the hydrogens in our calculations are too big (or hard), and other evidence which independently suggests that the carbons in Schleyer's calculations are too big (or hard). Thus it seems likely that the two force-fields will largely converge upon one another as they are further refined in the future.

While the ultimate decision regarding the importance of the "*gauche* effect" in determining conformational properties remains to be seen, the early ideas of conformational analysis have been dealt a real blow when it comes to the idea of the additivity of conformational energies. The compound 2,3-dimethylbutane will illustrate the point. This compound exists in two conformations *anti* [2] and *gauche* [3], of which the latter is *dl*, while the former is not. Calculations indicate that the enthalpies of the two conformations

---

* We are indebted to Professor Schleyer for supplying us with unpublished results obtained using his force-field.

are essentially identical. Since there are only two *gauche* interactions between two methyls in [2], and there are three in [3], one would predict an enthalpy difference favoring [2] by 0.7 kcal mole^{-1} or so. The calculations show that since the enthalpies are the same and entropy favors [3] by about 0·4 kcal mole^{-1}, the first order conclusion is highly inaccurate.

One must of course ask why is [3] predicted to have an enthalpy which is quite similar to [2] if there are different numbers of *gauche* interactions in these conformations as indicated. The diagram shows the dihedral angles involved. Because of symmetry, [2] cannot relieve the *gauche* repulsions, while [3] can relieve them somewhat by means of rotations. Because of the flattening of the tertiary groups, note that the dihedral angles between the methyls is not 60°, but only 54° in [2] (Allinger *et al.*, 1968a).

Whatever the interpretation placed on the interactions involved may ultimately be, in one respect the results are clear. That is, the simple idea of the additivity of conformational energies fails badly, even in this relatively simple case. Thus, although the idea of additivity of conformational energies has been widely used, it can be regarded as only a very crude approximation. Unfortunately for the bench chemist, this means that much of conformational analysis will have to be taken away from the man with the slide rule and given to the computer, if reliable and accurate results are required.

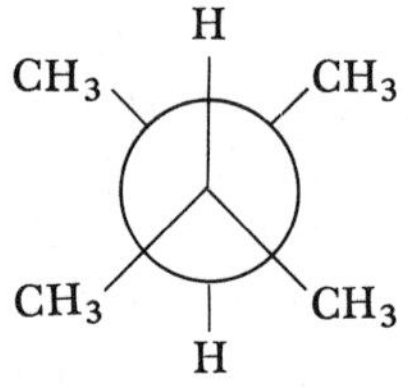

$$\text{[2]} \qquad \text{[3]}$$

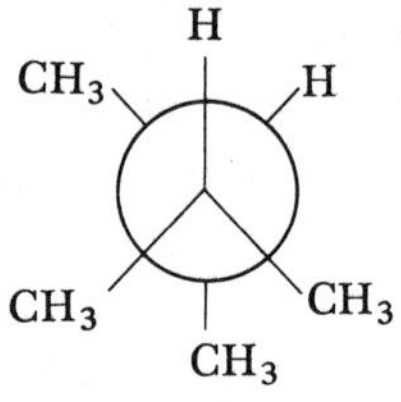

Bond stretching, with its large force constants, is not very much dependent on the other deformation modes. On the other hand, torsion, bending, and non-bonded interactions can be dealt with in a variety of ways. Changes in one of these quantities can be compensated for by changes in another quantity to a large extent, and in addition, because of the different kinds of terms (cross terms, or Urey–Bradley, etc.) in different force-fields, it is here that the force-fields show apparent great diversity. Nonetheless, the calculated results are generally quite similar.

For simple molecules the bond lengths are remarkably similar. For example, the bond lengths in ethane, propane, isobutane, and neopentane are according to the best current values as follows: 1·534 (Bartell and Higginbotham, 1965), 1·532 (Iijima, 1972), 1·535 (Hilderbrandt and Wieser, 1973), 1·541 (Beagley et al., 1969). Only neopentane appears different from the other by more than experimental error. In still more congested molecules, the bond lengths increase noticeably; thus 2,2,3,3-tetramethylbutane has a central bond-length of 1·580 Å (Jacob et al., 1967), while tri-t-butylmethane has a central bond length of 1·611 Å (Bartell and Burgi, 1972; Burgi and Bartell, 1972). Molecules having bond lengths in the range of the last two mentioned are sufficiently distorted that the harmonic approximation for them is noticeably in error. It is possible to correct for this in any force-field by an "add on" correction (Bartell and Burgi, 1972; Burgi and Bartell, 1972; for other examples of this approach, see Schubert et al., 1973, 1974; Hilderbrandt et al., 1973), or by building it into the force-field.

The tri-t-butylmethane molecule presented an interesting problem in structure determination. The molecule is too large for electron diffraction to give a unique structure. On the other hand, it is sufficiently deformed that it was uncertain if a molecular mechanics calculation would give a very accurate structure either. The problem was therefore tackled by using both the electron diffraction information, and the force-field calculations, together with the vibrational spectra, to try and arrive at the structure of the molecule. This procedure seems to have been successful and suggests a general way to extend the electron diffraction technique to molecules which were previously unapproachable, because of large size and asymmetry.

In some cases, good calculational results have been obtained prior to good experimental results. Thus, for example, in the case of cyclohexane, the best electron diffraction value for some years was 1·520 Å, later corrected to 1·528 (Buys and Geise, 1970; Geise et al., 1971), but apparently shorter than the corresponding values found in alkanes (about 1·533 Å). Force-field calculations indicated a longer bond, similar to that found in alkanes, and it was very satisfying when still further electron diffraction work finally yielded a value [1·536 Å (Bastiansen et al., 1973)] in better agreement with that calculated.*

* For example, the following calculated values are found: 1·532 (corrected from the microwave value) (Hendrickson, 1961), 1·530 (Lifson and Warshel, 1968; Warshel and Lifson, 1970), 1·534 (Wertz and Allinger, 1974), 1·530 (Engler et al., 1973).

Four-membered rings were discussed earlier. Five-, six- and seven-membered rings are well dealt with by all force-fields, since they deviate rather little from the strainless alkanes. The seven-membered ring was dealt with in a definitive way by Hendrickson (1961), and his general conclusions were that cycloheptane has two conformations, which we might call boat and chair, and each of these is flexible (i.e. can undergo pseudorotational motion with only low energy barriers separating the pseudorotational minima). There are thus a great many different equatorial-like positions for substituents to occupy. The important conclusion was drawn that different stereoisomers would in most cases have nearly equal energies in substituted cycloheptanes, in contrast to the corresponding cyclohexanes. What experimental data are available bears out this conclusion very well.

Bicyclic compounds are of considerable interest, because of the variations from normal geometry which they exhibit. Some representative structures calculated by our 1973 force-field (Wertz and Allinger, 1974) are shown in Figure 3. They are generally in good agreement with experimental information, with the exception of bicyclo[1.1.1]pentane [Fig. 3(c)] (Almenningen et al., 1971). There are two problems with the calculated structure here. The first has to do with the bond angles. The calculated angle at the secondary carbon is too small by about $2°$, while that of the tertiary carbon is correspondingly large. Clearly, the range of applicability of the bending function has been exceeded, which is perhaps not surprising when the bond angle is only $73°$. The other problem concerns the bond length, which is calculated to be about $0·03$ Å too long. It would seem that the stretch–bend interaction function has also broken down here, again because of the very large angular distortions. This molecule brings home the lesson that we cannot indefinitely distort our molecule and expect the force-field to give us a reliable structure.

For molecules less distorted than the bicyclo[1.1.1]pentane, the calculations are in much better agreement with experiment. The cubane molecule [Fig. 3(a)], for example, gives a calculated bond length of $1·557$ Å, compared with an experimental value (Fleischer, 1964) of $1·551 \pm 0·003$ Å. The difference here is only twice the estimated standard deviation (esd) in the crystal measurements. Generally, 2 esd is taken to be the experimental error.

With less strained systems, such as bicyclo[1.1.2]hexane [Fig. 3(e)], norbornane [Fig. 3(b)], and bicyclo[2.2.2]octane [Fig. 3(f)],

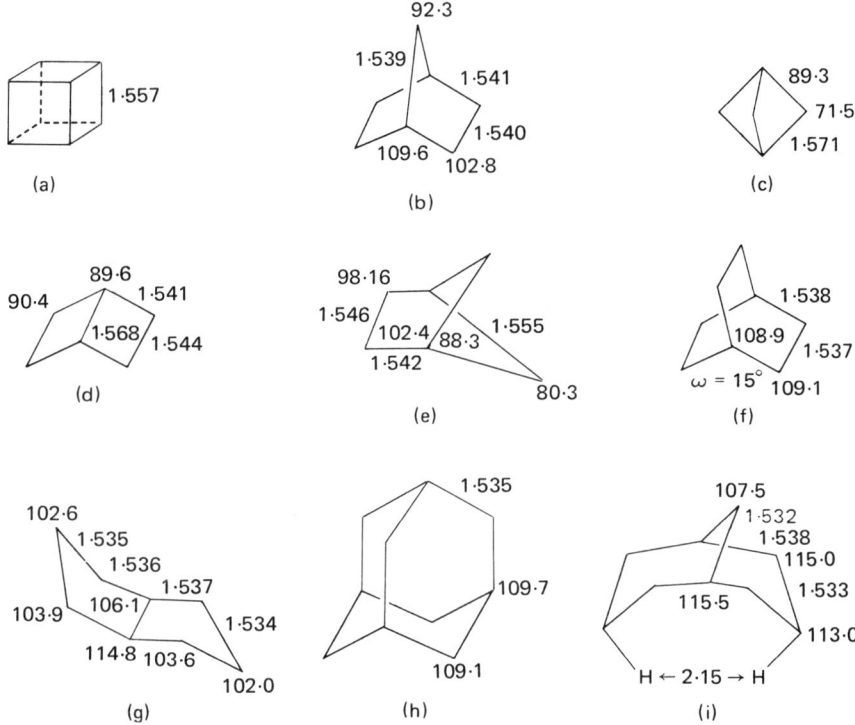

Figure 3. Bond lengths and bond angles in some representative bicyclic and related hydrocarbons calculated by the force-field method (Wertz and Allinger, 1974).

the agreement with experiment is expected to be good. For the first of these, there have been three experimental structures proposed (Dallinga and Toneman, 1967; Chiang, 1971; Almenningen et al., 1975), none of which is in very good agreement with the calculated structure. The most recent experimental structure gives fair agreement with the bond lengths (2-3 esd) except that between the secondary and tertiary carbons (reported 1·602 ± 0·004 Å). For norbornane, there have been several structure determinations, and the agreement is reasonable (± 1-2 esd), although the experimental errors are large. The most recent paper (Yokozeki and Kuchitsu, 1971) gives bond lengths as follows: bridge, 1·560 ± 0·010; $CH-CH_2$, 1·539 ± 0·005, CH_2CH_2, 1·557 ± 0·010; bridge angle 93·1 ± 0·7°. For bicyclo[2.2.2]octane, the calculations show the twisted (by 15°) D_3 structure is 0·3 kcal mole^{-1} lower in energy than the eclipsed (D_{3h}) one, while the experimental results (Yokozeki et al., 1970) fix these values at 0·1 kcal mole^{-1} and 7°. cis-Bicyclo[3.3.0]octane [Fig. 3] is calculated to have a boat-chair or C_s conformation, in agreement with experiment (Granger et al., 1971). Similarly, the bicyclo[3.3.1]nonane molecule is correctly calculated to have a double chair conformation of lowest energy (Dobler and Dunitz, 1964; Brown et al., 1965; Marvell and Knutson, 1970). A number of detailed studies have been carried out by Schleyer on adamantane and related compounds (homologs, dimers, isomers, etc.) (Farcasiu et al., 1974; Slutsky et al., 1973; Gund and Schleyer, 1973), which really show the power of the method in making thermodynamic predictions.

A number of studies have been carried out on large molecules, particularly in the steroid area. Accurate structures for a number of steroids are now becoming available through x-ray studies. We might consider the structure of androsterone [4] as an example of such an

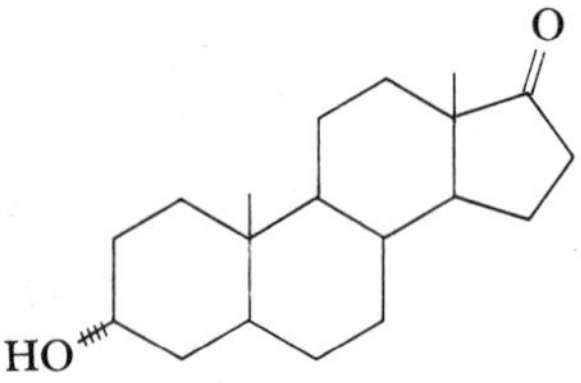

accurate structure (R = 0·04, esd 0·005 Å in bond length, 0·3° in angle, excluding hydrogens) (High and Kraut, 1966). When we compare our calculated structure with the x-ray structure, excluding those bonds around the alcohol group which we are not yet equipped to handle, we find that the average deviation between our calculated bond lengths and experimental ones is 0·006 Å for the C—C bonds, comparing favorably with the esd-values quoted. The average deviation in angles is 1·2°, about four times the quoted esd-values. On the average, then, our calculated structure is as good as the x-ray structure in terms of bond lengths, but not quite as good in terms of angles.* At this point we need to consider the difference between the structure observed in a crystal, and that calculated for an isolated molecule. Since the angle-bending constants are relatively small compared to the stretching constants, we might assume that any deformations due to crystal packing would be seen in terms of angular deformations, rather than in terms of bond-length deformations. Are the differences seen here between the calculated and observed values for the angles due to crystal packing forces, or are

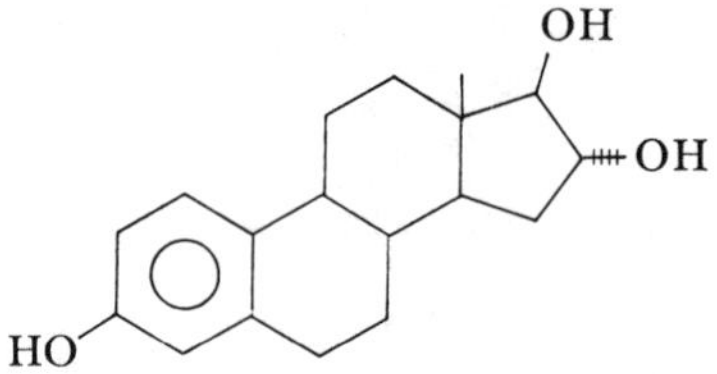

---

* A similar calculation was carried out earlier by C. Altona (personal communication).

they simply due to inaccuracy in the calculations? While we cannot answer that question with absolute certainty, we can try to get some idea of the magnitudes of the deformations which are to be expected as a result of crystal packing forces in molecules of this sort. For that purpose, the estriol molecule [5] is informative. This molecule exists in the crystal with two nonequivalent molecules per unit cell (Cooper et al., 1969). Clearly the two molecules would be identical if isolated, and any differences between them beyond experimental error can be blamed exclusively on crystal packing forces.

It is found that the estriol molecules have an average deviation in their C—C bond lengths of 0·014 Å, compared with an average esd of 0·007. Thus there appears to be no average bond length deformation beyond experimental error (2·0 esd). Only two of the bonds deviate from their counterparts by more than 3 esd (4·6 esd in both cases). The conclusion from this example is that crystal packing does not influence bond lengths in hydrocarbons much, if at all, at current levels of accuracy of measurement.

With bond angles the story is different. The average esd is 0·4°, while the average deviations between molecules is 0·9°, or 2·3 esd. This deviation is not sufficiently large to be given a definite meaning, as far as the average goes. However, here there are five angles that differ by more than 3 esd in the two molecules, the deviations being 9·8, 8·2, 6·0, 5·4 and 4·2 esd. Thus it would seem that deviations in bond angles of 2°, perhaps more, can sometimes be attributed to crystal packing forces. The differences in dihedral angle which correspond to differences in bond angle are often surprisingly large, especially in small ring compounds. In the estriol crystal, the different molecules have several dihedral angles which differ by as much as 12°. Of course, biphenyl is a classic case where a dihedral angle (the interplanar angle) changes markedly, from 0° in the crystal to about 40° in the gas.

We can thus conclude that while bond lengths will probably not be effected by packing forces in condensed phases, bond angles are likely to be by as much as a few degrees, and torsional angles may be completely different. This is an expected trend, because the force constants for deformation are usually large for stretching, intermediate for bending, and under proper circumstances, very small for torsion.

Turning now to some examples of energy calculations, the equilibrium between the C-20-epimers of the 5α-pregnane-3β,20-diols [6] was studied. The published proton spin coupling constants agreed well with the calculated results for each epimer, and the equilibrium constant was calculated to within an error of 0·5 kcal mole^{-1} (Altona and Hirschmann, 1970). The equilibrium between the 5α, 14α-androstane and its 5- and 14-epimers was similarly examined (Allinger and Wu, 1971). In this case, the four isomers were calculated to have stabilities qualitatively in the correct order, spanning a range of 1·9 kcal mole^{-1}, whereas the experimental values spanned a range of 2·7 kcal mole^{-1}, and seemed systematically somewhat larger. The epimerization at the 9-position for the 11-keto-androstane ring system [7] was also studied (Y. Yuh,

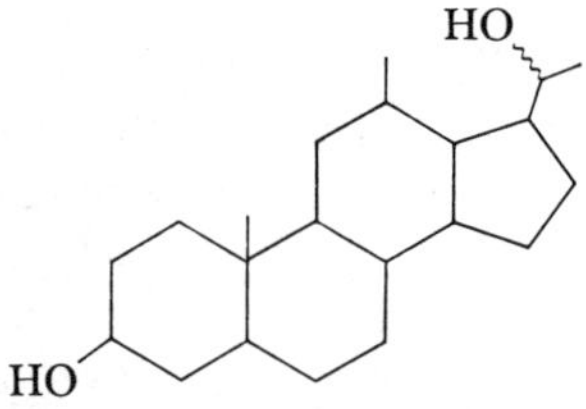

[6] 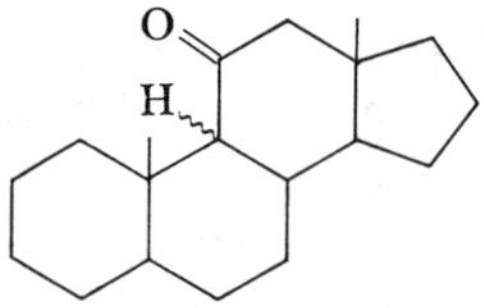   [7]

unpublished), since the experimental value (C. D. Liang, personal communication) (1·5 kcal mole^{-1}) favors the unnatural isomer (9β). It was calculated that the unnatural isomer was more stable by 0·5 kcal mole^{-1}. Since generally these steroidal systems do not contain any highly strained features, it seems their structures and energies can be fairly well calculated by existing methods.

Large rings seem to have been dealt with quite well by all of the current force fields which have been applied to studying them in detail. Thus, for example, if we take the rather well-studied molecule cyclodecane, we might examine the C–C–C angles which are calculated and observed experimentally. The overall ring structure found crystallographically is the same, regardless of the positions and nature of the substituents present, as long as they are small. Calculations have also been carried out for that structure, and the results are summarized in Table 3. The results indicate that for the carbon skeleton, the calculated structures are in good agreement both with one another, and with the experimental structures. Recently an electron diffraction study of the molecule has been carried out, but the results are not wholly convincing (Hilderbrandt et al., 1973). There is not enough information available from the diffraction pattern uniquely to specify the structure of the molecule, and, in addition, there appears to be more than one conformation present. Calculations by Schleyer (Engler et al., 1973), Wertz (1974),

TABLE 3
Comparison of Calculated and Experimental Cyclodecane Geometries

Structural parameter (degrees)	Exp	Schleyer	Allinger	Boyd
θ_{AV} CCC	117	118	118·6	116
ω_1	55	55	54	55
ω_2	−152	−149	−148	−153
ω_3	55	55	54	55
ω_4	66	65	65	67
ω_5	−66	−65	−65	−67

ω_6-ω_{10} are mirror images of ω_1-ω_5.

and with our 1973 force-field (Wertz and Allinger, 1974) indicate that the structure of lowest energy is not the same one that is found in crystals, and Hilderbrandt states (Hilderbrandt et al., 1973) that this is also true for Lifson's force-field. While not anticipated, this is not really an exceptional result, since the lattice forces between different conformations may vary by a kilocalorie or so, which is all that is needed to account for the observed results. It would be very useful to have an unequivocal gas phase structure for cyclodecane.

Recently, neutron diffraction data have given information on the location of hydrogens in the cyclodecane ring (Ermer and Lifson, 1973). The calculations have always indicated a "rocking" of the methylene groups where the repulsion was really severe, and this has now been borne out experimentally. However, in the case of our own force-field, the conclusion can be reached from a study of these calculations that the hydrogens retreat from one another further than they should, mainly by bending. This may result from the hydrogens being too hard, and the bending constants too small. The energy is reproduced correctly, suggesting that we have compensating errors. Of course, one cannot place too much weight on any one compound, but this is a particularly clear example of an error that appears from time to time in various places. Lifson's force-field gives a very good structure, including the hydrogen locations, but according to Hilderbrandt a very poor energy (Hilderbrandt et al., 1973).

Heats of Formation

After one has minimized the energy of the molecule and has the structure, one also has a good deal of information regarding the energy. One knows the "steric energy" of the molecule, that is, an energy relative to a hypothetical reference system. If one wants to know the difference in energies between conformations, say, this is directly given by a comparison of the calculated steric energies. However, if one wants to compare energies between molecules which differ in their bonding arrangements, or which differ in molecular size, then one needs some other way to compare energies besides the "steric energy" calculated. The easiest way to do this would seem to be to look at the heats of formation. These give calculated numbers, which may be directly compared with the corresponding experimental values.

Chemists have for years made use of bond energy tables, which permit one to calculate heats of formation by summing numbers and kinds of bonds in molecules (see, for example, Kalb *et al.*, 1966 and references therein). Such heat of formation tables work for strainless molecules, because the steric energy is proportional to the numbers of groups and bonds present. (With such a scheme, a strained grouping such as cyclopropane can be dealt with if a "group value" for the strained group is known.) All that is really needed for a general solution to the heat of formation problem is to add together bond energies and the "steric energies" from molecular mechanics calculations. Bond stretching, for example, gives a value for the energy relative to a hypothetical "natural" value, but the heat of formation will include the bond energy corresponding to the natural value. Obviously, we need to assign increments to the heat of formation for the numbers and kinds of bonds in a molecule, and perhaps for other things as well.

We can therefore see now two ways to approach this problem, either via statistical mechanics, or via bond energies. Let us consider the statistical mechanical approach first.

Here we know that the heat of formation of a compound (gas phase) will contain a contribution of $RT/2$ for each translational and rotational degree of freedom in the molecule, plus an additional RT if the measurements refer to a gas at constant pressure, rather than at constant volume. In addition, the vibrational contributions must be taken into account in some way. One needs to add the zero point energy, and these are very large numbers, to obtain the heat of formation at $0°K$. For higher temperatures, one must take into account the excitation of some of the molecules into higher vibrational and torsional states. This scheme has been used for a limited number of examples by Boyd, and by Lifson, and seemed promising. A thorough study was therefore carried out by Wertz (1974), who found that the heats of formation for a large group of compounds could be reproduced approximately, within the limits of accuracy of experimental measurement excluding, however, strained small-ring compounds. (There is no reason why the method cannot be extended to those compounds, but they involve some special problems and were not considered in his study.) This method requires calculation of the vibrational frequencies, which is straightforward once the force-field is specified.

The alternative bond energy approach, as modified by adding to it the steric energy terms calculated by molecular mechanics, has been used by Schleyer's group (Engler et al., 1973) and by ourselves, with a moderate degree of success. In order for such a bond energy scheme to work, it must be possible to include in the calculation, in

TABLE 4

Representative Zero Point and Thermal Energies[a] (kcal mole^{-1})

Compounds	Zero point energy	Thermal energy
Ethane	41·88	2·95
Neopentane	89·80	5·47
n-Pentane	90·39	5·54
Cyclohexane	96·53	4·76
trans-Decalin	149·85	7·15

[a] Wertz, 1974.

some average way, the zero point energies and the energies which result from the thermal occupation of higher states at room temperature. The reasonable success of the earlier bond energy schemes led us to optimism in this regard. Since we must have parameters for primary, secondary, tertiary, and quaternary carbon atoms, inherent differences in zero point energy that result from such substitution patterns can be taken into account in an average way. The question is whether or not this "average way" is going to be adequate in real cases.

Perhaps the hazards involved in assuming that the zero point energies and the thermal energies are additive functions of the number of bonds present can be best seen by looking at Table 4. The zero point energies are quite large, and so if they are additive to within a few percent, say, this is not going to be good enough to give us the kind of accuracy that we are seeking. Obviously, they are either additive to a much higher degree than that, or else the errors due to nonadditivity can be adequately compensated for in some other way. It is perhaps worth pointing out that in methylcyclohexane, the zero point energy difference favors the equatorial conformation by 0·5 kcal mole^{-1}, while the thermal energy difference favors the axial conformation by 0·1 kcal mole^{-1}. The net result is to raise the energy of the axial conformation by 0·4 kcal mole^{-1} from these two terms. It is not clear that this quantity can really be adequately accounted for without explicitly including these terms in the calculation.

In any case, it seems clear that the 3RT which is added to the heat content to account for the translational and external rotation of the molecule, plus the pressure-volume term, must also be added here. This was not done in earlier calculations, but is included in our 1973 force-field.

Of course, some molecules are a single conformation at 25°C (cyclohexane for example), while some are mixtures of quite a few conformations (n-hexane for example). These higher energy conformations raise the heat content, and this "conformational population increment" must be taken into account in individual cases. Finally, it was found that one additional modification was also necessary. Torsional degrees of freedom ordinarily involve vibrational frequencies which are very low, and for some molecules such torsional motion is quite free (n-hexane) while for others it is quite restricted (cyclohexane). Thus it would seem that there would be a systematic difference between compounds which can easily undergo torsional motion, as compared with those which cannot. If the frequency of such a motion is very low, the levels are closely spaced, and the higher levels will be occupied to a greater extent than otherwise. Perusal of a table (Pitzer and Gwinn, 1942) indicating how the heat content changes with the size of the torsional barrier led us to the conclusion that we could, to a good approximation, simply add 0·3 kcal mole^{-1} for each torsional degree of freedom present in the molecule, excluding methyl group rotation, where the torsional quantity would be absorbed into the primary carbon number. When allowance was made for this factor, in addition to the other factors discussed, we found that we could fit a wide variety of heats of formation for hydrocarbons with a standard deviation that does not exceed by very much the experimental errors reported for the measurements. In Table 5 is shown a list of typical hydrocarbons, chosen to be representative of organic molecules of interest, for

TABLE 5

Alkane Heat of Formation Data[a]

Compound	Wt	H_f° Calc	H_f° Exp[b]	Difference (Calc-Exp)	Reported probable errors
Methane	1	−17.82	−17.89	0.07	0.08
Ethane	2	−20.05	−20.24	0.19	0.12
Propane	9	−25.28	−24.82	−0.46	0.14
Butane	9	−30.26	−30.15	−0.11	0.18
Isobutane	9	−32.19	−32.15	−0.04	0.16
Pentane	7	−35.20	−35.00	−0.20	0.16
Isopentane	8	−36.62	−36.92	0.30	0.20
Neopentane	7	−41.06	−40.27	−0.79	0.25
Hexane	6	−40.14	−39.96	−0.18	0.19
Heptane	4	−45.09	−44.89	−0.20	0.19
n-Octane	3	−50.05	−49.82	−0.23	0.20
Hexamethylethane	5	−53.19	−53.95	0.75	0.29
2,3-Dimethylbutane	7	−42.16	−42.49	0.33	0.24
2,2,3-Trimethylbutane	6	−48.81	−48.95	0.14	0.27
Cyclobutane	2	6.16	6.38	−0.22	0.10
Cyclopentane	9	−18.02	−18.30	0.28	0.18
Cyclohexane	8	−30.08	−29.50	−0.58	0.17
Methylcyclohexane	7	−37.02	−36.99	−0.03	0.25
3,3-Diethylpentane	4	−55.41	−55.77	0.36	0.40
1,1-Dimethylcyclohexane	6	−44.02	−43.26	−0.76	0.46
cis-Dimethylcyclohexane	3	−41.73	−41.13	−0.60	0.27
trans-Dimethylcyclohexane	3	−43.06	−42.99	−0.07	0.27
Cycloheptane	7	−28.02	−28.22	0.20	0.26
Cyclooctane	4	−28.96	−29.73	0.77	0.33
Cyclodecane	3	−35.35	−36.29	0.94	1.00
trans-Decalin	3	−43.61	−43.54	−0.07	0.55
cis-Decalin	3	−41.11	−40.45	−0.66	0.55
cis-Hydrindane	4	−29.97	−30.41	0.44	0.47
trans-Hydrindane	4	−30.92	−31.45	0.53	0.50
Norbornane	7	−13.29	−12.40	−0.89	0.40
Cubane	1	149.18	148.70	0.48	1.00
Adamantane	3	−33.34	−32.96	−0.38	0.19
Congressane	1	−37.26	−36.64	−0.62	0.60
Bicyclo(2.2.2)octane	6	−23.81	−23.75	−0.06	0.30
cis-Bicyclo(3.3.0)octane	1	−21.43	−22.30	0.87	0.50
trans-Bicyclo(3.3.0)octane	1	15.01	15.90	0.89	0.60
trans-syn-trans-Perhydroanthracene	1	−57.22	−58.32	1.10	1.27
trans-anti-trans-Perhydroanthracene	1	−51.13	−52.93	1.80	1.47
Standard Deviation: 0.60 Correlation Coefficient: 0.999					±0.51

[a] A few new experimental values became available to us after the data in this table were assembled. The newer values have been included in the table together with the current difference between calculated and experimental values. However, the least squares fitting has not been repeated. If it were to be repeated, very small adjustments in the parameters would be expected, although no significant changes would result.

[b] The experimental values are generally taken from Cox and Pilcher, 1970 or API Tables, Project 44, Bureau of Standards, Washington, D.C.

which accurate experimental data are available, together with the heats of formation calculated and observed. Probably one can do a little better by a further small adjustment in the force-field, but clearly the results are very good, and the selection of compounds studied is sufficiently large that one can have confidence in the reliability of the method when applied to compounds which are similar to those in the table.

Using the statistical mechanical approach, Wertz was able to reproduce the heats of formation of 52 compounds with a standard deviation between calculated and experimental values of 0·41 kcal mole^{-1} (Wertz, 1974). The average experimental error claimed for the set of compounds he used is 0·39 kcal mole^{-1}.

Using the simplified method, not explicitly accounting for the zero-point and statistical mechanical energies, with a set of 38 compounds (including four-membered rings, but not three-membered rings) the standard deviation of the calculated from the experimental values was 0·60 kcal mole^{-1}, compared to a reported probable error average of 0·51 kcal mole^{-1}. Thus on the average, the effects of zero-point energy and of the higher vibrational levels on the heat of formation can be taken care of quite adequately by an averaging process. On the other hand, specific compounds can be expected to cause problems.

The data in Table 5 represent an updating (Wertz and Allinger, 1974) of our earlier work. A similar thorough study was independently carried out by Schleyer (Engler et al., 1973). His results with respect to the heats of formation of alkanes are on the average not significantly different from ours, yet his force-field is substantially different. If different force-fields give the same results, as far as they can be checked against experimental values, but different results in terms of internal details which are not experimentally accessible, it is clear that one cannot assign physical significance to the different sets of internal details. One conspicuous exception to the general statement above on the similarity in results of Schleyer's force-field and ours is the compound dodecahedrane, which is itself exceptional in being composed exclusively of C—H groups. (The only other similar molecule known is cubane, which is not easily compared with dodecahedrane because of the severe geometric differences.) The calculated heats of formation (gas, 25°C) from Schleyer's force-field and ours are respectively −0·22 and +40·88 kcal mole^{-1}. An experimental heat of formation would be most useful, but the compound has not yet been synthesized. Clearly though, the differences in these force-fields which exist are susceptible to comparison with experimental data, if and when the necessary data can be obtained.

Strain Energy

While numerical values for heats of formation are adequate for many purposes, if one wants to compare the "strain" in compounds which are not isomers, or which have different arrangements of branching, the heat of formation numbers are not directly informative. In the past it has been customary to relate strain energies to the normal alkanes. However, at ambient temperatures normal alkanes larger than propane are not really strainless in a sense, because they contain certain amounts of higher energy conformations. Thus one might want to refer strain to the stable (all *anti*) conformation alone. This definition would, however, have the disadvantage that even the *n*-alkanes would be strained. We therefore chose to define "inherent strain energies" in the following way. One can assign bond energies to the stable conformations of the *n*-alkanes in such a way as to fit the experimental heat of formation data, corrected from the conformational mixtures which actually exist at 25° to the stable alkane conformation. One can similarly obtain branching parameters from isobutane and neopentane. With the set of parameters thus developed, one can for any hydrocarbon calculate what the heat of formation would be for an isomer of that compound which is strainless, and exclusively in its minimum energy conformation. For the compound in question in its energy minimum conformation, one can calculate similarly a heat of formation, and the difference between those two numbers gives the "inherent strain" of the compound. These calculations do not depend on any experimental data for the compound in question. The only experimental data used is that pertaining to the simple and normal alkanes, when the parameter set was fixed. The advantage of this system is that an energy then becomes a purely calculated quantity, and can be determined without reference to experiment.

Schleyer defines strain energy somewhat differently, and uses experimental values for the compounds in question to evaluate it (Schleyer *et al.*, 1970). However, in practice, his calculated strain energies and ours differ by trivial amounts (perhaps with a rare exception). Using our scheme, the calculated values for "inherent strain" are as shown in Table 6.

TABLE 6

Calculated Inherent Strain Energies

Compound	Inherent strain (kcal mole^{-1})	Compound	Inherent strain (kcal mole^{-1})
Ethane	−0·1	Cycloheptane	8·4
Propane	0·0	Cyclooctane	13·3
Butane	0·0	Cyclononane	16·5
Pentane	0·0	Cyclodecane	17·8
Hexane	0·0	Cyclododecane	12·8
Isobutane	0·0		
Isopentane	0·9	Norbornane	15·1
Neopentane	0·0	Cubane	158·3
		trans-Decalin	1·5
2,3-Dimethylbutane	2·4	Bicyclo[2.2.2]octane	9·0
2,2,3-Trimethylbutane	4·7	Adamantane	3·5
2,2,4-Trimethylpentane	4·9	Bicyclo[2.1.1]hexane	41·1
2,2,3,3-Tetramethylbutane	9·2	Bicyclo[3.2.1]octane	10·3
tri-*t*-Butylmethane	43·3	Bicyclo[3.3.1]nonane	9·0
		cis-Bicyclo[3.3.0]octane	12·5
Cyclobutane	25·8	trans-Bicyclo[3.3.0]octane	19·0
Cyclopentane	7·5	Diamantane	5·3
Cyclohexane	1·0	Binorbornane	31·5

The normal and simple alkanes are calculated to have inherent strains of essentially zero. One can obtain from these numbers the actual strain energies, if desired, by adding the amount by which the enthalpy is raised due to admixture of the higher energy conformations. Such information is rarely of interest, however.

Note that the strain energy of isopentane is small, but that in 2,3-dimethylbutane is significant. This is a result of van der Waals repulsions, and distortions the molecule undergoes to relieve these repulsions. In hexylmethylethane (2,2,3,3-tetramethylbutane) and tri-*t*-butylmethane, the strain energies become increasingly large, and again this is a result of the van der Waals repulsions and the distortions they cause.

Turning to the ring compounds, cyclobutane is very strained, due mainly to angle strain, but also because of some unfavorable torsion. Cyclopentane suffers from similar problems to a lesser degree. Note that with this definition of strain, cyclohexane is not strainless. The medium rings through cyclodecane are calculated to be increasingly strained, in good agreement with experimental facts. Cyclododecane is calculated to be substantially strained, although less so than cyclodecane. The experimental results for cyclododecane are not completely clear.

Small ring compounds such as cubane and norbornane are strained for obvious reasons. Larger rings tend to be less strained, as perusal of Table 6 will show.

5. ALKENES

Structural Results

Alkenes have been studied fairly thoroughly by three groups, and the results seem to be about as good as they are for the alkanes (Allinger et al., 1968b; Allinger and Sprague, 1972, 1974; Ermer and Lifson, 1973; Favini et al., 1968; Favini and Nava, 1973; Zuccarello et al., 1971; Buemi et al., 1970). Bartell has also studied these compounds, using the alkene as a rigid substituent (Jacob et al., 1967), and results obtained in that way may be assumed to be good provided that the alkene is not further deformed about the double bond. The unsaturated carbon atom can clearly be considered to be different from the saturated one, and therefore all of the parameters (van der Waals, bending, stretching, etc.) involving it may also need to be different. To keep down the total number of parameters, some of them will often be assumed to be the same as for the saturated analog, unless something different is clearly required. In addition to these extra parameters the unsaturated carbon atom introduces an additional complication. Instead of measuring angular deformations as they are measured with saturated molecules, they must be factored into in-plane, and out-of-plane components. The reason for doing this can perhaps best be seen by looking at a molecule such as methylenecyclobutane [8]. Since the natural values for the bond

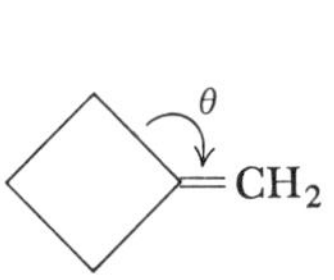

angles around the double bond are approximately 120°, if these angles are distorted by attaching a four-membered ring (for example), then the angles labeled θ in the figure obviously become much larger than 120°. Using the same approach as is used with saturated hydrocarbons, these angles will reduce their values to near 120° by bending the methylene group out of the plane of the four-membered ring. Naturally, that is not what happens. An out-of-plane disruption not only causes deformation in the σ-system, but also in the π-system. In other words, the forces leading to planarity are stronger than those leading to 120° bond angles. Hence, the molecule accepts whatever in-plane deformation is imposed upon it and retains the π-bond intact. The same factorization of bending must be applied whenever a tri-coordinate atom is present which has a preferred planar configuration (see, for example, Allinger et al., 1972; Winkler and Dunitz, 1971).

One additional complication takes place in principle with alkenes. With saturated hydrocarbons, most workers have been content to assume that the different atoms in the molecule are all electrically neutral, and no electrostatic factors have been considered in the calculation. An exception is one of the force-fields of Lifson (Lifson and Warshel, 1968; Warshel and Lifson, 1970), in which sizable atomic charges were used. These charges facilitated fitting to certain vibrational data, but the results are apparently about equally good with suitable parameterization when charges are omitted, and the simpler approach is to omit them. With alkenes, the neglect of atomic charges is less likely to be satisfactory, because alkenes have significant dipole moments. Electrostatic interactions of the part of the molecule around the double bond should therefore be taken into account. Again, it is not clear that any real improvement in the results can be obtained by including this interaction in the calculation. Our current feeling is that while one can probably safely ignore atomic charges for the alkenes themselves, if there are other polar groups in the molecule the dipolar interaction between the alkene and in the polar groups may be significant, and perhaps it is best to include the electrostatic effects in the alkenes themselves at the outset. (We may in the future wish we had felt this way about the alkanes!)

As with the alkanes, the parameters needed were obtained by fitting to simple molecules (Allinger and Sprague, 1972, 1974; Ermer and Lifson, 1973). The structures of ethylene, propene and the butenes are reproduced well by the two force-fields for which extensive studies have been reported. The structure of cyclobutene (Bak et al., 1969) was calculated fairly well, the single bonds being stretched as in other four-membered rings. The double bond is not excessively long, but the sp^2-sp^3 bond is (calc. 1·523, microwave 1·517 ± 0·003 Å), and so is the sp^3-sp^3 bond (calc. 1·560, microwave 1·566 ± 0·003 Å).

Cyclopentene has an envelope (C_s) conformation, and the planar form is higher in energy by 0·73 kcal mole^{-1} [experimental 0·66 kcal mole^{-1} (Laane and Lord, 1967)]. Cyclohexene exists preferentially in a half-chair (C_2) form, with the boat (C_s) being higher in energy by 6·31 kcal mole^{-1}. Recent calculations all indicate the boat is the saddle point on the chair ⇌ chair interconversion (Allinger and Sprague, 1972, 1974; see also Dashevskii and Lugovskoi, 1972; Bucourt and Hainaut, 1965), for which there is a low temperature nmr energy difference of 5·3 kcal mole^{-1} (Anet and Haq, 1965).

trans-Cyclohexene was calculated (Allinger and Sprague, 1972, 1974) to be 42 kcal mole^{-1} less stable than the *cis*-isomer, and separated from the latter by a barrier of only 12 kcal mole^{-1}. Because of the large distortions, these numbers are not expected to be accurate, but nonetheless, they give an idea of what the experimentalist is up against if he wishes to synthesize *trans*-cyclohexene.

Several of the larger cycloalkenes have been studied in considerable detail. They often have a number of stable conformations, and several pathways interconnecting these conformations. The details are complicated, and the interested reader is referred to the original literature for further information (Allinger and Sprague, 1972, 1974; Favini *et al.*, 1968; Favini and Nava, 1973; Zuccarello *et al.*, 1971; Buemi *et al.*, 1970).

A few non-conjugated polyenes have been studied by the force-field method. It was calculated that 1,4-cyclohexadiene is planar (D_{2h}), although one electron diffraction study (Oberhammer and Bauer, 1969) on the molecule indicated a boat form (C_{2v}). An independent electron diffraction work indeed gave a planar structure (Dallinga and Toneman, 1967). The problem with the former electron diffraction study seems to have been a misinterpretation of the observed 3,6-distance, a problem related to what is sometimes referred to as "shrinkage" (see Bartell and Kohl, 1963 and references therein). This effect occurs because the atomic nuclei are undergoing vibrational motion. Thus, one mode of vibration of 1,4-cyclohexadiene is as shown in eqn (11).

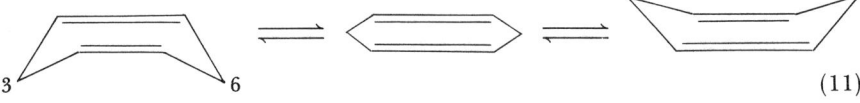

(11)

The energy minimum is clearly shown from vibrational spectra to be at the planar conformation (R. C. Lord, personal communication). However, the distance found between atoms 3 and 6 is not that corresponding to the planar form, but less. The reason for this is that as the molecule undergoes the wide amplitude vibration shown, the distance between the 3,6 atoms in the planar conformation is their *maximum* distance. Their average distance is less, corresponding to a shallow boat structure. If this "shrinkage" is not properly accounted for, incorrect structural conclusions can be reached.

A number of larger non-conjugated polyenes have also been studied (Allinger and Sprague, 1972; 1974; Favini et al., 1968; Favini and Nava, 1973; Zuccarello et al., 1971; Buemi et al., 1970).

Heats of Formation

The procedure here is similar to that used for the alkanes. There are parameters that need to be assigned to the different bonds and bonding arrangements found in alkenes and not found in alkanes (the numerical values for the alkane part of the molecule carry over, of course), and these were assigned by fitting to the available heat of formation data. Unfortunately, such data are severely limited. In Table 7 are given data for all of the compounds for which experimental values appear to be known with reasonable accuracy (Cox and Pilcher, 1970; API Tables, Project 44, Bureau of Standards, U.S. Govt. Printing Office, Washington, D.C.; Benson et al., 1969; Stull et al., 1969). For many alkenes, heats of formation are not known, but heats of hydrogenation in solution are (Turner et al., 1973 and earlier papers in this series). There are various approximations that can be applied to try to convert the heats of hydrogenation into heats of formation, but when this is done, one must anticipate errors in the solvation energy of the order of 1 kcal mole^{-1}. The reliability of the heat of formation calculations for alkenes is seriously limited by the lack of sufficient experimental data on strained alkenes.

6. ALKYNES

Proceeding as before, data for the alkynes were fitted (A. Y. Meyer, unpublished results) using parameters for structure and heat of formation (Table 8) based on experimental information which is even more limited than in the case of alkenes.

TABLE 7

Heats of Formation of Alkenes

Compound	Wt[e]	H_f° Calc	H_f° Exp[f]	Difference (Calc-Exp)
Ethylene	10	12·50	12·50	−0·00
Propene	9	4·85	4·88	−0·03
1-Butene	8	0·05	0·17	−0·12
cis-2-Butene	8	−1·66	−1·67	0·01
trans-2-Butene	8	−2·68	−2·67	−0·01
Isobutene	8	−4·11	−4·04	−0·07
cis-2-Pentene	7	−6·79	−6·71	−0·08
trans-2-Pentene	7	−7·58	−7·59	0·01
2-Methyl-2-butene	7	−10·16	−10·17	0·01
2-Methyl-2-pentene	7	−15·45	−15·98	0·53
2,3-Dimethyl-2-butene	6	−16·73	−16·68	−0·05
t-Butylethylene	6	−13·83	−14·51	0·68
cis-Di-t-butylethylene	0	−28·48	−32·52	4·04
trans-Di-t-butylethylene	0	−40·10	−42·82	2·72
Cyclobutene	0	39·66	37·45	2·21
Cyclopentene	8	8·05	8·56	−0·51
Methylcyclopentene	4	−0·92	−0·70	−0·22
Methylenecyclopentane	2	3·51	3·36	0·15
Cyclohexene	7	−1·38	−0·84	−0·54
Methylcyclohexene	5	−10·30	−10·34	0·04
Methylenecyclohexane	0	−8·30	−7·34	−0·96
Cycloheptene	5	−1·28	−2·19	0·91
cis-Cyclo-octene	4	−5·58	−6·45	0·87
trans-Cyclo-octene	0	4·84	4·10	0·74
Norbornene	0	18·21	21·73[a, c]	−3·52
Norbornadiene	0	51·27	57·71[b, d]	−6·44
Bicyclo[2,2,2]octene	0	6·94	5·50	1·44
Bicyclo[2,2,2]octadiene	0	37·75	34·46	3·29
Bicyclo[2,2,2]octatriene	0	69·80	73·06	−3·26

Standard Deviation: 0·37 (excluding compounds weighted zero)
Correlation Coefficient: 0·999

[a] Turner gives 20·73 (Turner et al., 1973), but stabilization from HOAc would correct to 21·73.
[b] Turner gives 55·71.
[c] Hall gives H_f° 15·12 ± 0·42 (gas) (Hall et al., 1973).
[d] Hall gives H_f° 50·59 ± 0·26 (Hall et al., 1973).
[e] A weight of zero means that the available number is suspect. Usually it is estimated from a heat of hydrogenation of acetic acid solution. An exception is cyclobutene, where the number is accurately known, but, because of the great strain in this molecule, a decision was made to fit to this value rather poorly so as to avoid distorting the results for less strained compounds.
[f] For references to experimental data, see Cox and Pilcher, 1970; API Tables, Project 44, Bureau of Standards, U.S. Govt. Printing Office, Washington, D.C.; Benson et al., 1969; Stull et al., 1969; Hall et al., 1973.

TABLE 8
Heats of formation of Alkynes

Compound	$H_f^°$ Calc[a]	$H_f^°$ Exp[b]	Difference (Calc-Exp)
Acetylene	54·20	54·34	−0·14
Propyne	43·98	44·39	−0·41
1-Butyne	39·45	39·49	−0·04
2-Butyne	34·57	34·71	−0·14
1-Pentyne	34·32	34·50	−0·18
3-Methyl-1-butyne	33·20	32·60	0·60
Standard Deviation: 0·36	Correlation Coefficient: 0·999		

[a] A. Y. Meyer, unpublished results.   [b] Ermer and Lifson, 1973.

A few interesting facts have been confirmed by the calculations. For example, the ethynyl group on a cyclohexane ring prefers to be equatorial rather than axial by 0·49 kcal mole^{-1} [experimental (Jensen et al., 1969), 0·41 kcal mole^{-1}].

The structure of cyclo-octyne has been calculated (A. Y. Meyer, unpublished results), and the agreement with the experimental electron diffraction structure (Haase and Krebs, 1971) is fair. In particular, the C-1-C-2-C-3 bond angle is calculated to be 161·5° (exp. = 158·5 ± 0·9°), but the calculated bond angles at the rear of the molecule are open too far (115–119°) compared to the experimental values (110 ± 1°), while the corresponding calculated bond lengths (1·550–1·555 Å), although long for ordinary aliphatic bonds, are shorter than those found experimentally (1·584 ± 0·009). While our force-field does show an error in bond angles which is systematic, letting them open perhaps 1–2° too much in a case like this, we believe that at least a good portion of the discrepancy is due to errors in the experimental values.

7. MOLECULES CONTAINING DELOCALIZED ELECTRONIC SYSTEMS

It should be clear that everything that has been said thus far applies to molecules for which a single Kekulé structure can be written. Thus atoms are either bound together, or not bound together, and there is no confusion on the point. With delocalized systems, the binding is much less easy to define.

Conjugated dienes were originally studied by a classical approach, and for simple polyenes the results seemed to be satisfactory (J. T. Sprague, unpublished results). Thus the essentially double bonds, for example the 1,2-bond in butadiene, are treated as ordinary double bonds, while the essentially single bonds (like the 2,3-bond in butadiene) are assigned a special set of parameters. It is known that the resonance interaction in a linear conjugated polyene is not very great, and, in any case, the approximation of taking the bond energies of such bonds as additive quantities appears to be perfectly adequate (Dewar, 1969). However, in the case of certain cyclic polyenes there are special problems because it is not a good approximation to consider such compounds as just polyenes. More elaborate calculations are therefore required.

Suppose we first consider benzene, and decide how we might deal with it in a force-field calculation. Can we find a natural bond length and a force constant for bond stretching, which would presumably be somewhere in between those for a carbon–carbon single bond and a carbon–carbon double bond, and which will adequately reproduce the structure of benzene? The answer is that we can, and simple benzene derivatives can then be dealt with using these parameters by the methods previously discussed. We only need to call the benzene bond a special kind of carbon–carbon bond with its own parameters.

Benzenoid compounds have been treated in this way by Boyd, who has looked at such molecules without considering the quantum mechanics of the electronic system (Boyd, 1968; Shieh et al., 1969; Boyd et al., 1971; Chang et al., 1970). He has treated a variety of paracyclophanes and related compounds with generally good results. Thus it seems clear that simple benzenoid hydrocarbons can be treated in the usual way, if a special set of parameters is assigned to the benzene ring.

What will happen, then, if we use these same parameters and try to calculate the geometry of naphthalene? We find that, on this basis, naphthalene has essentially all of the bond lengths equal (Allinger and Sprague, 1972, 1973), in contrast to the experimental situation where it is known that the individual bond lengths are proportional to their bond orders. The conclusion is that we cannot apply force-field calculations as discussed up to this point directly to delocalized systems, except in special cases like benzene where we are able adequately to pick the necessary parameters. If we want to deal in a general way with delocalized molecules, something more is needed.

Returning to naphthalene, there seems to be no calculational way to arrive at the correct structure without doing some kind of a quantum mechanical treatment of the π-system. The bond orders of the π-system determine what the structure will be, and it seems unavoidable. While a quantum mechanical calculation on naphthalene is a problem of sizable dimensions (48 valence orbitals or 58 total orbitals as a minimum basis set), the π-system of naphthalene contains but 10 orbitals. Before quantum chemists had computers, they studied in great detail the question of π–σ separation, and a great body of lore is available to us from those studies (see, for example, Flurry, 1968). It is well known just how to treat planar delocalized hydrocarbons by existing methods, to take advantage of the σ–π separation, and yet obtain accurate results. There are restrictions, however, planarity being a particularly important one.

To date there have been only two full-scale attempts at grappling with the problem of the use of force-field calculations to determine the structures of delocalized molecules. The approaches have been slightly different, and each has certain advantages. Warshel and Karplus (1972; see also Golebiewski and Parczewski, 1974) did a π system calculation, and determined forces and energies which were incorporated directly into the total calculation. Only a few compounds were studied, but the method seems to work well.

Our approach to the problem was somewhat different (Allinger and Sprague, 1972, 1973). Beginning in the 1930s and continuing up until the present time, a great deal of effort has been spent by quantum chemists on understanding the relationship between bond length (experimental quantity) and bond order (theoretical quantity) (for a summary, see Allinger and Graham, 1973; Graham, 1971). It would seem that in the context of the present type of calculation, the bond order–bond length relationship gives us a logical way to proceed. From the initial geometry, bond orders were calculated, and from these, force constants which were linearly related to bond orders were calculated in turn. The problem was thereupon reduced to the case of a molecule having a single Kekulé form; the quantum mechanical calculation on the electronic system was used only to ascertain the force constants needed. The planarity restriction imposed by σ–π separation was circumvented in the process because the force constants are calculated for the planar molecule, and its subsequent distortion does not affect the validity of the approximation of σ–π separation.

A number of numerically different bond order–bond length

relationships exist in the literature (for a summary, see Allinger and Graham, 1973; Graham, 1971), which were developed using different calculational procedures and parameters. For most structural work the simple Hückel method or SCF methods give approximately equivalent results. We have used the slightly more complicated VESCF method (Allinger and Tai, 1965; Allinger et al., 1967b), because with the orbitals obtained for the final structure one can carry out a configuration interaction calculation and obtain the electronic spectrum of the molecule.

Both the Karplus–Warshel and the Allinger–Sprague methods for delocalized molecules work quite well. In principle, they are applicable to π-electronically excited states, as well as the ground state, and this may be a fruitful area for study in connection with photochemistry.

Perhaps the situation here may be summarized by saying that one can do about as well with the calculation of the structure of a delocalized molecule (in its ground state) as one can with a localized molecule. The Karplus and Warshel (1972) calculations also give heats of formation directly, and these seem to be good, although the number of compounds so far studied has been small. Our method does not give this information directly, and a procedure for obtaining it has not yet been devised in detail. However, using the electronic energy of the π-system, together with the total "steric energy", there is no doubt that such a procedure can be worked out following the methods of Lo and Whitehead (1968a,b; see also Lindner, 1974 for a force-field extension of this approach).

Some of the structural problems involving delocalized molecules which have been examined are as follows. Butadiene and benzene are well reproduced as regards structure, and in the case of the former, also the energy difference between *cis*- and *trans*-conformations, as well as the barrier separating them.* These compounds were used in the parameterization, so naturally a good fit is anticipated. Planar aromatic systems such as naphthalene and anthracene have their structures accurately calculated (Allinger and Sprague, 1972, 1973). Phenanthrene, perylene, and quadricyclane had their structures fairly well calculated overall, but there were occasional sizable discrepancies in bond lengths. After calculational work (Allinger and

---

* Some recent work has been interpreted as indicating that *cis*-butadiene is non-planar (Lipnick and Garbisch, 1973), but the low frequency region of the Raman Spectrum shows with certainty that it is planar (Carreira, 1975). The calculations (Allinger and Sprague, 1972, 1973, 1974) fit the experimental potential function very well.

Sprague, 1972, 1973) was completed, refined structures for phenanthrene (Kay *et al.*, 1971) and quadricyclane (J. C. Speakman and A. Kerr, private communication) became available, and the agreement between the calculations and experiment is now quite good. The errors appear to have been in the original experimental values, not in the calculations.

The structure of azulene was well calculated, while that for fulvene deviated somewhat from the experimental one. Again, a better experimental structure of the latter (Baron *et al.*, 1972; Sunram and Harmony, 1973) has led to better agreement between the calculations (Allinger and Sprague, 1972, 1973) and experiment.

The structures of nonplanar compounds such as 1,3-cyclohexadiene (Allinger and Sprague, 1972, 1973, 1974) and 1,3-cyclooctadiene (Viskocil, 1974) are calculated in fairly good agreement with experiment (Traetteberg, 1970).

Cyclo-octatetraene presents an interesting calculational problem. The molecule exists in a tub conformation, the structure of which is well established experimentally (Traetteberg, 1966). Two barriers to inversion have also been determined. The molecule can invert (or reach a planar conformation) with or without a bond shift, so that the double bonds become single, and the single bonds become double. The barriers to both processes are known. The simple mechanical inversion can be treated adequately by force-field methods, and that barrier has been calculated to be 15·1 kcal mole^{-1} (Allinger *et al.*, 1973) [experimental (Anet, 1962; Anet *et al.*, 1964) 13·7 kcal mole^{-1} for $\Delta G^{\ddagger}$]. Because of quantum mechanical complications, the force-field method is not suitable in its present form for calculation of the barrier to bond shifting.

For methylated cyclo-octatetraenes, the inversion barriers become larger with increasing degrees of methylation (Allinger *et al.*, 1973), essentially because the methyls are mashed together in the planar transition state. The experimental barrier to inversion in 1,3,5,7-tetramethylcyclo-octatetraene is 22·5 kcal mole^{-1}, compared with the calculated value of 24·3 kcal mole^{-1}. For octamethylcyclo-octatetraene, the calculated barrier is 94 kcal mole^{-1} (no experimental value). The tetrabenzo- derivative of cyclo-octatetraene was originally reported to have a very low barrier to inversion (Figeys and Dralants, 1971), but molecular mechanics and CNDO calculations both indicated that this could not be (Rosdahl and Sandstrom, 1972; Finder *et al.*, 1972), and the calculations were later borne out by experiment (Gust *et al.*, 1972; Senkler *et al.*, 1972).

A great many annulenes and bridged annulenes have also been studied, and generally where comparison can be made, the agreement between calculation and experiment is good (Allinger and Sprague, 1972, 1973). One compound which might be specifically mentioned is [18]annulene. Crystallographic studies have indicated (Bregman et al., 1965) that the molecule has the general conformation shown (approximately D_6 symmetry), and is somewhat puckered from planarity. The C—C bond lengths are almost constant around the molecule, with a slight difference between the "inner" and "outer" bonds, shown light and heavy, respectively in structure [9]. The

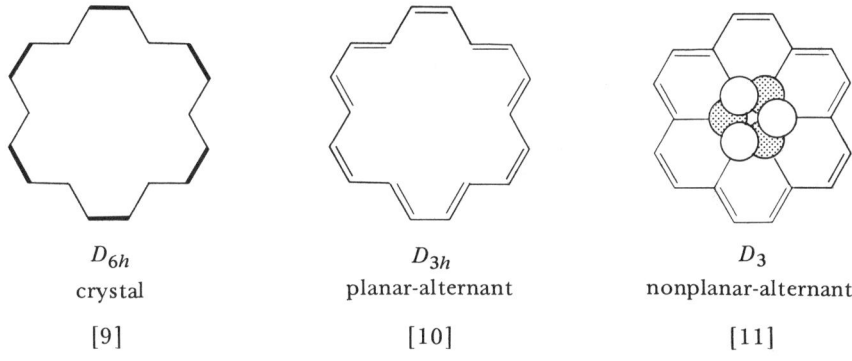

D_{6h} D_{3h} D_3
crystal   planar-alternant   nonplanar-alternant
[9]   [10]   [11]

calculations (Allinger and Sprague, 1972, 1973; Van-Catledge and Allinger, 1969) say that the isolated molecule should indeed be a little puckered all right, because of the repulsions between the inner hydrogens, but that the double bonds should be alternating long and short (D_3 symmetry). The electronic spectrum calculated (Van-Catledge and Allinger, 1969) from the alternating-bond model is in good agreement with experiment, while that calculated from the x-ray structure is not. MINDO calculations have also been reported (Dewar et al., 1974), and these support the alternating bond structure (planar). It would seem that the alternating bond structure is probably correct, at least in fluid phases. The x-ray structure obtained may be correct for the crystal, or there may be a problem with crystal disorder; the latter possibility was considered in the crystallographic work, specifically looked for, but not found.

Other compounds which might be mentioned, because they illustrate applications of the method to distinctly non-planar systems, are the bridged annulenes bicyclo[4,4,1]undecapentaene [12] and the isomeric[5,3,1]compound [13]. The x-ray structure is known for [12] and the calculated values are in quite good agreement (Allinger and Sprague, 1972, 1973). The conjugated system is reasonably flat, with average and maximum deviations from planarity (dihedral angles) of the polyene system of 19° and 30° respectively. The peripheral bond lengths are practically all the same.

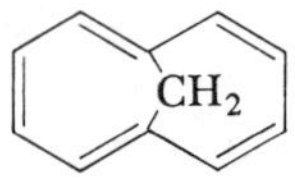

[12] 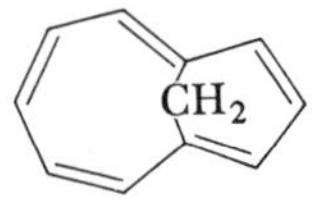   [13]

In contrast, [13] shows alternating long and short bonds, and considerable deviation from planarity arranged so that the short bonds are practically flat (14 and 28°), and the torsion is large about the single bonds (32 and 54°). No experimental data on [13] are yet available.

8. OTHER HYDROCARBONS

Classes of hydrocarbons discussed in previous sections include all but two types. These are cyclopropanes and allenes. The latter class has so far been ignored because of its limited significance and the equally limited available data. They could be dealt with if there were reason to do so. Cyclopropanes have been studied in a cursory manner (Jacob et al., 1967; Engler et al., 1973; R. Greengard and J. Kao, unpublished results). While we prefer to consider cyclobutane and other molecules which contain bond angle deformations not exceeding 20° or so, simply as distorted alkane derivatives, alkenes are clearly to be considered as a separate class of compounds (rather than as two-membered rings). Apparently everyone who has so far considered cyclopropanes has likewise regarded them as a separate class of compounds. In view of the bonding arrangements present in such molecules, this seems the most reasonable approach.

Our experience with cyclopropanes has been limited. There is not very much data, but, in this case (in contrast to the alkenes and alkynes), it is difficult to parameterize to that which is available. Because of the very fragmentary nature of the results, they will not be discussed further here.

9. MOLECULES CONTAINING HETEROATOMS

For compounds which contain other atoms in addition to carbon and hydrogen, the amount of *systematic* study to date has been comparatively small. There have been dozens of papers by various workers which have dealt superficially with different classes of compounds. While the results obtained have generally been good enough for the purposes of those who carried out the calculations, there has been little or no demonstration of the generality and limits of applicability of these force-fields, and this work will not be discussed in detail here.*

Silanes and Related Compounds

The silanes constitute a class of compounds which are, from the molecular mechanics point of view, only a slight step beyond hydrocarbons. They are comparatively non-polar, and the silicon preferentially has a tetrahedral geometry. There seems to be no difficulty in calculating structural information which is in good agreement with experiment (Tribble and Allinger, 1972; Ouellette et al., 1972) and a number of predictions have been made in this area. The calculations have also been extended to analogous compounds of germanium and tin, and again seem to work well (Ouellette, 1972). Values of the heats of formation for silanes are uncertain, however, due to the experimental difficulties met with in their combustion (Tribble and Allinger, 1972).

Some of the interesting features in silanes which are well reproduced by the calculations include the rotational barriers (1·7 kcal mole^{-1} in methylsilane as compared with 2·9 kcal mole^{-1} in ethane, and correspondingly low numbers for other methyl silanes). While the torsional function about the 2,3-bond in 1-silabutane looks almost like that in butane, the corresponding function

---

* Leading references, grouped according to compound type, are as follows: *acetylcholine* (Froimowitz and Gans, 1972); *alcohols* (Fournier and Waegell, 1973b); aromatic hydrocarbons (Kitaigorodskii and Dashevskii, 1968; Gleicher, 1967; Dashevskii and Kitaigorodskii, 1967; Bright et al., 1973); *carbohydrates* (Sundararajan and Marchessault, 1972; Sundararajan and Rao, 1968); *conjugated polyenes* (Gavezzotti et al., 1972; Sekigawa, 1970; Wettermark and Schor, 1967; Dodziuk, 1974; Bromberg and Muszkat, 1972; Fischer-Hjalmars, 1963); *enones* (Dodziuk, 1974); *halides* (Abraham and Loftus, 1974; Goursot-Leray and Bodot, 1971; Abraham and Parry, 1970; Fournier and Waegell, 1972); *heteroaromatics* (Ferre et al., 1974; Ollis et al., 1974); *heterocycles* (Blackburne et al., 1973; Lugovskoi et al., 1973; Wilson, 1974; Lee and Hamori, 1974); *hydrocarbons* (Fournier and Waegell, 1973a; Montaudo and Finocchiaro, 1972; Faber and Altona, 1971; Altona and Sundaralingam, 1970; Fournier and Waegell, 1970; Bucourt and Cohen, 1970; Cohen, 1971; Kitaigorodskii, 1961); *inorganic complexes* (Brubaker and Euler, 1972; Dwyer and Searle, 1972; Geue and Snow, 1971; Buckingham and Sargerson, 1971); *ketones* (Fournier and Waegell, 1972; Altona et al., 1971); *nucleotides* (Stellman et al., 1973; Lugovskoi and Kitaigorodskii, 1973; Yathindra and Sundaralingam, 1973; Sasisekharan and Lakshminarayanan, 1969; Usher et al., 1965); *phospholipids* (McAlister et al., 1973); *polyoxyethylene* (Mark and Flory, 1965); *terpenes* (White and Sim, 1973).

for 2-silabutane shows barriers of less than 2 kcal mole^{-1}. This can be attributed mainly to the longer C—Si bond length.

For silacyclohexane, the chair conformation is calculated to be the lowest in energy, as with cyclohexane, but because of the rather small torsional barrier about C—Si bond, the C_2 twist-boat conformer is only 3 kcal mole^{-1} higher (no experimental value). The barrier to the chair–chair inversion is also lower than in the hydrocarbon (for 1,1-dimethylsilacyclohexane, the calculated (Tribble and Allinger, 1972; Ouellette, 1974) and observed (Jensen and Bushweller, 1968) values are respectively 5·46 and 5·5 kcal mole^{-1}).

Carbonyl Compounds

Carbonyl compounds, partly because of their wide occurrence and the suitability of such compounds to other kinds of studies such as optical rotatory dispersion and circular dichroism (Crabbe, 1965), seemed to be a good target for study once an adequate force-field was available for hydrocarbons. For monoketones, electrostatic interactions were neglected, and this neglect did not appear to cause any difficulty. The carbonyl oxygen atom was taken to be spherical and assigned van der Waals properties accordingly. This is not really correct, of course. The electron density is greater around the oxygen in the plane of the σ-system than it is in the π-system, and this is even more true at the carbonyl carbon atom. The carbonyl carbon was assigned the same van der Waals properties as an olefinic carbon in the absence of other information. This is not really correct either, since the electron density is quite low here. However, these approximations seem to have worked well in all cases studied so far (Allinger et al., 1972; Allinger and Tribble, 1972).

The carbonyl compounds constituted the first thoroughly studied group in which there was a highly polar atom. As long as there is only one carbonyl group, and alkanes are taken to be non-polar, the electrostatic problem does not arise. If there is a second polar group in the molecule, electrostatic interaction needs to be considered. The problem of induction, on the other hand, is certainly present even in the monocarbonyl compound. *Ab initio* or semi-empirical calculations show that the presence of a polar group such as a carbonyl in a molecule causes sizable fluctuations in the charge densities at the various atoms. Just how serious this will be is not yet very clear. It means in principle, however, that not only must one worry about interactions of charges, but, perhaps more seriously, the presence of charges will change the van der Waals characteristics of the various atoms by induction. More charge density tends to make the atom both larger and more polarizable, while less charge density will have the opposite effect. So far these problems have been ignored, but as we attempt further to refine the results, the problem may have to be faced.

As usual, structural parameters were deduced by fitting the data for simple compounds, and then the heat of formation parameters were picked to reproduce the available thermochemical data. There are quite a lot of data available on the conformations of simple carbonyl compounds. Some of it is contrary to what might be expected from a knowledge of the conformational properties of hydrocarbons. For example, the stable conformation of acetaldehyde has a hydrogen on the methyl group eclipsing oxygen (Eliel et al., 1965, p. 20), and the stable conformation of propionaldehyde has the methyl eclipsing the carbonyl oxygen. The calculated (Allinger et al., 1972; Allinger and Tribble, 1972; S. Profeta, unpublished results) rotational potential function of the sp^2-sp^3 bond for propionaldehyde is shown in Fig. 4. The experimental curve (H. M. Pickett

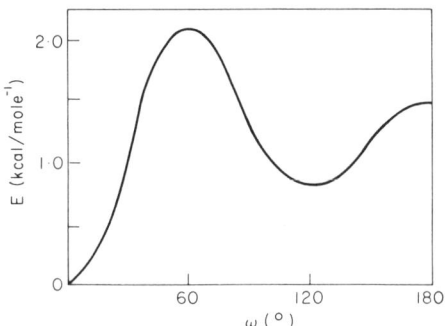

Figure 4. The calculated rotational potential function for propanal.

and D. G. Scroggin, personal communication) is comparable, with the respective maxima and minima in order from the left having values of 0·00, 2·1, 0·9, and 1·5 kcal mole^{-1}, respectively. Since these barriers are not very large, they can be overcome by steric effects rather easily. Thus t-butylacetaldehyde has the t-butyl group gauche to the carbonyl oxygen in the ground state (Karabatsos and Hsi, 1965; Karabatsos and Fenoglio, 1970). One concludes that the steric interaction between the bulky t-butyl group and the oxygen atom is sufficient to change the stability order for the conformations relative to propionaldehyde. Quite a lot of quantitative conformational information of this kind is available on aldehydes, largely from an nmr study of the coupling constants of the aldehyde hydrogens. These data are rather well reproduced by the calculations.

With ketones, one may start with acetone, the structure of which has long been accurately known except for the conformational arrangement of the methyls at the energy minimum (Nelson and Pierce, 1965). Opposing assumptions had been made by earlier workers, either that the hydrogens eclipse the carbonyl, or that they were staggered with respect to the carbonyl. No compelling experimental evidence existed to differentiate the two cases. *Ab initio* calculations, however, gave the correct magnitude of the rotational barrier and established that the eclipsed conformation was the more stable of the two (Allinger and Hickey, 1972). This is an example of something alluded to earlier, namely, that one can use an *ab initio* calculation to deduce force constants or other parameters and these can then be used in a force-field calculation. With this final parameter in hand, simple ketones were treated without difficulty (Allinger *et al.*, 1972; Allinger and Tribble, 1972).

2-Butanone and other straight chain homologs were examined, and it was calculated that the carbon skeleton prefers an all *anti* conformation; gauche forms are higher in energy, but as with hydrocarbons, not by very much. For 2-butanone, both the structure and the conformational equilibrium were well reproduced (Abe *et al.*, 1969; Pierce *et al.*, 1969).

Cyclobutanone has been studied by microwave spectroscopy, and the experimental structure is shown in Fig. 5 (Scharpen and Laurie, 1968; see also, d'Annibale and Lunazzi, 1973). The corresponding calculated numbers are given in parentheses. The experimental

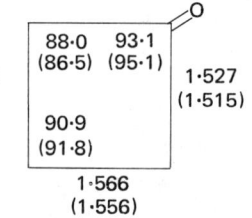

Figure 5. Experimental and calculated structure for cyclobutanone.

spectrum shows that the molecule has a very low barrier at the planar configuration, and the first vibrational level is above that barrier. Hence the molecule is effectively planar, vibrating back and forth through the planar configuration with a puckering motion of large amplitude. The calculated structure is also planar. The bond lengths are unusually long, which is typical for a four-membered ring. The angle at the carbonyl group is larger than the other angles in the four-membered ring, as one would expect from the hybridization. These features are reproduced in a semiquantitative way by the calculations, but the agreement is not impressive. Cyclopentanone was calculated to have two stable conformations, of which the half-chair (C_2) was more stable by 3·2 kcal mole^{-1} than the envelope (C_s). Only the half-chair form was detected experimentally (Kin and Gwinn, 1969; Geise and Mijlhoff, 1970), and this situation is to be contrasted with that found with cyclopentane itself, where the corresponding conformations are of identical energy.

Cyclohexanone has been studied in some detail. Early hand calculations on this molecule suggested (Eliel et al., 1965, p. 186) that the relative energies of the boat forms and the barrier to the chair-chair inversion would be smaller in cyclohexanone than in

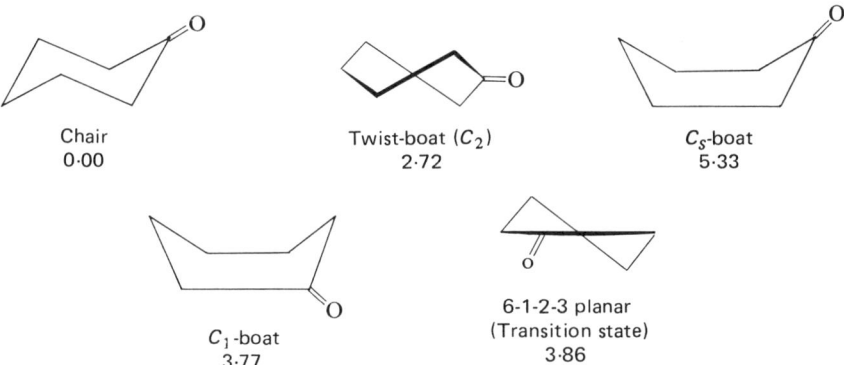

Figure 6. Relative energies (kcal mole^{-1}) of cyclohexanone by molecular mechanics calculations (Allinger et al., 1972; Allinger and Tribble, 1972).

cyclohexane, because the torsional barriers about the bond between the carbonyl carbon and the α-carbon are much lower in carbonyl compounds than in saturated molecules. These conclusions were borne out by the molecular mechanics calculations, which gave the energies shown in Fig. 6 (Allinger et al., 1972; Allinger and Tribble, 1972). Experimental data gave the free energy for the non-chair form of cyclohexanone as 3·3 kcal mole^{-1} above that of the chair (Allinger and Karkowski, 1965; Allinger et al., 1966). The barrier to inversion was measured by low temperature nmr methods and has a value of 4·9 kcal mole^{-1} (Anet et al., 1973), compared with a calculated value of 3·9 kcal mole^{-1}.

A considerable amount of calculation has been devoted to examining the various conformations of a number of medium-ring ketones (Allinger et al., 1972; Allinger and Tribble, 1972). These studies have been somewhat fragmentary, however, because in general a medium-ring hydrocarbon has accessible to it a substantial number of conformations, and when the hydrocarbon is converted to the ketone, the carbonyl group may occupy any of several non-equivalent positions for each of the basic hydrocarbon conformations. Thus the number of possible stable conformations for any one compound can be quite large. Typically, a few of the more likely conformations have been examined, and the one of lowest energy has been taken to represent the compound. It is quite possible that other structures of even lower energy have been overlooked, however. The conformations of medium ring ketones are of special interest because they are novel structures, about which there is available a considerable amount of information, mainly from the elegant low-temperature nmr work of Anet (Anet et al., 1974).

Carbonyl derivatives of steroidal systems presented a number of intriguing puzzles in the early days of conformational analysis. These have now been fairly well tracked down and sorted out, and the results are understood (for a review of the literature up to about 1965, see Eliel et al., 1965, pp. 113, 170).

To explain available conformational data, Klyne (1956) proposed the existence of a "2-alkyl ketone effect", and a "3-alkyl ketone effect", in which, to use the 2-methyl- and 3-methylcyclohexanones as examples, the conformational energy of the methyl group differs from what it would be in the corresponding hydrocarbon. The 2-alkyl ketone effect was subsequently shown to be nonexistent for a methyl group by experimental work, but the 3-alkyl ketone effect was real. For the latter, the facts are that a methyl group in an axial position on a cyclohexanone ring at carbon-3 has a somewhat smaller conformational energy than it would in the corresponding hydrocarbon (about 0.5 kcal mole^{-1} less). Klyne interpreted this as due to a lack of repulsion between the methyl and the *syn*-axial hydrogen which was present in the hydrocarbon but absent in the ketone. The force-field calculations give numbers close to those found experimentally, and verify Klyne's original interpretation.

In the course of the above studies, it was noted that the conformational energy of a methyl group at the 2-position of cyclohexanone was calculated to be slighly greater if a 6-equatorial methyl group was present (1.89 kcal mole^{-1} versus 1.82 kcal mole^{-1} for 2-axial methylcyclohexanone itself). The calculations showed that this difference was due to a small deformation introduced by the equatorial methyl group, which pushed the geminal hydrogen atom further into the ring, where it interfered more with the *syn*-axial methyl group (Allinger *et al.*, 1967). Although the effect is quite small, it had previously been detected experimentally (Eliel and Brett, 1965).

1,4-Cyclohexanedione is an interesting molecule, for which it has been shown that the ring prefers a non-chair conformation.* The calculations indicate that the chair and twist forms are of nearly equal enthalpy, so presumably the preponderance of the twist form results from an entropy effect. The calculation definitely indicates that for the isolated molecule the symmetrical (D_2) form is of lower energy than the C_1 structure which is found in the crystal. All of the available experimental data may also be interpreted as being consistent with this conclusion.

Returning to the steroids, the decalone and hydrindanone systems have been studied in some detail both experimentally and by calculation, in the steroids themselves, and also in model compounds (Allinger *et al.*, 1974). *trans*-1-Decalone exists in a single conformation [14], while *cis*-1-decalone is a mixture of the two conformations [15] and [16]. The calculated $\Delta G°_{298}$ for the reaction

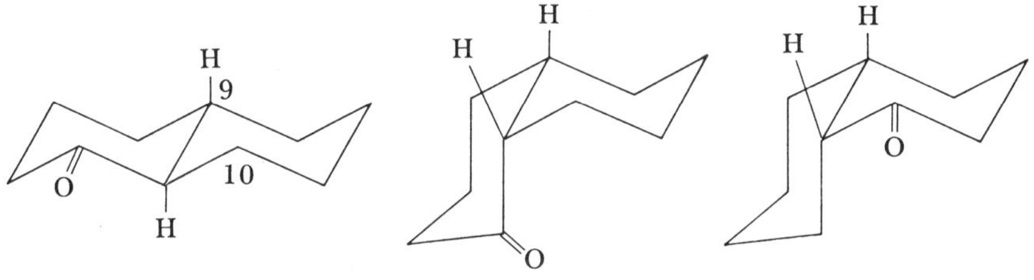

[14]   [15]   [16]

trans ⇌ *cis* is 1·85 kcal mole^{-1}, while experimental values range from about 1·1 to 3·1 kcal mole^{-1}. The presence of a bridgehead methyl group greatly alters the situation. For the 10-methyl-1-decalone system, the *trans*-isomer is slightly favored at equilibrium, while for the 9-methyl-1-decalone system, the *cis*-isomer is favored. These results are well reproduced by a molecular mechanics calculation (Allinger *et al.*, 1974). The analogous equilibria for the 4-keto- and the 6-keto-steroids have also been measured experimentally, and the values are again reproduced by the calculations (Allinger *et al.*, 1974).

* The literature regarding the structure of this molecule is massive and in part contradictory. For an up-to-date review, see Allinger and Wertz, 1973.

With the decalones, the simple qualitative ideas of conformational analysis lead to good predictions regarding the equilibria. For the hydrindanones, the situation is more complicated. Here the flexibility of the five-membered ring makes a simple interpretation of the results impossible. The calculations again reproduce the experimental data very well, however. For example, consider the 15-keto-steroids with a β-substituent at C-17 which is in turn hydrogen, methyl, or isopropyl. For these compounds the *cis*-isomer is the more stable experimentally by 1·2, 0·3 and −0·7 kcal mole^{-1}, respectively (van Horn and Djerassi, 1967). The calculations indicate that the *cis*-isomer has a lower enthalpy by 2·3, 1·4, and −0·7 kcal mole^{-1}, respectively (Allinger *et al.*, 1972; Allinger and Tribble, 1972). Thus the trend is predicted, although the numbers are not quantitatively accurate. From the details of the calculations, one can see exactly what interactions are bringing about the observed results. The interpretations based on the calculations are complicated, and they are different from what had been concluded earlier from an examination of models (van Horn and Djerassi, 1967).

Thiols and Thioethers

Extensive thermodynamic and spectroscopic studies have been carried out on the low molecular weight aliphatic thiols and thioethers, and a great deal of information is available concerning their conformations, conformational equilibria, rotational barriers, and heat contents. A moderate amount of structural information is also available, so that it was possible to deduce parameters to fit these compounds quite well (Allinger and Hickey, 1972; Hickey, 1973). As long as only one sulfur atom was present, the electrostatic problem was ignored. No explicit account of the lone pairs on sulfur was made, and this seems like a reasonable approximation here. In this case the structural and energetic results were quite good, comparable to those obtained for hydrocarbons. The calculations were also extended to include disulfides (Allinger and Hickey, 1975; Hickey, 1973). For example, the polycyclic disulfide shown in Fig. 7 was calculated to have the structure (a), which is what is found crystallographically (Wahl *et al.*, 1975) rather than the alternative structure (b). Since it is known that the torsional angle preferred about the S—S bond is approximately 90°, it had been supposed that structure (b), where the torsional angle is 100°, would be favored over (a),

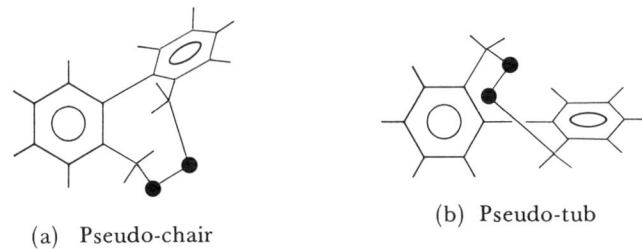

Figure 7. Conformations of 5H,8H-dibenzo(d,f)(1,2)dithiocin.

where this angle is 54°. According to the calculations, while it is true that the torsion about the S—S bond does in fact favor (b), much more serious is the torsion about the C—S bonds, which combined with the unfavorable torsion about the biphenyl bond leads to (a) being more stable than (b) by a calculated value of 1.34 kcal mole^{-1}.

It is known that the axially protonated form of thiacyclohexane is the stable conformation and the same is true in the corresponding selenium case (Allinger and Hickey, unpublished; Hickey, 1973). The that the axial proton is favored because of attractions between it and the axial hydrogens in the 3- and 5-positions.

When the calculations were carried out on these compounds, it was indeed found that the axial conformation is preferred, by 0·27 kcal mole^{-1} in the sulfur case, and 0·57 kcal mole^{-1} in the selenium case (Allinger and Hickey, unpublished; Hickey, 1973). The calculations do not indicate any attraction, however, between the axial hydrogens; rather, there is appreciable repulsion (0·2 and 0·4 kcal mole^{-1} for the selenium and sulfur compounds respectively). The main source of energy which is tending to stabilize the axial conformation comes from the hydrogens vicinal to the proton in question which exert a substantial repulsion on the latter. When the proton is in the equatorial position, there are four such repulsions, and only two when it is axial. This appears to be an example of the "gauche effect" (Wertz and Allinger, 1974) discussed earlier, and it very well accounts for the observed facts in this case. The driving force tending to push the proton axial is greater in the sulfur compound than in the selenium compound because of the shorter bonds. However, the *syn*-axial protons are closer and push back harder in the case of the sulfur compound, and the balance of these interactions is as indicated.

Halides

Preliminary studies indicated that the structure of the monohalides (fluorides, chlorides, bromides, and iodides) could all be reproduced at least roughly without undue difficulty (Allinger et al., 1969). More systematic and detailed studies have now borne out this conclusion (A. Y. Meyer, unpublished results). There is, however, a systematic error which persists when the calculated *gauche/anti* ratio of halides is compared with that found experimentally. The calculated enthalpy is always a little smaller for the *anti*-conformation than for the *gauche*, although with simple compounds the *gauche* is a *dl* mixture, while the *anti* is not, and the calculated equilibrium constants favor slightly the *gauche* conformation. Experimentally, the *gauche* conformation seems to be preferred, apparently by a somewhat larger amount than is calculated, although the data are not very precise. However, carrying this over to the cyclohexyl system, it is found experimentally that the cyclohexyl halides prefer the equatorial conformation by $0·2-0·5$ kcal mole^{-1} and there is no systematic trend with the size of the halogen (Jensen et al., 1969; Hirsch, 1967). The calculations indicate that the equatorial conformation is preferred by $0·3-1·0$ kcal mole^{-1}, with the number growing larger with increasing halogen size. The errors are not very big, but they are quite definite and systematic. There seem to be no simple adjustments of the halogen parameters which will improve the situation. This leads us to believe that the inconsistency is due either to an error inherent in the hydrocarbon calculation as it carries over to the calculation of the halide, or to improper halide–hydrogen interactions, probably as a result of neglect of the electrostatics. More work will be needed to resolve the problems here, but, in the meantime, the results are pretty good in any case.

There is a moderate amount of heat of formation data available for simple halides (Cox and Pilcher, 1970) and while it is in general less accurate than that for hydrocarbons, it is still reasonably good. There is no difficulty in fitting these data, and the heats of formation of monohalides can be calculated about as accurately as they can be measured experimentally.

Other Compounds

In addition, a number of miscellaneous compounds have been studied to solve specific problems. These include sulfoxides, nitriles, and a number of acid derivatives (esters, amides) (Allinger et al., 1969; M. T. Tribble, unpublished results). Sufficient structural information on most classes of compounds seems to be available in the literature so that, together with some reasonable guesses, one can reproduce fairly well the structures of most functional groups, as long as the molecules are not highly strained. If they are highly strained, then one really needs to have some experimental data on force constants, torsional barriers, or whatever may be appropriate for the particular case at hand. Such data are often difficult to obtain experimentally, but they are probably available with sufficient accuracy from *ab initio* calculations, at least for compounds which contain only first and second row elements. Thus extension of these methods to many other functional groups has been and should continue to be straightforward.

There are two classes of compounds to which attention should be called, because they do *not* appear in Table 2. These are the amines and the ethers (including alcohols). While a cursory treatment seemed to indicate that these compounds could be studied without any special difficulty (Allinger et al., 1969), more detailed studies showed that this was not true (D. Y. Chung, unpublished results). The electron density about the nitrogen or oxygen atom in these compounds is clearly not well described by a sphere centred on the nucleus (Robb et al., 1973). It would seem, at least within the context of our own force-field, that lone pairs must be included and given explicit treatment if one is adequately to fit available experimental data on these compounds. There is certainly ample precedent for such a procedure*, and preliminary studies indicate that this approach probably will be adequate for the calculation of the structures and energies of these classes of molecules. Because of the incomplete nature of the work, however, it will not be further discussed here.

Electrostatic Effects

The electrostatic effects present in polar molecules clearly lead to pronounced conformational changes and sizable energy differences, relative to the corresponding molecules without such polar interactions. Thus, to take a simple example, 1,2-dichloroethane is a

---

* The explicit consideration of lone pairs is a significant part of the so-called X-α method (see, for example, Kaufman, 1973).

mixture of *gauche* and *anti* conformations (for a summary, see Mizushima, 1954). The composition of the mixture is highly dependent upon phase and solvent, since in general the solvent with the higher dielectric constant tends to stabilize the molecule with the more unfavorable charge interactions (*gauche*) more than it does the other one (*anti*).

There are two simple methods which have so far been used in the literature as a starting point to allow for electrostatic interactions in force-field calculations. These are the point-charge interaction model, and the dipole–dipole interaction model (see, for example, Lehn and Ourisson, 1963; Collins and Kirk, 1970). These models are both approximations which will be accurate in the limit in which the distance between charges is large compared to the diameter of the atom (or the extent of charge distribution). Thus the approximation should be quite good if the charges are separated from one another by several atomic diameters, but poorer when the charges are on atoms that are close together in space. Just how good the approximations will be in cases of practical interest is not certain. We have studied several cases by both methods, and the results were substantially the same in each case; there are indications in the literature that this is not always true, however (Collins and Kirk, 1970). In particular, the results of the calculations are sometimes sensitive to the positioning of the point charges or dipoles, and this is a rather arbitrary matter. The current situation can be summarized as follows. If there are only two or three dipoles in a molecule, and if they are reasonably far apart (separated by at least two or three atoms) the methods described are probably adequate for the most part. As the dipoles increase in number, or become closer together, such approximations are suspect at best, and may break down completely. A further complication in the study of polar molecules is that experimental studies are generally made in solution, whereas the calculations are for the isolated molecule in the gas phase; the effects of solvation must somehow be explicitly considered. Solvation makes some difference even in non-polar molecules, but the difference is sufficiently small that it can be ignored in most cases (Ford and Allinger, 1970; Ouellette and Williams, 1971). This is certainly not true with highly polar molecules. Carrying polarity to an extreme, we come to the case of a hydrogen bond which differs from ordinary bonds in the present context. While the latter may stretch or compress only slightly in the ground states of real molecules, hydrogen bonds come apart easily if they are extended. Therefore, while Hooke's law is adequate for ordinary bonds, one needs a Morse potential, or something equivalent, to describe a hydrogen bond. Some studies have been directed toward this problem, but they have so far been fairly limited (see, for example, McGuire et al., 1972).

We might consider 1,4-dichlorocyclohexane as an example of the seriousness of the electrostatic interaction which may arise in a simple case (Abraham and Rossetti, 1972; see also, Wood and Woo, 1967). The equatorial chlorine in chlorocyclohexane is more stable than the axial by about 0·5 kcal mole^{-1}; hence, neglecting electrostatic effects, *trans*-1,4-dichlorocyclohexane would be predicted to be more stable in the diequatorial conformation than in the diaxial by 1·0 kcal mole^{-1} (gas phase). The experimental value (corrected to the gas phase) shows the diequatorial form to be *less* stable than the diaxial by 0·8 kcal mole^{-1}, leaving about 1·8 kcal mole^{-1} to be accounted for. The dipoles are fairly distant from one another here, so that one might expect that either the point charge model or the dipole–dipole model would give a good account of the electrostatics. In fact, either of the two treatments gives an electrostatic interaction which favors the diaxial form by about 0·3 kcal mole^{-1}, still leaving about 1·5 kcal mole^{-1} to be accounted for. Thus the simple approximations mentioned are inadequate, and one must look further to understand the electrostatics in this molecule.

Abraham has carried out CNDO calculations on 1,4-dichlorocyclohexane, and has found that the axial chlorine at C-1 induces positive charges on the axial hydrogens at C-2 and C-6 (Abraham and Rossetti, 1972). These induced dipoles are close to and oriented so as to oppose the C–Cl dipole at C-4. Thus, in the diaxial conformation each halogen induces a large charge separation in the vicinal axial C–H bonds, and these induced charge distributions have a strong and favorable interaction with the other C–Cl dipole which stabilizes that conformation. The magnitude of this effect as calculated by the CNDO method is about right, and it suggests that induction will be an important factor in the conformational analysis of polar molecules.

As it turns out, a method for treating this problem of induction was developed more than 20 years ago (Smith et al., 1951; Smith and Mortenson, 1956), but it has fallen into relative obscurity in more recent years. The original method, developed before the availability of computers, contained simplifications to make it convenient for hand calculations. The method has more recently been extended to eliminate these simplifications, and has been applied to a number of representative molecules.* Preliminary results indicate that this kind of modification will very much improve our ability to deal with polyfunctional molecules.

Obviously, charges (and hence dipoles) can be calculated by the *ab initio* method and its simplified modifications. A very simple procedure exists (Del Re, 1958) which, because of parameterization, essentially reduces to the classical case.

Additionally, if one wishes to study polar molecules in solution, some treatment of solvation energy is required. The early simple approach of adjusting the "effective dielectric constant" of the medium is pretty crude (Allinger and Allinger, 1958). The more sophisticated consideration of dipole and quadrupole terms would appear to be better (Abraham and Cooper, 1967), and in fact works well in at least one case where the simple method fails to predict correctly even the direction of relative changes in conformational energies with solvent changes (Viskocil, 1974). Studies utilizing these methods have been only preliminary.

10. REACTION RATES

One important area of research in organic chemistry in recent years has been concerned with rates of chemical reactions. In order to predict and understand reaction rates, one must understand both the starting and transition states of the reaction. The calculations we have discussed so far have applied to starting states. If we can apply similar calculations, modified as appropriate, to transition states, presumably we can learn something about chemical reactions and in particular about reaction rates. Indeed, Westheimer's initial application of the molecular mechanics method was to such a problem; the rates of racemization of some *ortho*-substituted biphenyls.

We might begin by thinking about a very simple "reaction", in which bonds are neither made nor broken. The conformational inversion of a cyclohexane ring can be taken as an example of such a "reaction". Calculations like those described for the starting state may be also carried out for the transition state. In some cases the

---

* The current method was developed by M. T. Wuesthoff at Wayne State University, 1968-9, and has been utilized in further conformational studies by Dosen-Micovic, 1973. See also, Mark and Sutton, 1972.

transition state is defined or at least approximated by a geometric constraint, and the calculation is straightforward, the only change from the usual situation being that the energy is not allowed to minimize with respect to certain constrained degrees of freedom. The inversion of cyclohexane falls into this category. For a more general case, where simple geometric constraints are inadequate, a clever device has been developed (Wiberg and Boyd, 1972). This begins at the starting state, and systematically drives a chosen dihedral angle through a range of values, such that the molecule goes through the transition state and ends up as the product. This is done by using an artificially large dihedral angle potential to calculate the geometry, and then the true potential to calculate the energy. The artificially large potential is moved incrementally along the reaction co-ordinate so that repetition of the calculation gives points (energies and corresponding geometries) at intervals along the reaction coordinate. The highest point on the curve gives the structure and energy of the transition state. This device has been applied in a number of cases, and works very well.

If the reaction is of the more usual type, in which one or more bonds are made or broken, the situation is more complicated and additional information is necessary. Perhaps the most straightforward situation arises if one wants to compare the same reaction for a series of compounds, where certain geometric features of the transition state can be assumed to be similar throughout the series, and in which the rates are affected only by steric interactions. Such a case was investigated as long ago as the 1950s. Ingold and co-workers studied $S_N 2$ reactions on a series of alkyl halides, essentially carrying out force-field calculations by hand, but making use of a large number of simplifying assumptions in order to keep the arithmetic manageable (Ingold, 1957; de la Mare et al., 1955). Their conclusion was that they could calculate very well indeed relative rates for a series such as methyl, ethyl, isopropyl, t butyl, neopentyl. Obviously the same can be done with much greater ease using the more automated and less approximate procedures now available. It is not certain that the results will be any better than they were with the hand calculations, however. The reason for this is that with only a small set of numbers to calculate, a lot of error can be soaked up in the approximations. In a more rigorous force-field calculation, many of these approximations would not be justifiable, and one would have to use the parameters that one uses for ground state molecules. In any case, this should be a very fruitful area for future work.

So far there have been a few calculations aimed along these lines. Schleyer has studied rates of solvolyses by calculating the relative energies of alkanes and the related carbonium ions (Fry et al., 1972; Lenoir et al., 1974; Gleicher and Schleyer, 1967). The interaction energy of one of the hydrogen atoms in a molecule is ordinarily rather small. If that hydrogen were replaced by a halogen atom, the difference in interaction energy between the hydrogen and the halogen would be relatively small (a fraction of a kcal), and, more important, it would be very nearly a constant for a series molecules. This would simply multiply the reaction rates for the series of molecules by a constant factor, and thus it would not affect relative rates. (A tosylate would similarly have an effect that would cancel out, provided that it could always orient so as to avoid serious steric interactions.) The carbonium ion is believed to be a fair model for the transition state in a solvolysis reaction, and therefore one would hope that the calculated steric energy differences between hydrocarbons and the corresponding carbonium ions would correlate well with observed rate differences. One might worry that because solvation energies are so large for carbonium ions, if they were in any way erratic, they would destroy such a correlation. One would also have to worry about possible complications due to some reactions going through different stages of ion pairing. Thus it was not clear *a priori* that one would be able to find a good correlation between the desired quantities. As Schleyer has shown, however, such a correlation is indeed good, and it is applicable over a much wider range than one might have suspected. In a series of studies, rate constants spanning 25 orders of magnitude were calculated with an average deviation of only a factor of two.

A second example of reaction rates recently studied concerned "conformational transmission"; specifically the relative rate of benzylidene formation when variously substituted 3-keto steroids were allowed to react with benzaldehyde (Barton et al., 1957). Here there are grave problems concerning the rate determining step and transition state, but the assumption was made that the Δ^2-olefin was a good model for the transition state, and the corresponding saturated compound was a good model for the starting state. Again, a fairly good correlation was observed between the calculated energy differences and the observed rates (Allinger and Lane, 1974).

Wipke has used an adaptation of the general method to study relative rates of attack (stereoselectivity) on ketones by hydride, and has found the experimental facts can be reproduced quite well (Wipke and Gund, 1974).

The Cope rearrangement of a diene to another diene has been examined in considerable detail by the molecular mechanics method for a number of different systems, with the full reaction coordinate being investigated. Generally satisfactory agreement with experiment was obtained, although the experimental data were scanty. Since in most cases small and medium rings were being opened and closed, and since a combined treatment of π- and σ-systems was necessary, the success obtained here is encouraging indeed (Simonetta et al., 1968). Finally, DeTar has studied the lactonization of hydroxy acids in bridged ring systems, with the aid of hydrocarbon models (DeTar, 1974 a,b). The calculated and experimental rates show a reasonable correlation.

The examples, together with Ingold's earlier study, seem sufficient to give one confidence that the use of molecular mechanics for the prediction to reactions rates will be a useful one. Such things as neighboring group participation will not, of course, be accounted for by such calculations. The test for unusual phenomena of that sort will be the failure of the calculations to predict the rate. The force-field method gives us a potential method for calculating the steric effects on reactions. When the calculations can be carried out reliably, the discrepancy between the calculated and observed rates will be a clear measure of the "non-steric" phenomena, and this use of the method has been exploited by Schleyer (Fry et al., 1972; Lenoir et al., 1974; Gleicher and Schleyer, 1967).

11. CONCLUSIONS

In a paper published in 1967 (Allinger et al., 1967) we stated that it was already possible at that time to determine the structures of many hydrocarbons by molecular mechanics calculations with an accuracy competitive with that obtainable by existing experimental methods. By 1968 (Allinger et al., 1968a) we had concluded that the heats of formation and relative energies of most hydrocarbons were also susceptible to accurate prediction by the same calculations. At that time, the range of applicability of the available force-fields was somewhat limited, and this range has now greatly increased. In addition, the amount of testing which has been carried out has yielded a much improved understanding of the reliability of the method. The amount of accurate structural data available from calculations on hydrocarbons now probably exceeds that which has been obtained by all experimental methods combined to date. The amount of data on heats of formation of hydrocarbons is approaching that which has been determined experimentally. A considerable amount of calculational information is now available on monofunctional compounds, and, with a few exceptions, it would seem that calculations in this area will be reasonably straightforward.

With polyfunctional compounds, serious problems remain to be solved. Electrostatic interactions, induction, and solvation are sticky problems for which general solutions are at present lacking, or at best, of questionable reliability. These are areas of much current excitement and interest.

So far, almost all of the applications of the method have been to the ground states of molecules. Various kinds of excited states are equally susceptible to study. Transition states are likewise open to attack by these methods, although there are of course limitations. Considerable advances here are anticipated in the near future. These areas seem potentially even more useful ones for molecular mechanics calculations than the area of ground states, because ground states can be accurately studied by several experimental methods. The structures of excited states and transition states seem virtually unassailable experimentally, however. They can be attacked by *ab initio* methods, but the same considerations of general usefulness for routine calculations apply here as discussed at the beginning of this article for ground states.

We continue to believe that the force-field method offers a rapid, convenient and reliable method for the determination of molecular structures and energies. While there are limitations to the method, as there are with each of the experimental methods, the usefulness of this technique now seems generally appreciated. We can forsee only a continuing expansion of the development and applications of force-field calculations in many areas of chemistry.

REFERENCES

Abe, M., Kuchitsu, K. and Shimanouchi, T. (1969). *J. Mol. Struct.* **4**, 245.
Abraham, R. J., and Rossetti, Z. L. (1972). *Tetrahedron Lett.* 4965.
Abraham, R. J., and Cooper, M. A. (1967). *J. Chem. Soc.* (B) 202.
Abraham, R. J., and Loftus, P. (1974). *Chem. Comm.* 180.
Abraham, R. J., and Parry, K., (1970). *J. Chem. Soc.* (B) 539.
Allinger, J., and Allinger, N. L. (1958). *Tetrahedron* **2**, 64.
Allinger, N. L., and Graham, J. C. (1973). *J. Amer. Chem. Soc.* **95**, 2523.
Allinger, N. L., and Hickey, M. J. (1972). *Tetrahedron* **28**, 2157.

Allinger, N. L., and Hickey, M. J. (1975). *J. Amer. Chem. Soc.* **97**, 5167.
Allinger, N. L., and Karkowski, F. M. (1965). *Tetrahedron Lett.* 2171.
Allinger, N. L., and Lane, G. A. (1974). *J. Amer. Chem. Soc.* **96**, 2937.
Allinger, N. L., and Sprague, J. T. (1972). *J. Amer. Chem. Soc.* **94**, 5734.
Allinger, N. L., and Sprague, J. T. (1973). *J. Amer. Chem. Soc.* **95**, 3893.
Allinger, N. L., and Sprague, J. T. (1975). *Tetrahedron* **31**, 21.
Allinger, N. L., and Tai, J. C. (1965). *J. Amer. Chem. Soc.* **87**, 2081.
Allinger, N. L., and Tribble, M. T. (1972). *Tetrahedron* **28**, 1191.
Allinger, N. L., and Wertz, D. H. (1973). *Rev. Latinamericana de Quimica* **4**, 127.
Allinger, N. L., and Wu, F. Z. (1971). *Tetrahedron* **27**, 5093.
Allinger, N. L., Blatter, H. M., Freiberg, L. A., and Karkowski, F. M. (1966). *J. Amer. Chem. Soc.* **88**, 2999.
Allinger, N. L., Miller, M. A., Van-Catledge, F. A., and Hirsch, J. A. (1967a). *J. Amer. Chem. Soc.* **89**, 4345.
Allinger, N. L., Tai, J. C., and Stuart, T. W. (1967b). *Theoretica Chim. Acta* **8**, 101.
Allinger, N. L., Hirsch, J. A., Miller, M. A., Tyminski, I. J., and Van-Catledge, F. A. (1968a). *J. Amer. Chem. Soc.* **90**, 1199.
Allinger, N. L., Hirsch, J. A., Miller, M. A., and Tyminski, I. J. (1968b). *J. Amer. Chem. Soc.* **90**, 5773.
Allinger, N. L., Szkrybalo, W., and Van-Catledge, F. A. (1968c). *J. Org. Chem.* **33**, 784.
Allinger, N. L., Hirsch, J. A., Miller, M. A., and Tyminski, I. J. (1969). *J. Amer. Chem. Soc.* **91**, 337.
Allinger, N. L., Tribble, M. T., Miller, M. A., and Wertz, D. H. (1971). *J. Amer. Chem. Soc.* **93**, 1637.
Allinger, N. L., Tribble, M. T., and Miller, M. A. (1972). *Tetrahedron* **28**, 1173.
Allinger, N. L., Sprague, J. T., and Finder, C. J. (1973). *Tetrahedron* **29**, 2519.
Allinger, N. L., Lane, G. A., and Wang, G. L. (1974). *J. Org. Chem.* **39**, 704.
Almenningen, A., Andersen, B., and Nyhus, B. A. (1971). *Acta Chem. Scand.* **25**, 1217.
Almenningen, A., Andersen, B., Nyhus, B. A., and Srinivasan, R., *in press* (1975).
Altona, C. (1971). In "Conformational Analysis. Scope and Present Limitations", G. Chiurdoglu (ed.), Academic Press, New York, p. 1.
Altona, C., and Hirschmann, H. (1970). *Tetrahedron* **26**, 2173.
Altona, C., and Sundaralingam, M. (1970). *Tetrahedron* **26**, 925.
Altona, C., de Graaff, R. A. G., Leeuwestein, C. H., and Romers, C. (1971). *Chem. Comm.* 1305.
Andrews, D. H. (1930). *Phys. Rev.* **36**, 544.
Anet, F. A. L. (1962). *J. Amer. Chem. Soc.* **84**, 671.
Anet, F. A. L., and Haq, M. Z. (1965). *J. Amer. Chem. Soc.* **87**, 3147.
Anet, F. A. L., Bourn, A. J. R., and Lin, Y. S. (1964). *J. Amer. Chem. Soc.* **86**, 3576.
Anet, F. A. L., Chmurny, G. N., and Krane, J. (1973). *J. Amer. Chem. Soc.* **95**, 4423.
Anet, F. A. L., St. Jacques, M., Henrichs, P. M., Cheng, A. K., Krane, J., and Wong, L. (1974). *Tetrahedron* **30**, 1629.
Bak, B., Led, J. J., Nygaard, L., Rastrup-Andersen, J., and Sørensen, G. O. (1969). *J. Mol. Struct.* **3**, 369.
Baron, P. A., Brown, R. D., Burden, F. R., Domaille, P. J., and Kent, J. E. (1972). *J. Mol. Spectrosc.* **43**, 401.
Bartell, L. S. (1960). *J. Chem. Phys.* **32**, 827.

Bartell, L. S., and Burgi, H. B. (1972). *J. Amer. Chem. Soc.* **94**, 5239.
Bartell, L. S., and Higginbotham, H. K. (1965). *J. Chem. Phys.* **42**, 851.
Bartell, L. S., and Kohl, D. A. (1963). *J. Chem. Phys.* **39**, 3097.
Barton, D. H. R. (1950). *Experientia* **6**, 316.
Barton, D. H. R., Head, A. J., and May, P. J. (1957). *J. Chem. Soc.* 935.
Bastiansen, O., Fernholt, L., Seip, H. M., Kambara, H., and Kuchitsu, K. (1973). *J. Mol. Struct.* **18**, 163.
Beagley, B., Brown, D. P., and Monaghan, J. J. (1969). *J. Mol. Struct.* **4**, 233.
Benson, S. W., Cruickshank, F. R., Golden, D. M., Haugen, G. R., O'Neal, H. E., Rodgers, A. S., Shaw, R., and Walsh, R. (1969). *Chem. Rev.* **69**, 279.
Bingham, R. C., and Dewar, M. J. S. (1972). *J. Amer. Chem. Soc.* **94**, 9107.
Blackburne, I. D., Duke, R. P., Jones, R. A. Y., Katritzky, A. R., and Record, K. A. F. (1973). *J.C. S. Perkin II*, 332.
Bondi, A. (1964). *J. Phys. Chem.* **68**, 441.
Boyd, R. H. (1968). *J. Chem. Phys.* **49**, 2574.
Boyd, R. H., Sanwal, S. N., Shary-Tehrany, S., and McNally, D. (1971). *J. Phys. Chem.* **75**, 1264.
Bregman, J., Hirschfeld, F. L., Rabinovich, D., and Schmidt, G. M. J. (1965). *Acta Crystallog.* **19**, 227.
Bright, D., Maxwell, I. E., and de Boer, J. (1973). *J.C.S. Perkin II*, 2101.
Bromberg, A., and Muszkat, K. A. (1972). *Tetrahedron* **28**, 1265.
Brown, W. A. C., Martin, J., and Sim, G. A. (1965). *J. Chem. Soc.* 1844.
Brubaker, G. R., and Euler, R. A. (1972). *Inorg. Chem.* **11**, 2357.
Buckingham, D. A., and Sargerson, A. M. (1971). *Topics in Stereochem.* **6**, 219.
Bucourt, R., and Cohen, N. C. (1970). *Bull. Soc. Chim. France* 2015.
Bucourt, R., and Hainaut, D. (1965). *Bull. Soc. Chim. France* 1366.
Buemi, G., Favini, G., and Zuccarello, F. (1970). *J. Mol. Struct.* **5**, 101.
Burgi, H. B., and Bartell, L. S. (1972). *J. Amer. Chem. Soc.* **94**, 5236.
Buys, H. R., and Geise, H. J. (1970). *Tetrahedron Lett.* 2991.
Carreira, L. (1975). *J. Chem. Phys.* **62**, 3851.
Chang, S., McNally, D., Shary-Tehrany, S., Hickey, M. J., and Boyd, R. H. (1970). *J. Amer. Chem. Soc.* **92**, 3109.
Chiang, J. F. (1971). *J. Amer. Chem. Soc.* **93**, 5044.
Clark, R. G., and Stewart, E. T. (1970). *Quart. Rev.*, **24**, 95.
Cohen, N. C. (1971). *Tetrahedron* **27**, 789.
Collins, L. J., and Kirk, D. N. (1970). *Tetrahedron Lett.*, 1547.
Cooper, A., Norton, D. A., and Hauptman, H. (1969). *Acta Cryst.* **25B**, 814.
Cox, J. D., and Pilcher, G. (1970). "Thermochemistry of Organic and Organometallic Compounds", Academic Press, New York.
Crabbé, P. (1965). "Optical Rotatory Dispersion and Circular Dichroism in Organic Chemistry", Holden-Day, San Francisco.
Dallinga, G., and Toneman, L. H. (1967a). *Rec. Trav. Chim.* **86**, 171.
Dallinga, G., and Toneman, L. H. (1967b). *J. Mol. Struct.* **1**, 117.
d'Annibale, A., and Lunazzi, L. (1973). *J.C.S. Perkin II*, 1908.
Dashevskii, V. G., and Kitaigorodskii, A. I. (1967). *Theor. Eksp. Khim.* **3**, 22.
Dashevskii, V. G., and Lugovskoi, A. A. (1972). *J. Mol. Struct.* **12**, 39.
de la Mare, P. B. D., Fowden, L., Hughes, E. D., Ingold, C. K., and Mackie, J. D. H., (1955). *J. Chem. Soc.* 3200.
Del Re, G. (1958). *J. Chem. Soc.* 4031.
DeTar, D. F. (1974a). *J. Amer. Chem. Soc.* **96**, 1254.
DeTar, D. F. (1974b). *J. Amer. Chem. Soc.* **96**, 1255.
Dewar, M. J. S. (1969). "The Molecular Orbital Theory of Organic Chemistry", McGraw-Hill, New York.

Dewar, M. J. S., Haddon, R. C., and Student, P. J. (1974). *Chem. Comm.* 569.
Dobler, M., and Dunitz, J. D. (1964). *Helv. Chim. Acta* **47**, 695.
Dodziuk, H. (1974). *J. Mol. Struct.* **20**, 317.
Dosen-Micovic, L. (1973). Ph.D. Thesis, University of Georgia.
Dwyer, M., and Searle, G. H. (1972). *Chem. Comm.* 726.
Eliel, E. L., and Brett, T. J. (1965). *J. Amer. Chem. Soc.* **87**, 5039.
Eliel, E. L., Allinger, N. L., Morrison, G. A., and Angyal, S. J. (1965). "Conformational Analysis", Wiley-Interscience, New York.
Engler, E. M., Andose, J. D., and Schleyer, P. von R., (1973). *J. Amer. Chem. Soc.* **95**, 8005.
Ermer, O., and Lifson, S. (1973). *J. Amer. Chem. Soc.* **95**, 4121.
Faber, D. H., and Altona, C. (1971). *Chem. Comm.* 1210.
Farcasiu, D., Wiskott, E., Osawa, E., Thielecke, W., Engler, E. M., Slutsky, J., Schleyer, P. von R., and Kent, G. J., (1974). *J. Amer. Chem. Soc.* **96**, 4669.
Favini, G., and Nava, A. (1973). *Theor. Chim. Acta* **31**, 261.
Favini, G., Buemi, G., and Raimondi, M. (1968). *J. Mol. Struct.* **2**, 137.
Ferre, Y., Vincent, E.-J., Metzger, J., Samat, A., and Guglielmetti, R. (1974). *Tetrahedron* **30**, 787.
Ferro, D. R., and Hermans, J. (1970). In "Liquid Crystals and Ordered Fluids", J. F. Johnson and R. S. Porter (eds.), Plenum, New York.
Figeys, H. P., and Dralants, A. (1971). *Tetrahedron Lett.* 3901.
Finder, C. J., Chung, D., and Allinger, N. L. (1972). *Tetrahedron Lett.* 4677.
Fischer-Hjalmars, I. (1963). *Tetrahedron* **19**, 1805.
Fleischer, E. B. (1964). *J. Amer. Chem. Soc.* **86**, 3889.
Flurry, R. L. (1968). "Molecular Orbital Theories of Bonding in Organic Molecules", Dekker, New York.
Ford, R. A., and Allinger, N. L. (1970). *J. Org. Chem.* **35**, 3178.
Fournier, J., and Waegell, B. (1970). *Tetrahedron* **26**, 3195.
Fournier, J., and Waegell, B. (1972). *Tetrahedron* **28**, 3407.
Fournier, J., and Waegell, B. (1973a). *Bull. Soc. Chim. France* 436.
Fournier, J., and Waegell, B. (1973b). *Bull. Soc. Chim. France* 1599.
Froimowitz, M., and Gans, P. J. (1972). *J. Amer. Chem. Soc.* **94**, 8020.
Fry, J. L., Engler, E. M., and Schleyer, P. von R. (1972). *J. Amer. Chem. Soc.* **94**, 4628.
Gavezzotti, A., Mugnoli, A., Raimondi, M., and Simonetta, M. (1972). *J.C.S. Perkin II*, 425.
Geise, H. J., and Mijlhoff, F. C. (1971). *Rec. Trav. Chim.* **90**, 577.
Geise, H. J., Buys, H. R., and Mijlhoff, F. C. (1971). *J. Mol. Struct.* **9**, 447.
Geue, R. J., and Snow, M. R. (1971). *J. Chem. Soc. (A)* 2981.
Gleicher, G. L., (1967). *Tetrahedron* **23**, 4257.
Gleicher, G. J., and Schleyer, P. von R. (1967). *J. Amer. Chem. Soc.* **89**, 582.
Golebiewski, A., and Parczewski, A. (1974). *Chem. Revs.* **74**, 519.
Goursot-Leray, A., and Bodot, H. (1971). *Tetrahedron* **27**, 2133.
Graham, J. C. (1971). Ph.D. Dissertation, Wayne State University.
Granger, R., Bardet, L., Sablayrolles, C., and Girard, J. P. (1971). *Bull. Soc. Chim. France*, 4454.
Gund, T. M., and Schleyer, P. von R. (1973). *Tetrahedron Lett.* 1959.
Gust, D., Senkler, G. H., and Mislow, K. (1972). *Chem. Comm.* 1345.
Haase, J., and Krebs, A. (1971). *Z. Naturforsch.* **26a**, 1190.
Hall, H. K., Smith, C. D., and Baldt, J. H. (1973). *J. Amer. Chem. Soc.* **95**, 3197.
Hariharan, P. C., and Pople, J. A. (1974). *Mol. Phys.*, **27**, 209.
Hendrickson, J. B. (1961). *J. Amer. Chem. Soc.* **83**, 4537.
Hendrickson, J. B. (1962). *J. Amer. Chem. Soc.* **84**, 3355.

Hendrickson, J. B. (1964). *J. Amer. Chem. Soc.* **86**, 4854.
Hendrickson, J. B. (1967). *J. Amer. Chem. Soc.* **89**, 7036.
Hickey, M. J. (1973). Ph.D. Dissertation, University of Georgia.
High, D. F., and Kraut, J. (1966). *Acta Cryst.* **21**, 88.
Hilderbrandt, R. L., and Wieser, J. D. (1973). *J. Mol. Struct.* **15**, 27.
Hilderbrandt, R. L., Wieser, J. D., and Montgomery, L. K. (1973). *J. Amer. Chem. Soc.* **95**, 8598.
Hill, T. L. (1948). *J. Chem. Phys.* **16**, 399.
Hirsch, J. A. (1967). *Topics in Stereochem.* **1**, 199.
Hirschfelder, J. O., Curtiss, C. F., and Bird, R. B. (1954). "The Molecular Theory of Gases and Liquids", Wiley, New York.
Hoffmann, R. (1963). *J. Chem. Phys.* **39**, 1397.
Iijima. T. (1972). *Bull. Chem. Soc. Japan* **45**, 1291.
Ingold, C. K. (1957). *Quart. Rev.* **11**, 1.
Jacob, E. J., Thompson, H. B., and Bartell, L. S. (1967). *J. Chem. Phys.* **47**, 3736.
Jensen, F. R., and Bushweller, C. H. (1968). *Tetrahedron Lett.* 2825.
Jensen, F. R., Bushweller, C. H., and Beck, B. H. (1969). *J. Amer. Chem. Soc.* **91**, 344.
Kalb, A. J., Chung, A. L. H., and Allen, T. L. (1966). *J. Amer. Chem. Soc.* **88**, 2938.
Karabatsos, G. J., and Fenoglio, D. J. (1970). *Topics in Stereochem.* **5**, 167.
Karabatsos, G. J., and Hsi, N. (1965). *J. Amer. Chem. Soc.* **87**, 2864.
Kaufman, J. J. (1973). *Int. J. Quant. Chem.* **7**, 369.
Kay, M. I., Okaya, Y., and Cox, D. E. (1971). *Acta Cryst.* **B27**, 26.
Kim, H., and Gwinn, W. D. (1969). *J. Chem. Phys.* **51**, 1815.
Kitaigorodskii, A. I. (1960). *Tetrahedron* **9**, 183.
Kitaigorodskii, A. I. (1961). *Tetrahedron* **14**, 230.
Kitaigorodskii, A. I., and Dashevskii, V. G. (1968). *Tetrahedron* **24**, 5917.
Klyne, W. (1956). *Experientia* **12**, 119.
Kuchitsu, K., Fukuyama, T., and Morino, Y. (1967-8). *J. Mol. Struct.* **1**, 463.
Laane, J., and Lord, R. C. (1967). *J. Chem. Phys.* **47**, 4941.
Lambert, J. B., Mixan, C. E., and Johnson, D. H. (1972). *Tetrahedron Lett.* 4335.
Lee, Y. N., and Hamori, E. (1974). *Biopolymers* **13**, 77.
Lehn, J.-M., and Ourisson, G. (1963). *Bull. Soc. Chim. France*, 1113.
Lennard-Jones, J. E. (1931). *Proc. Phys. Soc.* **43**, 461.
Lenoir, D., Hall, R. E., and Schleyer, P. von R. (1974). *J. Amer. Chem. Soc.* **96**, 2138.
Lide, D. R., Jr. (1962). *Tetrahedron* **17**, 125.
Lifson, S., and Warshel, A. (1968). *J. Chem. Phys.* **49**, 5116.
Lindner, H. J. (1974). *Tetrahedron* **30**, 1127.
Lipnick, R. L., and Garbisch, E. W., Jr. (1973). *J. Amer. Chem. Soc.* **95**, 6370.
Lo, D. H., and Whitehead, M. A. (1968a). *Canad. J. Chem.* **46**, 2027.
Lo, D. H., and Whitehead, M. A. (1968b). *Canad. J. Chem.* **46**, 2041.
Lugovskoi, A. A., and Kitaigorodskii, A. I. (1973). *Konform. Izmen. Biopolim. Rastvorakh. Vses. Sov.* 86.
Lugovskoi, A. A., Dashevskii, V. G., and Kitaigorodskii, A. I. (1973). *Tetrahedron* **29**, 287.
Mansson, M., Rapport, N., and Westrum, E. F., Jr. (1970). *J. Amer. Chem. Soc.* **92**, 7296.
Mark, J. E., and Flory, P. J. (1965). *J. Amer. Chem. Soc.* **87**, 1415.
Mark, J. E., and Sutton, C. (1972). *J. Amer. Chem. Soc.* **94**, 1083.

Marvell, E. N., and Knutson, R. S. (1970). *J. Org. Chem.* **35**, 388.
McAlister, J., Yathindra, N., and Sundaralingam, M. (1973). *Biochemistry* **12**, 1189.
McCullough, R. L., and Lindenmeyer, P. H. (1972). *Kolloid, Z.* **250**, 440.
McGuire, R. F., Momany, F. A., and Scheraga, H. A. (1972). *J. Phys. Chem.* **76**, 375.
Mizushima, S. (1954). "The Structure of Molecules and Internal Rotation", Academic Press, New York.
Montaudo, G., and Finocchiaro, P. (1972). *J. Amer. Chem. Soc.* **94**, 6745.
Nelson, R., and Pierce, L. (1965). *J. Mol. Spect.* **18**, 344.
Nyburg, S. C., and Szymanski, J. T. (1968). *Chem. Comm.* 669.
Oberhammer, H., and Bauer, S. H. (1969). *J. Amer. Chem. Soc.* **91**, 10.
Ollis, W. D., Stoddart, J. F., and Sutherland, I. O. (1974). *Tetrahedron* **30**, 1903.
Ouellette, R. J. (1972). *J. Amer. Chem. Soc.* **94**, 7674.
Ouellette, R. J. (1974). *J. Amer. Chem. Soc.* **96**, 2421.
Ouellette, R. J., and Williams, S. H. (1971). *J. Amer. Chem. Soc.* **93**, 466.
Ouellette, R. J., Baron, D., Stolfo, J., Rosenblum, A., and Weber, P. (1972). *Tetrahedron* **28**, 2163.
Pasternak, R., and Meyer, A. Y. (1972), *J. Mol. Struct.*, **13**, 201.
Pauling, L. (1960). "The Nature of the Chemical Bond", Cornell University Press, Ithaca, N.Y., 3rd Edition.
Pierce, L., Chang, C. K., Hayashi, M., and Nelson, R. (1969). *J. Mol. Spect.* **32**, 449.
Pitzer, K. S., and Gwinn, W. D. (1942). *J. Chem. Phys.* **10**, 428.
Pople, J. A., and Beveridge, D. L. (1970). "Approximate Molecular Orbital Theory", McGraw-Hill, New York.
Ramachandran, G. N., and Sasisekharan, V. (1968). *Adv. Protein Chem.* **23**, 283.
Robb, M. A., Haines, W. J., and Csizmadia, I. G. (1973). *J. Amer. Chem. Soc.* **95**, 42.
Robiette, A. G. (1973). In "Molecular Structure by Diffraction Methods", Vol. 1, The Chemical Society, London.
Rosdahl, A., and Sandstrom, J. (1972). *Tetrahedron Lett.* 4187.
Sasisekharan, V., and Lakshminarayanan, A. U. (1969). *Biopolymers* **8**, 505.
Schaefer, M. F. (1972). "The Electronic Structure of Atoms and Molecules", Addison-Wesley, Reading, Mass.
Scharpen, L. H., and Laurie, V. W. (1968). *J. Chem. Phys.* **49**, 221.
Scheraga, H. A. (1968). *Adv. Phys. Org. Chem.* **6**, 103.
Schleyer, P. von R., Williams, J. E., and Blanchard, K. R. (1970). *J. Amer. Chem. Soc.* **92**, 2377.
Schubert, W. K., Southern, J. F., and Schaefer, L. (1973). *J. Mol. Struct.* **16**, 403.
Schubert, W., Schaefer, L., and Pauli, G. H. (1974). *J. Mol. Struct.* **21**, 53.
Sekigawa, K. (1970). *Tetrahedron* **26**, 5395.
Senkler, G. H., Gust, D., Riccobono, P. X., and Mislow, K. (1972). *J. Amer. Chem. Soc.* **94**, 8626.
Shieh, C. F., McNally, D., and Boyd, R. H. (1969). *Tetrahedron* **25**, 3053.
Simonetta, M., Favini, G., Mariani, C., and Gramaccioni, P. (1968). *J. Amer. Chem. Soc.* **90**, 1280.
Slutsky, J., Engler, E. M., and Schleyer, P. von R. (1973). *Chem. Comm.* 685.
Smith, R. P., and Mortensen, E. M. (1956). *J. Amer. Chem. Soc.* **78**, 3932.
Smith, R. P., Ree, T., Magee, J. L., and Eyring, H. (1951). *J. Amer. Chem. Soc.* **73**, 2263.

Snyder, R. G., and Schachtschneider, J. H. (1965). *Spectrochim. Acta* 21, 169.
Stellman, S. D., Hingerty, B., Broyde, S. B., Subramanian, E., Sato, T., and Langridge, R. (1973). *Biopolymers* 12, 2731.
Stull, D. R., Westrum, E. F., and Sinke, G. C. (1969). "The Chemical Thermodynamics of Organic Compounds", Wiley, New York.
Sundararajan, P. R., and Marchessault, R. H. (1972). *Biopolymers* 11, 829.
Sundararajan, P. R., and Rao, V. S. R. (1968). *Tetrahedron* 24, 289.
Suenram, R. D., and Harmony, M. D. (1973). *J. Chem. Phys.* 58, 5842.
Traetteberg, M. (1966). *Acta Chem. Scand.* 20, 1724.
Traetteberg, M. (1970). *Acta Chem. Scand.* 24, 2285.
Tribble, M. T., and Allinger, N. L. (1972). *Tetrahedron* 28, 2147.
Turner, R. B., Mallon, B. J., Tichy, M., Doering, W. von E., Roth, W. R., and Schroder, G. (1973). *J. Amer. Chem. Soc.* 95, 8605.
Usher, D. A., Dennis, E. A., and Westheimer, F. H. (1965). *J. Amer. Chem. Soc.* 87, 2320.
Van-Catledge, F. A., and Allinger, N. L. (1969). *J. Amer. Chem. Soc.* 91, 2582.
Van Horn, A. R., and Djerassi, C. (1967). *J. Amer. Chem. Soc.* 89, 651.
van't Hoff, J. H. (1875). *Bull. Soc. Chim. France*, [2], 24, 295.
Viskocil, J. (1974). Ph.D. Dissertation, University of Georgia.
Wahl, G. H., Bordner, J., Harpp, D. N., and Gleason, J. G. (1972). *Chem. Comm.* 985.
Warshel, A., and Karplus, M. (1972). *J. Amer. Chem. Soc.* 94, 5612.
Warshel, A., and Lifson, S. (1970). *J. Chem. Phys.* 53, 582.
Wertz, D. H. (1974). Ph.D. Dissertation, University of Georgia.
Wertz, D. H., and Allinger, N. L. (1974). *Tetrahedron* 30, 1579.
Westheimer, F. H. (1956). In "Steric Effects in Organic Chemistry" M. S. Newman, ed., Wiley, New York.
Wettermark, G., and Schor, R. (1967). *Theor. Chim. Acta* 9, 57.
White, D. N. J., and Sim, G. A. (1973). *Tetrahedron* 29, 3933.
Wiberg, K. B. (1965). *J. Amer. Chem. Soc.* 87, 1070.
Wiberg, K. B., and Boyd, R. H. (1972). *J. Amer. Chem. Soc.* 94, 8426.
Wiberg, K. B., and Lampman, G. M. (1966). *J. Amer. Chem. Soc.* 88, 4429.
Wilson, E. B., Decius, J. C., and Cross, P. C. (1955). "Molecular Vibrations", McGraw-Hill, New York.
Wilson, G. E., Jr. (1974). *J. Amer. Chem. Soc.* 96, 2426.
Williams, D. E. (1966). *J. Chem. Phys.* 45, 3770.
Williams, D. E. (1967). *J. Chem. Phys.* 47, 4680.
Williams, J. E., Stang, P. J., and Schleyer, P. von R. (1968). *Ann. Rev. Phys. Chem.* 19, 591.
Winkler, F. K., and Dunitz, J. D. (1971). *J. Mol. Biol.* 59, 169.
Wipke, W. T., and Gund, P. (1974). *J. Amer. Chem. Soc.* 96, 299.
Wood, G., and Woo, E. P. (1967). *Canad. J. Chem.* 45, 2477.
Wright, J. S. and Salem, L. (1972). *J. Amer. Chem. Soc.* 94 322.
Yathindra, N., and Sundaralingam, M. (1973). *Biochim. Biophys. Acta* 308, 17.
Yokozeki, A., and Kuchitsu, K. (1971). *Bull. Chem. Soc. Japan* 44, 2356.
Yokozeki, A., Kuchitsu, K., and Morino, Y. (1970). *Bull. Chem. Soc. Japan* 43, 2017.
Zuccarello, F., Buemi, G., and Favini, G. (1971). *J. Mol. Struct.* 8, 459.

Protonation and Solvation in Strong Aqueous Acids

EDWARD M. ARNETT[1] and GIANFRANCO SCORRANO[2]

[1] *Department of Chemistry, University of Pittsburgh, Pittsburgh, Pennsylvania 15260, U.S.A.*
[2] *Istituto di Chimica Organica, Centro CNR Meccanismi di Reazioni Organiche, Università di Padova, Italy.*

1. Introduction	84
The Purpose of this Article	84
The Importance of Aqueous Acid Systems	85
2. Standard Free Energies of Proton Transfer and Solvation in Aqueous Acid	88
General Description	88
Determination of Ionization Ratios at High Dilution . . .	89
Activity Coefficients	97
Empirical Approaches to Individual Acidity Functions . . .	101
3. Measured Heats of Ionization in Various Acidic Media . . .	106
General Comments on Thermodynamic Properties for Ionization and Solution	106
Strong Bases in Water	109
Thermodynamics of Ionization for Weakly Basic Amines in Strong Aqueous Acids	117
Thermodynamics of Ionization of Oxygen Bases in Water . .	119
Calorimetric Heats of Protonation in Aqueous Acids . . .	119
Heats of Ionization in HSO_3F	124
4. Heats of Solvation	131
Calculating the Heats of Solvation of Ammonium and Oxonium Ions in HSO_3F and H_2O	131
Comparison of Heats of Solvation in Water and HSO_3F. . .	139
5. Acidity Functions and Solvation	142
Standard Free Energies of Hydration of Ammonium Ions and Oxonium Ions	142
Hydration Energies and ϕ Values	143
6. Summary	146
Acknowledgements	147
References	148

1. INTRODUCTION

The Purpose of this Article

The proton transfer properties of organic molecules in strong aqueous acids is a major area of physical organic chemistry with many practical and theoretical implications. Research in this field has been documented in at least 2000 separate publications over the past two decades and the results have been collected in critical reviews, (Paul and Long, 1957; Arnett, 1963; Deno, 1964; Boyd, 1969; Rochester, 1970; Liler, 1971), several of which have appeared quite recently. It is not our intention here to provide yet another. Instead, we hope to take advantage of the accessibility of the data to approach the subject from a somewhat different angle, so that this article will complement the material presented elsewhere with the least possible redundancy or overlap. We intend to limit our tabulations to material related directly to our principal aim, which is an interpretative one.

For our purposes we will consider strong aqueous acids as those systems whose pH values are too negative to be measured directly with a pH meter, thus covering the traditional H_0 scale from 1 M acid to pure sulfuric acid, for example. We will define weak organic bases as those compounds which require strong acids for measurable conversion to their conjugate acids.

We will attempt to: (a) tabulate the available measured and derived thermodynamic data for the protonation of important representative bases in aqueous acids, superacids, and the gas phase; (b) show how these data may be used to estimate the ionization ratios of the different bases in aqueous acid media, including their pK-values* in water at 25° and at different temperatures; (c) estimate the solvation energies of the 'onium ions of the protonated bases; (d) relate the acidity function and hydration behavior of the different classes of bases in order to provide the necessary data for a practical theory of acidity functions.

We will *not* attempt a comprehensive review of aqueous acid systems since this has been done adequately by others; nor will we consider the important questions of kinetics and mechanism of acid-base catalysis in these systems. Likewise, the behavior of

---

* Throughout, pK will be understood to represent the pK_a-value of the conjugate acid BH^+ corresponding to the base B under consideration. When comparison is made with a second base, usually of a different type, it will be referred to as X and its conjugate acid as XH^+.

weak acids in strong bases (Bowden, 1966; Rochester, 1966; Jones, 1973) will not be dealt with here. Although that topic is conceptually continuous with the present problem, there are certain differences and we do not wish to include it. Likewise, we will not discuss non-aqueous acid systems (Jander *et al.*, 1963; Waddington, 1965; Lagowski, 1966; Davis, 1968, 1970) except for gas-phase proton transfer and the behavior of several superacid systems which may be considered as reasonable extrapolations of the aqueous strong acid continuum. Detailed discussion of activity coefficients will be found in three excellent reviews, one very recent (Edward, 1964; Boyd, 1969, Yates and McClelland, 1974), and there have been several tabulations (Paul and Long, 1957; Arnett, 1963; Pal'm *et al.*, 1966; Collumeau, 1968; Rochester, 1970; Liler, 1971) of pK_a estimates derived largely on the assumption, *now known to be erroneous*, that all weak bases follow the original Hammett acidity function.

Finally, we will reduce to an absolute minimum any repetition of material presented in two recent reviews (Arnett, 1973; Arnett and Jones, 1974) which relate proton transfer in the gas phase to that in aqueous acid-base systems. We hope that the reader of this chapter will find that it is composed primarily of new material with a minimum of recycled old straw.

Our cut-off date for literature coverage is July 1974. Since a number of the areas treated below are developing at a rapid rate, some of the data may need revision in due course. However, we believe that the accumulated weight of consistent evidence at this time supports our basic conclusions.

The Importance of Aqueous Acid Systems

Of all pure chemicals water must surely be considered the essence of terrestrial life. Of Empedocles's original four elements it is the only one which we would still recognize as being a single pure discrete species, although of course, at the molecular level rather than at the atomic. Water is not only essential to biochemical processes, but is of primary importance to industrial and economic chemistry. There is thus some practical justification for the habitual aquocentricity of classical solution chemistry which chooses high dilution in water as the standard state for most studies. The economic importance of sulfuric and hydrochloric acids as heavy chemicals for many acid-catalysed processes makes the understanding of catalytic processes in strong aqueous acids of particular practical importance to the kineticist and chemical engineer.

Chemicals have been classified as acids or bases since the days of the alchemist. However, the role which hydrogen plays in determining acidity was not resolved until the middle of the 19th century (Davy, 1816; Liebig, 1838). The fact that organic molecules were ionized by hydrogen transfer only became related to acidity as the Arrhenius notion of ionization developed through conductivity studies (Arrhenius, 1887; Ostwald, 1888). It was recognized by 1900 that amines were converted to ionized species in dilute acid but that most other organic compounds remain neutral under these conditons. The work of Hantzsch and Oddo (Hantzsch, 1907, 1908, 1909a, 1909b, 1922, 1930; Oddo and Scandola, 1908, 1909a, 1909b, 1910; Oddo and Casalino, 1917, 1918) demonstrated that concentrated sulfuric acid was capable of ionizing most of the organic compounds which are too weakly basic to be ionized in dilute aqueous acids. It remained for Hammett (1928) to relate Hantzsch's and Oddo's work to the Brönsted theory (1923) of protolysis to provide our present notions of a continuum of acidities through aqueous sulfuric acid solutions. Hammett's use of nitrated anilines with the newly developed methods of quantitative colorimetry (Hammett and Deyrup, 1932), provided the practical and theoretical basis for the quantitative discussion of acid-base equilibria and catalysis in aqueous acid systems. Hammett's insight into this complicated problem was so perceptive that his formalism is still the standard of discussion after almost 50 years. His assumptions and their possible shortcomings were clearly recognized by him in the beginning, so that most of the development which has occurred in the field has been an elaboration of his original analysis of the problem.

Acidity scales are energy scales, and they are therefore arbitrary both with regard to the reference point and to the magnitude of units chosen. There is no measurable inherent property that can be identified as "basicity" which would allow us to correlate the interaction of a series of bases with a variety of acids in a consistent and predictable manner. We have recommended (Arnett et al., 1974) therefore that the proton affinity in the gas phase at 25° should be the operationally defined property for defining basicity. Yet, most proton transfer reactions of interest to ordinary chemists occur in solution where there is no close correlation between basicity and proton affinity. Even in the gas phase there is no general relationship between proton affinity and the basicity of a series of compounds against non-protonic Lewis acids. Although these comments might suggest that the problem of understanding acid–base interactions is hopelessly complex, we believe that the prospects are brighter than they have ever been for gaining a thorough understanding of the factors which go into base strength in aqueous acid and base systems. Such optimism is only possible because the principal problems have been identified and ways have been developed to deal with them.

Thanks to a number of new mass spectrometric methods (Baldeschwieler and Woodgate, 1971; Beauchamp, 1971; Kebarle, 1972; Howard *et al.*, 1972; Böhme *et al.* 1972) we can now determine the relative energies of proton transfer for many types of organic molecules and functional groups in the gas phase. By means of high pressure mass spectrometry and variants of the flowing afterglow technique it is possible to determine the energies for clustering one, two, three, and more molecules of solvent, including water, stepwise around 'onium ions formed in the gas phase. Some of these energies are related directly to the solvation of 'onium ions in strongly acidic condensed phases, and from there it is possible to go step by step through a series of binary aqueous acid systems to extrapolate the behavior which the weakly basic substance might show towards protonation in water itself at infinite dilution. We have demonstrated elsewhere (Arnett *et al.*, 1972a; Arnett and Jones, 1974) how a complete thermodynamic analysis can be applied to determine the thermodynamic properties for hydration of ammonium ions from the gas phase to water. The data which would allow such a complete analysis for most bases weaker than the amines, such as ethers, sulfides, esters, amides, or ketones, for example, are not yet directly available. We will describe here how they might be obtained and will present several examples of complete solutions.

On the basis of past interest in the field, we have hopes that our conclusions will be of value not only to physical chemists concerned with the understanding of structure-reactivity problems, but to the industrial chemist and chemical engineer who desires a higher predictability for acid catalysed systems and to the biochemist who is concerned with proton transfer properties in enzymatic catalysis.

2. STANDARD FREE ENERGIES OF PROTON TRANSFER AND SOLVATION IN AQUEOUS ACID

General Description

The fact that most organic compounds are reversibly protonated in pure strong acids but are neutral to water is a qualitative fact of enormous importance for setting the limits on a scale of Brönsted basicities. With these boundaries in mind it is possible to attack the following questions.

(*a*) Is there a continuous scale of acidities which may be generated by mixing superacids with water or with each other?

(*b*) How strong is each acid on this scale compared, let us say, to 1 M aqueous mineral acid?

(*c*) At what points on this scale are the different common organic functional groups protonated?

(*d*) What are the fundamental chemical and physical factors which determine the range of protolysis energies for organic bases?

(*e*) How does the scale and basicity behavior respond to changes in conditions such as temperature, the presence of cosolvents or added solutes?

There are many continuous scales of physical properties, such as density, refractive index, dielectric constant, which can be generated by mixing water with strong acids. The simplest of these scales are the stoichiometric compositions expressed in different concentration terms (e.g. molarity, mole fraction), but they have no general relationship to each other nor to the protonation behavior of organic molecules in these solutions. Hammett (Hammett and Deyrup, 1932; Hammett, 1970) provided a means of answering all the questions listed above through the operational approach of calibrating acidity scales in terms of the observable protonation behavior of nitroaniline indicators. This choice was a sensible one because the scale was related directly to its most probable uses. Unfortunately, it has turned out that the original scale is precisely applicable only to the compounds used to generate it. The same could be said if any other series of closely related compounds had been used. In fact, we now know that every family of compounds with a common functional group (e.g. ethers, sulfides, ketones) generates its own protonation scale and at the limit of precision a separate scale is needed for each compound (Arnett and Mach, 1964; Scorrano, 1973).

The differences in protolytic behavior in response to changing the composition of aqueous acid solutions can be enormous. For example, ethanol in 70 wt % aqueous sulfuric acid (Lee and Cameron, 1971) is of equal basicity to N-methyl-2,4-dinitro-4'-bromodiphenylamine (Arnett and Mach, 1964). However, if one uses their individual acidity scales to predict their Brönsted basicities in water, it is easy to show that ethanol is nearly a million times more basic than this substituted diphenylamine.*

In order to make such a comparison of the strengths of Brönsted bases a number of serious difficulties have to be surmounted. In terms of a simple Brönsted protolysis equilibrium (1), our problem is to

$$B + HS^+ \rightleftharpoons BH^+ + S \qquad (1)$$

determine the concentration ratio of $[B]/[BH^+]$ in one or more mixtures of aqueous acid and convert this to a pK-value at the standard state of infinite dilution in water at 25°C. Chemists who are used to dealing with such equilibria for strong organic acids and bases such as aliphatic amines within the pH range of acidity may not be prepared for the difficulties which arise at every step in trying to acquire and treat the data for the protonation of weak bases. In fact these problems are so formidable that there is still doubt for some types of bases whether a single unique Brönsted equilibrium constant can be isolated which is applicable to a broad range of protonation phenomena.

Determination of Ionization Ratios at High Dilution

Analytical problems

It has been shown (Bartlett and McCollum, 1956) that the presence of organic impurities or cosolvents in aqueous sulfuric acid can have very powerful effects on the observed concentration ratio for nitroaniline indicators. It is not, therefore, surprising that if B [eqn (1)] is present in concentrations much greater than 10^{-2} M, it may begin to affect its own ionization ratio through solute–solute interactions. Hammett anticipated and avoided this problem by exploiting the high molar absorptivities of his nitroanilines in order to keep them at very low concentration where their influence on

---

* pK(amine) − pK(alcohol) = −7.7 − (−1.9) = −5.8.

the medium was negligible. Unfortunately, many of the simple aliphatic compounds of interest to physical organic chemists or chemical engineers do not have strong electronic transitions and cannot be studied at high dilution by uv-visible spectroscopy. For many such prototype weak bases (e.g. acetone, ethanol, tetrahydrofuran, dimethyl sulfoxide) there is an abundant literature documenting the attempts by many ingenious and careful workers who have attempted to determine the ionization ratio by different methods. Discrepancies between these estimates for the most heavily studied compounds often cover many powers of ten (Arnett, 1963; Arnett et al., 1970; Scorrano, 1973) and give the distinct impression that different methods may be measuring different equilibria. However, in many cases these studies have been carried out in solutions that are so concentrated with respect to the base that the base itself may possibly be contributing a serious medium effect to its own ionization. This, of course, will interfere with any ultimate extrapolation to its behavior at infinite dilution in water. It is possible that most of the many discrepancies between uv, nmr, conductance, extraction and Raman data could be removed if all such measurements were carried out at solute concentration $<10^{-4}$ M. However, present methods are not sensitive enough to test this point for many of the most interesting compounds. In principle this problem of solute–solute interactions could intrude on the application of acidity functions to the kinetics of acid-catalysed reactions where the solute is present in concentrations high enough to be of synthetic or industrial utility.

Special medium effects

Even if the solute base is effectively at high dilution in aqueous acid, special medium effects may complicate the interpretation of the results of analytical methods used to distinguish between B and BH^+ as the composition of the binary aqueous acid solvent is changed. Thus the detection of an ionization ratio for protonating aromatic carbonyl bases (and even some nitroanilines) by electronic spectra cannot be handled simply by the usual isosbestic point method. Although the relative heights of absorption maxima for B and BH^+ change as a function of acid, the wavelengths of either or both maxima also change. Special corrections must then be made to estimate how much of the absorbance change for B or BH^+ is due to conversion to its conjugate acid and how much is due to a lateral medium shift of λ_{max}. There is no agreement on how these corrections are to be made (Flexser et al., 1935; Gold and Hawes, 1951; Davis and Geissman, 1954; Arnett and Wu, 1960; Katritzky et al., 1963). Most of the organic basic groups are fairly good hydrogen-bond acceptors (Mitchell, 1972; Arnett et al., 1974), and an aqueous acid binary system is a good hydrogen bond donor whose ability doubtless increases as the acidity increases. Likewise the protonated base BH^+ is a hydrogen bond donor and the acidic medium contains many species, water, the acid HA and its conjugate anion all of which are good hydrogen bond acceptors. The analytical determination of the conversion of B to BH^+ shown in eqn (1) really requires the ability to follow the process $B-HS^+ \rightleftarrows BH^+-S$, where S represents principal solvent species at a given acidity. One may imagine that the actual structures and compositions of the two hydrogen-bonded species are changing steadily as a function of the acidity. In many cases the species are so similar that it may be quite difficult to distinguish between their interconversion and solvent induced changes in hydrogen-bonding with the solvent, the acid, or its counter ion, as the acidity increases.

Conductance studies of Haldna and Pal'm (1960) suggest that in terms of this probe discrimination between the strengths of weak base disappears at relatively low acidities where hydrogen bonded oxonium complexes, $B-HS^+$, migrate to the cathode. By this method one might be led to suppose that even nitromethane (Haldna et al., 1964) is completely protonated in 38% aqueous sulfuric acid even though cryoscopic (Hantzsch, 1909a; Gillespie, 1950) and spectral (Deno et al., 1966) measurements show almost no ionization for this compound in 100% H_2SO_4. Again, in terms of distribution measurements, one might be misled into assigning an unrealistically high Brönsted basicity to nitrobenzene and some other very weak bases as a result of strong hydrogen bonding in aqueous acids of moderate strength, yet far too weak to produce protonation. In view of these problems it is hard to exaggerate the recent contribution which high resolution proton magnetic resonance spectrometry, capable of observing simple organic bases and their conjugate acid cations at 10^{-2} M, has made to this field.

The review by Olah, White and O'Brien (1970) documents extensive evidence for the formation of various protonated heteroatom bases in very strong acids. The identification of very weak bases, such as nitro-compounds, halides or sulfones, which require acids stronger than pure H_2SO_4 is easy with pmr. However, it is only recently that high resolution pmr has been used successfully to determine titration curves such as that for methyl isopropyl sulfide (see Fig. 1). Early attempts by Arnett and Bothner-By (1963), and Edward *et al.* (1962) to get such curves for ethers and alcohols probably failed because the studies did not cover a wide enough range of acidity and the authors did not realize fully the inappropriateness of using the primary nitroaniline H_0 scale for these simple oxygen bases. In the past few years, Haake (Haake and Hurst, 1966; Haake *et al.*, 1967; Haake and Cook, 1968), Moodie (Armstrong and Moodie, 1968), Liler (1969), Modena and Scorrano (Landini *et al.*,

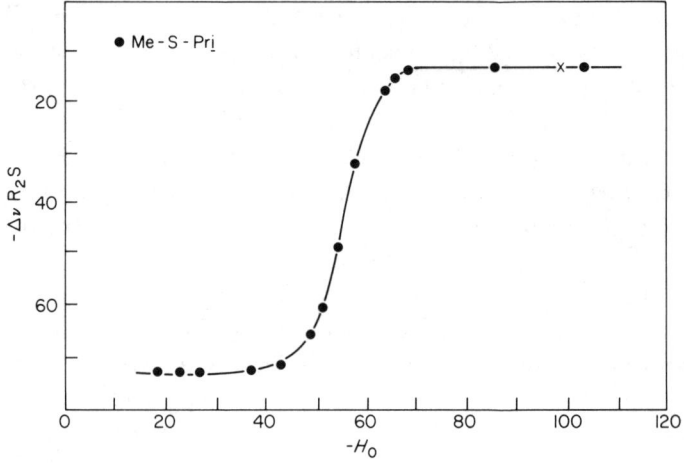

Figure 1. Sigmoid plot of $-\Delta\nu$ from nmr shifts vs $-H_0$ to determine protonation of methyl isopropyl sulfide.

1969; Bonvicini *et al.*, 1972a; Curci *et al.*, 1973; Bonvicini *et al.*, 1973; Levi *et al.*, 1974), Levy (Levy *et al.*, 1970), Lee and Cameron (1971, 1972) have applied pmr to protonation in dilute solutions of amides, phosphinates, phosphine oxides, ketones, alcohols, ethers, sulfoxides, and sulfides. In several cases it has been shown that the nmr results correspond exactly to those from uv spectroscopy and circular dichroism (Landini *et al.*, 1969; Bonvicini *et al.*, 1972b). Thus, for some of these compounds, we are finally at the point where there is good general agreement between several methods in detecting the exact range of acid where B undergoes complete conversion to BH^+. We have high hopes that, as the sensitivity of nmr equipment increases, this method may be used to settle many of the remaining analytical problems in determining protonation equilibria of weak bases.

Apart from the possibility of monitoring protonation equilibria of very simple organic bases, experience has shown two important experimental advantages in using the nmr technique. Since it involves chemical shift measurements, usually of protons bonded to carbon atoms near the protonation center relative to an internal standard, there is no need to use exactly equal amounts of substrate in each measurement. Moreover, even if decomposition of the base occurs it will only diminish the intensity of the signal and will not change its position, provided that complete decomposition is slow relative to the time for the experiment. Hence, measurements are reliable even in these cases.

TABLE 1

Protonation Equilibria[a] of Diethyl Ether and Ethanol at 60°

		pK_{BH^+}[b]	$(H_0)_{1/2}$[c]
Et$_2$O	$(\nu_{CH_2} - \nu_{ref})$	−2·43	−6·13
	$(\nu_{CH_2} - \nu_{CH_3})$	−2·45	−5·99
EtOH	$(\nu_{CH_2} - \nu_{ref})$	−1·97	−6·79
	$(\nu_{CH_2} - \nu_{CH_3})$	−1·99	−7·04

[a] Modena et al., 1974.
[b] This is the correct value at a standard state in water using the appropriate acidity function for the base in question.
[c] This is the strength of acid in the H_0 scale at which the base in question is half protonated.

However, even with this technique we must take into account the possibility of medium effects on chemical shifts. So far two techniques have been used to compensate for this effect: one procedure uses the intramolecular shift between two groups (e.g. CH$_2$ and CH$_3$ in an ethyl derivative), and the other the intermolecular shift relative to a suitable chosen reference (in most cases the trimethylammonium ion) as an internal reference. The first method has the advantage that it does not require the introduction of any other substances into the solutions, but it suffers from the disadvantage that the relative intramolecular shifts are small. The use of an internal standard, whose chemical shift is also affected by the solvent, may alter the acidity of the solution, although the use of small amounts (0·05 M) of trimethylammonium sulfate should give negligible changes.

In Table 1 we compare results obtained for diethyl ether and ethanol at 60° by using both methods of correcting for solvent effect.

The data show that both techniques yield consistent results. However, different ionization ratios have been evaluated at the same acidity for methyl phenyl phosphinate by following the CH_3 or OCH_3 proton resonance chemical shifts with respect to the trimethylammonium ion (Curci et al., 1973). This evokes a word of caution in the use of internal standards, although in many cases (see Fig. 1) very good results have been obtained.

Extrapolation problems

Assuming that a method has been found for detecting the conversion of B to BH^+ at high dilution and that "irrelevant" medium effects have been expunged, we are finally in a position to attack the principal aim of protonation studies. This is to determine the acidity where B is half-ionized and to find an appropriate extrapolation function which will describe how its ionization ratio changes across the entire spectrum of acidities down to the aqueous standard state. Work during the past decade has strengthened our position here so much that we believe the problem is solved for most practical purposes.

Again the reference point for discussion is Hammett's acidity function treatment (Hammett and Deyrup, 1932; Hammett, 1970). We write the Brönsted acid–base equilibrium in terms of proton loss from BH^+ as in eqn (2), for which the thermodynamic equilibrium

$$BH^+ \rightleftharpoons B + H^+ \qquad (2)$$

constant is given in eqn (3). Since none of these activities can be measured directly for most compounds over a wide range of aqueous

$$K = \frac{a_{H^+} a_B}{a_{BH^+}} \qquad (3)$$

acid concentrations, we usually isolate the directly measured concentration ratio from the more elusive activity coefficient to get eqns (4)

$$K = a_{H^+} \frac{\gamma_B}{\gamma_{BH^+}} \frac{[B]}{[BH^+]} \qquad (4)$$

and (5). The activity coefficients γ_B and γ_{BH^+} are taken on the molarity scale and are referred to a standard state of infinite dilution

$$pK = -\log\left[a_{H^+}\frac{\gamma_B}{\gamma_{BH^+}}\right] + \log\frac{[BH^+]}{[B]} \qquad (5)$$

in water where they approach unity and $(-\log a_{H^+})$ approaches the pH acidity. The entire first term on the right is called H_0 (Hammett, 1970; Paul and Long, 1957) for primary nitroaniline indicators and corresponds, with minor modification, to Hammett's original colorimetric acidity scale. The resulting equation (6) becomes the

$$H_0 = pK + \log\frac{[B]}{[BH^+]} \qquad (6)$$

familiar Henderson-Hasselbalch equation (Henderson, 1908; Hasselbalch, 1913) within the pH range of acidity and implies (as does that equation) a linear relationship with unit slope between the acidity function H_0 (or pH) and the logarithm of the indicator ratio, $\log([B]/[BH^+])$, henceforth referred to as log I. The pK is that which should be found (if it could be measured directly) for dissociation of BH^+ at high dilution in water, the standard state.

It is now common knowledge among physical organic chemists that eqn (6) is only completely applicable to a small percentage of organic bases.

In fact, the basic assumption used in developing the first acidity function, and in applying it to other bases, (Paul and Long, 1957) was that the activity coefficient ratios, at a given acid concentration, of a base and its conjugate acid (γ_B/γ_{BH^+}) be the same, no matter what the nature of the base B. It was only with the passage of a good many years after the definition of the H_0 acidity function, that other acidity scales began to be reported based on the conversion of arylcarbinols to carbonium ions (H_R) (Westheimer and Kharasch, 1946; Lowen et al., 1950; Murray and Williams, 1950; Gold and Hawes, 1951; Deno et al., 1955) dialkylnitroanilines (H_0''') (Arnett and Mach, 1966) indoles (H_I) (Hinman and Lang, 1964) and amides (H_A) (Yates et al., 1964). These findings indicated clearly that different classes of bases follow different acidity functions and that the original assumption of constant activity coefficient ratios is valid only within a narrow range of structurally similar bases (Arnett and Mach, 1964). As a matter of fact, the difference in logarithms of activity coefficient ratios [eqn (7)] has a finite value, which represents, in numerical terms, the "failure" of the original acidity function H_0 which was generated chiefly from primary anilines to represent the behavior of other types of weak base, X (Arnett and Mach, 1966).

It is implicit in eqn (7) that each single base actually has its own acidity function, being thus defined through a ratio of individual

$$\log \frac{\gamma_X}{\gamma_{XH^+}} - \log \frac{\gamma_B}{\gamma_{BH^+}} = H_0 - H_X \qquad (7)$$

activity coefficients. We will return to this point later.

The need for a special acidity function for each class of base in order to interpret acid-base equilibria in concentrated acids led to increasing disenchantment with this field. However, two empirical treatments have been reported in recent years which provide a means for returning the area to a sound basis. These equations have been proposed by Bunnett and Olsen (1966) and by Yates and McClelland (1967). Yates and McClelland (1967) recognized that there is a relation (8) between H_0 and other acidity functions. Substitution of

$$H_X = mH_0 \qquad (8)$$

(8) in (6) gives the pK_X as the value of H_0 at half protonation multiplied by m.

$$pK_X = mH_0 + \log [XH^+]/[X] \qquad (9)$$

According to Bunnett and Olsen (1966) and Hammett (1970), the correlation between acidity functions is of the type shown in eqn (10), and this, together with eqn (6), gives (11). Thus it is possible to

$$H_X + \log c_{H^+} = (1 - \phi)(H_0 + \log c_{H^+}) \qquad (10)$$

$$H_0 - \log I_X = \phi(H_0 + \log c_{H^+}) + pK_X \qquad (11)$$

estimate pK_X-values as the intercept of plots of $(H_0) - \log I_X)$ vs $(H_0 + \log c_{H^+})$.

The essential difference between eqns (8) and (10) lies in the presence of the $\log c_{H^+}$ term in (10). It seems, therefore, that a choice between the two methods might be made by examining the behavior of two acidity functions in the range of acidity in which $\log c_{H^+}$ is changing most rapidly. Two very careful sets of measurements have been reported recently by Kresge et al. (1972) on nitroanilines (H_0) and triarylcarbinols (H_R) in perchloric acid (0 to 24%) and in hydrochloric acid (0 to 14%).

In these ranges of acidity, the Bunnett-Olsen plot is linear whereas the Yates plot is clearly curved, as expected since all acidity function should merge, by definition, into the same pH scale.

Data of comparable quality are not available for more concentrated solutions. However, by using the accepted data for H_R, H_I, H_0''' and H_A, it is possible to show, as did Hammett (1970), that eqn (11) is followed to a quite good approximation even in more concentrated solutions, giving in every case a zero intercept. This is important, since one of the purposes of these correlations is to extrapolate data evaluated in concentrated solution down to the range of acidity where acidity functions and the pH scale have the same values.

From a practical point of view, the two treatments usually give similar pK-values. As a matter of fact (10) reduces to (8) when the term log c_{H^+} is small compared with H_0, as happens in the range of high acid concentration. This implies also that $(1 - \phi)$ be equal to m. As pointed out by Greig and Johnson (1968) this is practically the case: for H_R, $(1 - \phi) = 2 \cdot 24$, $m = 2 \cdot 05$; H_I, $(1 - \phi) = 1 \cdot 43$, $m = 1 \cdot 31$; H_A, $(1 - \phi) = 0 \cdot 62$, $m = 0 \cdot 61$.

However, there are two important points in favor of the Bunnett-Olsen relationship. First, its range of validity extends down to dilute acid concentrations where the Yates treatment fails. Second, by expressing eqn (10) in terms of activity coefficients we may obtain (12). This free energy relationship draws attention to ϕ which is a

$$\log \gamma_{H^+} \gamma_X / \gamma_{XH^+} = (1 - \phi) \log \gamma_{H^+} \gamma_B / \gamma_{BH^+} \quad (12)$$

diagnostic tool similar to the ρ-value of the Hammett $\rho\sigma$ treatment (Hammett, 1970) or to m of the Grunwald-Winstein equation (Grunwald and Winstein, 1948; Leffler and Grunwald, 1963). The internal consistency of eqn (12) has been confirmed by a number of independent measurements whereby ϕ-values have been estimated both through ionization ratios and through activity coefficients. We will develop this point after a brief discussion of the relevant activity coefficients.

Activity Coefficients

One approach toward analysing the origin of discrepancies between acidity functions lies in dissecting each acidity function into its component parts, e.g. log a_{H^+}, log γ_B, and log γ_{BH^+} [see eqn (5)]. Since log a_{H^+} is a component common to all acidity functions, the problem reduces to that of estimating the other two activity coefficients. Comparison of these factors does not of itself define the particular molecular interactions which are responsible for different ϕ-values, but it does immediately establish the relative importance of changing molecular structure on the free energy of transfer ($-RT \ln \gamma$) for B and BH^+ from water across a range of aqueous acid solutions.

The experimental estimation of these activity coefficients has been discussed in two recent reviews (Boyd, 1969; Yates and McClelland, 1974) both of which provide extensive tabulations of data and authoritative interpretations.

The determination of γ_B or γ_{BH^+} amounts to measuring the solubility of the species in water and the series of acid solutions under study. If, as is often the case, B is too soluble in these aqueous media, the necessary condition of high dilution can usually be approached by measuring its distribution constant with an immiscible organic phase (Long and McDevit, 1952; Long and McIntyre, 1954; Arnett et al., 1962).

Boyd (1963) solved the problem of comparing γ_{BH^+} through the use of slightly soluble salts of polycyanoalkanes. For example, pentacyanopropane (PCPH) is a very strong organic acid whose salts are highly colored so that their concentrations at saturation are determined easily by colorimetry. Since he used a common anion, the relative mean activity coefficients of the cations could be compared. There is obviously an unresolvable dilemma in any attempt to measure γ_B and γ_{BH^+} independently across the same wide range of acidities since as one passes through the half protonation point one species becomes predominantly replaced by the other. The best one can do is to compare the behavior of well chosen model compounds or ions in order to extrapolate into the solvent region where B and BH^+ coexist.

A minimum requirement for the consistency of these dissections is that the combination of γ_X and γ_{XH^+} terms for a series of compounds should reproduce exactly their acidity function relative to that for primary anilines. A number of studies confirm this (Boyd, 1963, 1969; Sweeting and Yates, 1966; Yates and McClelland, 1974). Furthermore, by using the rearranged equation (13), where

$$\log a_{H^+} = \log \gamma_{XB^+} - \log \gamma_X - H_X \tag{13}$$

H_X is the appropriate acidity function for bases of the type X, Yates has obtained remarkably consistent values for a_{H^+} from five independent acidity functions in H_2SO_4 and $HClO_4$ (Yates and

McClelland, 1974). These values are also in agreement with proton activities derived from electrometric measurements (Jánata and Jansen, 1972; Yates and McClelland, 1973). The assumption must be made that variations in a_{H^+} are independent of the counterion. As Yates and McClelland (1973) point out, the ability to use log a_{H^+} with directly measured values of log γ_X and H_X opens the way for estimating the free energy of solvation for many kinds of BH^+ 'onium ions in media where direct measurement would be difficult or impossible.

Some important general conclusions from studies of γ_B and γ_{BH^+} (or γ_X and γ_{XH^+}) are as follows.

For non-electrolytes

(a) Dilute aqueous acids behave in many ways like other aqueous electrolytes towards non-electrolyte solutes. Thus, most neutral solutes are "salted out" by aqueous mineral acids (up to about 30% by weight in the case of H_2SO_4). The use of the term "salted out" is appropriate since the Setschenow equation log $S = K_S C_E$ (where C_E is the molar concentration of electrolyte and K_S is the salting constant), describes the relationship between solubility S and acid concentration quite accurately (Long and McDevit, 1952). Beyond this, Arnett and Mach (1966) noted the wide ranges over which acidity functions are linear against acid concentration implying a considerable degree of parallel behavior for salt effects on γ_B and γ_{BH^+}.

(b) At high acidities (>80 wt.% of H_2SO_4), the solubility of all neutral solutes increases rapidly as the acid concentration increases.

(c) Behavior at intermediate acidities depends mainly on the presence of electronegative groups such as $-NO_2$, $-COOH$, $-OH$, $-COR$ or CN. When present they cause the inversion from salting-out to salting-in to occur at lower acidities. Compounds containing many other functional groups are already converted to their conjugate acids in 50% sulfuric acid. If more of the above groups are present there is little change in γ_B over quite a range of intermediate acid strengths.

(d) Arnett and Mach (1966), Burke (1966), and Yates and McClelland (1974) have called attention to the effect that molecular volume can have on γ_B, larger molecules tending to be salted out more readily than analogous smaller ones.

Cation behavior

(a) Like the neutral bases there is considerable variety in the response of solubilities of different types of cations to changes in acidity. Compared to tetraethylammonium (the standard ion) all large organic cations which do not have acidic hydrogens behave very much like the non-electrolytes described above. Thus triarylcarbonium, sulfonium, oxonium, diarylchloronium and iodonium salts are first salted out and then salted in as the acidity rises. Cycloheptatrienyl cation behaves almost exactly like tetraethylammonium ion.

(b) Among anilines the salting out behavior is clearly differentiated in terms of the number of acidic protons on the ammonium group. Thus trimethylanilinium ion behaves almost exactly like tetraethylammonium ion but anilinium ion is progressively more salted out by increasingly strong acid.

General Comments

Yates and McClelland (1974), Burke (1966), Arnett (Arnett et al., 1974) and many other workers have considered the role which electrostatic, internal pressure and hydrogen bonding may play in activity coefficient behavior of different species. Although many of the necessary data to investigate these factors are not available, the information at present on hand does not suggest that any one of them alone is the key to predicting acidity function behavior accurately. As we shall see the most powerful approach towards understanding these systems is a thorough knowledge of hydrogen bonding.

It is important to realize that the free energy scale for most comparisons of activity coefficients in aqueous acids is too limited for precise interpretation in terms of the crude resolving power of modern theories of the behavior of large, complicated molecules in solution. A factor of two or three kilocalories per mole at 25°C is rather small on the scale of absolute hydration energies, cavity energies or proton affinities, yet it is enough to change the solubility of a species by a factor of 10^2 to 10^3. However, from a practical viewpoint it is advisable to keep in mind Louis Hammett's aphorism, "... a particularly happy aspect of the existence of linear free energy relationships has been the proof it supplies that one need not suppose that the behaviour of nature is hopelessly complicated merely because one cannot find a theoretical reason for supposing it to be otherwise" (Chapman and Shorter, 1973).

Empirical Approaches to Individual Acidity Functions

We are now in the position to use measured activity coefficients to check the Bunnett-Olsen equation and to consider its meaning in more detail (Bonvicini et al., 1973; Levi et al., 1974). By rearranging eqn (12) we obtain (14).

$$\log \frac{\gamma_X}{\gamma_{XH^+}} \frac{\gamma_{BH^+}}{\gamma_B} = \phi(\log \gamma_{BH^+}/\gamma_B \gamma_{H^+}) = \phi(H_0 + \log c_{H^+}) \quad (14)$$

Values of ϕ obtained from plots according to (14) (see Table 2) show very good agreement with those evaluated through ionization-ratio measurements, confirming in this way the validity of (10) and (11).

TABLE 2

ϕ Values from Activity Coefficients (Eqn. 14) and from Ionization Ratios (Eqn. 11) for Four Acidity Functions in Sulfuric Acid[a]

Acidity function	ϕ Values	
	Equation 14[b]	Equation 11[c]
H_A	0·38(±0·03)	0·42 to 0·55
H_0'''	−0·47(±0·04)	−0·33 to −0·48
H_I	−0·61(±0·03)	−0·26 to −0·46 and −0·67 to −0·85
H_R	−1·11(±0·09)	−1·02 to −1·59

[a] From 0 to 70%.
[b] Values in parentheses are standard deviations as evaluated from least-squares analysis; activity coefficient values have been taken from Yates and McClelland, 1974.
[c] From Bunnett and Olsen, 1966.

We may further rearrange (12) to obtain eqn (15).

$$\log \gamma_{H^+} - \log \frac{\gamma_{XH^+}}{\gamma_X} = (1 - \phi) \left(\log \gamma_{H^+} - \log \frac{\gamma_{BH^+}}{\gamma_B} \right) \quad (15)$$

From the definition (Edward, 1964; Boyd, 1969; Yates and McClelland, 1974) of activity coefficients and of their standard states (infinite dilution in water), the $\log \gamma$ values of eqns (12)–(15) are proportional to the difference in standard chemical potentials in two media, water and the acid solution of interest. $\log \gamma$ is, therefore, proportional to the free energy change in transferring one mole of the indicated species from its infinitely dilute solution in water to that in the acid solution. Since in this case the most important interaction which causes the departure of the solute from ideal behavior is that with the solvent (Yates and McClelland, 1974), the activity coefficients considered in connection with acidity functions are essentially "medium effect" activity coefficients (Yates and McClelland, 1974). Hence, eqn (15) correlates the effect of changing medium on the equilibrium (16) with the solvent effect on the

$$X + H_3O^+ \rightleftharpoons XH^+ + H_2O \tag{16}$$

similar equilibrium involving the reference base B, where H_3O^+ represents the solvated proton in the media under discussion. This, once again, points out the importance of ϕ as a guide to solvent effects in aqueous acid.

We may further speculate on the meaning of (15). It is known (Edward, 1964; Boyd, 1969; Yates and McClelland, 1974) that activity coefficients decrease in the series $\gamma_{H^+} > \gamma_{XH^+} > \gamma_X$. This is quite reasonable since the increase in acid concentration is accompanied by a reduction in the availability of water, an excellent medium for solvating hydronium ions. This decrease will obviously tend to increase the free energy of the cationic species in particular. The extent of this increase will depend, in turn, on the ability of the individual cation to disperse the positive charge through the residues linked to the protonated atom by hydrogen bonding with the solvent.

It follows that both sides of eqn (15) are positive with the actual magnitude depending on the value of the ratio γ_{XH^+}/γ_X. Two limiting cases might be considered: (i) γ_{XH^+} is of magnitude similar to γ_X. This is found (Edward, 1964; Boyd, 1969; Yates and McClelland, 1974) when the positive charge is buried within a large and polarizable molecule and interactions with the solvent are therefore not much different for the base and its conjugate acid. In this case the ratio γ_{XH^+}/γ_X is small and the ϕ-values are negative. (ii) γ_{XH^+} is much greater than γ_X. This situation is found in those cases where the introduction of a positive charge greatly enhances the differences between X and XH^+, that is when the proton is bound to a small and not very polarizable molecule so that $X-H^+ \cdots OH_2$ bonds are strong. In these cases the ratio γ_{XH^+}/γ_X will be large and the ϕ-values positive since the difference $\log \gamma_{H^+} - \log \gamma_{XH^+}/\gamma_X$ is small. We may even set (Bonvicini et al., 1973) a higher limit for ϕ. In (15),

γ_{H^+} really stands for $\gamma_{H_3O^+}/\gamma_{H_2O}$; this implies that when X is H_2O, ϕ must be unity. It is significant therefore that ϕ for methanol is +0·85 (Bonvicini et al., 1973).

The above conclusions on the behavior of activity coefficients are confirmed by the experimental data. In Fig. 2 the values of log γ_{XH^+}/γ_X are plotted as a function of the acid concentration for a few classes of bases.

This ratio clearly varies in the order triaryl carbonium ion < indoles < dialkylanilines < anilines < amides, as expected.

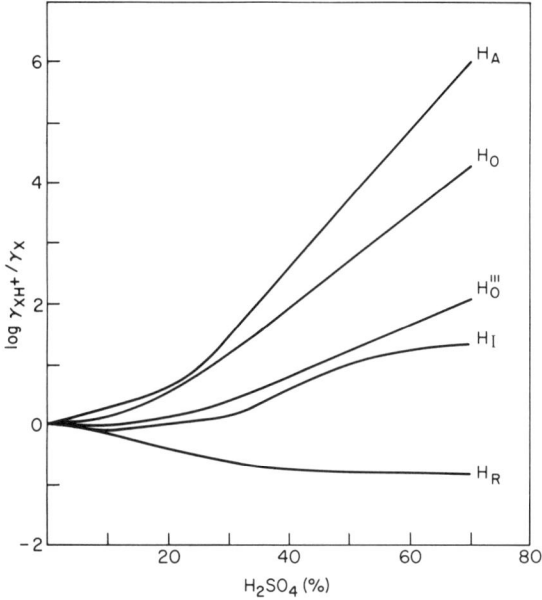

Figure 2. Logarithm of activity coefficient ratio as a function of acid concentration for various acidity functions.

There are also available several ϕ-values for compounds whose activity coefficients have not yet been evaluated. Some are reported in Table 3. The ϕ-values decrease going from oxonium to ammonium to carbonium ion bases. This is expected since the magnitude of ϕ depends on the solvation requirements of the protonation equilibrium. We expect stronger hydrogen bonds between XH^+ and water in those cases when X is more electronegative and, therefore, the positive charge in XH^+ is more localized on the hydrogen. We do not know at the present time of any quantitative information which would allow us to compare the strength of hydrogen bonds between charged species and water. In the absence of facts we would simply expect the same order, OH $\cdots$ O > NH $\cdots$ O > CH $\cdots$ O, found by Pimentel and McClelland (Vinogradov and Linnel, 1971) for uncharged compounds.

The data in Table 3 allow consideration of the importance of solvation of the free base. The acidity functions H_R and H_C are defined for ionization of triarylcarbinols and of diarylethylenes, respectively. The conjugate acid is, in both cases, the arylcarbonium ion. The ratio γ_{XH^+}/γ_X is different since the solvation requirements of the free bases are different, with the carbinol being obviously more

TABLE 3

ϕ-Values for Several Types of Bases in Aqueous Sulphuric Acid

Acidity function	Base	ϕ^a	References
H_R	Triarylcarbinol/triaryl carbonium ion	$-1 \cdot 02$ to $-1 \cdot 59$	Deno et al., 1955
H_I	Indoles	$-0 \cdot 26$ to $-0 \cdot 46$ and $-0 \cdot 67$ to $-0 \cdot 85$	Hinman and Long, 1964
H_C	Azulenes, diphenylethylenes	$-0 \cdot 70^b$	Reagan, 1969
H_0'''	Tertiary anilines	$-0 \cdot 33$ to $-0 \cdot 48$	Arnett and Mach, 1966
H_T	Thioamides	$-0 \cdot 36^b$	Tissier and Tissier, 1972
H_S	Dialkyl sulfides	$-0 \cdot 26$ to $-0 \cdot 29$	Bonvicini et al., 1972a
H_0	Primary anilines	0	Hammett, 1970
H_A	Amides	$+0 \cdot 42$ to $+0 \cdot 55$	Yates et al., 1964
H_{ROR}	Dialkyl Ethers	$+0 \cdot 75$ to $+0 \cdot 82$	Bonvicini et al., 1973
H_{ROH}	Alcohols	$+0 \cdot 85$	Bonvicini et al., 1973; Lee and Cameron, 1971
H_{H_2O}	Water	$+1 \cdot 00$	Bonvicini et al., 1973

^{a} From $(H_0 - \log I)$ vs $(H_0 + \log c_{H^+})$ plots, if not otherwise indicated.
^{b} From $(H_X + \log c_{H^+})$ vs $(H_0 + \log c_{H^+})$ plots.

solvated than the alkene. Hence the ratio is smaller for the carbinol and ϕ more negative.

Among a series of derivatives containing the same basic site, Y, the amount of positive charge on the proton will depend on the nature of the groups linked to Y and on their ability to conjugate with the atom bearing the positive charge. This is exemplified in the series of oxonium ions where the ϕ-values decrease (Levi et al., 1974) from acetone (+0·75) to ethyl acetate (+0·40) to methyl benzoate (+0·17). Other things being equal, the effectiveness of charge dispersal will increase with the number of hydrogens linked to the protonation site, since this increases the number of hydrogen bonds to the solvent. This is found in the ammonium ion series. (Arnett, 1973; Trotman-Dickenson, 1949.)

It is clear from the above discussion that each individual base defines its own acidity function. Experience has shown that ϕ-values for compounds containing the same functional group lie in most cases in a narrow range, possibly within experimental error. There is a recent analysis of the behavior of ketones, which shows, however, a variety of ϕ-values (see Table 4). The trends in Table 4 are in conformity with these principles since, obviously, an alkyl group has less ability to disperse a positive charge than a cyclopropyl, vinyl or phenyl group. Differences in ϕ-values between alkyl and phenyl esters or amides are also found (see Table 4), although of smaller

TABLE 4

Acid Base Equilibria of Some Ketones, Esters, Amides[a]

Compound	pK	ϕ
Acetone	−2·85	+0·75
Methyl Cyclopropyl Ketone	−3·27	+0·55
2-Cyclohexen-1-one	−3·02	+0·55
Acetophenone	−4·36	+0·40
4-Methoxybenzophenone	−4·18	+0·26
2,4,4'-Trimethoxybenzophenone	−3·89	−0·11
Ethyl Acetate	−4·61	+0·40
Methyl Benzoate	−7·05	+0·17
n-Butyramide	−0·74	+0·59
Benzamide	−1·50	+0·42

[a] See Levi et al., 1974 for references to original papers.

magnitude than in the case of ketones. This is expected since in these two cases the positive charge is also dispersed through conjugation with the OR- and NH_2-groups.

The examples of ketones, esters, and amides confirm the hypothesis of individual acidity functions. Moreover, they also show the reason why, in several cases, it has been possible to build acidity scales by using a series of bases to cover the entire range of acid concentration. Major variations in ϕ-values with the nature of the groups linked to the cationic center occur only when strong resonance interactions are present. The choice of nitroanilines to evaluate the first acidity function is proved, once again, to have been fortunate and, from this point of view, the "worst" acidity function for variation of ϕ should be the H_R function defined through the equilibrium triarylcarbinol ⇌ triarylcarbonium ion. In fact, ϕ-values increase from the most basic (4,4'-dimethoxytriphenylmethanol, ϕ = $-1 \cdot 16$) to the least basic substrate (4,4'-dinitrotriphenylmethanol, ϕ = $-0 \cdot 79$). (Bunnett and Olsen, 1966). Moreover, the relation $H_R + \log c_{H^+} = (1 - \phi)(H_0 + \log c_{H^+})$ shows the greatest curvature (Hammett, 1970) of any acidity function. In conclusion, it appears necessary to define the acid-base equilibria of each base with at least two parameters, namely, the pK, referred as usual to infinitely dilute aqueous solution as standard state, and the ϕ-value. The latter allows us to gain an insight into the solvation requirements of X and XH$^+$ a matter to which we will return on p. 142.

3. MEASURED HEATS OF IONIZATION IN VARIOUS ACIDIC MEDIA

General Comments on Thermodynamic Properties for Ionization and Solution

Much has been written about the relative merits of standard free energies, enthalpies* and entropies as fundamental properties to elucidate chemical processes (see, for example, Taft, 1956; Leffler and Grunwald, 1963; Hepler, 1963; Larsen and Hepler, 1969; Wells, 1968; Exner, 1964a, b; Hammett, 1970; Bell, 1973). In our opinion this question can only be answered in terms of the use to which the data will be put. Since $\Delta G°$, $\Delta H°$ and $\Delta S°$ at room temperature all contain kinetic energy (partition function) terms, none of these properties corresponds exactly to the potential energy. Physical organic chemists are not put off much by this fact since they are usually more concerned with how properties change in response to systematic variation of molecular structure or solvent than they are in particular properties of individual compounds.

Since free energy is the property which has been specifically designed to express chemical driving force, escaping tendency, or reactivity, it is almost tautological to say that the standard free energy, $\Delta G°$, is the property of primary interest to chemists. The remarkable precision of extrathermodynamic linear free energy

---

* For processes of the type which are discussed in the chapter the difference between the molar enthalpy change and the molar heat of reaction is so small that we use the terms interchangeably.

correlations to store and predict rates and equilibrium constants in terms of substituent or solvent parameters has enormous practical value and considerable theoretical interest.

It is also widely appreciated that extrathermodynamic linear correlations between enthalpy and entropy changes are widespread in nature. The interrelations of $\Delta G°$, $\Delta H°$ and $\Delta S°$ were probably first discussed at length by Evans and Polanyi (1936, 1937). They were explored in much greater detail twenty-seven years later by Leffler and Grunwald (1963). Particular discussion has revolved around the value (or lack thereof) of so-called isokinetic correlations between enthalpy and entropy changes for various processes. Two questions in particular have prevented the development of widespread interest in isokinetic plots.

The first of these is, "can I believe that there really is a genuine correlation between $\Delta H°$ and $\Delta S°$ for a set of kinetic or thermodynamic data, or is a linear isokinetic plot just an artifact produced by common errors in deriving both $\Delta H°$ and $\Delta S°$ from the same set of data through the van't Hoff, Arrhenius, or Eyring equations?" Thanks to the rapidly expanding use of solution calorimetry, enthalpy changes ($\Delta H°$) for many thermodynamic processes are now often measured directly and quite independently of $\Delta G°$. This allows realistic separation of errors in $\Delta H°$ and $\Delta G°$ and hence a realistic assignment of errors in $\Delta S°$.

The second important question about isokinetic plots is "what are they good for?" This question has been considered by Leffler and Grunwald (1963), Hammett (1970), Hepler (1963) and Lumry and Rajender (1970), to mention but a few authors, and we will not elaborate it further here. For our purposes, the most important fact is that changes in ΔH and ΔS are often the primary guides to solvation changes and that they have a remarkable tendency to compensate each other. This compensation (now referred to as Lumry's Law by some biochemists) is the physical phenomenon which is expressed by an isokinetic linear relationship of ΔH and ΔS. Thanks to this compensation, large changes in solvation which affect the ordering ($\Delta S°$) of many solvent molecules so as to reduce the energy ($\Delta H°$) of a solute may have very little effect on $\Delta G°$. This cancellation of solvation factors is especially pronounced in water and highly aqueous binary systems. Indeed, the ability of aqueous systems to dampen thermal and chemical changes which would have a devasting effect in most other media is probably a major reason why this peculiar solvent is so essential to terrestrial life.

For the orientation of the reader to the rest of this chapter we now wish to direct his attention to a much less commonly recognized extrathermodynamic relationship—that between free energy changes and the corresponding changes in enthalpy. A moment's reflection on eqn (17) is sufficient for appreciating the fact that if a true

$$\Delta G° = \Delta H° - T\Delta S° \qquad (17)$$

proportionality exists between $\Delta H°$ and $\Delta S°$ then either property is related linearly to $\Delta G°$. In view of the widespread occurrence of linear free energy and isokinetic correlations, it is then not surprising to find that linear correlations of free energy changes and enthalpy changes for similar processes are common. Indeed, in the limiting case where entropy changes are negligible, $\Delta G°$ not only becomes equal to $\Delta H°$, but as Hammett (1970) has shown, both become equal to the potential energy change such as would be calculated by molecular quantum mechanics.

Final verification of Hepler's (1963) assignment of large entropy changes to solvation for processes such as $BH^+ + B' \rightleftarrows B + B'H^+$ in solution is found in the practical equivalence of $\Delta G°$ and $\Delta H°$ for such processes in the gas phase (Arnett et al., 1972a; Taagepera et al., 1972; Briggs et al., 1972).

For our purposes the existence of a number of good linear correlations between carefully determined heats and standard free energies of ionization is of crucial importance. In foregoing sections of this article we have considered the difficulties in obtaining reliable estimates of relative strengths of bases in terms of pK ($\Delta G_i°$) when a large range of base strengths are involved. No such problem exists in determining the corresponding enthalpy change (ΔH_i) by solution chemistry provided that an acidic system can be found in which both bases are completely converted to their protonated forms. If there is reason to believe that the corresponding free energy of ionization is equal or proportional to ΔH_i, the latter becomes a useful guide to the relative "basicity".

Justification for this approach can be found in many sources, although it must always be regarded with caution in view of what we have said about systems where large changes in ΔH are almost completely compensated by ΔS. Thus for over fifty compounds a rather good (see p. 132) linear correlation of ΔH_i *in fluorosulfuric* acid is found with the corresponding pK *in water* (Arnett et al., 1970a, 1970b; Mitchell, 1972). Again we find an exact equivalence within experimental error between $\Delta G_i°$ for thirty weak acids in

DMSO solution over a range of twenty pK units (Arnett et al., 1973). Even in water, very precise values of ΔH_i of aryl carboxylic acids are found to correlate closely with pK (Bolton et al., 1972; Matsui et al., 1974) although such a correlation does not always hold in this medium (Christensen et al., 1967).

For our present purposes we will justify the widespread use of heats of ionization in HSO_3F in the following sections on several grounds:

(a) it allows us immediately to compare the proton affinities in solution of many compounds whose pK-values are at present inaccessible or uncertain;

(b) it is very likely, in view of the successful ΔH_i vs pK plots already alluded to, that changes in ΔH_i in HSO_3F are almost exactly equivalent to changes in ΔG_i° in the same medium if it were only possible to measure this property;

(c) ΔG_i° and ΔH_i can now be determined to within 0·3 kcal mole^{-1} in the gas phase. Using ΔH_i in HSO_3F and ΔH for vaporization, the heats of solvation of many BH^+ ions in HSO_3F can now be calculated exactly. This seems worth doing to us and is the aim of much of the rest of this article;

(d) the range of ΔH_i-values is so large (about 60 kcal mole^{-1} for compounds with a single basic site) that minor failures (1 kcal mole^{-1}) of individual compounds to fit an exact proportionality between ΔH_i and ΔG_i° will have little effect on our overall conclusions.

We shall use heats of ionization in HSO_3F later to compare the protonation energies of bases and the solvation energies of their anions in this medium to the corresponding values in water. This will provide us with a clear quantitative picture of ionization and solvation factors at the two limiting ends of the aqueous acid domain.

Strong Bases in Water

Our ultimate concern in this article is to compare protonation of various bases in the gas phase and strong aqueous acids and to relate them to the values which they would have in water if we could measure them directly in that solvent. We shall approach this task by first considering the data presently available for thermodynamics of

TABLE 5

Thermodynamics of Ionization for Substituted Pyridines in Water at 25°C

(a) pK-Values and Gibbs Free Energies

Substituent	pK				ΔG_i° (kcal mole^{-1})		
4-Ac	3·62	3·51^a			5·18 ± 0·25	4·79	
3-Ac	3·26	3·26^a			4·48 ± 0·15	4·45	
2-Ac		2·64^a				3·62	
4-NH$_2$	9·68	9·11^b	9·29		12·91 ± 0·31	12·43	12·45 ± 0·11
3-NH$_2$	6·03					8·26	
2-NH$_2$	6·71					9·17	9·17
4-Br	4·05	3·68			5·36 ± 0·11	5·02	
3-Br		2·72	2·91			3·71	3·91 ± 0·16
4-Cl	4·11				5·54 ± 0·27		
3-Cl	2·88				3·96 ± 0·16		
4-CN	2·26	1·48			2·95 ± 0·09	2·02	
3-CN	1·59	1·17			2·21 ± 0·20	1·60	
4-NO$_2$		1·23^c				1·70	
3-NO$_2$		0·79^c				1·11	
2-NO$_2$		−2·06^c				−2·67	
4-NMe$_2$	9·47		9·71		13·19 ± 0·16		13·01 ± 0·43
4-NHMe			9·66				12·96 ± 0·43
4-CO$_2$H	4·61			6·79	6·35 ± 0·19		
2,4-Me$_2$	6·68			6·51	8·93 ± 0·11		9·26
2,5-Me$_2$	6·25			6·75	8·60 ± 0·30		8·88
2,6-Me$_2$	6·81				9·32 ± 0·12		9·21
3,4-Me$_2$	6·23				8·58 ± 0·08		
3,5-Me$_2$	6·03			6·18	8·24 ± 0·07		8·43
4-SH	1·56				2·06 ± 0·20		
4-CH$_2$NH$_2$	4·41				5·93 ± 0·27		
3-CO$_2$H	3·98				5·45 ± 0·13		

	1	2	3	4	5
2-Me	5.91	5.95			8.14	8.13
3-Me	5.68	5.66^d		8.09 ± 0.13	8.12	7.75	7.66
4-Me	6.64	6.00^d	5.79	7.80 ± 0.20	7.72	8.21	8.15
4-CHO	4.86			8.30 ± 0.06	8.18		
3-CHO				6.61 ± 0.08	6.17		
2-CHO				—	5.03		
					5.13		
H	5.31	5.21^e	5.17	7.10 ± 0.25	7.11	7.05	7.12
3-Et			5.27				
3-Pr^i			5.80				
2,3,5,6-Me$_4$			5.88		7.07 ± 0.24		
4-NH$_2$-3-Me			7.88		7.77 ± 0.38		
4-NH$_2$-3-Et			9.43		7.89 ± 0.05		
4-NH$_2$-3-Pr^i			9.51		10.57 ± 0.48		
4-NH$_2$-3,5-Me$_2$			9.54		12.64 ± 0.57		
4-NH$_2$-2,3,5,6-Me$_4$			10.58		12.75 ± 0.49		
4-NH$_2$-3-Br			7.04		12.79 ± 0.02		
4-NHMe-3-Me			9.83		12.79 ± 0.23		
4-NHMe-3-Et			9.90		14.19 ± 0.34		
4-NHMe-3-Pr^i			9.96		9.44 ± 0.23		
4-NHMe-3,5-Me$_2$			9.43		13.18 ± 0.33		
4-NHMe-2,3,5,6-Me$_4$			10.06		13.27 ± 0.19		
4-NHMe-3-Br			7.47		13.36 ± 0.58		
4-NMe$_2$-3-Me			8.68		12.65 ± 0.91		
4-NMe$_2$-3-Et			8.66		13.49 ± 0.19		
4-NMe$_2$-3-Pr^i			8.27		10.02 ± 0.49		
4-NMe$_2$-3,5-Me$_2$			8.15		11.64 ± 0.27		
4-NMe$_2$-3-Br			6.52		11.62 ± 0.53		
					11.09 ± 0.38		
					10.93 ± 1.00		
					8.75 ± 0.14		
Quinoline				4.80		6.55	
Isoquinoline				5.07		6.91	
Piperidine				11.13		15.18	
Pyrrolidine				11.11		15.15	
References	1	2	3	4			
					1	2	3	4	5

TABLE 5—continued

(b) Enthalpies and Entropies

Substituent	ΔH (kcal mole⁻¹)				−ΔS°ᵢ (cal mole⁻¹ deg⁻¹)				
4-Ac	3·71 ± 0·23	3·56			4·03	4·13			
3-Ac	2·40 ± 0·20	2·62			6·99	6·13			
2-Ac		3·50			—	0·38			
4-NH₂	5·23 ± 0·20	11·21	10·88 ± 0·08		24·30	4·09	5·35 ± 0·27		
3-NH₂		7·37			—	2·97			
2-NH₂		8·29			—	2·95			
4-Br	2·38 ± 0·14	3·31			10·0	5·73			
3-Br		1·77	1·85 ± 0·11		—	6·52	7·03 ± 0·37		
4-Cl	2·57 ± 0·17				9·97				
3-Cl	1·63 ± 0·17				7·82				
4-CN	3·04 ± 0·13	−0·19			−0·30	7·40			
3-CN	5·59 ± 0·17	−0·75			−1·13	7·87			
4-NO₂		−1·08				9·33			
3-NO₂		−2·02				10·5			
2-NO₂		−6·81				13·8			
4-NMe₂	9·53 ± 0·3		10·75 ± 0·31		1·23		7·71 ± 1·06		
4-NHMe			11·02 ± 0·30				6·62 ± 1·02		
4-CO₂H	8·14 ± 0·18				−0·60				
2,4-Me₂	6·91 ± 0·11		7·16 ± 0·01		6·79			7·00	
2,5-Me₂	5·59 ± 0·80		6·81 ± 0·01		10·11			6·90	
2,6-Me₂	7·93 ± 0·70		7·23 ± 0·02	6·15 ± 0·11	4·49			6·60	10·11
3,4-Me₂	4·11 ± 0·18				15·00			—	
3,5-Me₂	7·82 ± 0·51	5·29 ± 0·57	6·36 ± 0·01		1·41		10·44 ± 1·94	6·95	
4-SH	−3·01 ± 0·08				17·01				
4-CH₂NH₂	−0·31 ± 0·04				20·93				
3-CO₂H	3·34 ± 0·30				7·07				
2-Me	7·14 ± 0·46	7·57	5·99 ± 0·02	6·95 ± 0·40	2·40	1·85		7·20	3·95
3-Me	4·87 ± 0·13	7·05	5·64 ± 0·01	6·70 ± 0·22	9·84	2·24	10·64 ± 1·09	7·05	3·28

	1	2	3	4	5
4-Me	6.54 ± 0.43	7.03			
4-CHO	4.82 ± 0.02	6.24			
3-CHO	—	5.56			
2-CHO	—	6.95			
H	4.00 ± 0.11	4.35			
3-Et			4.37 ± 0.17		
3-Pr^i			5.30 ± 0.27		
2,3,5-Me$_4$			5.57 ± 0.03		
4-NH$_2$-3-Me			8.08 ± 0.40		
4-NH$_2$-3-Et			11.66 ± 0.40		
4-NH$_2$-3-Pr^i			10.89 ± 0.34		
4-NH$_2$-3,5-Me$_2$			11.28 ± 0.01		
4-NH$_2$-2,3,5,6-Me$_4$			10.48 ± 0.16		
4-NH$_2$-3-Br			10.35 ± 0.24		
4-NHMe-3-Me			7.69 ± 0.16		
4-NHMe-3-Et			10.87 ± 0.23		
4-NHMe-3-Pr^i			11.54 ± 0.14		
4-NHMe-3,5-Me			11.93 ± 0.41		
4-NHme-2,3,5,6$_2$-Me$_4$			12.06 ± 0.64		
4-NHMe-3-Br			9.95 ± 0.14		
4-NMe$_2$-3-Me			9.01 ± 0.35		
4-NMe$_2$-3-Et			9.02 ± 0.19		
4-NMe$_2$-3-Pr^i			9.15 ± 0.38		
4-NMe$_2$-3,5-Me$_2$			8.76 ± 0.27		
4-NMe$_2$-3-Br			9.83 ± 0.71		
			6.36 ± 0.13		
Quinoline				5.36 ± 0.01	
Isoquinoline				5.92 ± 0.01	
Piperidine				12.19 ± 0.06	
Pyrrolidine				12.37 ± 0.06	

	1	2	3	4	5
4-Me	5.91		6.02 ± 0.02		
4-CHO	6.02		7.03 ± 0.13		
3-CHO	—				
2-CHO	—				
H	10.39		5.70 ± 0.30		
3-Et				3.87	
3-Pr^i				−0.22	
2,3,5-Me$_4$				−2.32	
4-NH$_2$-3-Me				−6.09	
4-NH$_2$-3-Et				9.25	
4-NH$_2$-3-Pr^i					
4-NH$_2$-3,5-Me$_2$					
4-NH$_2$-2,3,5,6-Me$_4$					
4-NH$_2$-3-Br					
4-NHMe-3-Me					
4-NHMe-3-Et					
4-NHMe-3-Pr^i					
4-NHMe-3,5-Me					

	3	4	5
4-Me		7.35	3.77
H	9.21 ± 0.58	7.55	4.76
3-Et	8.42 ± 0.92		
3-Pr^i	7.91 ± 0.10		
2,3,5-Me$_4$	8.49 ± 1.36		
4-NH$_2$-3-Me	3.34 ± 1.94		
4-NH$_2$-3-Et	6.34 ± 1.19		
4-NH$_2$-3-Pr^i	5.15 ± 0.03		
4-NH$_2$-3,5-Me$_2$	7.88 ± 0.54		
4-NH$_2$-2,3,5,6-Me$_4$	13.09 ± 0.82		
4-NH$_2$-3-Br	5.97 ± 0.55		
4-NHMe-3-Me	7.88 ± 0.78		
4-NHMe-3-Et	5.90 ± 0.48		
4-NHMe-3-Pr^i	4.88 ± 1.40		
4-NHMe-3,5-Me	2.01 ± 2.18		
4-NHme-2,3,5,6$_2$-Me$_4$	12.07 ± 0.48		
4-NHMe-3-Br	3.44 ± 1.19		
4-NMe$_2$-3-Me	8.94 ± 0.65		
4-NMe$_2$-3-Et	8.42 ± 1.30		
4-NMe$_2$-3-Pr^i	7.95 ± 2.42		
4-NMe$_2$-3,5-Me$_2$	3.75 ± 2.42		
4-NMe$_2$-3-Br	8.15 ± 0.48		
Quinoline		3.95	
Isoquinoline		3.30	
Piperidine		10.00	
Pyrrolidine		9.30	

References

1. Chakrabarty et al., 1973; 2. Bellobono and Monetti, 1973; 3. Essery and Schofield, 1961; 4. Sacconi et al., 1960a, b; 5. Mortimer and Laidler, 1959.
^{a} Cabani and Conti, 1965; ^{b} Bates and Hetzer, 1960; ^{c} Bellobono and Diani, 1972; ^{d} Perkampus and Prescher, 1968; ^{e} Bellobono and Beltrame, 1969.

TABLE 6: Thermodynamics of Ionization

Substituent	pK			ΔG_i^0 (kcal mole^{-1})				
H	4·60	4·596^a		6·27	6·28	6·27^a		6·27 ± 0·0
p-OCH$_3$	5·34	5·357^c		7·31	7·28	7·31^c	7·278	
p-OCH$_2$CH$_3$	–	–	5·25^7	7·16	–	–	7·155	
p-CH$_3$	5·07	5·083^a		6·94	6·92	6·94^a	6·930	6·93 ± 0·0
p-CH$_2$CH$_3$	–	–	5·00^8	6·82			–	
p-F	4·65			6·34			6·338	
p-Cl	3·98	3·982^c		5·44	5·44	5·43^c	5·425	
p-Br	3·88	3·888^c		5·30	5·29	5·30^c		
p-I	3·79	3·812^c		5·21	5·17	5·20^c		
p-CN	–	–	1·74^9	2·37	–	–		
p-NO$_2$	1·00	1·019^c		1·39	1·37	1·39^c		
m-OCH$_3$	4·23	4·204^b		5·73	5·76	5·73^b	5·765	
m-OCH$_2$CH$_3$	–	–	4·18^6	5·68	–	–	5·684	
m-CH$_3$	4·72	4·721^a	–	6·43	6·44	6·43^a	6·447	6·44 ± 0·0
m-CH$_2$CH$_3$	–	–	4·70^6	6·41		–	–	
m-F	3·57	–		4·89		–	4·893	
m-Cl	3·52	3·521^b		4·80	4·81	4·80^b	4·798	
m-Br	3·53	3·527^b		4·81	4·81	4·81^b		
m-I	3·59	3·583^b		4·89	4·89	4·89^b		
m-CN	–	–	2·75^9	3·75	–	–		
m-NO$_2$	2·46	2·460^b		3·35	3·36	3·35^b		
o-OCH$_3$	4·52							
o-CH$_3$	4·44	4·447^a			6·06	6·07^a		6·07 ± 0·0
o-F	3.20							
o-Cl	2·64							
o-Br	2·53							
o-I	2·55							
o-NO$_2$	−0·26							
References	1	2	6, 7, 8, 9	3	1	2	4	5

1. Biggs, 1961; 2. Bolton and Hall, (a) 1967, (b) 1968, (c) 1969; 3. Liotta et al., 1973; 8. de Courville and Peltier, 1967; 9. Fickling et al., 1959.

ionization of a number of Brönsted bases which are strong enough to be protonated within the pH range of acidity. These are presented for three classes of amines—in Tables 5, 6, and 7. Similar results for pyridine-N-oxides are listed in Table 8. In all cases the conventional process is ionization of the 'onium ion, the reverse of the protonation process (2). These data have been collected and evaluated elsewhere (Bell, 1973; Larsen and Hepler, 1969; Brown et al., 1955; Perrin, 1965; Albert and Serjeant, 1962; Jones, 1971; Arnett and Jones, 1974; Hall and Sprinkle, 1932) and we shall not examine them in detail here. For present purposes they provide a reference point for considering the data which will be used subsequently to

Substituted Anilinium Ions in Water at 298°K

	ΔH_i° (kcal mole^{-1})				ΔS° (cal mole^{-1} deg^{-1})				
3	6·50	7·38^b		7·24 ± 0·1	3·89	0·74	3·72^a		8·3 ± 0·3
1	7·56	8·34^c	8·210		4·02	0·94	3·44^c	3·1	
3	—	—	8·153		4·60	—	—	3·3	
6	6·97	8·06^a	7·592	7·60 ± 0·09	3·09	0·18	3·75^a	2·2	2·2 ± 0·2
4	—		—		3·09			—	
6	(6·93)		7·450		4·76			3·7	
7	6·47	6·63^c	6·420		4·13	3·44	4·01^c	3·3	
0	6·13	6·70^c			4·02	2·83	4·69^c		
0	6·05	6·55^c			4·33	2·95	4·51^c		
3	—	—			7·41	—	—		
2	3·10	3·42^c			9·16	5·83	6·79^c		
4	6·47	7·01^b	6·887		4·63	2·40	4·29^b	3·8	
5	—	—	6·660		4·60	—	—	3·3	
3	6·51	7·47^a	7·370	7·37 ± 0·06	3·86	0·23	3·48^a	3·1	3·1 ± 0·2
4	—				3·02			—	
5	(5·91)		6·226		5·60			4·5	
6	5·63	6·27^b	6·305		5·53	2·80	4·91^b	5·0	
3	5·55	6·25^b			5·97	2·50	4·82^b		
3	5·88	6·33^b			6·34	3·30	4·81^b		
	—	—			6·67	—	—		
	4·79	4·98^b			6·78	4·80	5·45^b		
	(6·72)								
	6·57	7·37^a		7·22 ± 0·09		1·7	4·35^a		3·9 ± 0·2
	(5·09)								
	(4·89)								
	(4·48)								
	(4·89)								
	(1·63)								
	1	2	4	5	3	1	2	4	5

4. Van de Poel and Slootmaekers, 1970; 5. O'Hara, 1968; 6. Bryson, 1960; 7. Whetsel, 1961;

estimate corresponding properties for weaker bases whose ionization cannot be studied directly in water.

A general examination of the data reveals that even for pK-values, which can be evaluated with high precision, there are discrepancies between values reported by different workers so that the accurate value for ΔG_i° may be in doubt by as much as ±0·5 kcal mole^{-1}, although usually there is agreement within ±0·2 kcal mole^{-1}. Disagreements about ΔH_i° values are greater, mainly reflecting difficulties in deriving them from the temperature dependence of ln$K$.

The range of ΔG_i° and ΔH_i° is restricted to about 10 kcal mole^{-1} as is required by the limited range of the pH scale itself within which

TABLE 7

Thermodynamics of Ionization of Aliphatic Ammonium Ions in Water at 25°C.

Amine, B	pK (molal)	ΔG_i^o (cal mole^{-1})	ΔH_i^o (cal mole^{-1})	$-\Delta S_i^o$ (cal mole^{-1} deg.$^{-1}$)	Ref.
Ammonia	9·2445	12611·5	12485	0·42	1, 2
Methylamine	10·6532	14533·3	13184	4·53	3
Dimethylamine	10·7788	14704·7	12040	8·94	5, 6
Trimethylamine	9·7977	13366·3	8819	15·25	5
Ethylamine	10·6784	14567·8	13710	2·88	4, 6
Diethylamine	11·0151	15027·0	12730	7·70	6
Triethylamine	10·7174	14620·9	10320	14·43	7, 6
Propylamine	10·5685	14417·8	13840	1·94	7, 6
Dipropylamine	11·00	15006	13170	6·16	8, 6
Tripropylamine	10·66	14543	10500	13·56	6
Isopropylamine	10·67	14556	13970	1·97	6
Diisopropylamine	11·20	15279	13550	5·80	6
Butylamine	10·6385	14513·3	13980	1·79	7, 6
Dibutylamine	11·25	15348	13660	5·66	8, 6
Tributylamine	9·93	13547			9
t-Butylamine	10·6837	14575·0	14354	0·74	10
Pentylamine	10·631	14503	13980	1·75	11,6

References
1. Bates and Pinching, 1949; 2. Vanderzee *et al.*, 1972; 3. Northcott, unpublished data cited in 4; 4. Van der Linde *et al.*, 1969; 5. Everett and Wynne-Jones, 1941; 6. Christensen *et al.*, 1969; 7. Cox *et al.*, 1968; 8. Girault-Vexlearschi, 1956; 9. Hall and Sprinkle, 1932; 10. Hetzer *et al.*, 1962; 11. Hoerr *et al.*, 1943.

TABLE 8

Standard Thermodynamic Quantities for the Protonation of Pyridine N-Oxide and its Methyl Substituted Derivatives in Aqueous Media[a]

Compound	pK 298·15°K	ΔG_{298}^o (cal mole^{-1})	$-\Delta H_{298}^o$ (cal mole^{-1})	$-\Delta S_{298}^o$ (cal mole^{-1} deg.$^{-1}$)
pyridine-N-oxide	0·686 ± 0·016	947 ± 3	1786 ± 53	9·17 ± 0·17
3-picoline-N-oxide	0·921 ± 0·009	1254 ± 1	1276 ± 20	8·49 ± 0·07
2-picoline-N-oxide	1·029 ± 0·008	1403 ± 1	1458 ± 20	9·59 ± 0·06
4-picoline-N-oxide	1·258 ± 0·006	1713 ± 1	1250 ± 21	9·94 ± 0·07
3,5-lutidine-N-oxide	1·181 ± 0·006	1604 ± 2	1086 ± 41	9·02 ± 0·13
2,5-lutidine-N-oxide	1·208 ± 0·005	1648 ± 1	825 ± 29	8·29 ± 0·09
2,6-lutidine-N-oxide	1·366 ± 0·009	1862 ± 1	1051 ± 15	9·77 ± 0·05
3,4-lutidine-N-oxide	1·493 ± 0·007	2038 ± 1	935 ± 15	9·97 ± 0·05
2,4-lutidine-N-oxide	1·627 ± 0·010	2228 ± 2	1002 ± 32	10·83 ± 0·10
2,4,6-collidine-N-oxide	1·990 ± 0·007	2715 ± 1	456 ± 18	10·64 ± 0·06

[a] Klofutar *et al.*, 1973.

these equilibria were all studied. Except for aromatic amines with steric encumbrance adjacent to the amino group, the correlation between amines of the same degree is fair. This is shown by the relative constancy of ΔS_i° for any series (recalling that a difference of 1 cal deg^{-1} at 25°C corresponds to $T\Delta S_i^\circ \sim 0{\cdot}3$ kcal mole^{-1}). For forty-three pyridines without *ortho*-substituents the correlation of ΔG_i° and ΔH_i° has a correlation coefficient of 0·885 (slope = 0·881) and for twenty-one anilines the corresponding values are 0·986 (slope = 1·344). It is quite possible in view of the recent re-examination of the pK-values of benzoic acids (Bolton *et al.*, 1972) that the correlation of *accurately* determined values for these anilines and pyridines is much better than we have found it to be from comparing scattered sources.

We may therefore conclude, within the somewhat crude, but fairly realistic, error limits of ±0·5 kcal mole^{-1} that variations in ΔH_i° are a fair guide to free energy changes produced by substituents which chiefly affect the electronic density on the basic site.

The pyridine N-oxides listed in Table 8 are of interest since their pK-values fall at the extreme acidic end of the pH scale at the interface between "strong" and "weak" bases. Unfortunately, the compounds listed represent only minor electronic changes in the pyridine nucleus by different substitution patterns of methyl groups. However, within the range of N-oxides shown, there is no obvious parallel between ΔG_i° and ΔH_i. Furthermore, the basic oxygen should be far enough removed from hindrance to solvation by groups on the 2- and 6-positions that steric factors can hardly be invoked.

Thermodynamics of Ionization for Weakly Basic Amines in Strong Aqueous Acids

Primary nitroanilines are the weak bases used for establishing the prototype H_0 scale. We have already discussed the problems involved in estimating pK (ΔG_i°) in water (Section 2). If we are to estimate the corresponding ΔH_i° and ΔS_i° we must repeat the whole process of setting up an acidity function at several temperatures and then use the temperature coefficients of pK. Here the indicator ratios are subjected to two major perturbations, namely, solvent change and temperature change. The accumulation of errors, especially for ΔH_i° and ΔS_i° of the weakest bases may be considerable.

The first study of this kind was that of Gel'bshtein *et al.* (1956). A few years later Arnett and Bushick (1964) studied the ionization of

TABLE 9

Thermodynamics of Ionization of Nitroanilinium Ions in Water at 25°C

Amine, B	pK		ΔG_i° kcal mole^{-1}		ΔH_i° kcal mole^{-1}		ΔS_i° cal mole^{-1} deg^{-1}	
4-Nitroaniline	1·00	1·00	1·37	1·37	3·09	3·25	5·75	6·32
2-Nitroaniline	−0·30	−0·26	−0·43	−0·35	1·68	3·34	7·06	12·37
4-Chloro-2-nitroaniline	−1·06	−0·97	−1·46	−1·32	0·84	2·17	7·70	11·70
2,5-Dichloro-4-nitroaniline	−1·75	−1·74	−2·41	−2·38	0·19	1·86	8·74	14·23
2-Chloro-6-nitroaniline	−2·38		−3·27		−0·06		10·74	—
2,4-Dichloro-6-nitroaniline	—	−3·01	—	−4·11	—	1·21	—	17·85
2,6-Dichloro-4-nitroaniline	−3·27		−4·47		−1·03		11·54	
2,4-Dinitroaniline	−4·27	−4·27	−5·82	−5·83	−3·16	0·21	8·91	20·27
2,6-Dinitroaniline	−5·39	−5·37	−7·34	−7·33	−4·87	−1·60	8·29	19·21
2-Bromo-4,6-dinitroaniline	−6·69	−6·46	−9·11	−8·82	−6·33	−3·56	9·33	17·64
3-Methyl-2,4,6-trinitroaniline	−8·33	−8·08	−11·31	−11·02	−7·59	−3·90	12·51	23·88
3-Bromo-2,4,6-trinitroaniline	−9·34		−12·82		−8·14		15·70	
2,4,6-Trinitroaniline	−10·03	−9·87	−13·75	−13·46	−9·43	−4·33	13·49	30·61
References	1	2	1	2	1	2	1	2

1. Johnson *et al.*, 1969; Bolton *et al.*, 1970.
2. Tickle *et al.*, 1970.

polyarylmethanols to the corresponding cations, a case where electrostatic factors might be most clearly revealed. In accordance with expectations from electrostatic theory, very large entropy differences between formation of the most stable and least stable ions were found.

The results in Table 9 from two recent studies for primary aniline Hammett indicators have enjoyed the full benefits of computer fitting of the data. As was observed previously in Tables 5 and 6, it is much easier to get agreement on ΔG_i° or pK than on the derived values ΔH_i° and ΔS_i°. For the stronger amines at the head of the list, where one is able to work close to the aqueous standard state, agreement on ΔH_i° is fairly good (± 1 kcal mole^{-1}). However, when we reach the compounds which require moderately strong acid (ca. 50% aqueous H_2SO_4), disagreements reach 3 kcal mole^{-1} and finally become 5 kcal mole^{-1} out of a possible 9 kcal mole^{-1} for the least basic compound. A general agreement of increasingly negative ΔH_i° and increasingly positive ΔS_i° is found paralleling earlier work but it is obvious that quantitative reliability is collapsing under the weight of accumulated errors.

Thermodynamics of Ionization of Oxygen Bases in Water

One of the aims of the present article is to interpret the differences between acidity functions in terms of solvation properties of the respective 'onium ions of the relevant bases. In order to apply such a complete analysis to weak oxygen bases in water the only presently available approach is to determine the pK-values, obtained through proton magnetic resonance, at a number of widely spaced temperatures and apply the van't Hoff equation in order to obtain ΔH_i° and ΔS_i°. Modena, Perdoncin and Scorrano (1974) have recently completed such a treatment for several important weak oxygen bases and the results are presented in Table 10. The structural variations and the analogies to corresponding nitrogen bases are so limited that we will forego discussion of any comparisons here.

Calorimetric Heats of Protonation in Aqueous Acids

In the past decade, thanks principally to the invention of the thermistor, solution calorimetry has become a convenient and inexpensive technique (Sturtevant, 1959; Arnett et al., 1965;

TABLE 10

pK (ϕ-Values) at Various Temperatures and Thermodynamic Quantities at 25° for Protonation of Oxygen Bases[a]

	25	40	60	90	ΔG_i° (kcal mole^{-1})	ΔH_i (kcal mole^{-1})	$T\Delta S_i$ (kcal mole^{-1})	ΔS_i (cal mole^{-1} deg^{-1})
Ethanol	−1.94 (0.80)	−1.96 (0.84)	−1.97 (0.85)		−2.65	+0.39 (±0.09)	3.04	10.2
Diethyl ether	−2.39 (0.78)	−2.39 (0.79)	−2.43 (0.74)	−2.48 (0.69)	−3.26	+0.73 (±0.14)	3.99	13.4
Acetone	−2.85 (0.75)	−2.87 (0.72)	−2.92 (0.69)	−2.94 (0.64)	−3.89	+0.73 (±0.12)	4.62	15.5
Dimethyl sulfoxide	−1.54 (0.58)	−1.49 (0.56)	−1.39 (0.56)	−1.24 (0.55)	−2.10	−2.33 (±0.20)	−0.23	−0.8

[a] Modena et al., 1974.

Wilhoit, 1967). Commercial instruments of very high quality and precision (LKB, Tronac, Beckman) are available. In addition, simpler, less expensive equipment (e.g., SKC, Tronac*) is available which can easily provide heats of ionization to ±0·1 kcal mole^{-1}, precision which is more than adequate for the resolution of most problems in physical organic chemistry.

The procedure for determining heats of ionization (ΔH_i) simply involves measuring the heat evolved when a small increment (10^{-5} to 10^{-3} mmole) of base is rapidly injected into about 200 ml of HSO_3F in a solution calorimeter. If the base is a liquid this may be done with a microsyringe. If it is a solid other devices are used (Arnett et al., 1965).

The heat of ionization conventionally refers to the enthalpy change for the process $BH^+ \rightarrow B + H^+$ where B and BH^+ are at high dilution in the solvent under discussion (in the present instance HSO_3F). In practice it is more convenient to measure the reverse process $B + HA \rightarrow BH^+ + A^-$ and reverse the sign. If we wished to determine the heat of ionization of a strong base such as an amine in water, we could do it easily by measuring the partial molar heat of solution of the liquid amine $\Delta \bar{H}_s(B)$ at high dilution in aqueous base of sufficient strength to repress completely (i.e. 99·99%) the hydrolysis of the amine. We would then repeat the experiment in aqueous acid, $\Delta \bar{H}_s(HA)$, of sufficient strength to convert the amine completely to its ammonium salt. After minor corrections for ionic effects (Jones, 1971) we can safely say that the only difference between the two measured partial molar heats is the heat of ionization [eqn (18)]. Values obtained in this way agree with those

$$\Delta \bar{H}_S(HA) - \Delta \bar{H}_S(B) = -\Delta H_i. \quad (18)$$

from thermometric titration of dilute aqueous amine with dilute aqueous acid or from the temperature coefficients of pK-values.

In view of the problems we have just seen which may arise in the determination of the heats of ionization of weak bases by the van't Hoff method, it is natural to consider whether these values might not be determined calorimetrically. Obviously, this cannot be done directly in dilute aqueous acid (the standard state) because weak bases are by definition not ionized in this medium. Unfortunately, there is no way that we know of to use free energy-enthalpy extrapolations across a range of aqueous acid solutions to estimate

* S.K.C., Inc., P.O. Box 8538, Pittsburgh, Pennsylvania 15220. Tronac, Inc., 1804 South Columbia Lane, Orem, Utah 84601.

ΔH_i° for a base in water, because highly aqueous media spoil extrapolations to the aqueous standard state.

The basis for this conclusion is an extensive study principally by Burke, Carter and Douty (Arnett et al., 1972b; Burke, 1966; Douty, 1965) of the heats of solution of a number of organic solutes of varying basicity in aqueous sulfuric acid. In some cases it was possible to combine the resulting heats of transfer from water with the corresponding activity coefficients and obtain the complete picture of how the standard free energy, enthalpy and entropy of solution for typical organic bases and related conjugate acids varied from dilute acid to pure H_2SO_4.

The most important facts, for the present discussion, which emerged from that survey are as follows, (a) Salts and very weak bases which were unprotonated across the entire range of acidity showed relatively small changes in their partial molar heats of solution ($\Delta \bar{H}_s$) on transfer from 10% to 96% H_2SO_4. (b) In contrast, bases which were protonated showed increasingly exothermic $\Delta \bar{H}_s$ as the acidity increased, the sizes of this increase being related closely to the strength of the base. (c) It was shown that the main contributor to the large increases in $\Delta \bar{H}_s$ for strong bases was the solvent effect on transferring H_2SO_4 from dilute to concentrated acid since process (19) is common to all these reactions and involves the

$$B + H_2SO_4 \rightarrow BH^+ + HSO_4^- \tag{19}$$

acid as well as the base and its conjugate acid salt. (d) For acid solutions stronger than 30% H_2SO_4 *entropies* of transfer for all species studied *remain nearly constant* so that a close parallel holds between free energies and enthalpies of solution; this even appears to extend to pure sulfuric acid and beyond that to fluorosulfuric acid and its mixture with SbF_5 ("magic acid"). (e) Comparison of the heats of solution in a series of aqueous sulfuric acid solutions for aniline, N-methylaniline, N,N-dimethylaniline and several sterically hindered pyridines and anilines suggested strongly that hydrogen-bonding from BH^+ to the solvent or counter-ion is a differentiating factor for ammonium ions (Arnett et al., 1972b).

Some of these data are presented in Table 11 where the enormous range of enthalpy effects is apparent. It is seen that, unlike ultraviolet spectroscopy, calorimetry is a technique which can be applied at high dilution to practically any kind of compound to give significant energy changes which provide a unique kind of acidity function for each compound.

TABLE 11

Partial Molar Heats of Solution ($-\Delta \bar{H}_S$ kcal mole^{-1}) of Several Amines in Aqueous Sulfuric Acid Solutions[a] at 25°C

%H$_2$SO$_4$	$-H_0$	$-(H_0 + \log [\text{H}^+])$	Aniline	N-Methylaniline	N,N-Dimethylaniline	2,6,N,N-Tetramethylaniline
0·00	—	—	−0·33 ± 0·07			
2·60	−0·04	0·20	7·81 ± 0·16	6·72 ± 0·09	6·33 ± 0·01	6·76 ± 0·40
9·93	0·43	0·39	7·17 ± 0·09	6·24 ± 0·05	6·16 ± 0·02	6·76 ± 0·04
19·12	1·04	0·69		6·08 ± 0·13	6·18 ± 0·04	
20·79	1·17	0·77	7·35 ± 0·09			
21·08	1·19	0·78				6·83 ± 0·07
30·09	1·82	1·25	7·64 ± 0·06	6·96 ± 0·08	7·06 ± 0·11	7·98 ± 0·17
38·71	2·45	1·74				9·84 ± 0·10
39·94	2·54	1·81	9·10 ± 0·11	8·39 ± 0·10	9·18 ± 0·11	
49·76	3·40	2·54				12·74 ± 0·16
52·07	3·61	2·74	11·24 ± 0·10			
52·09	3·61	2·74		10·74 ± 0·20	11·78 ± 0·20	14·86 ± 0·16
59·25	4·40	3·46				
61·92	4·77	3·79	13·85 ± 0·38	13·18 ± 0·10	13·91 ± 0·47	17·77 ± 0·63
70·09	5·93	4·88				
82·42	7·89	6·73	19·82 ± 0·24	20·46 ± 0·38	23·18 ± 0·33	
91·87	9·29	8·05	25·20 ± 0·34	26·83 ± 0·50	29·10 ± 0·56	
92·15	9·33	8·09	24·49 ± 0·18			
FSO$_3$H		12·22[b]	34·3	35·4	37·6	29·55 ± 0·19
"Magic Acid"		16·04[b]	43·5			

[a] Arnett et al., 1972b.
[b] Estimated from $\Delta \bar{H}_S$ vs $(H_0 + \log [\text{H}^+])$ plot for aniline.

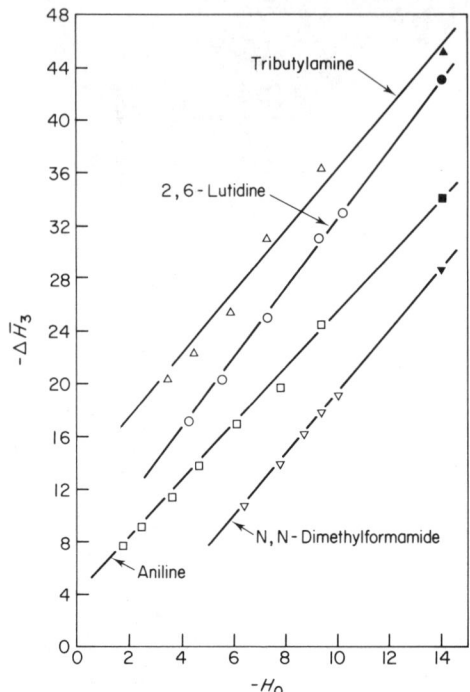

Figure 3. Plot of $\Delta \bar{H}_s$ vs H_0 for various nitrogen bases in aqueous H_2SO_4 (open points) and HSO_3F (black points) to estimate H_0 for the latter (Arnett et al., 1972b).

The facts given above and the close parallel between $\Delta \bar{H}_s$ values and the three acidity functions—$\%H_2SO_4$, H_0 and $(H_0 + \log [H^+])$ suggest that linear relationships may be found which would permit the estimation of otherwise inaccessible H_0 values or $\Delta \bar{H}_s$ values. This was indeed done in order to estimate the H_0 acidity of fluorosulfuric acid (Fig. 3) and HSO_3F–SbF_5 with independent corroboration of the former value within 1·3 H_0 unit (Arnett et al., 1972b). Unfortunately, we have not been able to find any objective treatment of such data which would yield reliable thermodynamic properties for ionization of weak bases in water as a standard state.

We will make considerable use of heats of ionization in HSO_3F below. The present section has drawn attention to the continuum of Brönsted acidity which links them to aqueous media.

Heats of Ionization in HSO_3F

The method and the "inert" solvent assumption

Fluorosulfuric acid is an extremely powerful proton donor, several orders of magnitude stronger than sulfuric acid. It presumably has a high dielectric constant (although it has never been reported), is fluid down to −90°C, and has much less tendency to oxidize compounds than does sulfuric acid. Thanks principally to Gillespie (1968) and Olah et al. (1970) we know through pmr studies that the vast majority of organic bases are converted instantly, cleanly and completely to their conjugate acids in this medium.

This fact was exploited in Arnett's laboratory during the years 1966–75 by Larsen, Quirk, Burke, Mitchell and Wolf who have used solution calorimetry to estimate heats of ionization in this medium. On p. 121 we described how ΔH_i could be obtained in water for strong Brönsted bases by measuring the heat of solution in aqueous base where it is completely unionized.

In the case of heats of ionization in pure liquid HSO_3F, we are faced with a separate problem in that the liquid base is converted instantly to $BH^+ SO_3F^-$ upon injection into the acid, and there is no way of measuring directly what the heat of solution of B would be in the absence of protonation.

We have handled this through an approximation which has been useful but which is certainly not above criticism. Instead of the inaccessible term $\Delta \bar{H}_s(B)$ which would be the heat of solution of the base in HSO_3F if no ionization occurred, we have used the partial molar heat of solution in an "inert" solvent represented as ΔH_s (inert); ΔH_i is then defined in eqn (20). The inert solvent is used to

$$-\Delta H_i = \Delta \bar{H}_S(HSO_3F) - \Delta H_S \text{ (inert)} \qquad (20)$$

level out idiosyncratic differences between the intermolecular interactions which different bases have as pure liquids or solids. These arise as the endothermic term which must be invested to separate the molecules of B in order to dissolve it in HSO_3F or any other medium. We use the term "inert" here in the sense of a solvent which exerts sufficiently strong intermolecular forces that it will dissolve nearly all kinds of bases, liquid or solid, which we wish to study in HSO_3F while keeping to a minimum specific interactions, such as strong hydrogen bonds or dipole–dipole forces, which will differentiate classes of bases in ways that have nothing to do with their proton affinities in solution. The demand for just the right degree of "inertness" as required here obviously cannot be perfectly met by any solvent. However, we expect that in no case is an error of more than 1 kcal mol^{-1} introduced in comparing the *relative* ΔH_i-values of two bases by this method compared to what might be obtained from the true, but at present unmeasurable, ΔH_i.

Justification for the use of ΔH_s (inert) for the approximation comes from several lines. First it is reasonable that in the absence of specific interactions such as hydrogen bonding, the solvation differences between a series of closely related bases (B) and their conjugate acid cations (BH$^+$) would be small. Each B and BH$^+$ pair will be of practically the same size so that internal pressure differences will be negligible. Furthermore, most non-basic non-polar liquids show negligible heats of solution in most other non-aqueous liquids, including HSO$_3$F (Arnett et al., 1970b).

The heats of transfer for several amines from different "inert" solvents to HSO$_3$F presented in Table 12 show only modest

TABLE 12

Heats of Solution and Condensation for Gaseous Amines into Various Liquids (determined by J. Wolf)

Compound	ΔH_S(kcal mole^{-1}) at 25°C				$\Delta H_v{}^a$
	FSO$_3$H	C$_6$H$_6$	CCl$_4$	o-C$_6$H$_4$Cl$_2$	
NH$_3$	−47·3 ± 0·7[b]	−3·52 ± 0·20		−4·45 ± 0·18	
CH$_3$NH$_2$	−51·7 ± 0·3[b]	−5·72 ± 0·18	−5·36 ± 0·4	−5·18 ± 0·10	5·996
CH$_3$CH$_2$NH$_2$	−52·9 ± 0·6	−5·84 ± 0·21	−6·08 ± 0·12	−6·29 ± 0·21	6·580
(CH$_3$)$_3$CCH$_2$NH$_2$	−46·5 ± 0·5	0·66 ± 0·09		0·37 ± 0·06	
(CH$_3$)$_2$NH	−53·8 ± 0·5[b]	−5·84 ± 0·14	−6·63 ± 0·19	−5·66 ± 0·12	5·779
(CH$_3$)$_3$N	−53·5 ± 0·7[b]	−5·69 ± 0·16	−6·47 ± 0·16	−5·87 ± 0·14	5·243

[a] Heat of vaporization (Jones, 1971).
[b] Arnett and Wolf, 1975.

differences compared to individual errors in measurement and, much more importantly, compared to the large magnitude of the ΔH_i values themselves. Finally, the close correlation of ΔH_i with other ionization properties presented both in this article and in other publications implies that large errors do not come from the "inert" solvent method as we have used it. Nonetheless, the method is based on the assumption that specific solvation differences for different bases are small in the solvents we have used—chiefly CCl$_4$ and o-dichlorobenzene. We expect that this will usually be true to within an error of ±0·5 kcal mol^{-1} but suggest that the reader abstain from interpreting differences of less than 1 kcal mol^{-1} in ΔH_i as being really significant.

The $\Delta H_i(HSO_3F)$ values which we will present are therefore simply heats of transfer of the solute from high dilution in an "inert" solvent to pure HSO_3F. These may be related to the gas phase through heats of vaporization or sublimation (Arnett and Oancea, 1975) which will be useful for calculating heats of solvation of BH^+.

We should like to emphasize in passing that for determining heats of ionization in solution it is more meaningful to use dilute solution in an inert solvent as the reference state for base B rather than the gas phase. This is because there is a dispersion force–cavity effect (Arnett and Carter, 1971) for condensing it from the gas phase into any solvent and this is dependent on the size of the molecule. This factor is cancelled out by using the "dilute solution in an inert solvent" reference state but will make large contributions of little significance to the ionization process if the heat of transfer from gas phase to HSO_3F is used instead of ΔH_i as we define it.

$\Delta H_i(HSO_3F)$ for Amines

We have used strongly basic amines to test relationships between processes in water and in other acidic media. In Table 13 we have collected all the data available for comparing heats of ionization of amines, both strong and weak, in water and HSO_3F. The data are plotted in Fig. 4 where it is seen that, despite a consistent trend towards correlation, there is a good deal of scatter. The straight line which we have drawn from corner to corner includes the entire range of primary amines including the weakly basic Hammett indicators (compounds 14–21). By this criterion, the data of Johnson, Katritzky and Shapiro (1969) are more consistent than those of Tickle, Briggs, and Wilson (1970), and we have therefore used the former. The pyridines (points 37–43) seem to define a separate line which extrapolates to a cluster of tertiary aliphatic amines (points 34–36). This again suggests some differentiation between primary, secondary and tertiary amines in either or both solvents.

It is instructive to compare Fig. 4 with Fig. 5 from a recent publication (Arnett et al., 1974) in which $\Delta H_i(HSO_3F)$ for 56 compounds representing many diverse classes of Brönsted bases are plotted against their pK-values in water. The data obey the linear equation (21) with a correlation coefficient of 0·986. The wide

$$-\Delta H_i = (1\cdot 77 \text{ p}K + 28\cdot 1) \text{ kcal mole}^{-1} \tag{21}$$

TABLE 13

Heats of Ionization (kcal mole^{-1}) of Amines in Water and HSO$_3$F; 25°C

	ΔH_i^o(water)[a]	ΔH_i^o(FSO$_3$H)[b]	Amines	ΔH_i^o(water)[a]	ΔH_i^o(FSO$_3$H)[b]
Substituted (X) Anilines					
X =					
1. H	7.43	34.0	24. C$_2$H$_5$–NH$_2$	13.71	46.8
2. 4-F	7.76	37.2	25. n-C$_3$H$_7$NH$_2$	13.84	46.2
3. 4-Cl	6.67	35.6	26. n-C$_4$H$_9$–NH$_2$	13.98	46.2
4. 4-Br	6.50	35.2	27. n-C$_5$H$_{11}$–NH$_2$	13.98	47.0
5. 4-I	6.50	34.5	28. i-C$_3$H$_7$–NH$_2$	13.97	49.3
6. 4-CH$_3$	7.86	36.9	29. $tert$-C$_4$H$_9$NH$_2$	14.35	48.8
7. 4-NO$_2$	3.09	31.1	30. (CH$_3$)$_2$NH	12.04	47.8
8. 3-Cl	6.45	34.2	31. (C$_2$H$_5$)$_2$NH	12.73	47.7
9. 3-NO$_2$	5.37	34.0	32. (n-C$_3$H$_7$)$_2$NH	13.17	48.6
10. 2-F	5.09	33.9	33. (n-C$_4$H$_9$)$_2$NH	13.66	46.4
11. 2-Cl	4.89	32.5	34. (CH$_3$)$_3$N	8.82	47.5
12. 2-I	4.89	32.4	35. (C$_2$H$_5$)$_3$N	10.32	49.2
13. 2-CH$_3$	6.57	35.6	36. (n-C$_3$H$_7$)$_3$N	10.5	48.1
14. 2-NO$_2$	1.68	26.8			
15. 2,4-(NO$_2$)$_2$	−3.16	21.5	Substituted (X) Pyridines		
16. 2,6-(NO$_2$)$_2$	−4.87	17.9	X =		
17. 2-NO$_2$-4-Cl	0.84	25.3	37. H	4.79	38.6
18. 2,5-Cl$_2$-4-NO$_2$	0.19	24.1	38. 4-CH$_3$	6.02	39.0
19. 2,6-Cl$_2$-4-NO$_2$	−1.03	21.8	39. 3-CH$_3$	5.64	39.1[a]
20. 2-Br-4,6-(NO$_2$)$_2$	−6.33	17.4	40. 3-Br	1.77	34.6
21. 2,4,6-(NO$_2$)$_3$	−9.73	13.2	41. 2-CH$_3$	5.99	39.6[a]
Amines			42. 2,6-(CH$_3$)$_2$	7.23	40.7
22. NH$_3$	12.48	43.3	43. Quinoline	5.36	37.0
23. CH$_3$–NH$_2$	13.18	46.3	44. Pyridine-N-oxide	−1.79	33.4

[a] See Tables 5, 6, 7, 9 and 10 for references.
[b] Anilines, Arnett et al., 1970a; Amines, Arnett and Wolf, 1975; Pyridines, Mitchell, 1972.

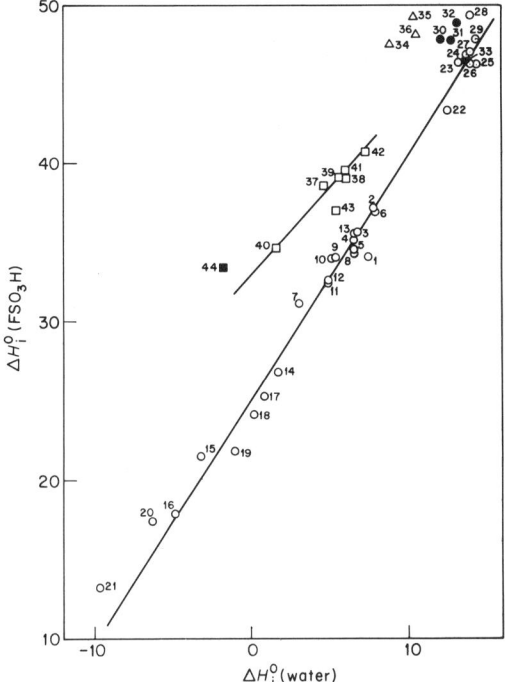

Figure 4. Heats of ionization in water vs heats of ionization in fluorosulfonic acid. Data points refer to Table 13.

differentiation between classes of amines seen in Fig. 4 is notably absent. We therefore conclude that it is $\Delta H_i(H_2O)$ which is producing the scatter in Fig. 4.

In Table 14 we have updated the pK and ΔH_i data on which Fig. 5 was based to include the best currently available values (August 1974). Several of the pK data used originally to plot Fig. 5 were compromised by the types of errors referred to earlier in this chapter.

Figure 6 compares the most reliable data for weak bases other than amines with the correlation lines for anilines (+) and pyridines (◊). The statistical data for the aniline line are: slope (1·57 ± 0·12), intercept (29·87 ± 0·64), correlation coefficient (0·9815). Corresponding data for the pyridines are (1·62 ± 0·068), (27·78 ± 0·64) and (0·9911).

Figure 7 portrays a corresponding plot omitting compounds (56–59) from Table 14 for reasons which are cited in that Table.

Comparison of Figs 5 and 7 shows, as was originally suggested,

TABLE 14

Comprehensive Tabulation of the Best Available Data for Anilines and Weak Bases: Heats of Ionization (ΔH_i) in HSO_3F and pK referred to Standard State in Water at 25°

Compound	pK	$-\Delta H_i(HSO_3F)$
1. 2,4,6-Trinitroaniline	−10·03	13·2
2. 2-Bromo-4,6-dinitroaniline	−6·69	17·4
3. 2,6-Dinitroaniline	−5·39	17·9
4. 2,4-Dinitroaniline	−4·27	21·5
5. 2,6-Dichloro-4-nitroaniline	−3·27	21·8
6. 2,5-Dichloro-4-nitroaniline	−1·75	24·1
7. 4-Chloro-2-nitroaniline	−1·06	25·3
8. 2-Nitroaniline	−0·30	26·8
9. 2,4,6-Tribromoaniline	0·8	25·3
10. 4-Nitroaniline	1·00	31·1
11. 2,4-Dichloroaniline	2·0	30·3
12. 3-Nitroaniline	2·46	34·0
13. 2-Iodoaniline	2·55	32·4
14. 2-Chloroaniline	2·64	31·1
15. 2-Fluoroalinine	3·20	32·5
16. 3-Chloroaniline	3·52	34·0
17. 4-Iodoaniline	3·79	34·2
18. 4-Bromoaniline	3·88	34·6
19. 4-Chloroaniline	3·98	35·6
20. 2-Methylaniline	4·44	35·2
21. Aniline	4·60	33·9
22. 4-Fluoroaniline	4·65	36·9
23. 4-Methylaniline	5·07	35·6
24. 2-Chloropyridine	0·6	31·7
25. 2-Bromopyridine	0·8	30·2
26. 3-Bromopyridine	2·72	34·5
27. Quinoline	4·81	37·0
28. Pyridine	5·31	38·6
29. 4-Methylpyridine	6·04	39·0
30. 2,6-Dimethylpyridine	6·81	40·7
31. 2,4,6-Trimethylpyridine	7·51	42·7
32. Pyridine-N-oxide	0·68	33·4
33. N,N-Dimethylacetamide	−0·36	32
34. N,N,-Dimethylbenzamide	−1·20	29·1
35. N,N-Dimethylformamide	−1·60	29·5
36. N-Methylformamide	−1·8	29·6
37. Dimethyl sulfoxide	−1·54	28·6
38. Tetramethylene sulfoxide	−1·34	29·5
39. Ethanol	−1·94	19·1
40. Diethyl ether	−2·39	19·1
41. Tetrahydrofuran	−2·32	19·6
42. Acetone	−2·85	19·1
43. Benzophenone	−4·03	16·9
44. 4-Chlorobenzophenone	−5·10	15·6
45. 4,4′-Dichlorobenzophenone	−5·29	14·9
46. Ethyl acetate	−4·61	17·4

TABLE 14—continued

Compound	pK	$-\Delta H_i(HSO_3F)$
47. Diethyl sulfide	−6·8	19·0
48. Tetrahydrothiophene	−6·84	19·7
49. N-Methyl-N-ethyl-n-butylamine	10·6	46·4
50. Dimethylamine	10·79	47·8
51. Diethylamine	11·01	47·7
52. Di-n-butylamine	11.25	46·4
53. Trimethylamine	9·797	47·5
54. Triethylamine	10·72	49·2
55. Tri-n-butylamine	9·93	45·2
56. Phenylmethylsulfoxide	−2·27	25·2[a]
57. 4-Methoxybenzophenone	−4·18	20·2
58. 4,4′-Dimethoxybenzophenone	−3·68	30·1[b]
59. 2,6-Dimethyl-γ-pyrone	−0·28	31· 2[c]

[a] Uncertain; slow reaction occurs with FSO_3H.
[b] Possibly too large; protonation of methoxy-groups may occur.
[c] Possibly too large; protonation of C=C may occur.

that refinement of pK and $\Delta H_i(HSO_3F)$ data would permit resolution of points scattered around the original single correlation line. The larger number of refined values upon which Fig. 7 is based now generates a series of nearly parallel lines for different basic functional groups. Using these it should be possible to estimate pK in water within 1 pK unit for any compound if the value of $\Delta H_i(HSO_3F)$ and the correlation line are known.

When these comparisons are added to the facts which were presented in pp. 106–109, much added strength is given to the general proposition that in aqueous solution pK is the best guide to potential energy changes, but in strong aqueous H_2SO_4 and pure HSO_3F, ΔH_i is an equally good measure of the potential energy of ionization. In Table 15 a few calorimetric values for protonating Hammet bases in 96% H_2SO_4 are presented for comparison.

4. HEATS OF SOLVATION

Calculating the Heats of Solvation of Ammonium and Oxonium Ions in HSO_3F and H_2O

Up to this point our entire discussion has been aimed at a thermodynamic analysis of the ways that different kinds of Brönsted

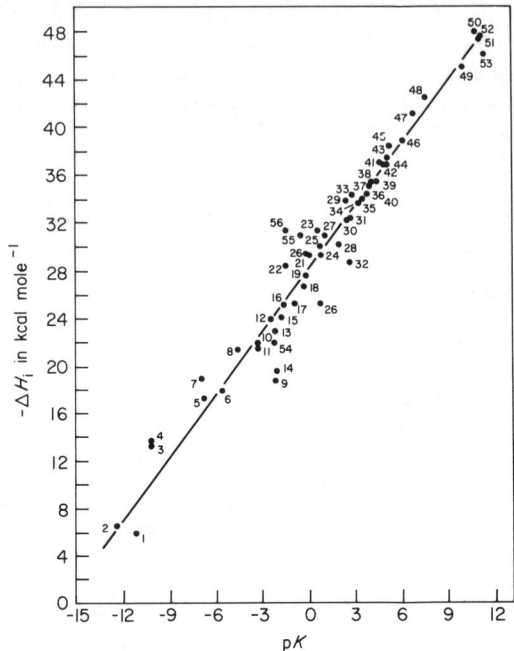

Figure 5. Plot of ΔH_i vs pK_a for the following bases: (1) benzoyl chloride, (2) nitrobenzene, (3) 2,4,6-trinitroaniline, (4) acetonitrile, (5) 2-bromo-4,6-dinitroaniline, (6) 2,6-dinitroaniline, (7) diethyl sulfide, (8) 2,4-dinitroaniline, (9) diethyl ether, (10) 2,6-dichloro-4-nitroaniline, (11) 1,4-dioxane, (12) diphenylcyclopropenone, (13) triphenylphosphine oxide, (14) tetrahydrofuran, (15) 2,5-dichloro-4-nitroaniline, (16) N,N-dimethylbenzamide, (17) 4-chloro-2-nitroaniline, (18) 2-nitroaniline, (19) N,N-dimethylacetamide, (20) N-methylformamide, (21) N,N-dimethylformamide, (22) dimethyl sulfoxide, (23) 2-chloropyridine, (24) pyridine N-oxide, (25) 2-bromopyridine, (26) 2,4,6-tribromoaniline, (27) 4-nitroaniline, (28) 2,4-dichloroaniline, (29) 3-nitroaniline, (30) 2-iodoaniline, (31) 2-chloroaniline, (32) triphenylphosphine, (33) 3-bromopyridine, (34) 2-fluoroaniline, (35) 3-chloroaniline, (36) 4-iodoaniline, (37) 4-bromoaniline, (38) 4-chloroaniline, (39) 2-methylaniline, (40) aniline, (41) 4-fluoroaniline, (42) quinoline, (43) N,N-dimethylaniline, (44) 4-methylaniline, (45) pyridine, (46) 4-methylpyridine, (47) 2,6-dimethylpyridine, (48) 2,4,6-trimethylpyridine, (49) tri-n-butylamine, (50) triethylamine, (51) quinolidine, (52) diethylamine, (53) di-n-butylamine, (54) phenyl methyl sulfoxide, (55) 2,6-dimethyl-γ-pyrone, (56) N-methyl-2-pyrrolidone. (Reproduced from Arnett et al., 1974).

bases and their conjugate acids behave relative to each other in response to changes of solvent. We have shown how it is possible to analyse this response by means of activity coefficients of the various charged and uncharged species. Furthermore, the results are consistent with the comparison based on the linear free energy relationships called acidity functions. Five years ago we would have had to leave the matter there; now we can go a step beyond and that is a very large step indeed.

Thanks mostly to ion-cyclotron resonance spectroscopy, it is possible to determine equilibrium processes for proton transfer such as eqn (22) within a small experimental error. This can be related to

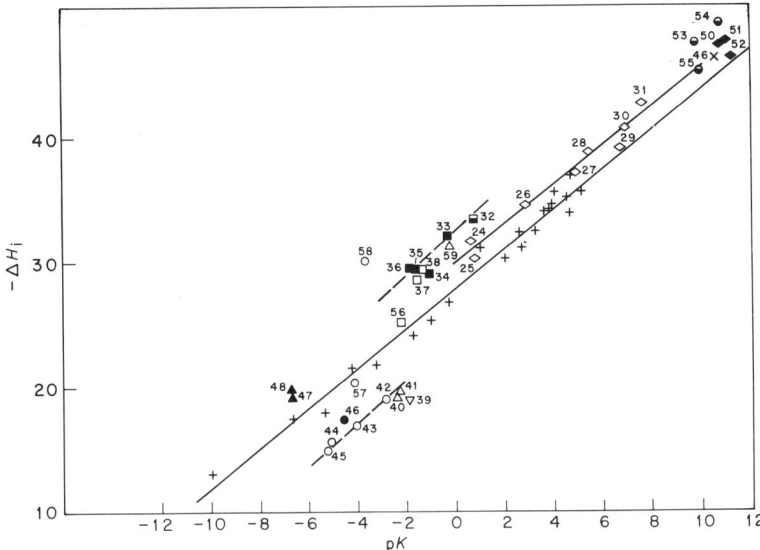

Figure 6. Correlation of Heat of Ionization in HSO_3F vs present best estimates of pK for weak bases. +, anilines; ◊, pyridines; x, RNH_2; ♦, R_2NH; ◉, R_3N; ○, ketones; ●, esters; △, ethers; ▲, sulphides; ▽, alcohols; □, sulfoxides; ■, Amides; ◒, pyridine-N-oxides. Data point numbers refer to Table 14.

the same process in solution, to give relative heats of solvation of the

$$BH^+ + B' \rightarrow B + B'H^+ \tag{22}$$

ionic species BH^+ or $B'H^+$. It is also possible to express the individual gas phase basicities as proton affinities and thus obtain an "absolute" solvation energy for each ion (see Scheme),

Scheme

gas phase $\quad BH^+(g) \xrightarrow{\text{i}} H^+(g) + B(g)$

$\quad\quad\quad\quad\quad\quad\quad\downarrow \text{ii} \quad\quad\quad\quad \downarrow \text{iii} \quad \downarrow \text{iv}$

aqueous solution $\quad BH^+(w) \xrightarrow{\text{v}} H^+(w) + B(w)$

$BH^+(g) \rightarrow H^+(g) + B(g)$	$\Delta P_i(g)$ (i)
$BH^+(g) \rightarrow BH^+(w)$	$\Delta P_s(BH^+)$ (ii)
$H^+(g) \rightarrow H^+(w)$	$\Delta P_s(H^+)$ (iii)
$B(g) \rightarrow B(w)$	$\Delta P_s(B)$ (iv)
$BH^+(w) \rightarrow B(w) + H^+(w)$	$\Delta P_i(w)$ (v)

from which

$$\Delta P_i(g) = \Delta P_i(w) - \Delta P_s(B) - \Delta P_s(H^+) + \Delta P_s(BH^+) \tag{vi}$$

where P represents any thermodynamic property ($G°, H°, S°$), g and w represent the gas phase and water medium, subscripts i and s represent ionization and solution processes, respectively. In principle, of course, this approach is almost as old as thermodynamics and has previously been applied in a few special cases (Bell, 1973). The

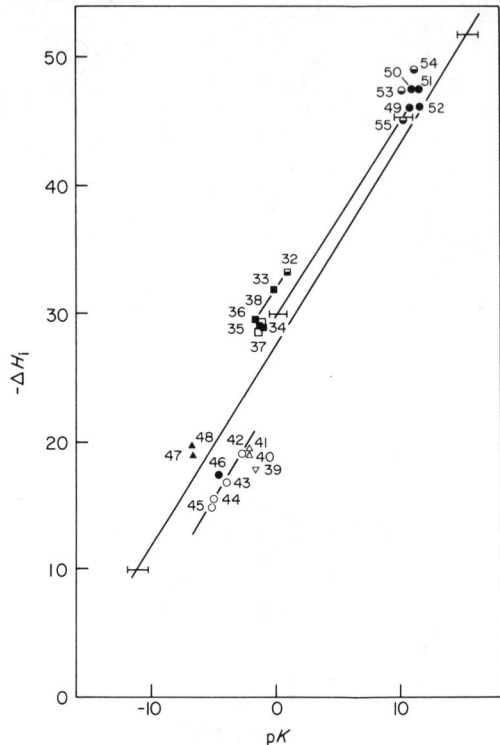

Figure 7. Data from Table 14 for weak bases with questionable data points (56–59) deleted.

TABLE 15

Protonation of Hammett Bases in Water, 96% H_2SO_4, and HSO_3F

X–Ar–NH$_2$	ΔH_i°(water)[a] kcal mole^{-1}	ΔH_i°(H_2SO_4)[b] kcal mole^{-1}	ΔH_i°(FSO_3H)[c] kcal mole^{-1}
4-NO_2	+3·09	20	31·1
2-NO_2	+1·68	17·4	26·8
4-Cl-2-NO_2	+0·84	16·2	25·3
2,5-Cl_2-4-NO_2	+0·19	14·9	24·1
2,6-Cl_2-4-NO_2	−1·03	12·2	21·8
2,4-(NO_2)$_2$	−3·16	9·9	21·5
2,6-(NO_2)$_2$	−4·87	8·8	17·9
2-Br-4,6-(NO_2)$_2$	−6·33	7·5	17·4
2,4,6-(NO_2)$_3$	−9·73		13·2

[a] From pK vs 1/T plots (Johnson et al., 1969).
[b] −[ΔH(H_2SO_4 96·4%) −ΔH_S° (inert)], (Burke, 1966).
[c] −[ΔH(FSO_3H) − ΔH_S° (inert)] (Arnett et al., 1970b).

difference in the present position is that the new gas phase data at last permit us to apply this long-awaited analysis to many common organic bases. Of particular relevance to this article is the use of such data to calculate the heats of solvation of ammonium and oxonium ions in HSO_3F with a probable accuracy of ±1 kcal mole^{-1}. Furthermore, using the free energies, enthalpies and entropies of ionization which have been presented above, we can estimate hydration energies and entropies of ammonium ions to ±1 kcal mole^{-1} and of oxonium ions to ±3 kcal mole^{-1} in water.

TABLE 16

Thermodynamic Data for Protonation of Amines in Gas Phase and HSO_3F (Determined by Wolf)

Compound	PA = ΔH_i(g)	ΔH_i(FSO_3H)	ΔH (inert)	ΔH_v^e
NH_3	207[a]	43·3 ± 0·9[c]	−4·0 ± 0·2[c]	
CH_3NH_2	216·1[a]	46·3 ± 0·5[c]	−5·4 ± 0·2[c]	
$(CH_3)_2NH$	222·5[a]	47·8 ± 0·7[c]	−6·0 ± 0·2[c]	
$(CH_3)_3N$	227·0[a]	47·5 ± 0·9[c]	−6·0 ± 0·2[c]	
$C_2H_5NH_2$	218·7[a]	46·8 ± 0·8	−6·1 ± 0·2	
$(C_2H_5)_2NH$	227·2[a]	47·7 ± 0·3[d]	−0·42 ± 0·06	7·45
$(C_2H_5)_3N$	233·7[a]	49·2 ± 0·3[d]	−0·64 ± 0·08	8·35
n-$C_3H_7NH_2$	220·1[a]	46·2 ± 0·5	0·48 ± 0·05	7·5
$(n$-$C_3H_7)_2NH$	229·5[a]	48·6 ± 1·0	0·19 ± 0·02	9·87
n-$C_4H_9NH_2$	220·6[a]	46·2 ± 0·7	1·13 ± 0·07	8·55
$(n$-$C_4H_9)_2NH$	230·3[b]	46·4 ± 0·6[d]	−0·30 ± 0·05	4·8
i-$C_3H_7NH_2$	221·1[a]	49·3 ± 0·8	0·46 ± 0·04	6·60
t-$C_4H_9NH_2$	223·1[a]	48·8 ± 0·5	0·44 ± 0·06	7·08

[a] Arnett et al., 1972a.
[b] Evaluated from data of Aue et al., 1972, by subtracting 2·4, the average difference between the values determined by Arnett et al., 1972a and Aue et al., 1972.
[c] Arnett and Wolf, 1973, 1975.
[d] Arnett et al., 1970a.
[e] Jones, 1971.

Some of the data needed for such a treatment of solvation of ions when they are transferred from the gas phase to HSO_3F are given in Tables 16 and 17. The proton affinities presented here are largely derived from work done in the laboratories of Aue, Beauchamp, McIver, McDaniel and Taft. Many have been difficult to establish and we expect that a few of the values will eventually prove to be seriously wrong. The value for H_2O is still in doubt and several other oxygen bases have been recently modified by nearly 10 kcal mole^{-1}. However, most of them have been subjected to so many

TABLE 17
Thermodynamic Data (kcal mole^{-1}) for Protonation of Weak Bases in Gas Phase and HSO$_3$F

Compound	Proton affinity = $\Delta H_i(g)$	$\Delta H_i(\text{FSO}_3\text{H})$	ΔH (inert)	ΔH_v
H$_2$O	165 ± 3^a, 179^e	16·5 ± 0·2^i	4·33 ± 0·15^{i,m}	10·5^i
CH$_3$OH	182 ± 3^b, 189^e	17·1 ± 0·4^i	2·74 ± 0·07^{i,n}	8·91 ± 0·02^q
C$_2$H$_5$OH	191^{c,d}, 193^e	18·7 ± 0·4^i	3·39 ± 0·12^{i,n}	10·11 ± 0·02^q
(CH$_3$)$_2$O	190 ± 5^a, 197^e	18·2 ± 0·7^i	−4·68 ± 0·26^{i,n}	
(C$_2$H$_5$)$_2$O	202^e	19·5 ± 0·7^i	0·09 ± 0·06^{i,n}	6·5^j
H$_2$S	170 ± 3^b, 183^e	5·3 ± 0·8^i	−5·1 ± 0·2^{i,p}	
CH$_3$SH	186^c, 193^e	19·2 ± 0·6^i	0 ± 0·3^{i,p}	
(CH$_3$)$_2$S	197^c, 203^e	18·1 ± 0·7^i	0 ± 0·03^{i,p}	6·6^r
PH$_3$	188^f, 194^e	14·0 ± 1·4^i	−2·9 ± 0·6^{i,p}	
(CH$_3$)$_3$P	228^f	44·6 ± 0·7^i	0 ± 0·03^{i,p}	6^s
(CH$_3$)$_2$CO	201^e	19·1 ± 0·1^k	0·79 ± 0·02^{k,o}	7·7^r
CH$_3$−CO−C$_3$H$_7$	199 ± 6^h	18·8 ± 0·2^k	0·43 ± 0·01^{k,o}	9·4^t
(CH$_2$)$_4$CO	202 ± 6^h	17·6 ± 0·1^k	0·32 ± 0·02^{k,o}	9·7^t
(CH$_2$)$_5$CO	204 ± 6^h	18·2 ± 0·1^k	0·09 ± 0·01^{k,o}	10·0^u
CH$_3$COOH	185 ± 2^h	13·9^l	0·76 ± 0·09^l	5·58^v
CH$_3$COOC$_2$H$_5$	205 ± 3^a, 204^e	17·4 ± 0·1^k	0·014 ± 0·004^{k,o}	8·40^q
CH$_3$CN	186^c	13·6 ± 0·2^k	1·8^{k,o}	7·87^s
(CH$_3$)$_2$SO	214^e	28·6^e	1·76^e	13·67^x

^{a} Long and Munson, 1973; ^{b} Haney and Franklin, 1969; ^{c} Beauchamp, 1971; ^{d} Estimated from differences in PA of ethanol and diethyl ether and PA of diethyl ether (ref. e); ^{e} Taft and Wolf, unpublished measurements; Taft, 1975; ^{f} Holtz et al., 1970; ^{g} Staley and Beauchamp, 1974; ^{h} Melby, 1971, ^{i} Arnett and Wolf, 1973; ^{j} Arnett et al., 1970a; ^{k} Arnett et al., 1970b; ^{l} Mitchell, 1972; ^{m} In sym-tetrachloroethane; ^{n} In benzene; ^{o} In carbon tetrachloride; ^{p} In toluene at −52°C; ^{q} Wadso, 1966; ^{r} Dreisbach, 1955; ^{s} Estimated from ΔH_v of similar compounds taken from Dreisbach, 1955; ^{t} Estimated from ΔH_v vs boiling point plots. Data taken from Weast, 1970; ^{u} Weast, 1970; ^{v} Konicek and Wadso, 1970; ^{w} Howard and Wadso, 1970; ^{x} Prueckner, 1963.

repetitions and cross checks in different laboratories that their relative values can be relied on to ±1 kcal mole^{-1}. All of the gas phase data cited here are taken relative to 207 kcal mole^{-1} for the proton affinity of NH$_3$. If (as seems quite possible) that value is modified it will change all of the other values for proton affinity, $\Delta H_i(g)$ and $\Delta G_i^\circ(g)$ but will leave unchanged all *relative* properties (preceded by δ) derived from them.

It has become a fairly common practice to refer to proton exchange equilibria and derived solvation energies to ammonia. The data from Tables 16 and 17 have been treated in this way to derive the relative solvation energies in HSO$_3$F for ammonium ions and a

TABLE 18

Evaluation of the Enthalpy of Transfer from Gas Phase to Water for Ammonium Ions[a]

Compound	$\Delta H_i^o(w)$	$\Delta H_S^o(B)$	$\delta_R \Delta H_i^o(g)$	$\delta_R \Delta H_i^o(w)$	$\delta_R \Delta H_S^o(B)$	$\delta_R \Delta H_S^o(BH^+)$
NH_3	12·48	−8·54	0	0	0	0
CH_3NH_2	13·18	−11·11	9·1	0·70	−2·6	5·8
$(CH_3)_2NH$	12·04	−13·26	15·5	−0·44	−4·7	11·2
$(CH_3)_3N$	8·82	−13·23	20·0	−3·66	−4·7	18·9
$C_2H_5NH_2$	13·71	−13·05	11·8	1·23	−4·5	6·0
$(C_2H_5)_2NH$	12·73	−15·31	20·2	0·25	−6·8	13·8
$(C_2H_5)_3N$	10·32	−16·76	26·7	−2·16	−8·2	20·6
$n\text{-}C_3H_7NH_2$	13·84	−13·37	13·1	1·36	−4·8	6·9
$(n\text{-}C_3H_7)_2NH$	13·17	−17·26	22·5	0·69	−8·7	13·1
$n\text{-}C_4H_9NH_2$	13·98	−14·11	13·6	1·50	−5·6	6·5
$(n\text{-}C_4H_9)_2NH$	13·66	−18·94	23·1	1·18	−10·4	11·5
$i\text{-}C_3H_7NH_2$	13·97	−13·37	14·1	1·49	−4·8	7·8
$t\text{-}C_4H_9NH_2$	14·35	−14·10	16·1	1·87	−5·6	8·6

[a] Jones 1971; Arnett and Wolf, 1975. All values are in kcal mole^{-1}.

TABLE 19

Heat of Transfer (kcal mole^{-1}) of Oxonium Ions from Gas Phase to Water

Compound	Proton affinity[a]	$\Delta H_i^\circ(w)$	$\Delta H_s^\circ(B)$[b]	$\delta_R \Delta H_i^\circ(g)$	$\delta_R \Delta H_i^\circ(w)$	$\delta_R \Delta H_s^\circ(B)$	$\delta_R \Delta H_s^\circ(BH^+)$
NH_3	207	12·48	−8·54	0	0	0	0
H_2O	179	0·0	−10·5	−28	−12·5	−1·96	−17
EtOH	193	+0·39	−12·6	−14	−12·09	−4·1	−6
Et_2O	202	+0·76	−11·1	−4	−11·72	−2·6	5
$(CH_3)_2CO$	201	+0·73	−10·2	−6	−11·75	−1·7	4
$(CH_3)_2SO$	214	−2·33	−18·2	7	−14·81	−9·7	13

[a] See Table 17 for references; [b] ΔH_V-values were taken from Table 17; ΔH_s for EtOH (−2·49), Et_2O (−4·6), and $(CH_3)_2CO$ (−2·44) are those reported by Arnett et al., 1972b, whereas that for $(CH_3)_2SO$ (−4·5) is from Corkill et al., 1969.

number of oxonium-, phosphonium- and sulfonium-ions. In Tables 18 and 19, an equivalent analysis is provided for some important ammonium and oxonium ions. In all these Tables the symbol δ_R is used as the structural operator to imply that the thermodynamic property which it modifies is taken relative to that for ammonia. Thus the term $\delta_R \Delta H_i^\circ(g)$ means "the heat of ionization in the gas phase relative to that for ammonia" and refers to the heat evolved (or absorbed) in the gas phase process (23). Similarly,

$$BH^+ + NH_3 \rightarrow B + NH_4^+ \qquad (23)$$

$\delta_R \Delta H_S^\circ(BH^+)_{H_2O}$ reads "the heat of solvation of BH^+ ion when it is transferred from the gas phase to water relative to NH_4^+", and refers to process (24). In view of the uncertainties referred to above we

$$BH^+(g) + NH_4^+(H_2O) \rightarrow BH^+(H_2O) + NH_4^+(g) \qquad (24)$$

have rounded off the proton affinities (and derived solvation energies) for oxygen bases to the nearest kcal mole^{-1}. Likewise, we have made no distinction between proton affinity, $\Delta H_i(g)$ and $\Delta G_i^\circ(g)$ although there are small formal entropy differences between 'onium ions with different numbers of acidic hydrogens.

Comparison of Heats of Solvation in Water and HSO_3F

In Table 20 we present the final results of our calculations, the solvation energies of a number of simple ammonium and oxonium ions relative to NH_4^+ in H_2O and HSO_3F.

Perhaps the most remarkable fact that emerges from these results is that the *relative* solvation energies of the ammonium ions in water are very close to those in the superacid, HSO_3F. Since both media have high dielectric constants, we might expect that on an electrostatic basis they would have comparable abilities to stabilize ions so that solvation energy should be strongly dependent on ionic radius. However, a moment's examination of Table 20 will reveal that the size of the ion, as reflected by the carbon number, seems to have much less influence on solvation energy than does the number and type of hydrogen bonds. Thus as we (Arnett et al., 1972a, 1974) and Taft et al. (1973) have noted there is a remarkable tendency for *most* primary ammonium ions so far reported to have heats of hydration about 7 kcal mole^{-1} less exothermic than NH_4^+, most

TABLE 20

Comparison of $\delta_R \Delta H_s^\circ(BH^+)$ of Transfer (kcal mole^{-1}) from Gas Phase to FSO_3H and H_2O

#	Compound (B)	$\delta_R \Delta H_s^\circ(BH^+)_w$	$\delta_R \Delta H_s^\circ(BH^+)_{FSO_3H}$[a]
1.	NH_3	0	0
2.	CH_3NH_2	5.8	4.7
3.	$(CH_3)_2NH$	11.2	9.0
4.	$(CH_3)_3N$	18.9	13.4
5.	$C_2H_5NH_2$	6.0	6.2
6.	$(C_2H_5)_2NH$	13.1	11.9
7.	$(C_2H_5)_3N$	20.6	15.4
8.	$n\text{-}C_3H_7NH_2$	6.9	7.0
9.	$(n\text{-}C_3H_7)_2NH$	13.1	11.8
10.	$n\text{-}C_4H_9NH_2$	6.5	7.1
11.	$(n\text{-}C_4H_9)_2NH$	11.5	12.1
12.	$i\text{-}C_3H_7\text{-}NH_2$	7.8	6.1
13.	$t\text{-}C_4H_9\text{-}NH_2$	8.6	8.8
14.	H_2O	−17	−1
15.	C_2H_5OH	−6	4
16.	$(C_2H_5)_2O$	4	18
17.	$CH_3\text{-}CO\text{-}CH_3$	4	16
18.	$CH_3\text{-}SO\text{-}CH_3$	13	13.8

[a] Data for amines are from Arnett and Wolf, 1975; [b] Estimated, see Table 19.

secondaries about 12 kcal mole^{-1} less exothermic and most tertiaries including pyridines and other diverse species to be about 19 kcal mole^{-1} less exothermic. To a first approximation, then, it is the number of hydrogen bonds which an ammonium ion can form with its environment which seems more than anything else to determine its relative hydration energy. Table 20 shows that this factor carries directly over to HSO_3F, a medium whose proton donating ability ($H_0 = -15$) is about twenty-two powers of ten greater than that of water ($H_0 = +7$).

Returning again to Table 20 and to Fig. 8 where the same data are plotted, we note that oxonium ions are far removed from the line for the ammonium ions. However, the slope of the line appears to be the same and we note that the displacement indicates that oxonium ions are relatively much better solvated in water than are ammonium ions if we use their behavior in HSO_3F as a guide. Thus in HSO_3F, $\delta_R \Delta H_s^\circ(BH^+)$ for $C_2H_5OH_2^+$ (4.0) is not far from that for $C_2H_5NH_3^+$ (6.2) [or $(CH_3)_2NH_2^+$ (9.0) if that is a better model]. In water however $C_2H_5OH_2^+$ is quite exothermically solvated (−6) while the

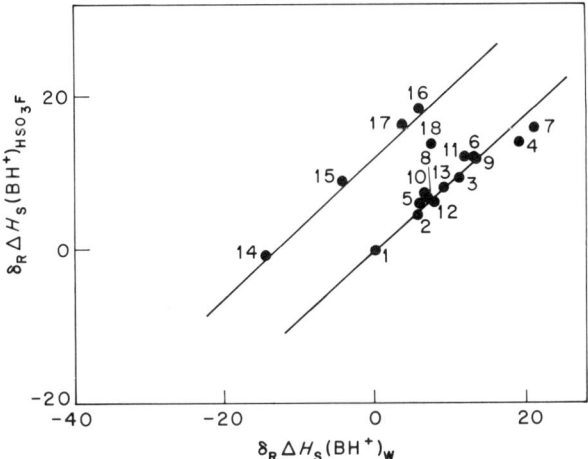

Figure 8. Correlations of heats of solvation from gas phase to HSO_3F vs these to water for the oxonium and ammonium ions listed in Table 20.

behavior of the ammonium ions has hardly changed. The same may be said for diethyl oxonium ion which changes its solvation energy also by about 14 kcal mole^{-1} from 18 to 4 while any comparable ammonium ion shows no similar effect. We remind the reader that our standard of comparison for both oxonium and ammonium ions is NH_4^+ and it is not surprising that alkyl ammonium ions should follow its behavior better than oxonium ions. The force of the comparisons in Table 20 is to show indeed how large the effect really is. This documents *strongly* that oxonium ions are much better solvated than are comparable ammonium ions in water whereas in at least one strong acid they are not nearly so differentiated.

One fact of great interest and significance is the very high exothermicity for solvating H_3O^+ relative to NH_4^+. The low proton affinity of H_2O (Table 17) shows that isolated water molecules are extraordinarily weak proton acceptors. In contrast, the ability of such weak acids as ammonium ions or carboxylic acids to generate H_3O^+ in aqueous solution proves that liquid water provides enough hydrogen bonding energy to make the formation of H_3O^+ a facile process in this medium. We see in Table 20 that of all the bases listed H_3O^+ shows the greatest relative decrease in solvation energy on transfer from water to HSO_3F. This factor is mitigated partially for ethanol in which part of the charge density on the oxonium ion can be delocalized internally into the ethyl group. In general one may conclude that, were it not for their excellent ability to form strong hydrogen bonds to H_2O and HSO_3F, the simple oxygen bases such as H_2O and Me_2O could not be protonated in even the strongest acid since on the average their proton affinities lie a full 30 kcal mole^{-1} below comparable nitrogen bases. It is perhaps not surprising that oxonium ions form much stronger hydrogen bonds to water than do ammonium ions since O—H···O bonds are surely stronger than are N—H···O ones. This leaves unanswered the question of why solvation energies for the two classes of ions should be so similar in HSO_3F which is also an oxygenated solvent.

5. ACIDITY FUNCTIONS AND SOLVATION

Standard Free Energies of Hydration of Ammonium Ions and Oxonium Ions

Acidity functions are free energy terms and their comparison through ϕ values (eqn 14) comes down to examination of the behavior of γ_X/γ_{XH^+} in aqueous acid solution. Because X and XH^+ are comparable in most respects, except for the presence of a charge and one or more acidic hydrogens, the response of γ_X/γ_{XH^+} should be largely dominated by the standard free energy of solvation of XH^+ since this is generally more sensitive to solvent change than is the value for X (Yates and McClelland, 1974; Arnett et al., 1972b). For a few oxygen bases it is feasible to use reliable new pK-values with Henry's Law data to calculate standard free energies of transfer both for X and XH^+ from the gas phase to water. They may then be compared with ϕ values for the same compounds and for analogous amines. In Table 20 such data are assembled for a few important oxygen bases, some of which are compared to analogous amines.

Again, as with heats of solvation, there is a general parallel between the effect of substitution of alkyl groups on oxygen and nitrogen bases and conjugate ions but the effects are larger for oxonium ions. This again suggests that hydrogen bonds from oxonium ions to water are stronger than from ammonium ions. Since oxygen is slightly larger than nitrogen, the stronger solvation of oxonium ions can scarcely be attributed to an ion size effect. Comparison of the last two columns in Table 20 reveals clearly the much greater variation in solvation energy for XH^+ than for the comparable free bases, X.

TABLE 21

Relative Standard Free Energy Terms (kcal mole^{-1}) for Ionization and Hydration of Oxygen Bases and Their Conjugate Acids Compared to Analogous Ammonium Ions

Compound (B)	$\delta_R \Delta G_i^o(g)$	$\delta_R \Delta G_i^o(w)$	$\delta_R \Delta G_S^o(B)$	$\delta_R \Delta G_S^o(BH^+)$	$\delta_R \Delta G_S^o(R_3N^+H)$
	a	b	c		d
NH$_3$	0	0	0		0
H$_2$O	−28	−15·16^e	−3·07	−16,	9·6
EtOH	−14	−15·26	−0·60^g	1,	18·0
Et$_2$O	−4	−15·87	2·34^g	13	26·0
Me$_2$C=O	−6	−16·50	1·73^g	12,	—
EtOAc	−3	−17·32^f	0·100	14	—

^{a} See Table 17 for references; ^{b} From Table 10; ^{c} Calculated from data in Frank and Evans, 1945; ^{d} Values for comparable ammonium ions of same number of acidic hydrogens and carbon number, i.e., EtNH$_3^+$, Et$_2$NH$_2^+$, Et$_3$NH$^+$, Arnett, 1975; ^{e} $\Delta G_i^o(w)$ estimated on the basis of $\phi = 1$ (see Bonvicini et al., 1973) and the assumption that water is half protonated in 84·5% H$_2$SO$_4$ (H$_2$O/H$_2$SO$_4$ = 1); ^{f} Lee and Sador, 1974; ^{g} From data collected by Hine and Weimar, 1965 and Guthrie, 1973, which were taken mainly from Timmermans, 1960.

In Table 22 all hydration properties are brought together for available oxonium ions and compared with corresponding values for representative ammonium ions of each class and with ϕ. Unfortunately, we have no ϕ values for the simple amines for which solvation data are available. The simple amines are far too basic to be studied in the H_0 range. Where such data are lacking they have been estimated from Tables 2 and 3.

Hydration Energies and ϕ Values

Armed with the data in Table 22 we are at last able to apply a rough test to the question of the relationship between acidity functions and hydration energies. In Fig. 9, a plot of relative standard free energies of hydration vs. ϕ is presented. A comparable plot of hydration enthalpies is given in Fig. 10 with the immediate warning that enthalpies and entropies of solution in water are much more complex than free energies. Furthermore, since ϕ is a free energy term it should be compared with a free energy term.

Even when allowance is made for conservative error bars, it is evident enough that the simple hydration energy from the gas phase to water is not sufficient to explain or predict acidity functions. Clearly there is a trend and the correlation is good enough to suggest that hydration is a major factor in acidity function behavior. As

TABLE 22

Thermodynamic Properties for Solvation of 'Onium Ions in Water at 25°

Compound (B)	$\delta_R \Delta G_S^\circ(BH^+)$ (kcal mole^{-1})	$\delta_R \Delta H_S^\circ(BH^+)$ (kcal mole^{-1})	$-\delta_R T\Delta S_S^\circ(BH^+)$ (kcal mole^{-1})	$\delta_R \Delta S_S^\circ(BH^+)$ (cal mole^{-1} deg.$^{-1}$)	ϕ
NH_3	0	0	0	0	0
CH_3NH_2	6·9	5·8	1·1	3·7	−0·3[a]
$(CH_3)_3N$	20·3	18·9	1·4	4·7	1·00
HOH	−16	−17	1	5·4	0·80
EtOH	1	−6	5	17	0·78
Et_2O	13	4	9	30	0·75
$(CH_3)_2CO$	12	13	8	27	0·58
$(CH_3)_2SO$					0·54
EtOAc	14				
$C_6H_5NH_2$	12[b]	7·2			0
Pyridine	+23·6	19·5	4·1	13·7	−0·40[a]

[a] Assuming that it behaves as a tertiary anilinium ion.
[b] Estimated from data in Arnett, 1975.

PROTONATION AND SOLVATION IN STRONG AQUEOUS ACIDS 145

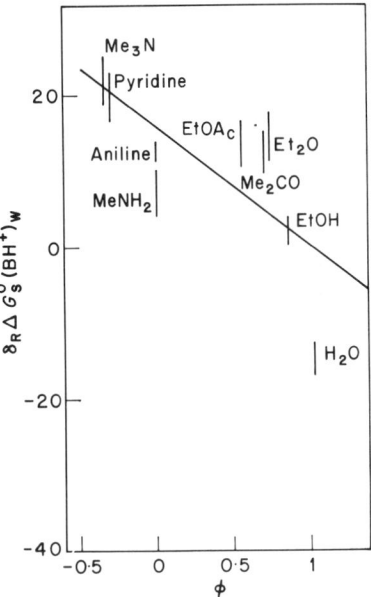

Figure 9. Correlation of ϕ with the standard free energies of hydration of ammonium and oxonium ions.

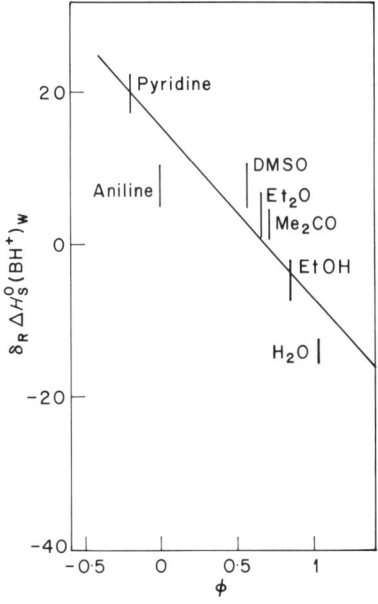

Figure 10. Correlation of ϕ with hydration enthalpies of ammonium and oxonium ions.

more and better data become available, it may become possible at last to make a thorough accounting for acidity function behavior in these terms.

We must conclude then that there is a rough correspondence between the hydration energy of an 'onium ion in water and the ease with which that hydration energy will change as the water content is reduced by addition of mineral acid. However, $\Delta G_S^o(BH^+)$ and $\delta G_S^o(BH^+)/\delta H_0$ (which is one way of expressing ϕ) show quite different response to solute structure. Careful examination of the detailed behavior of activity coefficients (Yates and McClelland 1974) and of the thermodynamics of transfer for weak bases and their ions in aqueous sulfuric acid (Arnett et al., 1972) supports this conclusion. However, it could not have been foreseen in the absence of accurate solvation energies.

The aqueous sulfuric acid system is highly complex with varying concentrations of H_2O, H_2SO_4, HSO_4^-, SO_4^{2-}, and $H_3SO_4^+$; less is known about other aqueous acids. The ability of X and XH^+ to respond to varying modes of interaction with these species can scarcely be expected to be neatly related to a single interaction such as $XH^+ \cdots OH_2$ and the above results show clearly that they cannot.

6. SUMMARY

We have approached the problem of the relationship between protonation, solvation, and acidity functions. First we have discussed the present status of the determination of pK and acidity functions for individual bases in terms of the Bunnett-Olsen equation.

Our thesis is summarized in the following statements.

(a) New spectroscopic methods, especially nmr, permit determination of the ionization ratio for most basic molecules in aqueous acids.

(b) Ultimately each compound has its own acidity function and the ϕ value for the Bunnett-Olsen equation (14) is a convenient means for comparing them.

(c) The magnitude of ϕ depends on the way the activity coefficient ratio γ_X/γ_{XH^+} varies with changing acidity.

(d) The ratio (γ_X/γ_{XH^+}) depends primarily on the behavior of γ_{XH^+} although variation of γ_X is also of some importance.

(e) Electrostatic solvation (Born charging) is probably significant for ions of very different size and shape but is probably a minor factor in differentiating among organic ammonium and oxonium ions.

(f) The principal differentiating term which causes variation in ϕ for oxygen and nitrogen bases is the balance between charge delocalization within the ion and hydrogen bonding by the ion to the medium.

(g) It is now possible in some cases to determine the relative free energies and enthalpies of solvation in water and some acidic media for oxonium, ammonium and sulfonium ions.

(h) There is a surprisingly close parallel between heats of solvation of ammonium and oxonium ions in water and fluorosulfuric acid suggesting that nearly parallel behavior obtains across the range of intermediate acid strengths.

(i) Oxonium ions are more strongly solvated than ammonium ions in water and fluorosulfuric acid. In both series the solvation energy depends sharply upon the number of acidic protons on the heteroatom which can hydrogen bond to the medium.

(j) For a small number of representative oxygen and nitrogen bases it is now possible to compare ϕ directly with the enthalpy and free energy of hydration. The correlation is not very good although the trend is clear. Variable solvation effects of γ_{BH^+} and γ_B must also play a role

(k) The value of ϕ is influenced by the ability of the rest of the molecule to delocalize positive charge. This will reduce the strength of hydrogen bonds from XH^+ to the solvent. Delocalization is maximized for arylcarbonium ions and their ϕ values change accordingly.

(l) The solvating ability of the medium depends on a complex combination of factors such as the activities of various water aggregates and anions. Their hydrogen bonding acceptor ability from XH^+ should be important but is not defined clearly at present.

ACKNOWLEDGEMENTS

Professor K. Yates generously made his important review on activity coefficients available to us long before its publication. Professor Robert Taft supplied valuable gas phase data. We also appreciate Professor G. Modena's many helpful suggestions and Dr. B. Chawla's and Dr. M. Stewart's care with proof corrections.

Our research in this field was supported by a Senior NATO Fellowship to G.S. and by grants, NSF #GP-6550-X and OSW #14-30-2570 to E.M.A.

REFERENCES

Albert, A., and Serjeant, E. P. (1962). "Ionization Constants of Acids and Bases", Methuen, London.
Armstrong, V. C., and Moodie, R. B. (1968), *J. Chem. Soc. B*, 275.
Arnett, E. M. (1963). *Progr. Phys. Org. Chem.* **1**, 223.
Arnett, E. M. (1973). *Accounts Chem. Res.* **6**, 404.
Arnett, E. M. (1975). "Proton Transfer Reactions", (E. F. Caldin and V. Gold, ed.) Chapman and Hall.
Arnett, E. M., and Bothner-By, A. A. (1963). Cited in Arnett (1963).
Arnett, E. M., and Bushick, R. D. (1964). *J. Amer. Chem. Soc.* **86**, 1564.
Arnett, E. M., and Carter, J. V. (1971). *J. Amer. Chem. Soc.* **93**, 1516.
Arnett, E. M. and Jones, F. M. III (1974). *Progr. Phys. Org. Chem.* **11**, 263.
Arnett, E. M., and Mach, G. W. (1964). *J. Amer. Chem. Soc.* **86**, 2671.
Arnett, E. M., and Mach, G. W. (1966). *J. Amer. Chem. Soc.* **88**, 1177.
Arnett, E. M., and Oancea, D. J. (1975) *J. Chem. Ed.*, **52**, 269.
Arnett, E. M., and Wolf, J. F. (1973). *J. Amer. Chem. Soc.* **95**, 978.
Arnett, E. M., and Wolf, J. F. (1975). *J. Amer. Chem. Soc.* **97**, 3262.
Arnett, E. M., and Wu, C. Y. (1960). *J. Amer. Chem. Soc.* **82**, 4999, 5660.
Arnett, E. M., Wu, C. Y., Anderson, J. N., and Bushick, R. D. (1962). *J. Amer. Chem. Soc.* **84**, 1674.
Arnett, E. M., Bentrude, W. G., Burke, J. J., and Duggleby, P. McC. (1965). *J. Amer. Chem. Soc.* **87**, 1541.
Arnett, E. M., Quirk, R. P., and Burke, J. J. (1970a). *J. Amer. Chem. Soc.* **92**, 1260.
Arnett, E. M., Quirk, R. P., and Larsen, J. W. (1970b). *J. Amer. Chem. Soc.* **92**, 3977.
Arnett, E. M., Jones, F. M. III, Taagepera, M., Henderson, W. G., Beauchamp, J. L., Holtz, D., and Taft, R. W. (1972a). *J. Amer. Chem. Soc.* **94**, 4724.
Arnett, E. M., Burke, J. J., Carter, J. V., and Douty, C. F. (1972b). *J. Amer. Chem. Soc.* **94**, 7837.
Arnett, E. M., Moriarity, T. C., Small, L. E., Rudolph, J. P., and Quirk, R. P. (1973). *J. Amer. Chem. Soc.* **95**, 1492.
Arnett, E. M., Mitchell, E. J. and Murty, T. S. S. R. (1974). *J. Amer. Chem. Soc.* **96**, 3875.
Arrhenius, S. (1887). *Z. Physik. Chem.* **1**, 631.
Aue, D. H., Webb, H. M., and Bowers, M. T. (1972). *J. Amer. Chem. Soc.* **94**, 4726.
Baldeschwieler, J. D., and Woodgate, S. S. (1971). *Accounts Chem. Res.* **4**, 114.
Bartlett, P. D., and McCollum, J. D. (1956). *J. Amer. Chem. Soc.* **78**, 1441.
Bates, R. G., and Hetzer, H. B. (1960). *J. Res. Nat. Bur. Stand.* **64A**, 427.
Bates, R. G., and Pinching, G. D. (1949). *J. Res. Nat. Bur. Stand.* **43**, 519.
Beauchamp, J. L. (1971). *Ann. Rev. Phys. Chem.* **22**, 527.
Bell, R. P. (1973). "The Proton in Chemistry", second edition, Chapman and Hall, London.
Bellobono, I. R., and Beltrame, P. (1969), *J. Chem. Soc. B.* 620.
Bellobono, I. R., and Diani, E. (1972). *J. C. S. Perkin II*, 1707.
Bellobono, I. R., and Monetti, M. A. (1973). *J. C. S. Perkin II*, 790.

Biggs, A. I. (1961). *J. Chem. Soc.* 2572.
Bolton, P. D., and Hall, F. M. (1967). *Aust. J. Chem.* 20, 1797.
Bolton, P. D., and Hall, F. M. (1968). *Aust. J. Chem.* 21, 939.
Bolton, P. D., and Hall, F. M. (1969), *J. Chem. Soc. B*, 259.
Bolton, P. D., Johnson, C. D., Katritzky, A. R. and Shapiro, S. A. (1970). *J. Amer. Chem. Soc.* 92, 1567.
Bolton, P. D., Fleming, K. A., and Hall, F. M. (1972). *J. Amer. Chem. Soc.* 94, 1033.
Bohme, D. K., Lee-Ruff, E., and Young, L. B. (1972). *J. Amer. Chem. Soc.* 94, 5153.
Bonvicini, P., Levi, A., Lucchini, V., and Scorrano, G. (1972a). *J. C. S. Perkin II*, 2267.
Bonvicini, P., Levi, A., and Scorrano, G. (1972b). *Gazz. Chim. Ital.* 102, 621.
Bonvicini, P., Levi, A., Lucchini, V., Modena, G., and Scorrano, G. (1973). *J. Amer. Chem. Soc.* 95, 5960.
Bowden, K. (1966). *Chem. Rev.* 66, 119.
Boyd, R. H. (1963). *J. Amer. Chem. Soc.* 85, 1555.
Boyd, R. H. (1969). In "Solute Solvent Interactions" (J. F. Coetzee and C. D. Ritchie eds.), Marcel Dekker, New York.
Briggs, J. P., Yamdagni, R., and Kebarle, P. (1972). *J. Amer. Chem. Soc.* 94, 5128.
Brönsted, J. N. (1923). *Rec. Trav. Chim.* 42, 718.
Brown, H. C., McDaniel, D. H., and Häfliger, O. (1955). In "Determination of Organic Structures by Physical Methods", (E. A. Braude and F. C. Nachod, eds.), Academic Press, New York.
Bryson, A. (1960). *J. Amer. Chem. Soc.* 82, 4858.
Bunnett, J. F., and Olsen, F. P. (1966). *Can. J. Chem.* 44, 1899.
Burke, J. J. (1966). Ph.D. Thesis, University of Pittsburgh.
Cabani, S., and Conti, G. (1965). *Gazz. Chim. Ital.* 95, 533.
Chakrabarty, M. R., Handloser, C. S., and Mosher, M. W. (1973). *J. C. S. Perkin II*, 938.
Chapman, N. B., and Shorter, J. (1972). "Advances in Linear Free Energy Relationships", Plenum Press, London.
Christensen, J. J., Izatt, R. M., and Hansen, L. D. (1967). *J. Amer. Chem. Soc.* 89, 213.
Christensen, J. J., Izatt, R. M., Wrathall, D. P., and Hansen, L. D. (1969). *J. Chem. Soc. A*, 1212.
Collumeau, A. (1968). *Bull. Soc. Chim. France*, 5087.
Corkill, J. M., Goodman, J. F., and Tate, J. R. (1969). *Trans. Faraday Soc.* 65, 1742.
Cox, M. C., Everett, D. H., Landsman, D. A., and Munn, R. J. (1968). *J. Chem. Soc. B* 1373.
Curci, R., Levi, A., Lucchini, V., and Scorrano, G. (1973). *J. C. S. Perkin II*, 531.
Davis, M. M. (1968). *Nat. Bur. Stand. Monographs* 105.
Davis, M. M. (1970). In "The Chemistry of Non-aqueous Solvents" (J. J. Lagowski, ed.), Academic Press, Inc.
Davis, C. T., and Geissman, T. A. (1954), *J. Amer. Chem. Soc.* 76, 3507.
Davy, H. (1816). In "The Collected Work of Sir Humphry Davy", Vol. 5, pp. 514–515.
de Courville, A., and Peltier, D. (1967). *Bull. Soc. Chim. France*, 2164.
Deno, N. C. (1964). *Survey Progr. Chem.* 2, 155.
Deno, N. C., Jaruzelski, J. J., and Schriesheim, A. (1955). *J. Amer. Chem. Soc.* 77, 3044.
Deno, N. C., Gaugler, R. W., and Schulze, T. (1966). *J. Org. Chem.* 31, 1968.

Douty, C. F. (1965). Ph.D. Thesis, University of Pittsburgh.
Dreisbach, R. R. (1955). "The Physical Properties of Chemical Compounds", A. C. S., Advances in Chemistry Series No. 15.
Edward, J. T. (1964). *Trans. Roy. Soc. Can.* **2**, 313.
Edward, J. T., Leane, J. B., and Wang, I. C. (1962). *Can. J. Chem.* **40**, 1521.
Essery, J. M., and Schofield, K. (1961). *J. Chem. Soc.* 3939.
Evans, M. G., and Polanyi, M. (1936). *Trans. Faraday Soc.* **32**, 1333.
Evans, M. G., and Polanyi, M. (1937). *Trans. Faraday Soc.* **33**, 166.
Everett, D. H., and Wynne-Jones, W. F. K. (1941). *Proc. Roy. Soc.* **177A**, 499.
Exner, O. (1964), *Coll. Czech. Chem. Comm.* **29**, 1094.
Exner, O. (1964). *Nature* **201**, 488.
Fickling, M. M., Fischer, A., Mann, B. R., Packer, J., and Vaughan, J. (1959). *J. Amer. Chem. Soc.* **81**, 4226.
Flexser, L. A., Hammett, L. P., and Dingwall, A. (1935). *J. Amer. Chem. Soc.* **57**, 2103.
Frank, H. S., and Evans, M. W. (1945). *J. Chem. Phys.* **13**, 507.
Gel'bshtein, A. I., Shcheglova, G., and Temkin, M. I. (1956). *Zh. Neorgan. Khim.* **1**, 506.
Gillespie, R. J. (1950). *J. Chem. Soc.* 2542.
Gillespie, R. J. (1968). *Accounts Chem. Res.* **1**, 202.
Girault-Vexlearschi, G. (1956). *Bull. Soc. Chim. France* 589.
Gold, V., and Hawes, B. W. V. (1951). *J. Chem. Soc.* 2102.
Greig, C. C., and Johnson, C. D. (1968). *J. Amer. Chem. Soc.* **90**, 6453.
Grunwald, E., and Winstein, S. (1948). *J. Amer. Chem. Soc.* **70**, 846.
Grunwald, E., Loewenstein, A., and Meiboom, S. (1957). *J. Chem. Phys.* **27**, 641.
Guthrie, J. P. (1973). *J. Amer. Chem. Soc.* **95**, 6999.
Haake, P., and Cook, R. D. (1968). *Tetrahedron Lett.* 427.
Haake, P., and Hurst, G. (1966). *J. Amer. Chem. Soc.* **88**, 2544.
Haake, P., Cook, R. D., and Hurst, G. H. (1967). *J. Amer. Chem. Soc.* **89**, 2650.
Haldna, U. L., and Pal'm, V. A. (1960), *Dokl. Akad. Nauk. SSSR* **135**, 667.
Haldna, U. L., Kuura, H. J., Laaneste, H. E., and Puss, R. K. (1964). *Russ. J. Phys. Chem.* **38**, 469.
Hall, N. F., and Sprinkle, M. R. (1932). *J. Amer. Chem. Soc.* **54**, 3469.
Hammett, L. P. (1928). *J. Amer. Chem. Soc.* **50**, 2666.
Hammett, L. P. (1970). "Physical Organic Chemistry", 2nd ed., McGraw Hill, New York.
Hammett, L. P., and Deyrup, A. J. (1932). *J. Amer. Chem. Soc.* **54**, 2721.
Haney, M. A., and Franklin, J. L. (1969). *J. Phys. Chem.* **73**, 4328.
Hantzsch, A. (1907). *Z. Physik. Chem.* **61**, 257.
Hantzsch, A. (1908). *Z. Physik. Chem.* **62**, 626.
Hantzsch, A. (1909a). *Z. Physik. Chem.* **65**, 41.
Hantzsch, A. (1909b). *Z. Physik. Chem.* **68**, 204.
Hantzsch, A. (1922). *Ber.* **55**, 953.
Hantzsch, A. (1930). *Ber.* **63**, 1782.
Hasselbalch, K. A. (1913). *Biochem. Bull.* **2**, 367.
Henderson, L. J. (1908). *J. Amer. Chem. Soc.* **30**, 954.
Hepler, L. G. (1963). *J. Amer. Chem. Soc.* **85**, 3089.
Hetzer, H. B., Robinson, R. A., and Bates, R. G. (1962). *J. Phys. Chem.* **66**, 2696.
Hine, J., and Weimar, R. D. (1965). *J. Amer. Chem. Soc.* **87**, 3387.
Hinman, R. L., and Long, J. (1964). *J. Amer. Chem. Soc.* **86**, 3796.

Hoerr, C. W., McCorkle, M. R., and Ralston, A. W. (1943). *J. Amer. Chem. Soc.* **65**, 328.
Holtz, D., Beauchamp, J. L., and Eyler, J. R. (1970). *J. Amer. Chem. Soc.* **92**, 7045.
Howard, C. J., Bierbaum, V. M., Rundle, H. W., and Kaufman, F. (1972). *J. Chem. Phys.* **57**, 3491.
Howard, P. B., and Wadso, I. (1970). *Acta Chem. Scand.* **24**, 145.
Janata, J., and Jansen, G. (1972). *J. C. S. Faraday Trans.* **1**, 1656.
Jander, G., Spandau, H., and Addison, C. C., eds. (1963-seq.). "Chemistry in Non-aqueous Ionizing Solvents", Vierveg-Interscience, Braunschweig and London.
Johnson, C. D., Katritzky, A. R., and Shapiro, S. A. (1969). *J. Amer. Chem. Soc.* **91**, 6654.
Jones, F. M. III (1971). Ph.D. Thesis, University of Pittsburgh.
Jones, J. R. (1973). "The Ionization of Carbon Acids", Academic Press, London.
Katritzky, A. R., Waring, A. J., and Yates, K. (1963). *Tetrahedron* **19**, 465.
Kebarle, P. (1972). In "Ions and Ion Pairs in Organic Reactions" (M. Szwarc, ed.), Interscience, New York.
Klofutar, C., Palik, S., and Kremser, D. (1973). *Spectrochimica Acta* **29A**, 139.
Konicek, J., and Wadso, I. (1970). *Acta Chem. Scand.* **24**, 2612.
Kresge, A. J., Chen, H. J., and Chiang, Y. (1972). *J. Chem. Soc. Chem. Comm.* 969.
Landini, G., Modena, G., Scorrano, G., and Taddei, F. (1969). *J. Amer. Chem. Soc.* **91**, 6703.
Lagowski, J. J., ed. (1966-seq.), "The Chemistry of Non-aqueous Solvents", Academic Press, New York.
Larsen, J. W., and Hepler, L. G. (1969). In "Solute Solvent Interactions" (J. F. Coetzee and C. D. Ritchie, eds.), Dekker, New York.
Lee, D. G., and Cameron, R. (1971). *J. Amer. Chem. Soc.* **93**, 4724.
Lee, D. G., and Cameron, R. (1972). *Can. J. Chem.* **50**, 445.
Lee, D. G., and Sadar, M. (1974). *J; Amer. Chem. Soc.* **96**, 2862.
Leffler, J. E., and Grunwald, E. (1963). "Rates and Equilibria of Organic Reactions", John Wiley, New York.
Levi, A., Modena, G., and Scorrano, G. (1974). *J. Amer. Chem. Soc.* **96**, 6585.
Levy, G. C., Cargioli, J. D., and Racela, W. (1970). *J. Amer. Chem. Soc.* **92**, 6238.
Liebig, J. von (1838). *Ann. Chem.* **26**, 113.
Liler, M. (1969). *J. Chem. Soc.* B 385.
Liler, M. (1971). "Reaction Mechanisms in Sulphuric Acid", Academic Press, New York.
Liotta, C. L., Perdue, E. M., and Hopkins, H. P. (1973). *J. Amer. Chem. Soc.* **95**, 2439.
Long, F. A., and McDevit, W. F. (1952). *Chem. Rev.* **51**, 119.
Long, F. A., and McIntyre, D. (1954). *J. Amer. Chem. Soc.* **76**, 3243.
Long, J., and Munson, B. (1973). *J. Amer. Chem. Soc* **95**, 2427.
Lowen, A. M., Murray, M. A., and Williams, G. (1950). *J. Chem. Soc.* 3318.
Lumry, R., and Rajender, S. (1970). *Biopolymers* **9**, 1125.
Matsui, T., Ko, H. C., and Hepler, L. G. (1974). *Can. J. Chem.* **52**, 2906, 2912.
Melby, E. G. (1971). Ph.D. Thesis, University of Cincinnati.
Mitchell, E. J. (1972). Ph.D. Thesis, University of Pittsburgh.
Modena, G., Perdonein, G., and Scorrano, G. (1974). Unpublished results.
Mortimer, C. T., and Laidler, K. J. (1959). *Trans. Faraday Soc.* **55**, 1731.

Murray, M. A., and Williams, G. (1950). *J. Chem. Soc.* 3322.
Oddo, G., and Casalino, A. (1917), *Gazz. Chim. Ital.* **47 II**, 200, 232.
Oddo, G., and Casalino, A. (1918). *Gazz. Chim. Ital.* **48 I**, 17.
Oddo, G., and Scandola, E. (1908). *Z. Physik. Chem.* **62**, 243.
Oddo, G., and Scandola, E. (1909a). *Z. Physik. Chem.* **66**, 138.
Oddo, G., and Scandola, E. (1909b). *Gazz. Chim. Ital.* **39 II**, 1, 44.
Oddo, G., and Scandola, E. (1910). *Gazz. Chim. Ital.* **40 II**, 163.
O'Hara, W. F. (1968). *Can. J. Chem.* **46**, 1965.
Olah, G. A., White, A. M., and O'Brien, D. H. (1970). *Chem. Rev.* **70**, 561.
Ostwald, W. (1888). *Z. Physik. Chem.* **2**, 36.
Pal'm, V. A., Haldna, U. L., and Talvik, A. J. (1966). In "The Chemistry of the Carbonyl Group" (S. Patai, ed.), Interscience, New York.
Paul, M. A., and Long, F. A. (1957). *Chem. Rev.* **57**, 1.
Pauling, L. (1928). *Proc. Nat. Acad. Sci.* **14**, 359.
Pauling, L. (1931). *J. Amer. Chem. Soc.* **53**, 1367.
Perkampus, H. H., and Prescher, G. (1968). *Ber. Bunsengesellschaft Phys. Chem.* **72**, 429.
Perrin, D. D. (1965). "Dissociation Constants of Organic Bases in Aqueous Solution", Butterworths, London.
Prueckner, H. (1963). *Erdoel. Kohle* **16**, 188.
Reagan, M. T. (1969). *J. Amer. Chem. Soc.* **91**, 5506.
Rochester, C. H. (1966). *Quart. Rev.* **20**, 511.
Rochester, C. H. (1970). "Acidity Functions", Academic Press, New York.
Sacconi, L., Paoletti, P., and Ciampolini, M. (1960a). *J. Amer. Chem. Soc.* **82**, 3828.
Sacconi, L., Paoletti, P., and Ciampolini, M. (1960b). *J. Amer. Chem. Soc.* **82**, 3831.
Scorrano, G. (1973). *Accounts Chem. Res.* **6**, 132.
Staley, R. H. and Beauchamp, J. L. (1974). *J. Amer. Chem. Soc.* **96**, 1604, 6252.
Sturtevant, J. M. (1959). In "Physical Methods of Organic Chemistry" (A. Weissberger, ed.), 3rd ed., Vol. I, Interscience, New York.
Sweeting, L. M., and Yates, K. (1966). *Can. J. Chem.* **44**, 2395.
Taagepera, M., Henderson, W. G., Brownlee, R. T. C., Beauchamp, J. L., Holtz, D., and Taft, R. W. (1972), *J. Amer. Chem. Soc.* **94**, 1369.
Taft, R. W., Jr. (1956). In "Steric Effects in Organic Chemistry" (M. S. Newman, ed.), John Wiley and Sons, New York.
Taft, R. W. (1975). "Proton Transfer Reactions" (E. F. Caldin and V. Gold ed.), Chapman and Hall, London.
Taft, R. W., Taagepera, H., Summerhays, K. D., and Mitsky, J. (1973). *J. Amer. Chem. Soc.* **95**, 3811.
Tickle, P., Briggs, A. G., and Wilson, J. M. (1970). *J. Chem. Soc.* B 65.
Timmermans, J. (1960). "The Physico-Chemical Constants of Binary Systems in Concentrated Solutions", Vol. 4, Interscience, New York.
Tissier, C., and Tissier, M. (1972). *Bull. Chem. Soc. France* 2109.
Trotman-Dickenson, A. F. (1949). *J. Chem. Soc.* 1293.
Van de Poel, W., and Slootmaekers, P. J. (1970). *Bull. Soc. Chim. Belg.* **79**, 223.
Van der Linde, W., Northcott, D., Redmond, W., and Robertson, R. E. (1969). *Can. J. Chem.* **47**, 279.
Vanderzee, C. E., King, D. L., and Wadso, I. (1972). *J. Chem. Thermodynamics* **4**, 685.
Vinogradov, S. N., and Linnel, R. H. (1971). "Hydrogen Bonding", Van Nostrand Reinhold, New York.

Waddington, T. C., ed. (1965). "Non-aqueous Solvent Systems", Academic Press, New York.
Wadso, I. (1966). *Acta Chem. Scand.* **20**, 544.
Weast, R. C., ed. (1970). "Handbook of Chemistry and Physics", The Chemical Rubber Co., Cleveland.
Wells, P. R. (1968). "Linear Free Energy Relationships", Academic Press, London.
Westheimer, F. H., and Kharasch, M. S. (1946). *J. Amer. Chem. Soc.* **68**, 7871.
Whetsel, K. B. (1961). *Spectrochimica Acta* **17**, 614.
Wilhoit, R. C. (1967). *J. Chem. Educ.* **44**, A571, A629, A685, A688, A853.
Yates, K., and McClelland, R. A. (1967). *J. Amer. Chem. Soc.* **89**, 2686.
Yates, K., and McClelland, R. A. (1973). *J. Amer. Chem. Soc.* **95**, 3055.
Yates, K., and McClelland, R. A. (1974). *Progr. Phys. Org. Chem.* **11**, 323.
Yates, K., Stevens, J. B., and Katritzky, A. R. (1964). *Can. J. Chem.* **42**, 1957.
Yates, K., Wai, H., Welch, G., and McClelland, R. A. (1973). *J. Amer. Chem. Soc.* **95**, 418.

Formation, Properties and Reactions of Cation Radicals in Solution

A. J. BARD[1], A. LEDWITH[2], and H. J. SHINE[3]

[1] *Department of Chemistry, University of Texas, Austin, Texas 78712, U.S.A.*

[2] *Donnan Laboratories, University of Liverpool, Liverpool L69 3BX, U.K.*

[3] *Department of Chemistry, Texas Tech University, Lubbock, Texas 79409, U.S.A.*

1. Formation of Cation Radicals 156
   Oxidations in Brönsted Acid Media 157
   Oxidation by Lewis Acids 164
   Oxidation by Halogens 167
   Use of Metal Salts 169
   Formation via Charge Transfer Complexes 175
   Light-induced Formation 179
   Formation on Catalytic Surfaces 188
   Isolable Cation-Radical Salts 192
2. Electrochemical Methods of Formation and Investigation of Cation Radicals. 197
   Electrochemical Techniques 198
   Examples of Electrochemical Studies 203
3. Physical Properties of Cation Radicals 210
   Dimerizations 210
   Disproportionations 214
   Ion-Pairing 218
4. Electron Transfer Reactions of Cation Radicals . . . 218
   Exchange Electron Transfer Reactions 220
   Chemiluminescent and Electrogenerated Chemiluminescent Reactions 222
5. Reactions of Cation Radicals with Nucleophiles . . . 226
   Reaction with Water 227
   Reaction with Alcohols 230

                Reaction with Cyanide Ion 232
                Reaction with Halide Ions 234
                Reaction with Amines 238
                Acetoxylation 247
                Reaction with Aromatics 249
                Reaction with Olefins 250
                Reaction with Ketones 252
        6. Cation Radicals from Bipyridylium Salts 254
                Electron-transfer scavenging of Organic Radicals 258
        7. Concluding Remarks 264
                Acknowledgements 265
                References 265

1. FORMATION OF CATION RADICALS

Cation radicals are formed by the removal of one electron from a neutral molecule. The result is the formation of a species which is at the same time a cation (the positive charge caused by the loss of an electron) and a radical (the remaining unpaired electron). One-electron oxidations may be achieved with a variety of chemical oxidants, e.g. concentrated sulfuric acid, by physical means, e.g. photoionization, pulse radiolysis, and electron impact (mass spectrometry), and by anodic oxidation. In this section we shall confine ourselves mainly to chemical oxidations, to oxidations on acidic surfaces, and to photoionizations. Anodic oxidation is dealt with more fully in section 2 and in Volume 12 of this series (Eberson and Nyberg, 1976). Formation by electron impact ionization will not be discussed here.

In proposing that a cation radical is an intermediate in any particular oxidation process it is useful to know if, for example, the neutral reactant will give up one or more electrons in the accessible potential range, and if the possible products will survive the potential necessary to oxidize the reactant. The ability of a neutral organic molecule to give up an electron is governed by the energy of the highest occupied molecular orbital (HOMO) and may be calculated (approximately) or, more conveniently, estimated experimentally by measurements of gas phase ionization potential (I.P.) (e.g. by photoelectron spectroscopy) and/or solution phase oxidation potential ($E_{1/2}$). Since HOMO energies govern both I.P. and $E_{1/2}$ values, it is to be expected that they will show a correlation (Miller *et al.*, 1972), although it is obvious that both specific and more general

TABLE 1

Oxidation and Ionization Potentials (Miller et al., 1972)

Compound	I.P. eV[a]	$E_{\frac{1}{2}}$V[b]
Ethylene	10·51	2·90
1,3-Butadiene	9·07	2·03
Cyclohexene	8·95	1·98
1,4-Dioxane	9·13	1·97
Pyridine	9·27	1·82
Benzene	9·24	2·04
Dimethyl sulfoxide	8·84	1·73
Thiophene	8·86	1·70
Dimethyl sulfide	8·69	1·26
Mesitylene	8·39	1·53
Phenol	8·50	1·04
Anisole	8·22	1·40
Biphenyl	8·27	1·48
Naphthalene	8·12	1·34
Hexamethylbenzene	7·85	1·20
Triethylamine	7·50	0·79
Diphenylamine	7·40	0·53
Anthracene	7·23	0·84
N,N-Dimethylaniline	7·14	0·45

[a] Assumed to be vertical ionization potentials.
[b] Measured in CH_3CN at a Pt electrode vs $Ag/AgNO_3$.

solvation phenomena will also affect values of $E_{1/2}$. Table 1 gives representative data for I.P. and $E_{1/2}$ values of typical organic molecules.

The most easily oxidized molecules are those containing π-electrons and heteroatoms with unshared pairs. Consequently, most of our treatment concerns aromatic and heteroaromatic cation radicals. Cation radicals of saturated molecules are, of course, easily but fleetingly made in the mass spectrometer, and even the anodic oxidation of alkanes is achievable and describable in terms of cation radical chemistry (Clark et al., 1973).

Oxidations in Brönsted Acid Media

Concentrated sulfuric acid

Many aromatic and heterocyclic compounds dissolve in concentrated sulfuric acid and become oxidized to their cation radicals. The acid is reduced to sulfur dioxide, and the complete reaction may be written as in eqn (1). Oxidation in sulfuric acid became a standard technique for characterizing cation radicals by esr (atmospheric

$$2ArH + 3H_2SO_4 \rightarrow 2ArH^{\cdot+} + 2H_2O + SO_2 + 2HSO_4^- \qquad (1)$$

oxygen does not broaden the esr signals in sulfuric acid) and electronic absorption spectroscopy. Recognition of the formation of cation radicals in this way has a curious history. During the early period of esr spectroscopy, about 20 years ago, colored solutions of aromatic and heterocyclic compounds in sulfuric acid were found to be paramagnetic, but it was not recognized that simple one-electron oxidation had occurred. Paramagnetism in solutions of bianthrone, for example, was attributed to the biradical form of protonated bianthrone (Hirschon et al., 1953), while, on the occasion of the first demonstration of esr of a hydrocarbon (perylene) solution, the phenomenon was attributed to the triplet state of perylene (Yokosawa and Miyashita, 1956) because the esr signal decreased as the temperature was lowered, a behavior now understood as the reversible dimerization (or pairing) of the cation radical.

Anion radicals of aromatic hydrocarbons had already been characterized spectroscopically when exploration of the sulfuric-acid reactions was under way. Theory (Vincow, 1968) predicts that the electronic absorption spectra of cation radicals and anion radicals of alternant aromatic hydrocarbons should be similar to each other. This led several groups of workers, particularly in Holland, to recognize from absorption spectra that sulfuric-acid solutions of hydrocarbons such as perylene, anthracene, naphthacene, and 3,4-benzopyrene contained the corresponding cation radicals (Hoijtink and Weijland, 1957; Kon and Blois, 1958; Aalbersberg et al., 1959a). Direct identification of cation radicals of anthracene, and of perylene and naphthacene by esr spectroscopy followed (Weissman et al., 1957; Carrington et al., 1959). In particular, the well-resolved spectra of the anthracene, perylene and naphthacene cation radicals allowed measurement of all of the ring-proton coupling constants and their comparison with those of the already-known anion radicals, Thus, by 1959 the formation of cation radicals in sulfuric acid had been shown without doubt.

The mechanism of oxidation of aromatic hydrocarbons by sulfuric acid is not known. Hydrocarbons which do not oxidize easily (e.g. naphthalene) are known to be protonated in sulfuric acid. A series of steps [eqns (2-4)] has been proposed in which the protonated aromatic is a key intermediate. These steps have been written as equilibria (Carrington et al., 1959), and if they are to be valid, the equilibrium constant for (3) must be very very small and the equilibrium constant for (4) very large. The reason for the first

$$ArH + H_2SO_4 \rightleftharpoons ArH_2^+ + HSO_4^- \tag{2}$$

$$ArH + ArH_2^+ \rightleftharpoons ArH_2\cdot + ArH^{+\cdot} \tag{3}$$

$$ArH_2\cdot + 2H_2SO_4 \rightleftharpoons ArH^{+\cdot} + 2H_2O + HSO_4^- + SO_2 \tag{4}$$

condition is that cation radicals cannot be detected when aromatics are dissolved in other strong protic acids (e.g. trifluoroacetic acid–boron trifluoride mixtures, hydrogen fluoride, and trifluoroacetic acid) unless oxygen is present (Hoijtink and Weijland, 1957; MacLean and van der Waals, 1957; Aalbersberg et al., 1961; Dallinga et al., 1958), while the reason for the second condition is that cation-radical conversion is normally complete in sulfuric acid. The situation is not quite clear, however, because solutions of methanesulfonic acid in nitrobenzene oxidize certain aromatics (e.g. naphthacene, pentacene, 9,10-diphenylanthracene) very nicely (Malachesky et al., 1966). This observation would support equilibrium (3) if the nitrobenzene played no part, but whether or not it does play a part is not known.

The attractiveness of the proposed steps is that eqn (2) accounts for the solubility of an aromatic hydrocarbon in sulfuric acid, and the radical $ArH_2\cdot$ in eqn (4) is expected to be more easily oxidized than the parent hydrocarbon, ArH. No direct evidence for the existence of the radical $ArH_2\cdot$ in sulfuric acid is known, however.

Polarographic reduction of aromatic cation radicals (e.g. perylene, naphthacene, anthracene, pyrene and others) in trifluoracetic acid–boron trifluoride leads to the formation of the radical $ArH_2\cdot$ [eqns (5)–(7)] and has shown that reduction of the cation radical (5) is

$$ArH^{+\cdot} + e^- \rightleftharpoons ArH \tag{5}$$

$$ArH + H^+ \rightleftharpoons ArH_2^+ \tag{6}$$

$$ArH_2^+ + e^- \rightleftharpoons ArH_2\cdot \tag{7}$$

easier than of the protonated aromatic (7) by values of 0·5 to 0·8 V (Aalbersberg and Mackor, 1960). This again indicates that equilibrium (3), if it exists in sulfuric and other strong acids, must lie on the left-hand side.

The summation of the situation is that cation radicals are formed in strong oxidizing acids, but we do not precisely know all of the steps. We shall see that this situation also prevails for other methods of forming cation radicals.

The use of sulfuric acid is accompanied by other difficulties. Most aromatic and heterocyclic compounds dissolve only slowly in concentrated sulfuric acid, and this leads to the possibility of condensation reactions on the surface of solid compounds or in the bulk of liquid ones. The decomposition of the cation radical in solution during slow dissolution of the parent compound can also be a problem. These difficulties are sometimes overcome by first dissolving the compound in a small amount of solvent such as hexane, dimethylformamide, acetonitrile, or acetic acid (Grace and Symons, 1959; Wheeler et al., 1966). The use of sulfuric acid in nitromethane gives exceptionally well-resolved esr spectra.

The most troublesome consequence of using concentrated sulfuric acid is in its sulfonating property. An aromatic molecule is likely to be more easily oxidized to a cation radical if it contains electron-donor substituents. But these substituents also make the molecule more susceptible to electrophilic attack, i.e. to sulfonation. This is the case, for example, with aromatic ethers and diarylsulfides. For example, 1,4-dimethoxybenzene undergoes 70-90% sulfonation within a few minutes in concentrated sulfuric acid, and only a low concentration of the cation radical can be obtained (Nishinaga et al., 1970). Diphenyl sulfide undergoes disulfonation very rapidly without being able to form the cation radical, while other diaryl sulfides give only low concentrations of cation radical (Shine et al., 1967). Fortunately, these problems can be avoided by using other oxidizing systems such as aluminum chloride-nitromethane solutions.

Although it took only a few years after the first esr measurements to recognize that cation radicals were formed in sulfuric acid, the knowledge that they are formed already lay in the literature. Kehrmann deduced over 50 years ago that phenothiazine lost both one and two electrons in sulfuric acid solution (Kehrmann et al., 1914; Kehrmann and Sandoz, 1918). The two oxidation states were described in the terminology of the times as semiquinoid and holoquinoid, respectively. Kehrmann characterized each state by ultraviolet and visible spectroscopy, but had no direct way of proving the odd-electron nature of the semiquinoid state—the cation radical. It was some 25-30 years later that Michaelis showed by potentiometric titration that phenothiazine underwent one-electron oxidation in reaction with bromine or lead tetraacetate, and wrote Kehrmann's semiquinoid state in the present way, that is, as the cation radical (Michaelis et al., 1941). The probability that aromatic hydrocarbons are also oxidized to cation radicals was explicitly stated by Weiss (1946). Thus, the background to cation-radical formation in sulfuric acid was in the literature, but the formation was re-discovered by present-day spectroscopists.

Benzene and its alkyl derivatives do not ordinarily form cation radicals in sulfuric acid. *p*-Xylene was once thought to give xylene cation radical in solutions of persulfuric-sulfuric acid, but is now known to give the semiquinone cation radical instead (8) (Brivati *et*

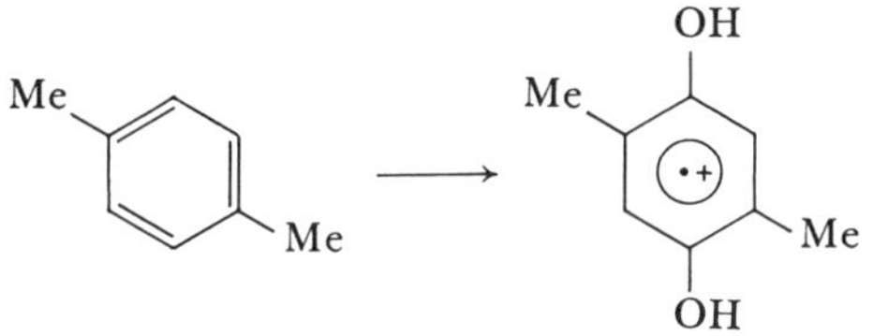

$$\text{Me-C}_6\text{H}_4\text{-Me} \longrightarrow \text{[Me-C}_6\text{H}_2\text{(OH)}_2\text{-Me]}^{\cdot+} \quad (8)$$

al., 1961; Bolton and Carrington, 1961). Benzene also gives its semiquinone cation radical in Caro's acid (Carter and Vincow, 1967a). Hexamethylbenzene and hexaethylbenzene are oxidized to their cation radicals in fuming sulfuric acid (Carter and Vincow, 1967b). The cation radicals are unstable and their esr signals soon decay. Indeed, hexamethylbenzene in sulfuric acid has a quite complicated, but now fairly well understood, chemistry. From reactions of hexamethylbenzene in fuming sulfuric acid, Hulme and Symons (1965a) detected the pentamethylbenzene cation radical. This indicates that either disproportionation or protolysis of hexamethylbenzene occurs. In line with this, it has also been found that on standing in sulfuric acid hexamethylbenzene gives the well-resolved esr spectrum of 4-methylene-1,1,2,3,5,6-hexamethylcyclohexadiene cation radical (Singer and Lewis, 1965). This apparently arises from the disproportionation of hexamethylbenzene, which gives the heptamethylbenzenium ion, and the latter undergoes oxidation (9).

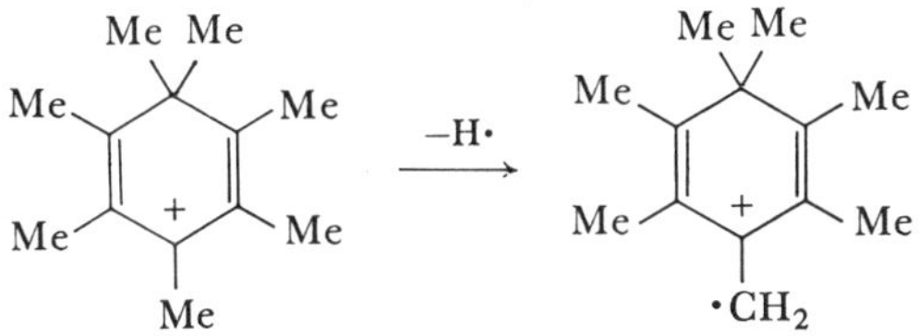

$$\text{heptamethylbenzenium}^+ \xrightarrow{-\text{H}\cdot} \text{[hexamethyl-methylenecyclohexadiene]}^{\cdot+}\text{-CH}_2 \quad (9)$$

It is evident that alkylbenzenes are too resistant to one-electron oxidation by sulfuric acid. However, cation radical formation can be achieved in sulfuric acid by ultraviolet irradiation. Hexamethylbenzene (Hulme and Symons, 1965b) and even benzene cation radical (Carter and Vincow, 1967a) have been made in that way. The latter is very unstable and disappears above about $-105°$. Disappearance is attributed to recapture of the photoejected electron. The formation of anthracene cation radical is enhanced by ultraviolet irradiation (Oster and Yang, 1973). It is most likely that in strong acids the photoejected electron is trapped—or even removed [eqn (10)]—as a hydrogen atom. Photoionization can also be achieved in

$$ArH_2^+ \xrightarrow{h\nu} ArH^{\cdot +} + H\cdot \qquad (10)$$

weaker acids, such as phosphoric (benzene) (Carter and Vincow, 1967a) and boric acid glasses (perylene, anthracene, naphthalene, naphthacene, phenanthrene, pyrene, diphenyl) (Vincow and Johnson, 1963; Bennema et al., 1959), chrysene and tetracene (Khan and Khanna, 1974) and in these cases trapping of the electron within the framework of the glass may occur, since decay on warming is accompanied by luminescence.

Aromatic cation radicals are often very stable in sulfuric acid solution. The major reason for this is that the solution is only weakly nucleophilic. The free water content is very small and the nucleophilicities of sulfate and bisulfate ions are too low for reaction with cation radicals. Cation radicals which have removable protons participate in acid-base equilibria with the corresponding neutral radicals and would be expected to be predominantly in the acid form in sulfuric acid. For example, the phenothiazine cation radical becomes the neutral phenothiazinyl radical on deprotonation and the neutral radical dimerizes to biphenothiazines (11) (Tsujino, 1968; Shine et al., 1972; Hanson and Norman, 1973).

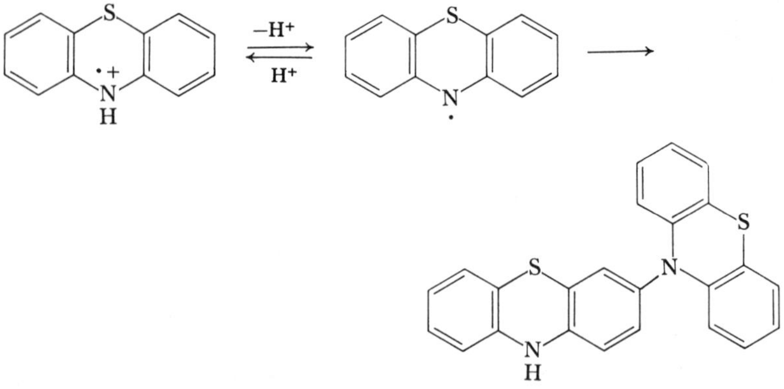

Other Brönsted acids

Reference has already been made to the use of methanesulfonic acid in nitrobenzene, and the question of how oxidation occurs was raised. Trifluoromethanesulfonic acid (Yang and Pohland, 1972) gives stable solutions of cation radicals which, in contrast, have short life in sulfuric acid (e.g. dibenzo-p-dioxin, anthracene, and 9,10-dimethylanthracene). In some cases, e.g. dibenzo-p-dioxin and 2,3-dichlorodibenzo-p-dioxin, an added oxidant was not needed in the trifluoromethanesulfonic acid, and it is thought that air, dissolved in the acid, was responsible for oxidation (Yang and Pohland, 1972). In other cases, e.g. with octachlorodibenzo-p-dioxin, cation-radical formation in trifluoromethanesulfonic acid occurred only if an oxidizing agent (potassium nitrate) was added or ultraviolet irradiation was applied.

Trifluoroacetic acid (TFA) has been used with considerable success in stabilizing cation radicals in anodic and, on some occasions, chemical oxidations. Solutions of thianthrene and 9,10-di-p-anisylanthracene cation radicals are stable when prepared anodically in TFA containing tetrabutylammonium fluoroborate as electrolyte (Hammerich et al., 1972). The redox equilibria of a number of cation radical systems have been measured in anodic oxidations in methylene chloride-TFA solutions (Svanholm and Parker, 1973), while the cation radicals of biphenylenes (Ronlan and Parker, 1974) and methoxybiphenyls (Ronlan et al., 1974) have been made anodically in similar solutions, sometimes containing trifluoroacetic anhydride.

At one time TFA itself was thought to cause the oxidation of thianthrene to the cation radical (Fava et al., 1957). Persistent efforts to purify the acid and to work in degassed solutions (Shine and Piette, 1962) were almost, but not entirely, successful in preventing its formation. Very low concentrations of thianthrene cation radical were obtained and preserved for years in sealed ampoules, suggesting that equilibrium (3) (p. 159) may have been responsible for cation-radical formation. In contrast with these several results, perylene and other polynuclear aromatics are only protonated in hydrofluoric acid and mixtures of trifluoroacetic acid–boron trifluoride, and cation-radical formation does not occur unless oxygen is admitted (Aalbersberg et al., 1959a). Analogously, tetramethoxybiphenylene forms a cation radical in TFA if air is admitted (Ronlan and Parker, 1974). Oxidation of aromatics by metal ions [e.g. Co(III), Tl(III)] in TFA is discussed later.

Perchloric acid oxidizes perylene (Matsunaga, 1961), thianthrene and phenoxathiin (Murata and Shine, 1969) to the cation radicals, but whether or not the acid serves as the oxidizing agent or catalyses oxidation by atmospheric oxygen has not been studied.

The overall view is that in strong, non-oxidizing acids aromatic compounds suffer no more than protonation. Where cation radicals are formed an oxidant must be added, and this may be, adventitiously, atmospheric oxygen. Further investigation of the validity of equilibrium (3) seems desirable.

Oxidation by Lewis Acids

A number of Lewis acids cause one-electron oxidation of aromatic compounds. Prominent among these are aluminum chloride and bromide, and antimony pentachloride. It was known in the early 1900s that solutions of some aromatic hydrocarbons in carbon tetrachloride became colored and others gave colored precipitates on the addition of antimony pentachloride (Meyer, 1910; Hilpert and Wolf, 1913). A solution of benzene became yellow, for example, while one of anthracene gave a green solid.

Analogously, solutions of aromatics in nitrobenzene-aluminum chloride were also known to be colored, ranging from the yellow-orange of toluene to deep red of hexamethylbenzene solution (Brown and Grayson, 1953).

With the advent of esr spectroscopy aromatic-antimony pentachloride precipitates were shown to contain the aromatic cation radical (Weissman et al., 1957), and this in turn accounted for the earlier discovery of paramagnetism in the salts obtained from reaction of aromatic amines with antimony pentachloride (Kainer and Hausser, 1933). Characterization of cation radicals in nitromethane and nitrobenzene solutions of antimony pentachloride by visible spectroscopy soon followed. Eventually, by working with degassed solutions of antimony pentachloride in dichloromethane at $-70°$ it was possible to obtain esr spectra of aromatic hydrocarbon cation radicals with extraordinarily well-resolved hyperfine patterns (Lewis and Singer, 1965, 1966). Similar success was obtained with alkyl aryl ether (Forbes and Sullivan, 1966), and organosulfur cation radicals in aluminum chloride–nitromethane solutions at $-50°$ (Shine and Sullivan, 1968; Sullivan, 1968). In this work, resolution of hitherto unparalleled definition was obtained, not only for proton couplings, but also for naturally abundant ^{33}S-couplings.

The way in which Lewis acids oxidize aromatic compounds is not known clearly. Aromatics which are not easily oxidized, such as benzene, alkylbenzenes, naphthalene, and which give colored solutions as noted above, undoubtedly form charge-transfer complexes with the Lewis acid (see p. 175). Aromatics which are oxidized easily undergo complete electron transfer and form the cation radical, but the final state of the electron acceptor is not too-well known. A Lewis-acid anion radical has never been detected in these systems. Although the initial reaction in oxidations by antimony pentachloride has been represented as in eqn (12), it is not

$$ArH + SbCl_5 \rightarrow ArH^{\cdot +} + SbCl_5^{\cdot -} \qquad (12)$$

the final state of reaction. Solid, paramagnetic salts obtained with some triarylamines and antimony pentachloride have the composition $ArH^{\cdot +}SbCl_6^-$ (Bell et al., 1969). The salt obtained with 9,10-diphenylanthracene has been similarly formulated (Sato et al., 1969), whereas salts obtained with phenothiazine, 10-methyl phenothiazine, thianthrene, perylene, 9,10-dimethyl-, and 9,10-dichloroanthracene have been formulated as $ArHSbCl_5$. Any implication that the latter group of salts must contain the $SbCl_5^{\cdot -}$ anion radical must surely be wrong, unless for unknown reasons the magnetic characteristics of this anion have evaded detection. The suggestion has been made that electrons are in fact paired, perhaps in a vacant orbital, in a dinegative $SbCl_5$ ion (Blomgren and Kommandeur, 1961). This possibility would, however, contradict known analytical data.

The formulation $ArH^{\cdot +}SbCl_6^-$ appears most reasonable, but how the hexachloroantimonate ion is formed has never been determined. It is proposed (Kainer and Hausser, 1933) that $SbCl_5^{\cdot -}$ disproportionates (13), and that where analysis indicates a pentachloro-

$$2SbCl_5^{\cdot -} \rightarrow SbCl_4^- + SbCl_6^- \qquad (13)$$

antimonate, the true situation is formation of a mixed salt, such as that shown in eqn (14). This proposal would account for all reports

$$2\left(MeO-\!\!\left\langle\!\!\bigcirc\!\!\right\rangle\!\!-\right)_2 NMe + 2SbCl_5 \rightarrow$$

$$2\left[\left(MeO-\!\!\left\langle\!\!\bigcirc\!\!\right\rangle\!\!-\right)_2 \overset{+\cdot}{N}Me\right], SbCl_4^-, SbCl_6^- \qquad (14)$$

of cation-radical salt compositions, and points up one of the failings in the analyses of salts, namely that the oxidation states of antimony are not determined, only total amounts of antimony and chloride ion.

As a final comment on the possible complexity of oxidation systems containing antimony pentachloride, it is to be noted that the hexachloroantimonate ion itself is an oxidizing agent and is thought to behave as in eqns (15) and (16) (Cowell et al., 1970).

$$SbCl_6^- + 2e^- \rightarrow (SbCl_6)^{3-} \tag{15}$$

$$(SbCl_6)^{3-} \rightarrow SbCl_4^- + 2Cl^- \tag{16}$$

Knowledge of how aluminum chloride oxidizes aromatics to cation radicals is practically non-existent. At one time it seemed that a nitro compound was a necessary co-acceptor (Buck et al., 1960) and that, whereas with mononuclear alkylaromatics, the Lewis acid-nitro compound pair formed only charge transfer complexes (Brown and Grayson, 1953), complete electron transfer occurred with more easily oxidized aromatics. But, cation-radical formation from perylene, anthracene, and chrysene was found to occur in carbon disulfide, chloroform, and benzene solutions, too (Rooney and Pink, 1961) and even occurs on warming anthracene and naphthacene with solid aluminum chloride (Sato and Aoyama, 1973). There is no doubt that a nitro compound enhances electron transfer, however (Sullivan and Norman, 1972). Cation radical formation in $AlCl_3$-nitromethane has been estimated as approximately 100% as compared with 1% in sulfuric acid oxidation of dialkoxybenzenes (Forbes and Sullivan, 1966). Unfortunately, aluminum halide salts have not been isolated and, therefore, even the beginnings of analytical data have yet to be collected. There is no definite knowledge of either the nature of the counter ion or the fate of the electrons in these cation-radical formations.

It was once thought that cation radicals played a direct part in aluminum chloride Friedel-Crafts reactions, because esr signals were detected in them (Adams and Nicksie, 1962). Later it was shown that the esr signals could be attributed to cation radicals (e.g. of anthracene) from polynuclear aromatics formed as byproducts (Banks et al., 1964).

Among other Lewis acids, sulfur trioxide in dimethyl sulfate solution oxidizes perylene to the cation radical. Naphthacene is so easily oxidized that the dication is formed (Aalbersberg et al., 1959b).

Boron trifluoride in liquid sulfur dioxide (de Boer and Praat, 1964), in nitromethane or in nitrobenzene (Aalbersberg et al., 1959b), and phosphorus pentafluoride in the last two solvents (Aalbersberg et al., 1959b) oxidize perylene, anthracene and tetracene to the cation radicals. The influence of solvent is again seen, because only small amounts of cation radical are observed with anthracene and tetracene, and hardly any with perylene in 1,2-dichloroethane. In fact, pumping off of the boron trifluoride (and phosphorus pentafluoride) reverses the reaction with these Lewis acids in 1,2-dichloroethane. On the other hand, irradiation of the solutions with ultraviolet light enhances cation-radical formation.

Once again, in none of these cases is the nature of the anion known.

Some aromatic hydrocarbons (e.g. perylene) give cation radicals in (neat) antimony trichloride solutions at 75° (Porter et al., 1970; Johnson, 1971). However, the cation radicals are not formed in the absence of oxygen. In fact, molten antimony trichloride can be used as a solvent at 99° for the anodic oxidation of perylene, naphthacene, and other polynuclear aromatics, provided that the electrolyte (e.g. KCl) is highly dissociated (Bauer et al., 1971). When a more covalent electrolyte (e.g. $AlCl_3$) is used, the solvent system itself becomes the oxidant [(17) and (18)].

$$AlCl_3 + SbCl_3 \rightleftharpoons AlCl_4^- \, SbCl_2^+ \qquad (17)$$

$$3SbCl_2^+ + 3ArH \rightleftharpoons 3ArH^{\cdot+} + 2SbCl_3 + Sb(s) \qquad (18)$$

Mixtures of some compounds (e.g. thianthrene, anthracene, naphthacene) with heavy metal halides (CdX_2, HgX_2, and PbX_2) give esr signals on heating, grinding or even compressing the mixture (Brown et al., 1973).

Oxidation by Halogens

Electron-transfer and the formation of charge-transfer complexes between a great variety of donors and halogen acceptors is widely documented in the literature, and is not within the scope of our considerations. The oxidation potentials of many molecules are low enough, however, to permit oxidation by the halogens and many oxidations have been identified at the cation-radical stage. In fact, the earliest examples of cation-radical formation were brought about with halogen oxidants, when aminium salts were isolated, although the nature of an aminium ion was not clearly understood. Among Wieland's classic researches in arylamine and tetra-arylhydrazine chemistry are found the controlled oxidations of tri-p-tolylamine and tri-p-anisylamine with bromine at low temperature, from which were obtained what we now recognize as the cation radical (or aminium ion) perbromides (19) (Wieland, 1907; Wieland and Wecker, 1910).

$$2Ar_3N + 3Br_2 \rightarrow 2Ar_3N^{\cdot+}Br_3^- \qquad (19)$$

It is interesting to note that Wieland observed that solutions of the perbromide in chloroform decomposed to give both tri-p-tolylamine and tri-bromotolylamine, in which we see the first signs of possible disproportionation reactions in cation-radical chemistry.

Wieland's oxidation was soon applied to heterocyclic nitrogen compounds, and since interest was concentrated on the unusual nature of the salts, emphasis was made on isolating and characterizing them. Phenothiazine perbromide was made (Pummerer and Gassner, 1913) and later characterized as a semiquinoid salt (Kehrmann and Diserens, 1915), although its structure was not understood until the work of Michaelis *et al.* (1941) on titrimetric oxidations with bromine and with lead tetraacetate. Oxidation of N-substituted p-phenylenediamines by bromine is quantitative and can be used in producing the cation radicals in solution for absorbance measurements (Fitzgerald *et al.*, 1971).

The direct use of iodine and chlorine is not as successful as the use of bromine. Iodine is not a good enough oxidant, while chlorine engages in electrophilic substitution.

Iodine, of course, forms charge-transfer complexes with a very large number of compounds. These complexes are mostly detectable in solution, but in some cases they can be isolated as solids and some of these exhibit esr spectra. For example, among alternant hydrocarbons perylene and pyrene form complexes $2ArH \cdot 3I_2$ and $ArH \cdot 2I_2$, respectively (Kommandeur and Hall, 1961; Singer and Kommandeur, 1961). Among non-alternant hydrocarbons, azulene, guaiazulene, and acepleiadylene form $ArH \cdot I_2$ complexes (Danyluk and Schneider, 1962). These complexes give broad, single-line esr spectra, attributable to the organic cation radical.

Use of iodine–silver perchlorate in ether gives solid triarylamine perchlorates (20), and the technique which was developed (Weitz and

$$2Ar_3N + I_2 + 2AgClO_4 \rightarrow 2Ar_3N^{\cdot+}ClO_4^- + 2AgI \qquad (20)$$

Schwechten, 1926, 1927) of extracting the perchlorate from a precipitated mixture of the salt and silver iodide is increasingly being put to use today (Bell *et al.*, 1969). When perchlorates were first made in this way it was thought that the effective oxidant was chlorine tetroxide. The need to use fresh solutions of iodine–silver perchlorate was known, but the complexity of the reactions of iodine with silver perchlorate was not. These reactions vary with the solvent, and a prominent (but not the only) reaction in ether is the formation of iodine triperchlorate (21) (Alcock and Waddington,

$$2I_2 + 3AgClO_4 \rightarrow 3AgI + I(ClO_4)_3 \qquad (21)$$

1962). Use of iodine–silver perchlorate may accomplish cation-radical formation before the oxidizing pair can themselves react. In modern usage, silver ion (as the perchlorate usually) is added to a solution of the substrate and iodine, and the complexity of the iodine–silver perchlorate system is avoided, provided that the substrate undergoes reasonably fast oxidation. Such is the case with perylene (Sato *et al.*, 1969; Ristagno and Shine, 1971) and phenothiazine, but not the case with diphenylanthracene and thianthrene (Shine *et al.*, 1972).

The interhalogens iodine bromide and iodine chloride form stronger charge-transfer complexes than iodine. Therefore, oxidation of suitable substrates should be feasible. In the case of thianthrene, oxidation is very rapid and a cation radical salt with a complex anion is formed (22) (Murata and Shine, 1969).

$$2C_{12}H_8S_2 + 6ICl \rightarrow 2C_{12}H_8S_2^{\cdot+}I_2 Cl_3^- + I_2 \qquad (22)$$

Use of Metal Salts

A variety of organic compounds are oxidized by salts of polyvalent metals. We have already discussed one-electron oxidations by some Lewis acids and noted that the mechanisms of these oxidations are unknown. In contrast, oxidations by some other metal salts have been studied quite thoroughly and much is known about the mechanisms. Of particular interest to us are cobalt(III), manganese(III), lead(IV), and cerium(IV) salts, and the most commonly used are the acetates. Another contrast with the Lewis acids is that cation radicals formed in oxidations by salts of polyvalent metals usually have short lifetimes since they are involved in further oxidation and, also, nucleophilic reactions with the anions of the metal salt.

Cobalt(III) acetate oxidizes thianthrene and 9,10-diphenylanthracene in acetic or trifluoroacetic acid to the cation radicals, as shown by esr (Heiba et al., 1969b). In flow systems, Co(III) acetate in trifluoroacetic acid has also given well-resolved esr spectra of polyalkylbenzene cation radicals (hexa-, penta-, and all tetramethylbenzenes, 1,3,5-tri-t-butyl-, and 1,4-di-t-butylbenzene) which are difficult to make in other ways (Dessau et al., 1970). When similar oxidations of alkylaromatics are carried out in static systems, further reactions occur, and these are sometimes of preparative value, since among the products are nuclear and side chain acetates. The reactions have been clarified most recently by the work of Heiba and Dessau.

Three pathways are open to a metal acetate, $M(OAc)_x$, in reactions with aromatics. One of these begins with one-electron transfer, to which we have already referred, and is illustrated in eqn (23). The other two begin with radical-forming decomposition of the

$$ArH + M^{x+} \rightleftarrows ArH^{\cdot +} + M^{(x-1)+} \tag{23}$$

$$M(OAc)_x \rightarrow M^{x-1}(OAc)_{x-1} + \cdot CH_2CO_2H \tag{24}$$

$$M(OAc)_x \rightarrow M^{x-1}(OAc)_{x-1} + CH_3^{\cdot} + CO_2 \tag{25}$$

salt, and are illustrated in eqns (24) and (25). Which pathway will be taken depends not only on the oxidizability of the aromatic, but also on the nature of the metal ion (M^{x+}) and on the temperature. The work of Norman and his collaborators with lead tetra-acetate has been particularly important in distinguishing between the various pathways (Norman and Thomas, 1970; McClelland et al., 1972; Norman et al., 1971, 1973a).

Among the metal acetates which have been studied recently, Co(III) acetate engages in electron transfer more easily than Pb(IV), Mn(III), and Ce(IV) acetates (Heiba et al., 1968a, 1969a, b; Heiba and Dessau, 1971). Kinetics have shown, both with Mn(III) and Co(III) acetates, that electron transfer is fast and reversible and is followed by a rate-determining step whose nature depends on the aromatic. Oxidation of 1-methoxynaphthalene in acetic acid at 100°, for example, leads to 1-acetoxy-4-methoxynaphthalene [eqns (26)–(28)]. Nuclear acetoxylation is the slow step (Andrulis and Dewar, 1966). It has been suggested that the slow step in oxidations of toluene by Co(III) acetate (Heiba et al., 1969b; Sakata et al.,

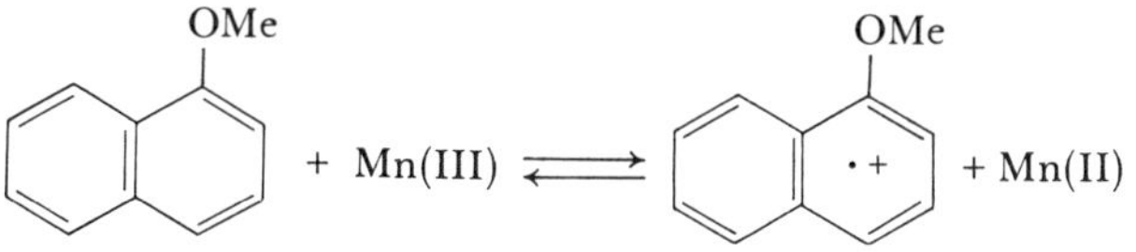

$$\text{1-methoxynaphthalene} + \text{Mn(III)} \rightleftarrows \text{[1-methoxynaphthalene]}^{\cdot+} + \text{Mn(II)} \quad (26)$$

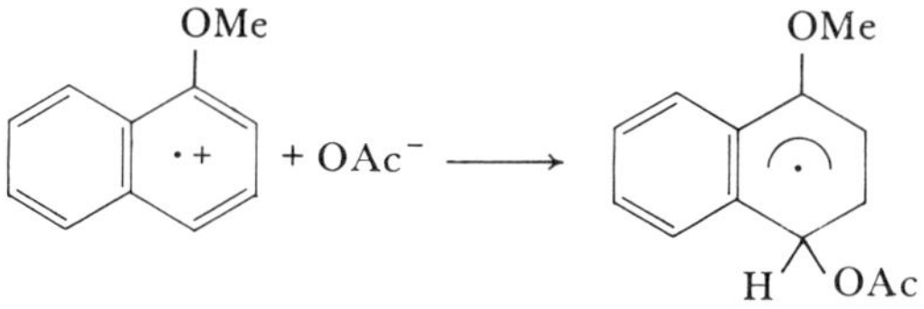

$$\text{[1-methoxynaphthalene]}^{\cdot+} + \text{OAc}^- \longrightarrow \text{4-OMe-1-H,OAc-dihydronaphthalene radical} \quad (27)$$

$$\text{radical intermediate} + \text{Mn(III)} \longrightarrow \text{1-OMe-4-OAc-naphthalene} + \text{H}^+ + \text{Mn(II)} \quad (28)$$

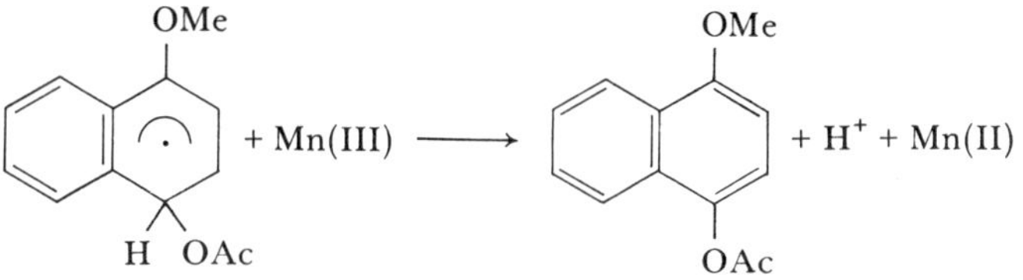

1969) and of *p*-methoxytoluene by Mn(III) acetate (Andrulis *et al.*, 1966) is proton abstraction from the cation radical, giving the benzyl radical (29; X = H or OMe), which is converted in the fast steps (30) and (31) to the benzyl acetate.

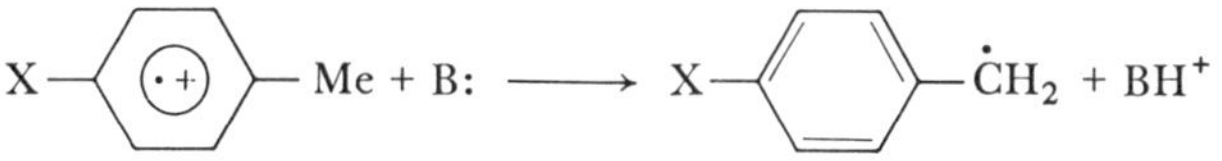

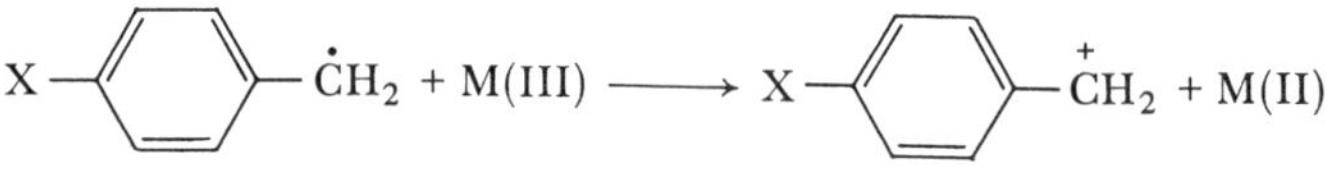

$$\text{X–C}_6\text{H}_4\text{–Me}^{\cdot+} + \text{B:} \longrightarrow \text{X–C}_6\text{H}_4\text{–}\dot{\text{C}}\text{H}_2 + \text{BH}^+ \quad (29)$$

$$\text{X–C}_6\text{H}_4\text{–}\dot{\text{C}}\text{H}_2 + \text{M(III)} \longrightarrow \text{X–C}_6\text{H}_4\text{–}\overset{+}{\text{C}}\text{H}_2 + \text{M(II)} \quad (30)$$

(M = Mn or Co)

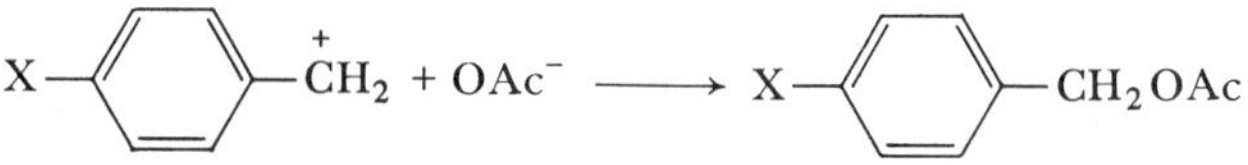

$$\text{X–C}_6\text{H}_4\text{–}\overset{+}{\text{C}}\text{H}_2 + \text{OAc}^- \longrightarrow \text{X–C}_6\text{H}_4\text{–CH}_2\text{OAc} \quad (31)$$

In these oxidations Co(III) is needed to oxidize toluene, whereas more easily oxidized aromatics, *p*-methoxytoluene and 1-methoxynaphthalene, react with Mn(III).

When decomposition of the metal salt acetate occurs more easily than electron transfer, the aromatic substrate is attacked by methyl radicals and/or carboxymethyl radicals. Both Pb(IV) and Ce(IV) acetates lead to methylation and carboxymethylation (Heiba *et al.*, 1968a; Heiba and Dessau, 1971), while Mn(III) acetate leads only to the latter (Heiba *et al.*, 1969a). Carboxymethylation produces not only arylacetic acids [(32) and (33)], but also benzyl acetates. The last compounds are formed from oxidative decarboxylation of the carboxylic acid (34) and subsequent reactions of the benzyl

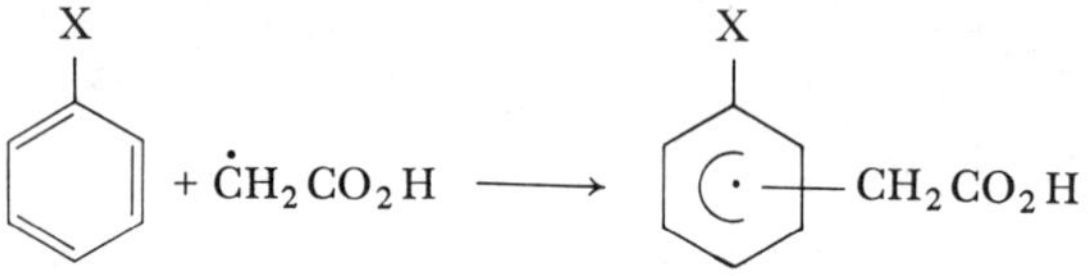

$$\text{X-C}_6\text{H}_5 + \dot{\text{C}}\text{H}_2\text{CO}_2\text{H} \longrightarrow [\text{X-C}_6\text{H}_5\text{-CH}_2\text{CO}_2\text{H}]^{\cdot+} \quad (32)$$

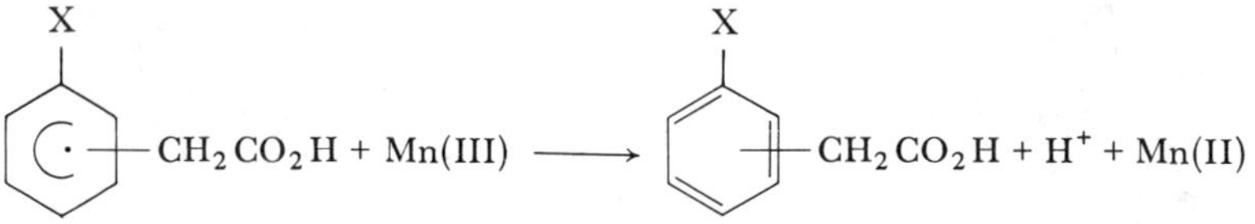

$$[\text{X-C}_6\text{H}_5\text{-CH}_2\text{CO}_2\text{H}]^{\cdot+} + \text{Mn(III)} \longrightarrow \text{X-C}_6\text{H}_4\text{-CH}_2\text{CO}_2\text{H} + \text{H}^+ + \text{Mn(II)} \quad (33)$$

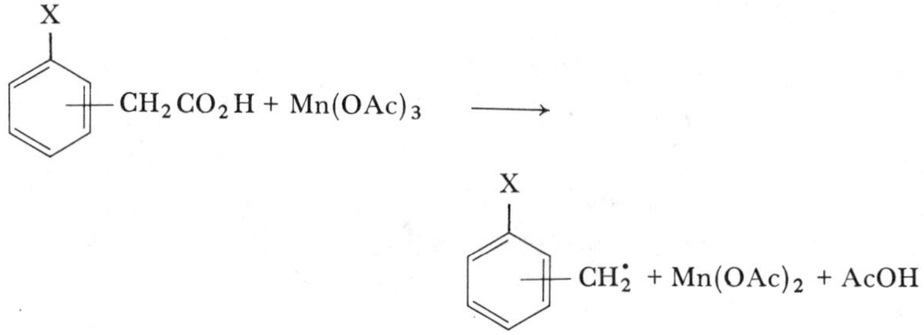

$$\text{X-C}_6\text{H}_4\text{-CH}_2\text{CO}_2\text{H} + \text{Mn(OAc)}_3 \longrightarrow \text{X-C}_6\text{H}_4\text{-CH}_2^+ + \text{Mn(OAc)}_2 + \text{AcOH} \quad (34)$$

radical [(30) and (31)]. In accordance with these free-radical reactions, chlorobenzene and toluene (van der Ploeg et al., 1968), and anisole (Heiba et al., 1968a) give large amounts of o-substituted benzylacetates.

The presence of strong acids makes cation-radical formation by Co(III), Mn(III) and Ce(IV) oxidation much easier, as Heiba et al. (1969b), have shown in flow systems. Kochi et al. (1973) have found that cobalt trifluoroacetate (CoTFA$_3$) in trifluoroacetic acid (TFA) or mixtures of TFA and its anhydride (TFAn) are also able to oxidize aromatics very easily. In low concentrations of the aromatic the cation radical formed in a slow oxidation step (23) reacts rapidly with TFA to give good yields of aryl trifluoroacetates (e.g. benzene, chlorobenzene) or benzyl trifluoroacetates (e.g. toluene). In high concentrations of the aromatic the cation radical tends to react with its parent to give biaryls. Rather similar results have been reported for oxidations by cerium(IV) trifluoroacetate (Norman et al., 1973b). The ability of TFA to stabilize certain cation radical solutions (Hammerich et al., 1972) in anodic oxidations has been mentioned; reaction of those cation radicals with TFA does not occur. Kochi et al. (1973) have commented on this, noting that some cation radicals are too stable to react with the poorly nucleophilic TFA. Dannenberg's (1975) explanation of the stabilization of carbenium ions by TFA is also to be noted.

Hanotier et al. (1973) have also found that in acetic acid made strongly acidic with sulfuric or chloroacetic acid (1-1.5 M), for example, electron transfer is so much easier that it occurs at room temperature. This is particularly notable for Mn(III) acetate since, at the higher temperatures ordinarily used (100°), reaction involves ·$CH_2 CO_2 H$ principally, whereas at room temperature electron transfer (23) prevails.

Oxidation of aromatics by thallium trifluoroacetate in TFA at 20° also gives the cation radicals of alkylaromatics, and it is thought, because of this, that thallation of aromatics may involve cation radical formation first (Elson and Kochi, 1973).

Olefins are also oxidized by metal salts. Whereas Pb(IV) and Mn(III) acetates lead to γ-lactones by free-radical addition routes (Heiba et al., 1968b, c), Co(III) in aqueous sulfuric acid at room temperature causes oxidative cleavage of a glycol which is formed by a cation-radical route [(35)-(38)] (Bawn and Sharp, 1957). The final products are ketones and carboxylic acids.

$$R_2C=CH_2 + Co(III) \rightleftarrows R_2\overset{+}{C}-\overset{\cdot}{C}H_2 + Co(II) \qquad (35)$$

$$R_2\overset{+}{C}-CH_2\cdot + 2H_2O \rightarrow R_2C(OH)-CH_2\cdot + H_3O^+ \qquad (36)$$

$$R_2C(OH)-CH_2\cdot + Co(III) \rightarrow [R_2C(OH)-CH_2^+] + Co(II) \qquad (37)$$

$$[R_2C(OH)-CH_2^+] + H_2O \rightarrow R_2C(OH)-CH_2OH + H^+ \qquad (38)$$

Phenols are oxidized easily to phenoxyl radicals. The reaction involves deprotonation of the first-formed cation radical, and once the phenoxyl radical is formed further oxidative dimerization often occurs with the formation of quinones. If oxidation is carried out [with Mn(III) acetylacetonate, for example] in the absence of oxygen and in neutral, aprotic solvents (carbon disulfide, acetonitrile), however, only the formation of phenol dimers (39) occurs (Dewar and Nakaya, 1968).

$$2 \text{ (2-naphthol)} + 2\text{Mn(III)} \longrightarrow \text{(1,1'-bi-2-naphthol)} + 2\text{H}^+ + 2\text{Mn(II)} \tag{39}$$

Oxidation of 1,2-diarylethanes by Ce(IV) ammonium nitrate produces substituted benzaldehydes, benzyl alcohols and nitrates, and is thought to begin by the formation of the substrate cation radical (Trahanovsky and Brixius, 1973).

The one-electron oxidations described briefly above are referred to as non-bonded (or outer-sphere) electron transfers (Littler, 1971). They differ from bonded (or inner-sphere) oxidations, such as the oxidation of alcohols by Cr(VI), in that a bond between the organic substrate and the metal ion or the complexed metal ion is not formed. Electron transfers of the non-bonded type may be very fast. Although in the equations above it has been convenient to represent the oxidant as the free ion, this cannot be so in solution, in which the ion must be solvated or complexed in some way. Clear cut cases of non-bonded or outer sphere oxidation can be seen in the use of the hexachloroiridate(IV) ion (40) (Littler, 1971) and the 12-tungstocobalt(III) ion (41) (Chester, 1970). In the latter example

$$\text{IrCl}_6^{2-} + \text{ArOH} \rightarrow \text{IrCl}_6^{3-} + \text{Ar}\overset{\cdot}{\text{O}}\text{H}^+ \tag{40}$$

$$\text{ArCH}_3 + \text{Co(III)}\text{O}_4\text{W}_{12}\text{O}_{36}^{5-} \rightarrow \text{ArCH}_3^{\cdot +} + \text{Co(III)}\text{O}_4\text{W}_{12}\text{O}_{36}^{6-} \tag{41}$$

electron transfer (or electron tunneling) occurs through the very complex ligand system to the central Co(III). The final product of oxidation, carried out in acetic acid, is the benzyl acetate, ArCH_2OAc, formed according to eqns (29)–(31).

Recently it has been shown that peroxydisulfate ion or, more correctly, $\text{SO}_4^{\cdot -}$ formed by homolytic fission of $\text{S}_2\text{O}_8^{2-}$, is a particularly convenient reagent for the oxidation of aromatics (including pyridine, benzene and chlorobenzene) in water or aqueous acetonitrile solutions (Ledwith and Russell, 1974a, b, c). These reactions have mechanistic significance in respect of the more complex oxidations outlined above and may be represented overall as in eqn (42). It is interesting that, generated in this manner, the arene cation

$$\text{ArH} + \text{SO}_4^{\cdot -} \rightarrow \text{ArH}^{\cdot +} + \text{SO}_4^{2-} \tag{42}$$

radical may be captured by direct reaction with chloride ion or copper(II) chloride to give nuclear chlorinated products or, for appropriate structures, may undergo side chain deprotonation producing corresponding benzyl radicals (43). Reaction mechanisms for

$$CH_3Ar + SO_4^{\cdot -} \rightarrow CH_3Ar^{\cdot +} + SO_4^{2-} \xrightarrow{-H^+} ArCH_2^{\cdot} \qquad (43)$$

these novel chlorination procedures are discussed more fully in section 5.

Metal-ion catalysed air oxidations of aromatic compounds are industrially important. Oxidation of a methyl to a carboxylic acid group, such as in the first stage of the manufacture of terephthalic acid from p-xylene, is believed to involve peroxidation of benzylic carbon [formed in eqn (29)] (Andrulis et al., 1966). The reactions leading to benzyl acetates [eqns (29)-(31) for example] must therefore be carried out in the absence of air. Oxidations of alkylaromatics in the presence of oxygen and involving cation radical intermediates have been reported (Onopchenko et al., 1972; Holtz, 1972; Scott and Chester, 1972).

Formation via Charge Transfer Complexes

Donor-acceptor interactions encompass a very wide variety of intermolecular combinations ranging from very weak to the very strong, readily isolated and characterized, complexes (Foster, 1969a; Mulliken and Person, 1969). The range here should include Lewis acid–Lewis base combinations, charge transfer complexes, and various kinds of ion pairs. The interaction of donors (D) and acceptors (A) may be represented, generally, as in eqn (44), and the

$$D + A \xrightleftharpoons{K} D, A \qquad (44)$$

strength of the association between D and A, although measurable and definable in several different ways, may be indicated by the value of the association constant (K). Whenever the donor possesses an unshared pair of electrons, as in the case of Lewis bases, and the acceptor an unfilled orbital of low energy, as in Lewis acids, the bonding between D and A is relatively strong and frequently represented by complete transfer of electron charge from D to A *in the ground state of the complex*, as shown, for example, in (45).

$$NH_3 + BF_3 \rightleftharpoons H_3\overset{+}{N} - \overset{-}{B}F_3 \qquad (45)$$

There are, however, a very large number of examples of donor-acceptor complexes which do not readily conform to the Lewis acid–Lewis base description, relying as it does on essentially localized filled and unfilled orbitals, and it was Mulliken who first proposed a theory to account for the bonding in these complexes. Intermolecular association in such combinations is often indicated by the appearance of absorption bands (frequently visible colors) not shown by either donor or acceptor alone, and it is this feature which led Mulliken, following a correct interpretation of their spectra, to describe them as "charge transfer complexes". According to Mulliken (1952), charge transfer complexes arise from the interaction (46) between donor molecules having high energy filled orbitals (i.e. low ionization potentials, I_D), and acceptors having low energy unfilled orbitals (i.e. high electron affinities, E_A). In contrast to Lewis

$$D + A \rightleftharpoons [D, A \longleftrightarrow D^{\cdot +}, A^{\cdot -}] \xrightarrow{h\nu} [D^{\cdot +}, A^{\cdot -} \longleftrightarrow D, A]$$

ground state excited state

acid–Lewis base adducts, the ground state of charge transfer complexes may be considered as a resonance hybrid with only a small contribution from the canonical structure representing electron transfer between D and A (i.e. $D^{\cdot +}, A^{\cdot -}$). A low lying electronically excited state of the complex would have a major contribution from the electron transfer canonical form ($D^{\cdot +}, A^{\cdot -}$) and explains the appearance of new absorption bands on reacting D and A. An important justification for this theory of charge transfer spectra is the general applicability of empirical relationships between I_D, E_A, and the energy of the charge transfer transition which may be generalized as

$$h\nu = I_D - E_A + C$$

where C is a constant representing mainly Coulombic forces. The concept of charge transfer phenomena has been widely, and sometimes uncritically accepted, and it cannot be overstressed that appearance of charge transfer absorption bands in reaction systems should *not* be taken as *a priori* evidence for the participation of complexes as reaction intermediates.

Kosower (1965, 1968) has critically reviewed the possibilities for involvement of charge transfer complexes in organic reactions (Scheme 1). The reactivity of the complex (D, A) will be influenced by several factors including the geometry of the complex, and the stabilization of a positive charge on D or a negative charge on A. In certain cases, it is quite possible that the ground state complex reacts

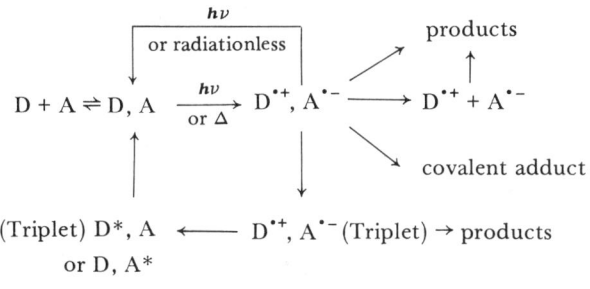

Scheme 1

completely, or promotes reactions of D or A, with added reagents. A most important feature of this generalized scheme is that formation of a charge transfer complex does not imply automatic generation of reactive ions or ion pairs for subsequent chemical reactions. In the vast majority of cases reactive intermediates are produced from charge transfer complexes only by thermal, photochemical, or catalysed activation.

Investigations of the esr spectra of solid charge-transfer complexes, in which electron transfer is thought to be complete, have been somewhat controversial. Some complexes show a two-line esr spectrum. In some cases this has been attributed to a triplet (e.g. in which the two electrons of the $D^{\cdot+}$, $A^{\cdot-}$ host are unpaired but magnetically coupled), while in others assignment has been made (seemingly in error) to the uncoupled spins of separate radicals, i.e. $D^{\cdot+}$ and $A^{\cdot-}$ (Foster, 1969a).

The phenomenon of complete electron transfer in solution is best illustrated with the now classic example of the N,N,N',N'-tetramethyl-p-phenylenediamine (TMPD)-chloranil reaction (Kainer and Wherle, 1955; Isenberg and Baird, 1962; Eastman et al., 1962). This pair forms a charge-transfer complex in solvents of low dielectric (e.g. dioxan, benzene, 1,2-dichlorethane) without complete electron transfer, whereas in acetonitrile the detectable, first-formed complex gives way to the separate cation and anion radicals, each of which is detectable by absorption and esr spectroscopy.

A similar situation is obtained with TMPD and tetracyanoethylene (TCNE), and both $TMPD^{\cdot+}$ and $TCNE^{\cdot-}$ are observed in acetonitrile solution (Liptay et al., 1962).

More recently, crystalline cation-radical: anion-radical salts of the donor tetrathiofulvalene with the acceptors tetracyanoquinodimethane (Phillips et al., 1973; Ferraris et al., 1973; Butler et al., 1974) and tetracyanomuconitrile (Wudl and Southwick, 1974) have been made. Analogous salts of tetraselenofulvalene-tetracyanoquinodimethane (Engler and Patel, 1975) and perylene bis(maleonitriledithiolene)-metal chelates have been made also (Alcacer and Maki, 1974). These salts exhibit unusually high electrical conductivity and therefore are of considerable interest.

Examples of the detection of both ion radicals, however, are rare. Even in the TMPD-chloranil case, the anion radical decays leaving TMPD$^{\cdot +}$ (Eastman et al., 1962). In most cases either a cation- or an anion-radical spectrum is obtained, but not both. A solution of TMPD and benzoic anhydride in acetonitrile, for example, exhibits both a charge-transfer and a TMPD$^{\cdot +}$ absorption band, but not an anion-radical band (Starkova et al., 1970).

Occasionally the absence of the signal of one of the ion radicals in the esr spectrum of a charge-transfer complex in solution is attributed to spin exchange between the ion radical and its precursor. The ion radical is thus assumed to be present in solution but undetectable by esr. This, in fact, may not be the case; the anion radical may be undetectable because it has disappeared. We have already seen in oxidations by antimony pentachloride that the first-formed $SbCl_5^{\cdot -}$ probably disproportionates.

It has already been noted, also, that even in the TMPD-chloranil reaction the chloranil anion radical disappears in time (although its fate is not known). Therefore, it seems highly likely that, in other cases in which only one of the two expected ion radicals is detectable by esr, the undetectable ion radical has itself ceased to exist.

However, it should be noted that studies of the effects of high pressures (several hundred kilobars) on the p-phenylenediamine-chloranil complex have established a complicated reaction sequence involving initial formation of the cation radical-anion radical pair, leading ultimately to tetrachlorohydroquinone and polyazophenylene (Sakata et al., 1974).

There may also be cases in which charge transfer may appear to be but is not in fact really responsible for radical ion formation, so that seeking the two ion radicals corresponding to a charge-transfer complex is in vain. For example, reaction between tetracyanoethylene and amines often leads to the TCNE$^{\cdot -}$ but no sign of the amine cation radical is obtained. When pyridine is used, a charge-transfer band is detectable in the ultraviolet (and is probably from an outer complex), but the reaction which leads to TCNE$^{\cdot -}$ is not electron donation by pyridine but the very complex reaction of hydroxide ion (generated by pyridine) with TCNE (Shine and

Goodin, 1970). Thus, there is no point in searching for the esr spectrum of the pyridine cation radical. Probably the same situation prevails in reactions of other amines with TCNE. For example, a variety of arylamines react with TCNE and form, finally, tricyanovinyl derivatives (Foster, 1974). The reaction is thought to involve complex formation first (Isaacs, 1966). The TCNE$^{\cdot\,-}$ is produced over a period of minutes, while the other anticipated radical, presumed to be an aminium ion, is not observed. Its absence is attributed to dimerization and proton loss, although the anticipated products (hydrazobenzene and its isomers) are not found. In these reactions, therefore, there can be little doubt that the TCNE$^{\cdot\,-}$ arises from reactions of TCNE with hydroxide ion and/or cyanide ion (formed in the tricyanovinylation reaction). Thus, in these cases the formation of an anion radical is not linked to electron transfer from an organic donor (Isaacs, 1966).

The overall picture is that solid charge-transfer complexes of the type $D^{\cdot\,+}$, $A^{\cdot\,-}$ can be made, and that in solution these may give the free ion radicals. The ion radicals are then able to take part in other reactions. Further, one-electron transfer can occur between a donor-acceptor pair in solution, and one of the ion radicals may persist in solution. Formation of an ion radical is, however, not necessarily confirmation of the anticipated donor–acceptor reaction.

Light-induced Formation

Reference has already been made to the photoionization of aromatic hydrocarbons in acid solutions (p. 161). The earliest recognized examples of light-induced cation-radical formation, however, were in neutral solution (Lewis and Lipkin, 1942). The solvent was a mixture of ether, isopentane, and ethanol (EPA) which forms a glass at low temperatures and has become standard usage for that reason. Irradiation of solutions of amines such as triphenylamine, tri-p-tolylamine, TMPD, and tetramethylbenzidine at liquid-air temperature gave the cation radicals. Heterocyclic compounds such as phenothiazine and phenoxazine, and analogous dyestuffs such as methylene blue and Capri blue, behaved similarly (Lewis and Bigeleisen, 1943). In this work Lewis recognized that cation radical formation had occurred and, because the cation radicals disappeared when the frozen glass was warmed, believed the photo-ejected electron to be trapped in the glass until it returned to the cation radical on warming. Lewis referred to this type of cation-radical formation as photo-oxidation, reserving the term photoionization for bond-breaking photolyses. The first indication of solvent effects on photoionization (i.e. electron ejection in the modern usage) were obtained also, because triphenylamine in isopentane-methylcyclohexane did not ionize.

Continuous irradiation of these frozen solutions gave cation radicals which were identifiable because they were of fairly stable types. Less easily formed and much shorter-lived cation radicals can be obtained by a high-intensity, short-duration flash technique. Here, unfortunately, the term flash photolysis, a general one for the flash technique, is often used. For our purposes the term connotes flash ionization. Simple phenol and arylamine (Land and Porter, 1963) cation radicals have been made in this way and characterized by absorption spectroscopy. The rapidity of the technique allows the use of liquid solvents at room temperature, such as water (the hydrated electron was detected by Joschek and Grossweiner, 1966; Feitelson and Hayon, 1973) and saturated hydrocarbons, as well as glasses at low temperature. Three primary processes may occur in the flash irradiation of arylamines [(47)-(49)], and which of them

$$ArNH_2 \rightarrow ArNH_2^{\cdot +} + e^- \qquad (47)$$

$$ArNH_2 \rightarrow ArNH\cdot + H\cdot \qquad (48)$$

$$ArNH_3^+ \rightarrow ArNH_2^{\cdot +} + H\cdot \qquad (49)$$

occurs depends on the nature of the solvent. In water and aqueous acids photoionization (47) occurs. The cation radical is formed also in strong acids by hydrogen atom ejection (49), a photolysis reaction mentioned earlier (p. 162). In neutral solvents (saturated hydrocarbons, water) the neutral radical is formed (48), a reaction which also prevails in flash irradiation of phenols.

Equation (47) suggests that photoionization is a simple process. In fact, the mechanistic details of photoionization are infrequently known and may be quite complicated (Albrecht, 1970; Kevan, 1973). For example, some ionizations are two-photon processes. Thermoluminescence (i.e. emission of light by recombination of cation radical and electron on warming frozen glasses) of toluene, aniline, perylene and other aromatics in 3-methylpentane is proportional to the square of the intensity of the ionizing light (Gibbons *et al.*, 1965). This suggests that photoionization is biphotonic, and is interpreted as involving excitation of the first-formed excited triplet

(^{3}ArH*) state [(50)-(53)]. Similar conclusions have been made from work with TMPD in 3-methylpentane (Cadogan and Albrecht, 1965),

$$\text{ArH} + h\nu \rightarrow {}^1\text{ArH*} \qquad (50)$$

$$^1\text{ArH*} \rightarrow {}^3\text{ArH*} \qquad (51)$$

$$^3\text{ArH*} + h\nu \rightarrow \text{ArH**} \qquad (52)$$

$$\text{ArH**} \rightarrow \text{ArH}^{\cdot+} + e^- \qquad (53)$$

and, interestingly, in the formation of the benzene cation radical by ultraviolet irradiation of benzene on a silica surface (Tanei, 1968). The first ionization potential of benzene is 9·2 eV, so that formation of the benzene cation radical is not achievable by one-photon excitation in the ultraviolet. Excitation to the singlet (4·84 eV) followed by intersystem crossing to the triplet state (3·7 eV) is, as it were, the first step up the ladder for ionization. Absorption of a second photon by triplet benzene raises the triplet state to a higher excited level (8·4 eV), from which level ionization is achieved by donation of an electron to the silica surface. The triplet-triplet transition for benzene has also been identified in irradiations in frozen hydrocarbon glass (Godfrey and Porter, 1966). In other examples, the biphotonic ionization of durene (Schwarz and Albrecht, 1973) and phenol (Feitelson et al., 1973) have been reported.

Kevan's group (Möckel et al., 1973) has shown, most interestingly, that, with TMPD in 3-methylpentane and 2-methyltetrahydrofuran, the distance traveled by the ejected electron of the biphotonic process depends on the energy of the second photon which causes the triplet-triplet transition (52). Recombination of TMPD$^{\cdot+}$ with the nearby ejected electron leads to fluorescence, and the rate of the isothermal fluorescence decay was found to be slower when ionization was initiated at 319 nm than at 365 nm. The interpretation of this result is that at 319 nm the ejected electron, because of its now higher energy, travels further away from the TMPD$^{\cdot+}$ than an electron ejected with 365 nm irradiation, and hence takes longer to return to recombination.

Photoionizations are likely to be mono-photonic if electron transfer from an excited state is made easy. This may occur if the solvent is a good electron acceptor or if charge-transfer complexation is involved. Each of these provides an easy path to electron transfer.

For example, even the well-explored TMPD ionization becomes monophotonic in acetonitrile solution, in which formation of a complex between TMPD·⁺ and MeCN is believed to precede ionization (Imura *et al.*, 1971).

A recent detailed investigation of the monophotonic ionization of TMPD reveals that the quantum yields are a maximum for irradiation at shorter wavelengths and are markedly dependent on solvent and temperature, e.g. the maximum quantum yield is 0·07 in tetramethylsilane and $\sim 10^{-5}$ in n-alkanes. It appears therefore that the observed values of quantum yields for the formation of TMPD·⁺ depend directly upon the relative values of cation-electron separation distances which are characteristic of any particular liquid/temperature combination (Holroyd and Russell, 1974).

Some photoionizations occur only in halogenomethane solvents whereas in other solvents (e.g. ethanol, benzene) only fluorescence occurs; in the halogenated solvents ionization is monophotonic. This is the case with di- and tetraphenyl-p-phenylenediamine in chloroform and carbon tetrachloride (Fitzgerald *et al.*, 1971), and with TMPD in chloroform, bromoform, methylene bromide and chloride (Meyer, 1970), and dimethylaniline in bromobenzene (Gradowski and Latowski, 1974; Pac *et al.*, 1972). Electron transfer may occur from the excited aromatic singlet to ground state solvent [(54) and

$$\text{ArH} + h\nu \rightarrow {}^1\text{ArH}^* \tag{54}$$

$$^1\text{ArH}^* + \text{CHCl}_3 \rightarrow \text{ArH}^{\cdot +} + \text{Cl}^- + \cdot\text{CHCl}_2 \tag{55}$$

$$\text{ArH, CHCl}_3 + h\nu \rightarrow \text{ArH}^{\cdot +} + \text{Cl}^- + \cdot\text{CHCl}_2 \tag{56}$$

(55)] or within a charge-transfer or contact complex (56). In either case, halide-ion ejection is the driving force in promoting transfer.

Equation (55) may also be an oversimplification in some cases. That is, there is fluorescence evidence that an excited aromatic compound may form an excited complex (an exciplex) with a halogenated solvent prior to dissociative electron transfer. Irradiation of triethylamine in chloro-, bromo-, and iodobenzene resulted in the quenching of triethylamine fluorescence, and in the appearance of a new fluorescence at higher wavelength (Tosa *et al.*, 1969). Irradiations of triethylamine and a halogenobenzene in methanol solution led to fluorescence quenching again but also to the formation of halide ion, biphenyl, and other free radical products. The several reactions are summarized in eqns (57)–(61), in which D represents

$$D + h\nu \rightarrow {}^1D^* \tag{57}$$

$$^1D^* + \text{PhX} \rightarrow {}^1(D, \text{PhX})^* \tag{58}$$

$$^1(D, PhX)^* \rightarrow (D, PhX) + h\nu \qquad (59)$$

$$^1(D, PhX)^* \rightarrow D^{\cdot+} + PhX^{\cdot-} \qquad (60)$$

$$PhX^{\cdot-} \rightarrow Ph^{\cdot} + X^- \qquad (61)$$

triethylamine. Equation (59) represents the new fluorescence, while eqn (60) represents electron transfer as it occurs in a polar solvent (methanol). In such case, of course, the new fluorescence is either diminished or not observed at all.

The generality of this excited charge-transfer phenomenon was first recognized by Weller (1968), who has shown also that an excited acceptor may promote electron transfer from a ground state donor (Davidson, 1969; Rao and Ramakrishnan, 1971; Bryce-Smith et al., 1971). Thus, excitation of triethylamine in a toluene solution of acceptor biphenyl follows eqns (58)–(60). On the other hand, excitation of acceptor anthracene (ArH) in a toluene solution of triethylamine (D) results in quenching of anthracene's fluorescence. The situation is summarized in eqns (62)–(65), and Fig. 1 (Weller,

$$ArH + h\nu \rightarrow {}^1ArH^* \qquad (62)$$

$$^1ArH^* + D \rightarrow {}^1(ArH, D)^* \qquad (63)$$

$$^1(ArH, D)^* \rightarrow (ArH, D) + h\nu \qquad (64)$$

$$(ArH, D)^* \rightarrow ArH^{\cdot-} + D^{\cdot+} \qquad (65)$$

1968). Again, ionization (65) is observed in a polar solvent (acetonitrile) and fluorescence (64) in a non-polar one (toluene).

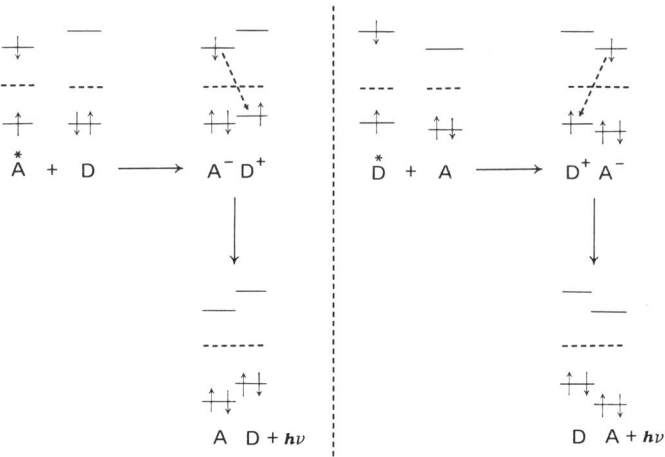

Figure 1. Molecular orbital scheme for interpretation of the complex emission (Weller, 1968).

The role of exciplexes in the wider compass of organic photochemistry has now been fully recognized, as evidenced by the proceedings of an International Meeting on Exciplexes (Gordon et al., 1975). Two recent studies are highly relevant in demonstrating the intervention of ion-radical intermediates.

Aryl ketones in their triplet states undergo fast reactions with alkylamines resulting in ketone reduction or quenching of triplet ketone without net chemical change. The efficiency of photoreduction varies greatly, depending upon the ketone–amine pair and upon the solvent (Cohen et al., 1973, Blanchi and Watkins (1974). A generalized reaction scheme for excited ketone–amine interactions is indicated in Scheme 2 and it may be seen that formation of the

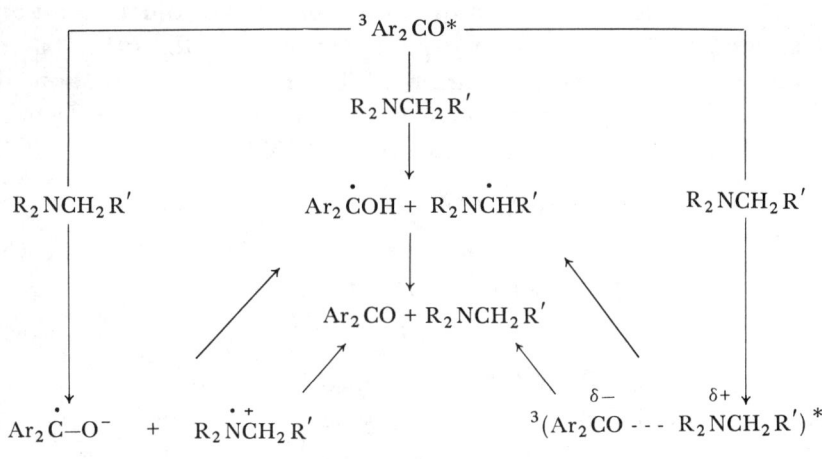

Scheme 2

neutral radicals, $Ar_2\dot{C}OH$ and $R_2N\dot{C}HR'$, required as intermediates to explain the products, may occur by hydrogen abstraction from the amine (R_2NCH_2R') by the triplet state ketone ($^3Ar_2CO^*$). However, much of the experimental data for such processes allow the primary photochemical step to be formulated as an electron transfer producing a pair of ion radicals, $Ar_2\dot{C}-O^-$ and $R_2\overset{\cdot+}{N}CH_2R'$. Proton transfer from $R_2\overset{\cdot+}{N}CH_2R'$ to $Ar_2\dot{C}-O^-$ would then produce the same neutral radical pair as that arising by direct hydrogen abstraction. A more general formulation involves the intermediacy of a ketone–amine exciplex, $^3(Ar_2CO\text{---}R_2NCH_2R')^*$, whose relaxation could be apportioned among three paths: decay to the ground state species (quenching), hydrogen atom transfer to form the neutral radical pair, and electron transfer to form the discrete radical ions.

An elegant study of the photolysis of p,p'-disubstituted benzophenones in the presence of diazabicyclo[2.2.2]octane (Dabco) utilizing nuclear spin polarization techniques revealed CIDNP effects for the ortho (enhanced absorption) and meta protons (net emission) (Roth and Lamola, 1974). These results provide clear evidence for electron transfer from the amine to the excited triplet ketones. Simultaneous nmr line broadening was observed for Dabco and p,p'-dichlorobenzophenone and attributed to degenerate electron transfer between the amine and its cation radical, and ketone and ketone anion radical respectively. This latter phenomenon is particularly relevant to the difficulties (noted above) in detecting both components of a pair of ion radicals generated by activation of charge transfer complexes.

A rather similar CIDNP study (Weinstein et al., 1975) of the photo-oxidation of phenylthioacetic acid by several aryl ketones leads to the conclusion that whilst several radical pairs may give rise to the observed polarization, there is no requirement for the intervention of ion-radical intermediates. It should be noted, however, that, in this case, any primary ion radical intermediates would be extremely short-lived because of facile decarboxylation, e.g. $PhSCH_2COOH \rightarrow PhSCH_2\cdot + \dot{C}OOH$.

Neutral radical intermediates formed in the photo-oxidation of amines etc. by aryl ketones are useful in initiation of free radical vinyl polymerization and may afford technological advantages over more conventional photoinitiation systems (Ledwith and Purbrick, 1973; Ledwith, 1975).

A more general photoionization phenomenon involving charge-transfer complexes is the irradiation of the charge-transfer complex itself. Quite a large number of examples are known, representative of which are acceptors TCNE (Ward, 1963; Ilten and Calvin, 1966) and pyromellitic dianhydride (Rao and Ramakrishnan, 1971) in solvent donor THF, acceptors TCNE, TCNQ in solvent donor acetonitrile (Kimura et al., 1973), and acceptors tetracyanobenzene and pyromellitic anhydride in donor solvents methyl tetrahydrofuran and benzene-dichloroethane (Shimada et al., 1973). In these cases the energy for ionization of the charge-transfer complex is supplied by light, whereas in ground-state ionization (pp. 175-9), the energy is supplied by solvation (Foster, 1969b). It is to be noted that in these photoionizations of charge-transfer complexes only one member of the pair of ion-radicals is usually observed. In particular, $THF^{\cdot+}$ is not observed and is thought to have decomposed very rapidly after formation.

There have been many reported instances of initiation of vinyl polymerization by photochemical (and thermal) activation of charge transfer complexes and the scope and reliability of these reports have been extensively reviewed (Hyde and Ledwith, 1974). It should be noted, however, that following the elegant studies of Ottolenghi (1973) and Mataga (1975), formation of a pair of ion radicals by photoexcitation of charge transfer complexes frequently involves excitation of, and to, higher excited states of the complex. A further generalization which may be made is that excitation of the ground

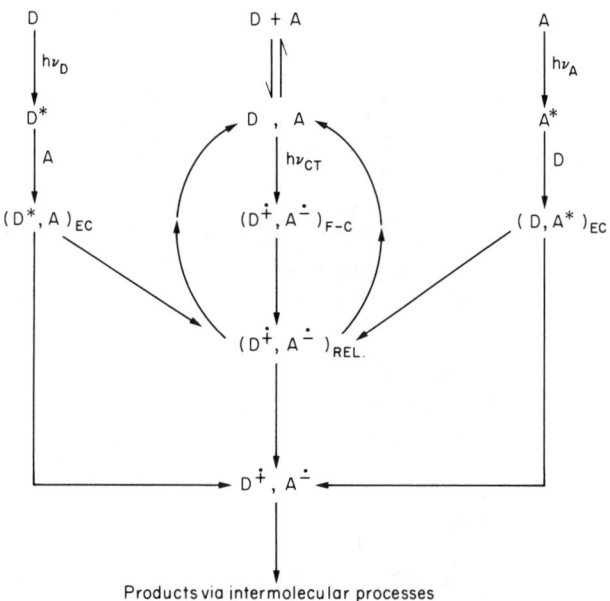

Figure 2. Charge transfer complex and encounter complexes. F-C implies a Franck–Condon state; REL implies a relaxed form of the F–C state; EC implies encounter complex (exciplex in appropriate cases).

state of a charge transfer complex produces initially a Frank–Condon excited state which, in appropriate circumstances, may relax to a lower energy excited state prior to reaction or degradation back to its ground state. Combining this with the previous discussion of ion radical production via exciplexes, the generalized scheme shown in Fig. 2 may be derived. Thus a pair of ion radicals, useful in subsequent chemical processes, may arise via excitation of a ground state charge transfer complex or by interactions of donor and acceptor respectively with locally excited states of the corresponding acceptor and donor. Experimental evidence for the fluorescence and chemical reactivities of exciplexes and charge transfer complexes formed from common donor–acceptor pairs is now available (Itoh, 1974; Ledwith, 1975.)

Finally, electron transfer within an excited charge-transfer complex may lead to products of reaction without the appearance of the ion radicals presumed to have been formed. Thus, irradiation of tetracyanobenzene in mesitylene at the system's charge-transfer band led to the formation of 2,4,5-tricyanodiphenylmethane in a reaction involving as the initial step the transfer of a proton from the cation radical of mesitylene to the tetracyanobenzene anion radical (Yoshino et al., 1971). Direct detection of the cation radical (and the anion radical) was not made.

Irradiation of phenol in chloroform gave, after work up, o- and p-hydroxybenzaldehyde, while irradiation of N,N-diethylaniline gave o- and p-diethylaminobenzaldehyde. The aldehydes are thought to have been formed by hydrolysis of the corresponding benzal chlorides, and these to have been formed in reactions of the dichloromethyl radical ($\cdot CHCl_2$) whose origin, of course, is the anion $CHCl_3^-$ (Hirao and Yonemitsu, 1972).

Ionization of aromatics has also been achieved with high-energy irradiation. Benzene and biphenyl gave the corresponding cation radicals on x-irradiation in boric-acid glass (Hughes et al., 1964), while γ-irradiation of benzene on silica and silica-alumina (Edlund et al., 1967), and on a molecular HY sieve (Komatsu and Lund, 1972), and of toluene on silica gel and Vycor glass (Komatsu et al., 1972) gave the aromatic cation radical.

In another radiation technique, pulse radiolysis, high-energy electrons formed in a Van de Graaff generator lead to cation radical reactions in solution with very short periods of irradiation. Solvent ions or radicals are formed which may next remove an electron from the solute. For example, TMPD and some triarylamine cation radicals have been made in carbon tetrachloride solution [(66) and (67)] by

$$CCl_4 + e^- \rightarrow CCl_4^{\cdot+} + 2e^- \tag{66}$$

$$CCl_4^{\cdot+} + Ar_3N \rightarrow CCl_4 + Ar_3N^{\cdot+} \tag{67}$$

$$\cdot OH + RSSR \rightarrow OH^- + RSSR^{\cdot+} \tag{68}$$

Burrows et al., (1972), and some disulfide cation radicals in water solution (68) by Möckel et al., (1974). Cation radicals of phenothiazines have also been made in this way in hydrocarbon solutions (Burrows et al., 1973). The technique is very useful in conjunction with rapid scan spectrophotometry in that the spectra of cation radicals and derived species can be recorded very soon after the radiolysis pulse.

Formation on Catalytic Surfaces

Cation radicals can be made from aromatic and some olefinic molecules by depositing them on solid surfaces of silica-alumina or alumina. Catalysts of this type are of great industrial importance, since they are used for large-scale hydrocarbon conversions in the petroleum industry. Therefore, formation of cation radicals on such surfaces aroused at one period considerable interest, particularly as to the mechanism.

Ordinarily, the catalyst is activated by heating in a current of air or oxygen for several hours and next heated under continuous pumping for several hours. We shall discuss the active sites on catalysts later, but note now that activation is necessary for cation-radical formation. The substrate is usually added to the catalyst as a solution, say in benzene, carbon disulfide, and carbon tetrachloride, and the solvent is removed by pumping. Occasionally the substrate is added neat, either as a liquid or in the vapor phase. One-electron oxidation is usually immediate.

Polynuclear aromatic hydrocarbons are the most easily oxidized on catalyst surfaces. Perylene and anthracene have been the most commonly used hydrocarbons, and their cation radicals have been characterized in such work by esr and absorption spectroscopy. Naphthalene (Rooney and Pink, 1962) and naphthacene (Flockhart et al., 1966) have been used, but their oxidation to the cation radical goes only poorly.

Benzene and alkylbenzenes (toluene, o-xylene, hexamethylbenzene) do not give cation radicals on silica-alumina, the most well known and widely used type for catalyst for hydrocarbon conversions (Rooney and Pink, 1962). However, more active catalysts can be made from zeolites, and use of these has given cation radicals of simple aromatics, although these cation radicals do not have the stability on the catalyst surface that, say, perylene cation radical has on silica-alumina. Benzene and pentamethylbenzene on calcium-ion-exchanged Y-zeolite gave weak, poorly resolved esr signals of the (presumed) cation radicals (Hirschler et al., 1965). Use of the commercial catalyst, hydrogen-zeolon, gave the cation radicals of durene and pentamethylbenzene, and of benzene too, although the latter appeared to be in equilibrium with the monocation dimer (benzene)$_2^{\cdot+}$ (Corio and Shih, 1970). The most potent catalyst found for simple aromatics is an ammonium-ion-exchanged zeolite. With this it was possible to make the cation radicals of benzene, toluene, o-xylene, p-xylene, 1,2,3-trimethylbenzene, durene, and pentamethylbenzene (Kurita et al., 1970). Only the last three compounds persisted as the monomeric cation radicals, however. Among the other hydrocarbons, p-xylene gave the dimer cation radical, (p-xylene)$_2^{\cdot+}$, benzene gave not only its dimer cation radical, (benzene)$_2^{\cdot+}$, but also the cation radical of biphenyl, while toluene and o-xylene gave only the corresponding biaryl cation radical. The formation of these biaryl cation radicals is a most interesting reaction, and must result from oxidative coupling, no doubt via cation radical intermediates.

Neither silica nor alumina itself is an effective catalyst. For example, the perylene cation radical was not obtained on silica gel and γ-alumina. By special treatment these solids can be made to act as catalysts, however; fluoridated alumina is a weak oxidant for perylene. But, whereas the perylene cation radical on silica-alumina is stable for at least a year, it decays in a matter of hours on activated alumina (Brouwer, 1962). Alumina which has been activated by heating at 900° will serve as an oxidant for perylene and anthracene, and also diphenylamine and benzidine provided oxygen is also present (Flockhart et al., 1966).

Both silica and alumina have served as a host for oxidation of benzene by ultraviolet irradiation, leading to the benzene dimer cation radical (benzene)$_2^{\cdot+}$ (Tanei, 1968). Photoionization here is thought to be biphotonic (p. 180). On the other hand, the formation of perylene and anthracene cation radicals on silica alumina is enhanced by ultraviolet irradiation, and the process is found to be monophotonic (Takimoto and Miura, 1972). The fate of the photoejected electron is, of course, not known, a state of ignorance which pertains to all of the cation-radical forming reactions on catalyst surfaces.

As yet, to the writers' knowledge, no simple olefin has been converted into a cation radical by one of these solid catalysts. Indeed, polymerization of propylene on silica-alumina has been used as a diagnostic tool in classifying protonic rather than one-electron oxidative states (Hodgson and Raley, 1965; Rooney and Pink, 1962), a matter which will be discussed later. Esr signals have been obtained from linear olefins (e.g. pentene-2, octene-1) on a calcium-ion-exchanged Y-zeolite, but could not be attributed reliably to cation-radical formation (Hirschler *et al.*, 1965). The only olefin for which the evidence of cation-radical formation has any degree of reliability (absorption spectroscopy) is 1,1-diphenylethylene, but even here there are conflicts in interpretation (Dollish and Hall, 1965).

At no time, of course, is the formation of cation radicals on catalyst surfaces likely to be claimed as being useful in itself. The great and legitimate interest which it commanded (but which has dropped off in recent years), lies in the bearing it may have on the nature of catalyst surfaces. Furthermore, at one time it was thought that an understanding of the formation of aromatic cation radicals on catalyst surfaces might provide some insight into undesirable reactions (e.g. coking) of aromatics in industrial processes.

Two major questions have arisen: at which sites does cation-radical formation occur, and does molecular oxygen have to be present?

It is generally agreed that silica-alumina surfaces contain two types of acidic sites, Lewis and Brönsted, and the concensus is that one-electron oxidation occurs at the Lewis-acid sites. This opinion has been developed using a number of diagnostic techniques. Brönsted (i.e. proton donating) sites can be altered by metal-ion (e.g. sodium, calcium) exchange, and this exchange diminishes the catalyst's capacity for proton-catalysed reactions. The cumene-cracking capacity of silica-alumina is drastically diminished by sodium-ion exchange, but its capacity to form perylene cation radicals is not much altered (Brouwer, 1962). Propylene is readily polymerized by silica-alumina catalysts. If sodium-ion exchange is carried out, the catalyst will no longer polymerize propylene and will no longer catalyse the alkylation of aromatics by propylene. Yet, the ion-exchanged catalyst will still oxidize perylene and anthracene to their cation radicals (Hodgson and Raley, 1965; Rooney and Pink, 1962). Furthermore, a silica-alumina catalyst which has been "poisoned" by an aromatic (perylene or anthracene) still retains its ability to polymerize propylene and to alkylate the aromatic with propylene (Hodgson and Raley, 1965; Shepard *et al.*, 1962). The last result suggests that the aromatic covers up Lewis sites and leaves Brönsted sites available for proton-catalysed reactions. Interesting results along this line with the use of water and ethanol have also been reported. Thus, the intensity of the esr signal of perylene cation radical on silica-alumina was diminished by the addition of-water or ethanol vapor. Curiously, the effect was reversible, since removal of the water by pumping led to regeneration of the esr signal intensity. The phenomenon is interpreted (69) as a competition between water

$$ArH^{\cdot +}, LA^- + H_2O \rightarrow H_2O:LA + ArH \qquad (69)$$

and perylene for a Lewis-acid (LA) site. One might have thought that the cation radical should also react with the water (p. 227), but information on this point is not at hand (Rooney and Pink, 1962).

Results of these kinds, then, lead to the aforementioned concensus that a cation radical is formed at a Lewis-acid site. There is no knowledge, however, about the fate of the abstracted electron. It has been suggested that the electron deposited initially at each site is spread over the surface or within the crystal lattice of the catalyst so that the esr signal becomes broadened and undetectable (Stamires and Turkevich, 1964). There is also a lot of evidence that molecular oxygen may be the true electron acceptor, and this possibility, in turn, is part of the much-debated question of the role of oxygen in cation radical formation.

There is no doubt that some catalysts cannot produce cation radicals in the absence of air or oxygen. One of these is activated alumina (Scott et al., 1964; Flockhart et al., 1966). It is thought that activated alumina possesses Lewis-acid sites with which aromatics can form weak charge-transfer bonding. These sites are not sufficiently strong acceptors for complete electron transfer. Molecular oxygen will also complex with these sites, and, having done so, provide a path for complete electron transfer from an aromatic (e.g. perylene) to the surface. This is proposed also for the oxygen-enhanced formation of the benzene dimer cation radical on mordenite (Tokunaga et al., 1971).

There is ample evidence to suggest that similar sites exist in silica-alumina surfaces, but there is also evidence which suggests that oxygen is not needed at all of them. The addition of small amounts of oxygen enhances cation-radical formation by silica-alumina catalysts. Furthermore, reduction of the surface with hydrogen diminishes, but does not entirely prevent, cation-radical formation. These results suggest that to a large extent cation-radical formation is promoted by electron transfer to chemisorbed oxygen. A problem about this suggestion is that when O_2^- is formed directly by x-irradiation of oxygen on silica-alumina, its esr signal can be detected, whereas no other signal but that of the aromatic cation radical is detectable in the hydrocarbon work (Dollish and Hall, 1967). In contrast, it has also been proposed that oxygen serves as a transfer agent from aromatic to a Brönsted site rather than a Lewis site (Hirschler, 1966), so that a proton becomes the electron acceptor. The proposal is based on the finding that the H_R strength of solid acids influenced the extent of formation of perylene cation radical (Hirschler and Hudson, 1964). This proposal, if correct, would account for the failure to detect the esr signal of O_2^-, but it still leaves the fate of the abstracted electron unknown. Certainly, it seems that the two opposing proposals—electron transfer at Lewis versus Brönsted sites—are not reconcilable. Hydride-ion transfer at Brönsted sites seems to have been accepted as initiating formation of diphenyl- and triphenylmethyl cations (Wu and Hall, 1967), and one might ask if, since hydride-ion transfer is feasible, electron transfer should not be feasible too, at a Brönsted site? The weight of metal-ion-exchange evidence tips the scales in favor of Lewis-acid sites, however, and as a corollary it appears that electron transfer to these sites *via* molecular oxygen is a prominent factor.

The story about oxygen's involvement is still not complete at this stage. The fact that lengthy heating in hydrogen gas does not eliminate cation-radical formation entirely suggests that some sites on acid surfaces are themselves good enough electron acceptors. It is also found that beyond certain limits molecular oxygen diminishes rather than enhances the esr signals of cation radicals on catalyst surfaces. To a large extent this is due to the well-known broadening of esr signals by paramagnetic molecular oxygen. This effect is reversed by pumping off the oxygen. There is also evidence, however, of reaction of anthracene cation radical with oxygen on silica-alumina (Roberts *et al.*, 1959).

Isolable Cation-Radical Salts

Crystalline cation-radical salts have been known for many years, the earliest examples—Wurster salts—having been prepared almost 100 years ago. Interest in the salts has quickened in the last decade or so because it is possible to examine their spectroscopic properties both in the solid state and in solution.

Stable salts are obtained from molecules which are oxidized reasonably easily. Delocalization of the unpaired electron and positive charge stabilizes the cation radical, and, if an appropriate anion such as perchlorate or bromide is available, crystallization may follow. Therefore, compounds which form stable salts are aromatic and often contain a heteroatom. Most commonly encountered classes among these are polynuclear aromatics, diaminobenzenes (which give Wurster salts), and triarylamines (which give aminium salts).

Wurster salts (Sidgwick, 1966) are usually obtained by oxidizing diaminobenzenes (*p*-phenylenediamines) with bromine in methanol-acetic acid solution. This leads to the Wurster bromide (Wurster and Sendtner, 1879), but if oxidation is carried out in the presence of a large amount of perchlorate ion the Wurster perchlorate is formed. This is the way in which Wurster's Blue [1] perchlorate (the most widely studied cation radical salt of all) and Wurster's Red [2] perchlorate are made (Michaelis and Granick, 1943). The chlorides and iodides are also known (Oohashi and Sakata, 1973; Sakata and Nagakura, 1970).

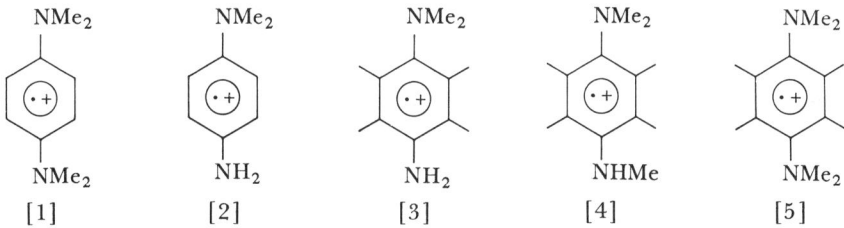

There are some pitfalls in the preparation of Wurster salts by this method. The cation radical of a *p*-phenylenediamine may readily undergo further oxidation to the dication (or the corresponding diimine). This, in turn, may be reactive in aqueous solution (as with *p*-phenylenediamine itself) and become hydrolysed. Further, steric inhibition of resonance may prevent stabilization of the cation radical (e.g. [5]). Thus, while the salts of cation radicals [1], [2], and [3] are readily isolable, those of [4] and [5] are not (Michaelis et al., 1939).

Wurster's Blue perchlorate has been the subject of many investigations. Its stability as a solid and in solution, its ease of preparation, and its symmetry make it an ideal model in spectroscopic and magnetic susceptibility studies. It was one of the first cation radicals to be shown to exist both as a monomer and as a dimer in solution (Hausser, 1956; Bolton et al., 1962) and in the solid state (Duffey, 1962; McConnell and Lynden-Bell, 1962), the dimeric form being obtainable reversibly by lowering the temperature. This phenomenon is widespread among cation radicals and is discussed in section 3.

Triarylaminium salts were also isolated in the early 1900s as perbromides ($Ar_3N^{\cdot+}Br_3^-$) and so-called pentachloroantimonates

(Wieland, 1907; Wieland and Wecker, 1910). Oxidation was carried out in an inert solvent such as benzene, from which the salt precipitated. The easier and more reliable preparation of perchlorate salts by the iodine-silver perchlorate method (Weitz and Schwechten, 1926, 1927), to which we have referred earlier (p. 168), allowed a much clearer understanding of the nature of triarylaminium ions to be obtained. Isolation of perchlorates permitted chemical studies, and easy reduction to the triarylamine by iodide ion, ferrous ion, etc., was consistent with the cation-radical view that was developed. The name aminium ion was coined by Weitz. Other salts were prepared such as tritolylaminium picrate (by oxidation of the amine with lead dioxide in the presence of picric acid), and it was also recognized that conjugate anions in salts obtained by oxidation with antimony pentachloride, phosphorus pentachloride and ferric chloride had to be complex anions rather than simple anionic radicals. This is a particularly pertinent point in antimony pentachloride oxidations (p. 165).

A variety of p-substituted triarylaminium perchlorates have been made (Walter, 1955, 1966). The p-substituent may be electron donating or withdrawing (MeO, Me, F, Cl, Ph, $CO_2 Me$), but oxidation of tri-(p-nitrophenyl)amine failed.

Tris-(p-bromophenyl)amine readily forms stable salts having anions such as $SbCl_6^-$, ClO_4^- (Bell et al., 1969) which are valuable as reagents for the relatively unambiguous one-electron oxidation of organic substrates. In this way it has been possible to investigate the reactivity of cation radicals from cycloheptatriene derivatives (Beresford and Ledwith, 1970), 1,2-di-9-carbazylcyclobutane (Beresford et al., 1970), methoxylated bibenzyls (Kricka and Ledwith, 1973), carbazoles and iminobibenzyls (Beresford et al., 1974), several types of molecule containing p-dimethylaminophenyl substituents (Bell et al., 1974), and hydrazine derivatives (Nelson et al., 1974). Cation radicals so generated undergo a variety of carbon–carbon and carbon–hydrogen scission processes in addition to the anticipated oxidative coupling reactions (carbon–carbon bond formation), and the results complement those from related anodic oxidation studies. A particularly interesting use of stable aminium salts is in the reported catalytic oxygenation of cisoid 1,4-dienes (Barton et al., 1972).

The stability of p-substituted triarylaminium ions which permits of their isolation as salts is in part resonance stabilization and in part due only to the blocking of the p-positions against benzidine-type coupling. Triphenylamine does not form a stable salt because the triphenylaminium ion undergoes coupling and further oxidation. Thus, oxidation of triphenylamine with iodine (Bruning et al., 1967) and at an anode (Seo et al., 1966) gives N,N,N',N'-tetraphenylbenzidine [eqn (70)]. Anodic oxidation of a variety of triarylamines

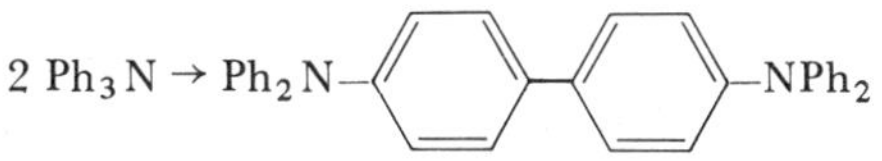

$$2\text{ Ph}_3\text{N} \rightarrow \text{Ph}_2\text{N}\text{—}\bigcirc\text{—}\bigcirc\text{—NPh}_2 \qquad (70)$$

has shown that if an unsubstituted p-position remains, benzidine formation occurs (Seo et al., 1966). At the same time it appears that a stable aminium ion will be formed even from an N,N-dimethylarylamine if the aryl group contains a good stabilizing group, such as methoxyl. Thus, in summary, it is not surprising that many triarylamines give stable aminium salts, and anodic oxidation indicates other amines which may do the same.

The stabilizing influence of p-substituents is seen also in the formation of a stable hexachloroantimonate from dianisylamine (Kainer and Hausser, 1953), and perchlorate from 2,2',4,4'-tetramethoxyldiphenylamine (Medzhikov et al., 1966), the latter having been oxidized by Frémy's salt in solution containing perchloric acid.

The bromide, picrate, perchlorate and disulfate of phenothiazine are well known (Sato et al., 1969; Kehrmann and Diserens, 1915; Iida, 1971). Oxidations of phenothiazines are usually direct, in the presence of the appropriate anion. An acidic solution is advisable so as to avoid dimerization of neutral phenothiazinyl radicals. Oxidation of 3,7-dimethoxyphenothiazine by Frémy's salt in a biphthalate-buffered solution led to the stable perchlorate salt (Medzhikov et al., 1966).

The most reliable way of preparing cation-radical salts of phenothiazine and its derivatives, however, seems to be the disproportionation of the phenothiazine and its 5-oxide in the appropriate acid (Billon, 1961; Shine et al., 1972; Matsunaga and Suzuki, 1973).

The same method has been used with thianthrene and thianthrene 5-oxide (Rundel and Scheffler, 1963), but Lucken's method of oxidizing thianthrene with perchloric acid in carbon tetrachloride-acetic anhydride is the better (Lucken, 1962; Murata and Shine, 1969).

Although many polynuclear aromatic hydrocarbons give stable cation radical solutions (e.g. in sulfuric acid), not many give stable salts. The best known salts are those from antimony pentachloride oxidations. These salts are very easy to prepare by direct oxidation in methylene chloride or carbon tetrachloride from which they precipitate almost instantly.

Anodic oxidation of perylene, pyrene, and azulene (Ristagno and Shine, 1971a, b; Reddoch et al., 1971) has given the aromatic perchlorate complexed with the aromatic in what appears to be an ArH . ArH$\cdot^+$ClO$_4^-$ salt. The salt deposits on the anode.

Anodic oxidations leading to stable salts are rare, no doubt because not only must the cation radical be resistant to further oxidation in the range of potential used but also must be reasonably insoluble in the anolyte. The perchlorate of 3,3'-dimethylbenzidine has been made anodically (Blount and Kuwana, 1970a).

Dibenzodioxin cation radical perchlorate has been made by anodic oxidation (Cauquis and Maurey-Mey, 1972). This salt undergoes further (unknown) reactions on the anode and therefore only small amounts can be deposited at a time (Shine and Shade, 1974). Anodic oxidation in trifluoroacetic acid can give trifluoroacetates or (if another anion is added after electrolysis) other salts (Hammerich et al., 1972).

We have already referred to oxidations by iodine-silver perchlorate (p. 168). This is an easy way of making the perchlorates of some cation radicals such as of perylene (Sato et al., 1969; Ristagno and Shine, 1971b). In appropriate circumstances, silver ion itself should be able to oxidize some compounds to their cation radicals, and this has been achieved with some 2-phenyl-3-arylaminoindoles and silver perchlorate in acetone or acetonitrile; a silver mirror is formed (Bruni et al., 1971).

Among the most stable cation radical salts are those of metalloporphyrins. Most commonly the zinc, and magnesium metalloporphyrins are used (M = Zn, Mg), but others, such as the cadmium porphyrin, are also used (Dolphin et al., 1973; Fajer et al., 1973; Dolphin and Felton, 1974). Tetraarylporphyrins [6, R = H; R_1 = aryl] or octaalkylporphyrins (R = alkyl, R_1 = H) are the most common parent molecules. Recently, some zinc tetraalkylporphyrin cation radicals were also reported (Fajer et al., 1974). The metalloporphyrins have low oxidation potentials (Fuhrhop and Mauzerall, 1969; Fuhrhop et al., 1973; Stanienda and Biebl, 1967), and their salts (e.g. perchlorates) can be made quite easily by anodic oxidation in the presence of an appropriate electrolyte. Chemical oxidation, by bromine or ferric ion for example, is also easily achieved, while

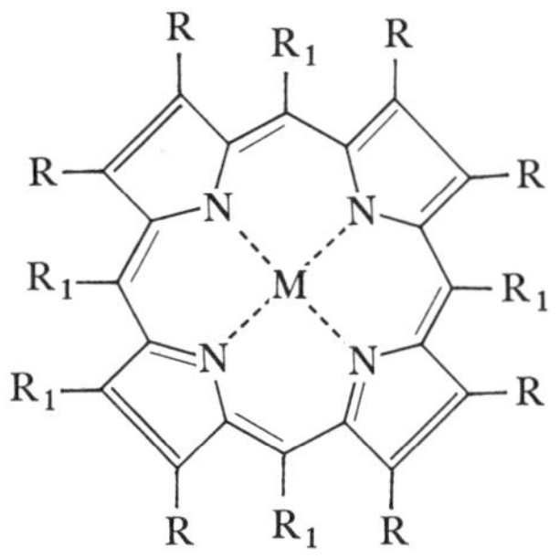

[6]

electron exchange with another cation radical (e.g. thianthrene cation radical perchlorate) is also a facile way of making metalloporphyrin cation radical salts.

2. ELECTROCHEMICAL METHODS OF FORMATION AND INVESTIGATION OF CATION RADICALS

Electrochemical methods have played an important role in the recognition of cation radicals as intermediates in organic chemistry and in the study of their properties. An electrode is fundamentally an electron-transfer agent so that, given the proper solvent system, anodic oxidation allows formation of the cation radical without any associated proton or other atom transfer and without the formation of a reduced form in the immediate vicinity of the cation radical. Moreover, because the potential of the electrode can be adjusted precisely, its oxidizing power can be controlled, and further oxidation of the cation radical can often be avoided. Finally, the electrochemical experiment can involve both production of the cation radical and an analysis of its behavior, so that information about the thermodynamics of its formation and the kinetics of its reaction can be obtained, even if the cation radical lifetime is as short as a few milliseconds. There are some limitations, however, in the anodic production of cation radicals. The choice of solvent is limited to those that show reasonable conductivity with a supporting electrolyte (e.g. tetra-n-butylammonium perchlorate, TBAP). Acetonitrile, methylene chloride and nitrobenzene have been employed as solvents, but other favorites, such as benzene and cyclohexane, cannot be used. The relatively high dielectric constant of the suitable solvents usually causes the cation radical salts produced to be fairly soluble, and separation of these from the supporting electrolyte salt may be a problem. Electrolysis methods require fairly elaborate cells and electronic equipment and large scale (i.e. greater than a few hundred milligrams) production of product is not very easy. Finally chemical methods will sometimes be capable of producing cation radicals for cases where further oxidation is possible at potentials of cation radical formation, while electrochemical methods may not. Consider the case of the methylhydrazines, for example. The cation radicals can be observed in sulfuric acid solutions by esr under the conditions of reaction of the methylhydrazine and cerium(IV) in a fast-flow mixing chamber (Atkinson and Bard, 1971). Here the relative amounts of the hydrazine and cerium(IV) can be controlled so that only one electron per hydrazine is extracted. The electrochemical oxidation of methylhydrazines under similar conditions (e.g. King and Bard, 1965) shows no evidence of cation radicals, since those produced at the electrode surface (and their reaction products) are immediately oxidized further with the electrode providing an unlimited sink for electrons at this potential.

This section will outline the experimental techniques that have been used to study cation radicals and the type of information that can be obtained about their behavior. Several examples of such investigations will be given, but, since several reviews have appeared which have discussed various aspects of this area (Peover, 1967, 1971; Hoijtink, 1970; Mann and Barnes, 1970; Baizer, 1973; Fry, 1972) and organic electrochemistry is the subject of an extensive chapter in this series (Eberson and Nyberg, 1976), an exhaustive review of electrochemical methods and studies involving the production or reactions of cation radicals will not be attempted.

Electrochemical Techniques

A large number of electrochemical techniques are available for producing and studying cation radicals (or other electrogenerated species). The instrumentation, techniques, and theory of most of these have been described (Cauquis and Parker, 1973; Adams, 1969; Sawyer and Roberts, 1974), so this discussion will only outline these methods to the extent necessary for an understanding of the cation radical chemistry in the examples described. Basically the electrochemical techniques can be separated into two classes: the microelectrode or voltammetric techniques and the bulk electrolysis or coulometric techniques. In the voltammetric techniques a small indicator electrode is employed and the electroactive substance (e.g. the cation radical) is generated in the immediate vicinity (i.e. within 0·4 mm) of the electrode surface. The bulk solution is essentially unchanged by the experiment and proof that a cation radical is formed and information about its properties are obtained from the nature of the current (i)-potential (E)-time (t)-behavior. In coulo-

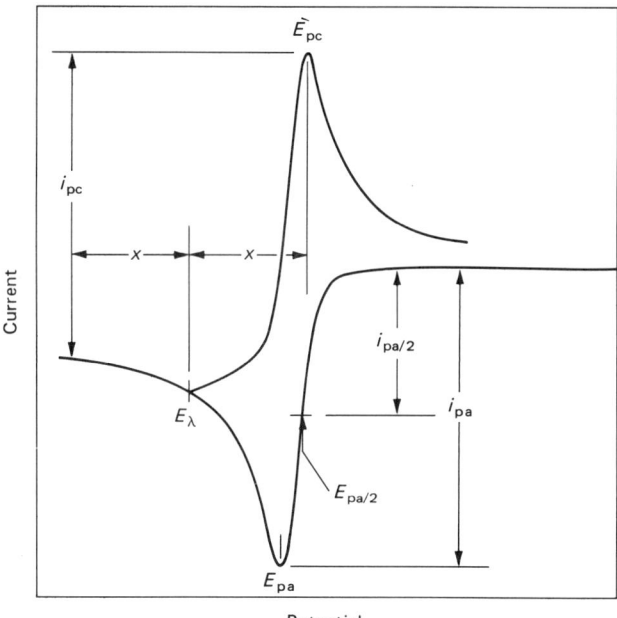

Figure 3. Typical cyclic voltammogram for nernstian oxidation of a species to a stable cation radical, showing parameters of interest: i_{pa}, maximum anodic current; $i_{pa/2}$, one-half maximum anodic current; i_{pc}, maximum cathodic current, measured from a single sweep curve at a point x past the switching potential E_λ; E_{pa}, potential of i_{pa}; $E_{pa/2}$, potential of $i_{pa/2}$; E_{pc}, potential of i_{pc}; E_λ, potential at which direction of potential sweep is reversed.

metric experiments exhaustive electrolysis of the solution occurs and analyses and isolation of products are sometimes possible.

The voltammetric technique used most frequently in studies of cation radicals is cyclic voltammetry (CV). CV involves the use of a small stationary electrode (usually of platinum but sometimes of gold or glassy carbon) in an unstirred solution whose potential is scanned linearly with time over a certain range [e.g. from 0 to +1·8 V vs a saturated calomel electrode (SCE)] and then scanned back again. The resulting current-potential (i–E) curve (Fig. 3) is called a cyclic voltammogram. The cyclic voltammogram in Fig. 3 is representative of the behavior of a substance which forms a cation radical which is stable within the time scan of the electrochemical experiment [e.g. 9,10-diphenylanthracene (DPA) in 0·1 M TBAP–CH$_3$CN]. On the initial scan from potentials where no electrode reaction occurs to positive potentials an anodic wave is observed for the oxidation of DPA to DPA$^{\cdot+}$. The parameters of interest on this forward scan are the anodic peak potential (E_{pa}) and the peak current (i_{pa}). At some potential past the peak, E_λ, the scan direction is reversed and a cathodic wave is observed for the reduction of any DPA$^{\cdot+}$ remaining in the vicinity of the electrode. For an electrode reaction which is nernstian or reversible (i.e. the rate of electron transfer at the electrode surface is large compared to the rates of mass transfer of reactants and products at the electrode) and the product of the electron transfer reaction is stable, the cyclic voltammogram is characterized by $|E_{pa} - E_{pc}| = 2\cdot3\, RT/nF$ (or at 25°, 59/n mV), $|E_{pa} - E_{pa/2}| = 2\cdot2\, RT/nF$ (or at 25°, 56·5/n mV), and $|i_{pa}| = |i_{pc}|$; the current function, $i_{pa}/v^{1/2} C_R$, (where v is the scan rate, e.g. in V sec^{-1}, and C_R is the bulk concentration of reactant) is constant at different concentrations and scan rates. For a given electrode, the current function, can be given as $i_p/v^{1/2} C_R = $ const. $n^{3/2} D_R^{1/2}$, where n is the number of electrons transferred per molecule of reactant in the electrode reaction and D_R is the diffusion coefficient, in cm^2 sec^{-1}, of the reactant. A cyclic voltammetric experiment in which a cation radical is formed which is stable for the time needed to complete the forward and reverse sweeps (ca 10 milliseconds to 10 seconds) thus provides evidence for the production and stability of the radical cation. The value of $E°$ for the half reaction, $R^{\cdot+} + e \rightleftarrows R$, can be obtained from the E_{pa}-value, since $|E_p - E_{1/2}| = 1\cdot109\, RT/nF$, where $E_{1/2}$ is the polarographic or voltammetric half wave potential, which is given by (71). To a

$$E_{1/2} = E° + (RT/2nF) \ln (D_R/D_{R^{\cdot+}}) \qquad (71)$$

good approximation, $E° = E_{pa} - 28\cdot5/n$(mV) at 25°C, measured against the reference electrode employed in the CV experiment, of course.

If the product of the electron transfer (e.g. R$^{\cdot+}$) is unstable and decays to form non-electroactive products, the reversal (cathodic) current decreases or is absent and E_{pa} is shifted towards less positive values. The extent of this shift depends upon the scan rate, the rate constant for the decay reaction, k, and the reaction order, m. For example, in the limit of large values of k/v, E_{pa} is given by (72).

$$E_{pa} = E_{1/2} + \frac{RT}{nF}(0\cdot 78) - \frac{RT}{(m+1)nF}\ln\left(\frac{RT}{nF}\frac{k}{v}\right) \qquad (72)$$

Thus, values of i_{pc}/i_{pa} less than one and shifts in E_{pa} with v demonstrate that the cation radical is unstable. In favorable cases these data can be employed to determine k. The current function for the anodic peak $(i_{pa}/v^{1/2}C_R)$ decreases slightly (a maximum of 10%) with increasing scan rate. If the product of the decay reaction is electroactive at potentials where the cation radical is formed, the reaction is said to be an ECE-reaction, e.g. eqns (73)-(75). For this

$$R - e \longrightarrow R^{\cdot+} \qquad (E) \qquad (73)$$

$$R^{\cdot+} + N\bar{u} \xrightarrow{k} RNu^{\cdot} \qquad (C) \quad (N\bar{u} \text{ nucleophile}) \qquad (74)$$

$$RNu^{\cdot} - e \longrightarrow RNu^{+} \qquad (E) \qquad (75)$$

case, not only does i_{pc}/i_{pa} decrease below one and E_{pa} shift with scan rate, but $i_{pa}/v^{1/2}C_R$ becomes a strong function of scan rate. At high scan rates, where the cyclic time is so short that appreciable decomposition of $R^{\cdot+}$ does not occur, the behavior approaches that of a reversible one-electron transfer wave. For slow scan rates, however, the experimental time becomes long enough that the second electron transfer reaction can occur to an appreciable extent, and $i_{pa}/v^{1/2}C_R$ approaches twice the value for fast scan rates. It is also possible to treat reaction schemes where the electron transfer reactions at the electrode are slow, although this rarely seems to occur for electrode reactions associated with cation radicals. A number of other reaction sequences involve single two-electron transfer steps, disproportionations, and variants of the above schemes; these are discussed in detail in the literature, especially in papers by Nicholson, Shain, Savéant, Vianello, and their co-workers (see, e.g. Nicholson and Shain, 1964, 1965; Olmstead et al., 1969; Kudirka and Nicholson, 1972; Savéant and Vianello, 1963, 1965, 1967, 1967a; Andrieux et al., 1970, 1973; Savéant et al., 1973; and references therein).

Precise quantitative measurements are sometimes difficult in CV, especially when the solution resistance is high (causing perturbation of the measured peak potentials because of the uncompensated iR drop), the nonfaradaic current associated with charging the double layer is appreciable (important at large values of v and small C_R-values), and adsorption or precipitation of reactants, intermediates or products occurs. Other voltammetric techniques, such as the double potential step-method, where the potential is stepped rather than scanned between two limiting potentials and the current or the coulombs passed are recorded, or rotating ring-disk electrode (RRDE) voltammetry may be useful complements to CV in these cases. The RRDE consists of a disk electrode and a concentric ring electrode electrically insulated from the disk and separated from it by a thin Teflon spacer. When this electrode is rotated, solution at the ring-disk surface is spun out in a radial direction and is replenished by bulk solution flowing to the RRDE in a normal direction. Products generated at the disk flow to the ring where they can be detected by their electrochemical behavior. For example, a cation radical generated at the disk, at a current i_d, will flow to the ring where it can be detected by its reduction back to parent, i_r. The ratio, i_r/i_d, analogous to i_{pc}/i_{pa} in CV, provides information about the stability of the cation radical. The variation of i_d and the disk potential with rotation rate also is useful in elucidating the electrode reaction mechanism. Details on the application of the RRDE in such studies are given in monographs by Albery and Hitchman (1971) and Pleskov and Filinovsky (1972) and in a number of recent papers (e.g. Prater and Bard, 1970, 1970a; Puglisi and Bard, 1972, and references therein).

In controlled potential coulometric studies exhaustive electrolysis of the solution is carried out at a large working electrode (e.g. platinum gauze) whose potential is maintained at a constant value with respect to a reference electrode by a potentiostat. The auxiliary electrode, used to complete the electrolysis circuit, is contained in a chamber separated from the working electrode compartment by one or more sintered-glass disks to prevent products formed at the auxiliary electrode from interfering with the electrolysis or reacting with products. Information about the electrode reaction and the stability of the products is obtained by observing the current- or coloumb-time behavior, the apparent number of faradays of electricity consumed per mole of reactant (n_{app}), and analysis of the resultant solution by electrochemical, spectroscopic or chromatographic methods. In general, coulometric methods involve much longer experimental time periods (several minutes to several hours) and are capable of studying much slower reactions of the electrogenerated product. Details on the application of coulometric methods to studies of reaction kinetics and mechanisms can be found in chapters by Bard and Santhanam (1970) and Meites (1960).

Examples of Electrochemical Studies

Numerous electrochemical studies of systems which form stable cation radicals exist. For example, the CV oxidation of aromatic hydrocarbons which have the sites of high electron density blocked, such as 9,10-diphenylanthracene (DPA) or rubrene, show voltammograms typical of nernstian one-electron systems (Phelps et al., 1967; Marcoux et al., 1967; Peover and White, 1967). Rotating disk and RRDE studies also provide evidence for their stability. Controlled potential coulometric oxidation of these show n_{app}-values of one and CV studies of the oxidized solution showed a cathodic peak for reduction of the cation radical at the same potentials and of the same height as that of the original solution. Similar electrochemical behaviour is observed with phenothiazines, thianthrenes, and other heterocyclic compounds in solvents suitably freed from nucleophilic impurities. Not only do the electrochemical results demonstrate the production of a stable cation radical, but the measured E_p- or $E_{1/2}$-values can be used as a measure of the energy required to abstract an electron from the molecule. Numerous correlations exist between the electrode potentials of nernstian systems and orbital energies calculated by different molecular orbital methods (Peover, 1967, 1971; Hoijtink, 1970). Correlations of $E_{1/2}$-values with optical parameters, such as long wavelength absorption bands or charge transfer bands, and ionization potentials also have been described (Peover, 1968; Gough and Peover, 1966; Miller et al., 1972). Anodic oxidation has also been employed to produce stable cation radicals (and sometimes dications) of metalloporphyrins (see, e.g. Stanienda and Biebl, 1967; Fajer et al., 1970; Wolberg and Manassen, 1970; Lexa and Reix, 1974; and references therein). Since tetraphenylporphin itself produces a stable cation radical in methylene chloride (Tokel et al., 1972), many of these metalloporphyrins (e.g. those with Mg, Zn, and Cu) can be considered as metal stabilized cation radicals, with the unpaired spin density located primarily on the ligand.

For many cation radicals electrochemical studies show evidence of reversible oxidation to stable dications. The CV of a typical EE reaction sequence is shown in Fig. 4. This kind of behavior is observed for 9,10-di(*p*-anisyl)anthracene (Hammerich and Parker, 1974), 2,3,4,5-tetra-aryl-pyrroles (Libert and Caullet, 1974; Libert *et al.*, 1972), tetrathioethylenes (Coffen *et al.*, 1971 and references therein), 2,2'-benzothiazolinone azines (Janata and Williams, 1972) and related violene-type systems (e.g. Hünig *et al.*, 1969; Happ *et al.*, 1972; Fritsch *et al.*, 1970; and references therein). Hammerich and Parker (1973) have pointed out that the purity of the solution is very important in observing reversible oxidation waves to the dications, and that, by performing the electrochemical studies in the presence of suspended alumina added directly to the solvent/supporting

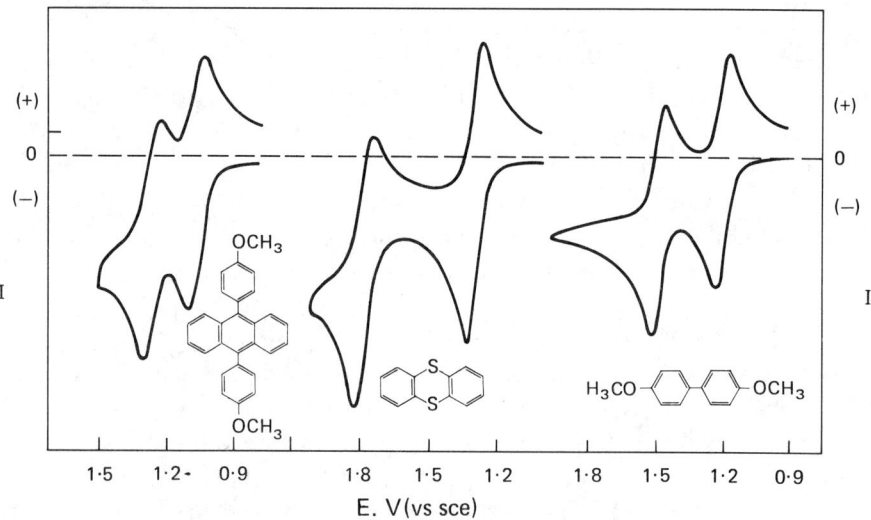

Figure 4. Cyclic voltammograms for several systems showing two nernstian one-electron oxidations. Systems in acetonitrile containing suspended alumina with 0·1 Mn-Bu$_4$N$^+$BF$_4^-$ as supporting electrolyte (from Hammerich and Parker, 1973).

electrolyte system in the cell, two reversible oxidation waves can be observed for thianthrene and 4,4'-dimethoxybiphenyl. The measured E_p-values of the two waves are useful in estimating the equilibrium constant for disproportionation of the cation radical. Attempts at measuring these constants when the second electron transfer step is irreversible, usually because of a rapid reaction of the dication, may lead to sizable errors, since the second wave will be shifted in a negative direction [see (72)]. For some molecules the second oxidation step occurs very near or even more easily than the first (i.e. disproportionation of the cation radical occurs to a very large extent). In these cases, two very closely spaced CV waves are observed, as in the case of tetra-*p*-anisylethylene (Bard and Phelps, 1970; Parker *et al.*, 1969), or only a single two-electron wave for direct oxidation to the dication, e.g. for tetrakis-(*p*-N,N-dimethylaminophenyl)ethylene (Bard, 1971; Bard and Phelps, 1970) or 1,1,4,4-tetrakis-(dimethylamino)butadiene (Fritsch *et al.*, 1970). Coulometric studies on the above systems attested to the long-term stability of the dications. In most cases the removal of a second electron from the molecule should be more difficult than removal of the first, since the electron repulsion, amounting to about 5 eV in the

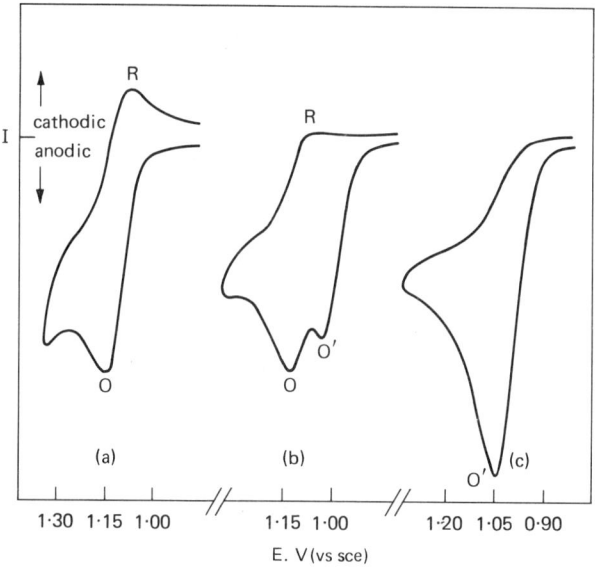

Figure 5. Cyclic voltammograms for the oxidation of 9-phenylanthracene (2.0 mM) in acetonitrile-0·1 M NaClO$_4$ in the presence of the following concentrations of pyridine: (a) 0; (b) 1·0 mM; (c) 2·0 mM. Note the growth and shift of the anodic wave and the disappearance of the reversal wave (from Jensen and Parker, 1973).

gas phase, assists in the first step. High solvation of the dication and a tendency to form stronger ion pairs or aggregates greatly decreases the potential for the second oxidation step, and 0·3 to 0·5 V appears to be the usual separation between the waves. However, geometric changes in the molecule after removal of an electron can alter the orbital energies; this is apparently the case for the closely-spaced or two-electron waves (Bard, 1971). This effect has been discussed in terms of entropy and enthalpy factors for the similar process of anion radical and dianion formation (Andrieux and Savéant, 1974).

Many electrochemical studies involve reactions of the electrogenerated cation radicals with added nucleophile or in coupling reactions; a number of these are discussed in the sections on reactions of cation radicals. Added nucleophile causes the disappearance of the CV cathodic (reversal) peak current and usually an increase in the forward (anodic) peak current because the RNu· species formed in the reaction of cation radical and nucleophile is more easily oxidized than the parent [see reactions (73)–(75)]. Moreover, the oxidation peak is shifted towards negative potentials [see (72)] and, when less than stoichiometric amounts of nucleophile are added, CV pre-peaks can be observed. The coulometric n_{app}-values will also show increases from 1 to 2. Typical behavior of this type is seen, for example, upon oxidation of 9-phenylanthracene in acetonitrile solutions containing increasing amounts of pyridine (Fig. 5) (Jensen and Parker, 1973). Numerous other studies involving methoxylations, acetoxylations and hydroxylations of hydrocarbon cation radicals, following similar routes, have been reported. RRDE studies of these kinds of reaction, for example that of p-methoxystyrene cation radical with pyridine (Landsberg and Müller, 1974) have also been reported. Cation radicals have also been proposed, in related studies, as intermediates in the reduction of products derived from nucleophile addition to the dication. For example, oxidation of DPA to the dication in benzonitrile leads to the appearance of a new reduction wave, at potentials about 0·8 V more negative than the DPA$^{·+}$/DPA wave, which corresponds to potentials where triphenylmethyl cation is reduced (Dietz and Larcombe, 1970). The wave was ascribed to reaction (76). The intermediate radical dissociates to form the cation radical which, at these potentials, is immediately reduced to parent DPA [(77) and (78)]. A similar ECE reaction

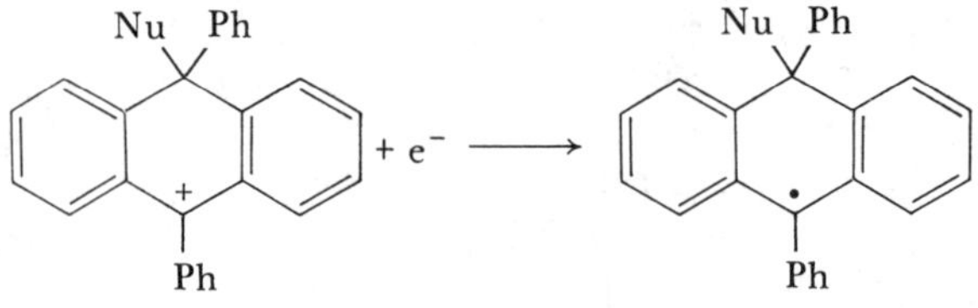

$$\text{(76)}$$

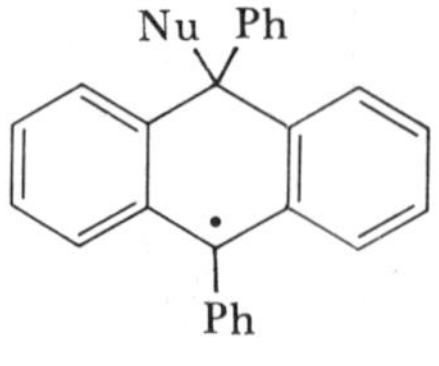

$$\text{(77)} \quad \longrightarrow \text{DPA}^{·+} + \text{N}\bar{\text{u}}$$

$$\text{DPA}^{·+} + e^- \longrightarrow \text{DPA} \quad \text{(78)}$$

sequence has been discussed by Hammerich and Parker (1974) for other 9,10-disubstituted anthracenes and nucleophiles. Some reactions of cation radicals, such as DPA$^{•+}$, with nucleophiles (e.g. n-butylamine) have been shown to occur with the emission of light, perhaps by an electron transfer reaction between the cation radical with the radical-nucleophile adduct (Sioda, 1968; Bolhuis et al., 1974) (see also section 5).

Coupling reactions of the electrogenerated cation radicals can also occur. If the parent molecule is a good nucleophile, this can be considered as a nucleophilic attack by parent on its own cation radical. The nature of the electrochemical responses in CV and coulometry depend upon whether the coupled product can undergo further oxidation. For example, in the oxidation of aromatic azahydrocarbons such as acridine (AcH), studied by Marcoux and Adams (1974), the CV was characterized by an irreversible one-electron wave and coulometry showed an n_{app}-value of one. An analysis of rotating disk voltammograms demonstrated that the reaction sequence (79)-(81) was more probable than that involving

$$AcH - e^- \rightleftarrows AcH^{•+} \tag{79}$$

$$AcH^{•+} + AcH \rightarrow (AcH)_2^{•+} \tag{80}$$

$$(AcH)_2^{•+} - e^- \rightarrow Ac_2H^+ + H^+ \tag{81}$$

coupling of two cation radicals to produce the same product. In many other cases, however, the coupled product can undergo further oxidation at the potentials of cation radical formation; for example, substituted-triphenylamine cation radicals form a benzidine which undergoes a further two-electron reaction as shown in eqns (82)-(84)

$$Ph_3R - e^- \rightleftarrows Ph_3R^{•+} \tag{82}$$

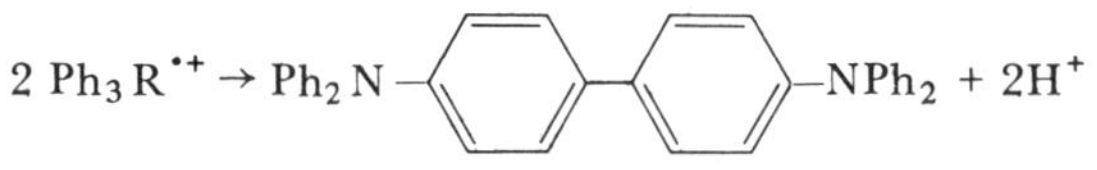

$$2\,Ph_3R^{•+} \rightarrow Ph_2N\text{–}\underset{}{\bigcirc}\text{–}\underset{}{\bigcirc}\text{–}NPh_2 + 2H^+ \tag{83}$$

$$Ph_2N\text{–}\underset{}{\bigcirc}\text{–}\underset{}{\bigcirc}\text{–}NPh_2 - 2e^- \rightleftarrows Ph_2N^+\text{=}\underset{}{\bigcirc}\text{=}\underset{}{\bigcirc}\text{=}N^+Ph_2 \tag{84}$$

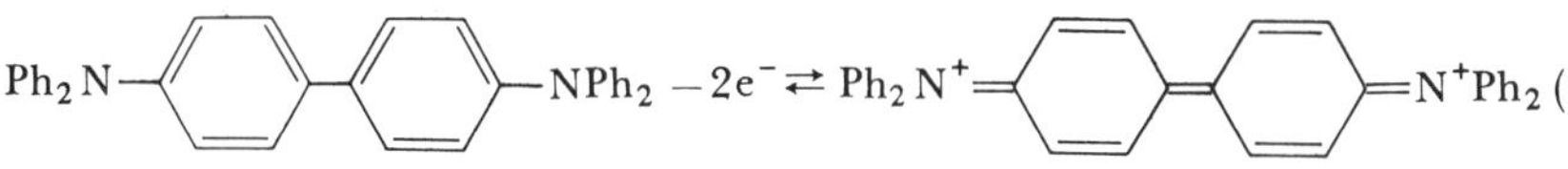

(Seo et al., 1966; Creason et al., 1972 and references therein). For these cases, the CV and coulometry are characterized by two-electron waves (i.e. two faradays per mole of parent). The nature of the coupling step (radical-radical or radical-parent) is not known with certainty, although the radical-radical route appears to be the preferred one (Nelson and Feldberg, 1969). Similar behavior is observed for the oxidation of phenazine-di-N-oxide (Stüwe et al., 1974), carbazoles (Ambrose and Nelson, 1968) and iminobibenzyls (Frank et al., 1975). In all of these cases electron and proton transfer reactions between the cation radicals and other intermediates and the parent molecule must be considered in determining the reaction mechanism.

Intramolecular coupling or nucleophilic substitutions sometimes occur upon oxidation of the cation radical; the new product frequently then undergoes an additional anodic electron-transfer reaction to yield a new cation radical. For example, tri-*p*-substituted triphenylamines form stable cation radicals upon oxidation at their first anodic wave. Reynolds et al. (1974) have shown, however, that further oxidation of the cation radicals leads to the appearance of a new reversible redox couple at potentials slightly positive with respect to the first anodic wave. This new couple was shown to occur at potentials where the corresponding carbazole is oxidized, so that the overall process involves conversion of the amine to the carbazole cation radical by reaction scheme (85).

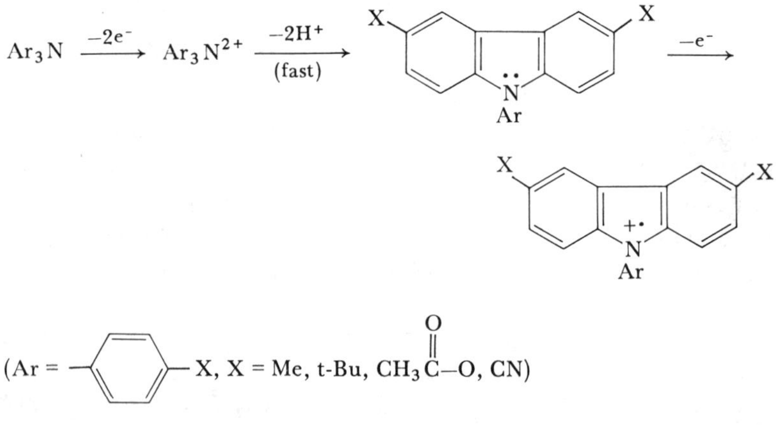

$$Ar_3N \xrightarrow{-2e^-} Ar_3N^{2+} \xrightarrow[\text{(fast)}]{-2H^+} \text{[carbazole]} \xrightarrow{-e^-}$$

$$\text{[carbazole cation radical]} \quad (85)$$

$(Ar = \text{—C}_6\text{H}_4\text{—X}, \ X = \text{Me, t-Bu, CH}_3\overset{\text{O}}{\overset{\|}{\text{C}}}\text{—O, CN})$

Side reactions of the dication (e.g. hydrolysis), and perhaps further oxidation of the carbazole at potentials where the dication forms, limits the yield of carbazole cation radical in some cases. A similar reaction sequence reported by Ronlan and Parker (1970) involves oxidation of 3,4,3',4'-tetramethoxybibenzyl [7a] which produces a cyclized product, the 9,10-dihydrophenanthrene [7b]. On the CV time scale [7b] forms a stable cation radical, but on coulometric oxidation it shows an n_{app}-value of two and produces the phenanthrene [7c]. A similar intramolecular coupling occurs upon oxidation of

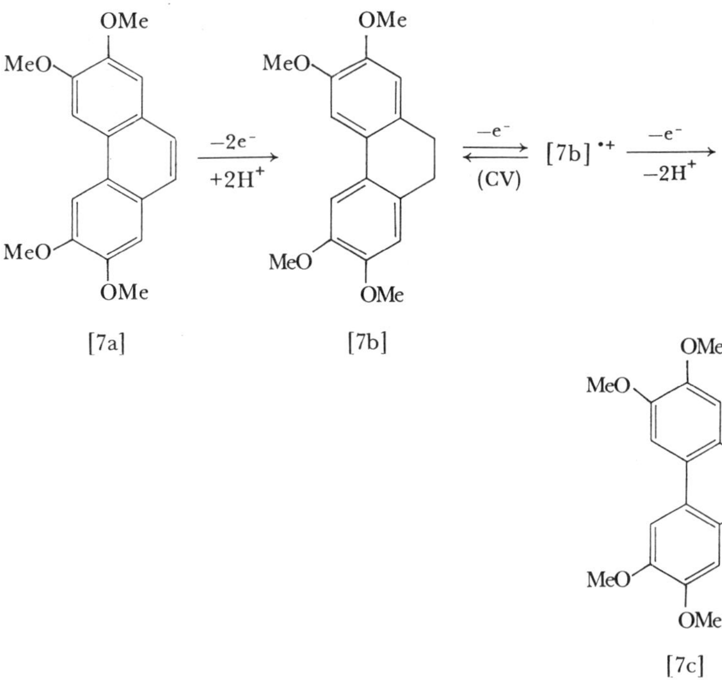

[Scheme showing conversion of [7a] → [7b] (−2e⁻, +2H⁺), [7b] ⇌ [7b]•⁺ (−e⁻, CV), [7b]•⁺ → [7c] (−e⁻, −2H⁺)] (86)

tetra-*p*-anisylethylene to the 3,6-dimethoxy-9,10-bis(*p*-anisyl)phenanthrene (Stuart and Ohnesorge, 1971). A more involved, intramolecular coupling route involves nucleophilic attack on the initially formed cation radical followed by further oxidation (an ECE-route) and intramolecular cyclization. Thus, the oxidation of 2,4,6-tri-t-butylaniline in the presence of pyridine (Cauquis and Cros, 1971; Cauquis *et al.*, 1971) leads to a benzimidazole by the following route:

(1) Oxidation, *via* the cation radical, in an ECE sequence (87) with pyridine acting as a base.

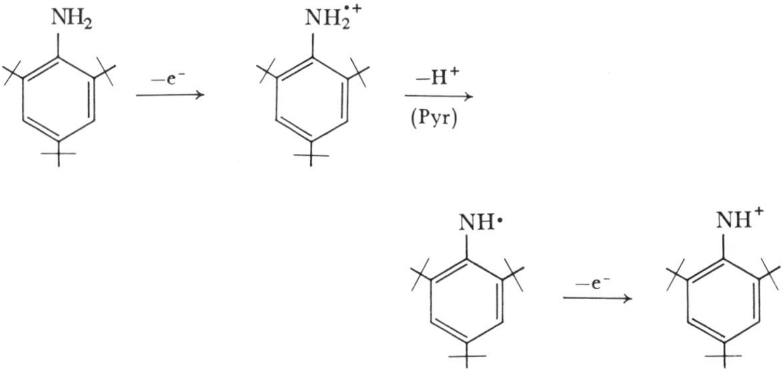

[Scheme: ArNH₂ —(−e⁻)→ ArNH₂•⁺ —(−H⁺, Pyr)→ ArNH• —(−e⁻)→ ArNH⁺] (87)

(2) Nucleophilic attack by pyridine, followed by cyclization, loss of t-butyl cation and further oxidation (88).

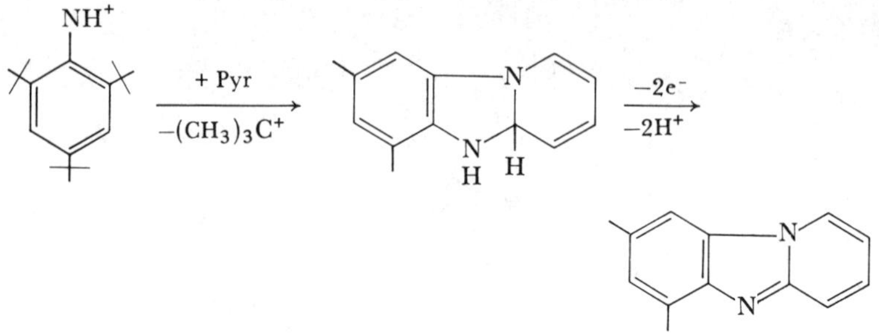

(88)

The change in the CV behavior of the aniline upon addition of pyridine clearly shows the intermediacy of the cation radical in the reaction sequence. A similar ECE sequence, *via* the cation radical, followed by intramolecular coupling occurs during the anodic oxidation of crowded 8-t-butyl-1-(2-pyridyl) naphthalenes (Popp, 1972).

3. PHYSICAL PROPERTIES OF CATION RADICALS

Cation radicals exhibit characteristic esr spectra. Many of the aromatic and heteroaromatic cation radicals are colored and therefore have characteristic electronic absorption spectra in the visible region. Detailed descriptions of these properties are outside the scope of this account. Changes in esr and absorption spectra are used to follow other physical and chemical changes of the cation radicals. Among the physical phenomena of interest to us are reversible dimerizations and disproportionation.

Dimerizations

Two types of reversible dimerization occur. Equation (89) shows the association of a cation radical with the parent molecule, while (90) shows the association of two molecules of cation radical. The former produces a paramagnetic dimer cation usually characterized

$$M^{\cdot +} + M \rightleftarrows M_2^{\cdot +} \tag{89}$$

$$2M^{\cdot +} \rightleftarrows M_2^{2+} \tag{90}$$

by an esr spectrum with twice as many lines and half the splitting constant of $M^{\cdot+}$, while the latter results in spin pairing and the disappearance of the esr signal.

Monocation dimers

The first definitive demonstration of $M_2^{\cdot+}$ formation was by Lewis and Singer (1965) in the oxidation of naphthalene by antimony pentachloride. The esr spectrum of $M_2^{\cdot+}$ was observed. Of historical interest is the fact that Brass and Tengler (1931) and Brass and Fanta (1936) oxidized perylene and other hydrocarbons (e.g. 1,2-benzanthracene) with $SbCl_5$ and isolated salts corresponding with the formula $2M,SbCl_5$. Allowing for possible error in the analyses, it may be that these workers had in fact isolated salts of the perylene cation dimer. The only other example of the isolation of a dimer salt we are aware of is also of (perylene)$_2^{\cdot+}$; the perchlorate was made anodically by Chiang and Reddoch (1970). Monocationic dimer formation by aromatic hydrocarbons in $SbCl_5$ solutions has also been observed for anthracene and pyrene as well as naphthalene by Howarth and Fraenkel (1970), while theoretical aspects of the dimerization (e.g. benzene, naphthalene, and 2,2'-paracyclophane) have been discussed by Badger and Brockelhurst (1970). Here it is argued that the distance separating $M^{\cdot+}$ and M, held in parallel planes, is controlled by the competing magnitudes of the energy of attraction and energy of π-electron repulsion. The former does not differ much between the benzene and naphthalene cases, while the latter increases with the number of filled π orbitals. Hence, separation of the planes is the greater in the naphthalene case.

The monocation dimer of coronene (Willigen et al., 1968) and pyrene (Cooper and Forbes, 1968) have been detected by esr spectroscopy in SO_3–BF_3 solution at low temperatures. Broad bands in the region of 500–600 nm in spectra of p-dihydroxy- and p-dialkoxybenzenes (e.g. hydroquinone, 1,4-dimethoxybenzene) in sulfuric acid at 77°K have been attributed to the presence of $M_2^{\cdot+}$ by Yamada and Kimura (1969), while Carter and Vincow (1967a) observed by esr (benzene)$_2^{\cdot+}$ in irradiated solutions of benzene in sulfuric acid at −103°. Using a flow-system, it was found by esr spectroscopy that the oxidation of mesitylene by cobalt(III) trifluoroacetate in TFA solution led to $M_2^{\cdot+}$ rather than $M^{\cdot+}$ signals (Shih and Dessau, 1971).

Monocation dimers have also been observed by esr in γ-irradiation of benzene (Edlund et al., 1967) and methylbenzenes (Nagai et al., 1971) adsorbed on silica surfaces at low temperatures. Irradiation of tetramethylethylene (TME) in 3-methylpentane glass with γ-rays gave $(TME)_2^{\cdot+}$, whose decay by internal proton transfer was monitored by esr spectroscopy (Ichikawa and Ohta, 1973). Pulse radiolysis (the use of high energy electron pulses) has also been used to make monocation dimers, for example, of perylene, naphthalene, and 9,10-dimethylanthracene, in benzonitrile solution at 17°. Even under these conditions, equilibrium constants (89) were quite high, being respectively 1·7, 5·2, and 29 x 10^2 M^{-1} (Kira et al., 1972).

Freed and Faulkner (1972) have deduced that the monocation dimer of N-methylphenothiazine (N-MP) affects the yield of triplet states in the chemiluminescent oxidation of fluoranthene anion radical (FA$^{\cdot-}$) by N-MP$^{\cdot+}$ [eqns (91) and (92)]. The triplet yields

$$FA^{\cdot-} + N\text{-}MP^{\cdot+} \rightarrow FA^{3*} + N\text{-}MP \qquad (91)$$

$$FA^{\cdot-} + N\text{-}MP^{\cdot+} \rightarrow FA + N\text{-}MP^{3*} \qquad (92)$$

were inversely proportional to the concentrations of N-MP. In contrast, when N-phenylphenothiazine cation radical (N-PP$^{\cdot+}$) was used, the triplet yields were independent of the concentration of N-PP. With the assumption that N-PP$^{\cdot+}$ and N-PP were sterically inhibited from association, the effect of N-MP was attributed to its causing a decrease in N-MP$^{\cdot+}$ concentration by formation of the dimer $(N\text{-}MP)_2^{\cdot+}$.

Examples of monocation dimer formation among aromatic hydrocarbons have been confined mostly to alternant hydrocarbons, and the dimer can be regarded as an association of two closed shell molecules which have lost an electron. Recently Paskovich and Reddoch (1972) made a new class of monocation dimers, in which an electron is missing from two associated open shell molecules. Oxidation of phenalene by oxygen led to the phenalenyl radical and, it is thought, to the phenalenyl cation, association of which gave the monocation dimer (93).

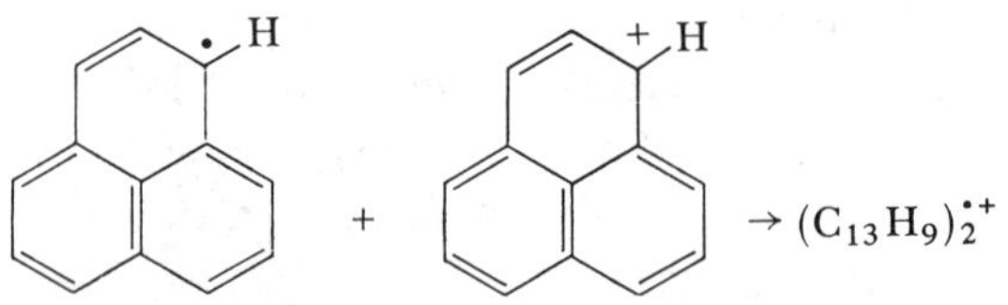

$$\text{[phenalenyl radical]} + \text{[phenalenyl cation]} \rightarrow (C_{13}H_9)_2^{\cdot+} \qquad (93)$$

Dication dimers

Formation of dicationic dimers (90) has been longer known than formation of the monocationic type (89), no doubt because of the availability of and interest in the very stable Wurster Salts. Hausser and Murrell (1957) proposed that the long wave-length absorption band (near 800 nm) of Wurster's Blue perchlorate in ethanol at $-90°$ was caused by two associated, cation radicals lying in parallel planes. Since that time a considerable number of workers have explored the dimerization of Wurster and analogous cation radicals, (e.g. Kawamori et al., 1966; Kimura et al., 1968). Not only does Wurster's Blue cation radical (i.e. TMPD$^{\cdot+}$) associate with itself, but it also forms a spin paired dimer with p-phenylenediamine cation radical (PD$^{\cdot+}$). In fact, Takimoto et al. (1968) conclude from absorption spectroscopy that solutions of TMPD$^{\cdot+}$ and PD$^{\cdot+}$ in ethanol-ether at $-195°$ contain (PD$^{\cdot+}$)$_2$ and (PD$^{\cdot+}$-TMPD$^{\cdot+}$)$_2$ but very little of (TMPD$^{\cdot+}$)$_2$. Dimerization of unlike cation radicals is known in other systems too. Perylene$^{\cdot+}$ and naphthacene$^{\cdot+}$ each forms an (M$^{\cdot+}$)$_2$ dimer in sulfuric acid at reduced temperatures (Kimura et al., 1971). Mixtures of the two cation radicals in sulfuric acid leads to a mixed dimer too, (Perylene$^{\cdot+}$, naphthacene$^{\cdot+}$), the heat of formation of which (-7.7 kcal mole^{-1}) incidentally, lies between that of the perylene$^{\cdot+}$ (-8.8) and naphthacene$^{\cdot+}$ (-5.6) dimers (Yamazaki and Kimura, 1972).

Dimerization of heterocyclic cation radicals was discovered in recent times. The well known thianthrene cation radical perchlorate forms a dimer in propionitrile solution. This association is, in fact, tetrameric (de Sorgo et al., 1972); that is, the anions are included in the aggregrate: (Th$^{\cdot+}$ClO$_4^-$)$_2$. Phenoxaselenine cation radical has also been found to form an (M$^{\cdot+}$)$_2$ dimer in sulfuric acid and nitromethane solutions (Cauquis and Maurey-Mey, 1973). It is not surprising that the very stable metalloporphyrin cation radicals should dimerize, but this was demonstrated only recently with zinc (Fuhrhop et al., 1972), and magnesium (Fajer et al., 1970) octaethylporphyrin (MOEP).

Dimerizations are highly solvent dependent. Thus, thianthrene cation radical does not dimerize as readily in acetonitrile as in propionitrile (de Sorgo et al., 1972). In acetonitrile at 20° a 10^{-5} M solution of ZnOEP$^{\cdot+}$ is 60% dimerized, while similar solutions in chloroform and methylene chloride are not dimeric at all. The cation radical of zinc hematoporphyrin was found to be 100% dimeric in water solution (Fuhrhop et al., 1972). Solvents with high dielectric constants reduce the repulsion between the two associated cations and are highly involved in solvation as illustrated by the large negative entropies or formation of these dimeric dications (Kimura et al., 1971; Fuhrhop et al., 1972).

Disproportionations

In disproportionation, two molecules of ion radical enter a redox reaction (94); the products are the neutral (parent) molecule and its

$$2M^{\cdot+} \rightleftharpoons M + M^{2+} \tag{94}$$

dication. The reaction is usually detected kinetically (that is its presence is deduced from the kinetics of reaction of a cation radical), spectroscopically, or by electrochemical techniques. Not many examples are known, except among a particular group of cation radicals, the violenes. These have been explored mostly in Hünig's laboratories (Hünig, 1967). Violenes are cation radicals derived from conjugated systems of the type [8], in which n may be zero and

$$X-(CH=CH)_n-X \quad X-(CH=CH)_n-\overset{+}{X}{}^{\cdot} \quad \overset{+}{X}(CH=CH)_n-\overset{+}{X}$$

[8]   [8$^{\cdot+}$]   [8^{2+}]

X may be NR_2, SR or OR. The name is derived from the violet color of the tetraphenylhydrazine cation radical [8$^{\cdot+}$; n = zero, $X = N(C_6H_5)_2$]. This, like all of the violenes is a markedly stable cation radical. In fact, the violenes are often made by oxidation of M by M^{2+}, and the equilibrium thus achieved is called comproportionation (95). Comproportionation constants (96) have been obtained

$$M + M^{2+} \rightleftharpoons 2M^{\cdot+} \tag{95}$$

for a number of compounds from the one- and two-electron oxidation potentials E_1 and E_2, which are obtained either polarographically or by cyclic voltammetry. The constants are quite large, indicating that the violenes predominate, and in fact do not enter

$$E_2 - E_1 = \frac{RT}{F} \log K_{\text{comp}} \tag{96}$$

much into disproportionation (94). The tetrathiofulvalene cation radicals mentioned earlier (p. 177) are among this class (see also

Pittman et al., 1972). Compounds [9]–[12] are a few examples. It should be noted that [11] is a bipyridylium ion which is often called

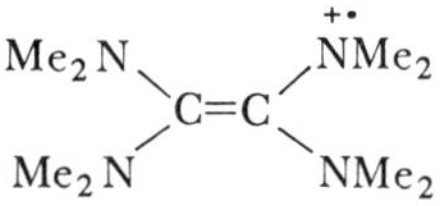

[9] $K_{comp} = 2 \times 10^2$ $(n = 1)$

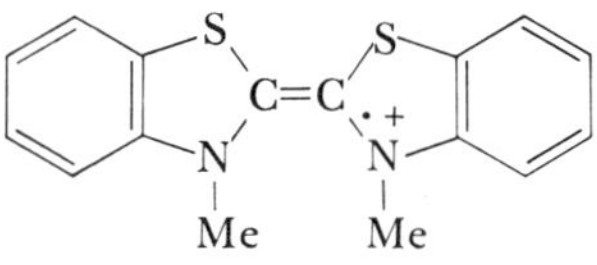

[10] $K_{comp} = 7 \times 10^4$ $(n = 2)$

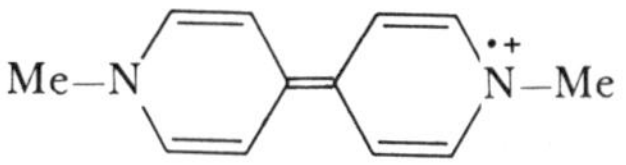

[11] $K_{comp} = 1 \times 10^6$ $(n = 3)$

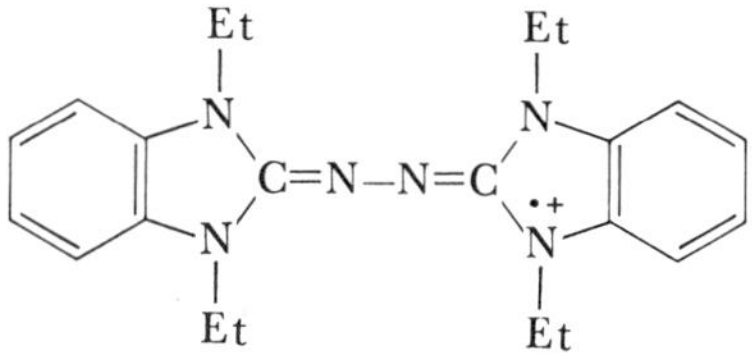

[12] $K_{comp} = 2 \times 10^9$

a "viologen", while [12] is in Hünig's terminology an "azaviolene". Comproportionation constants have been determined for violenes also by temperature jump ([13], Bernasconi et al., 1971), and by stopped-flow kinetics ([14], Bennion et al., 1972). In these cases it was possible to measure also the rate constants of the forward and

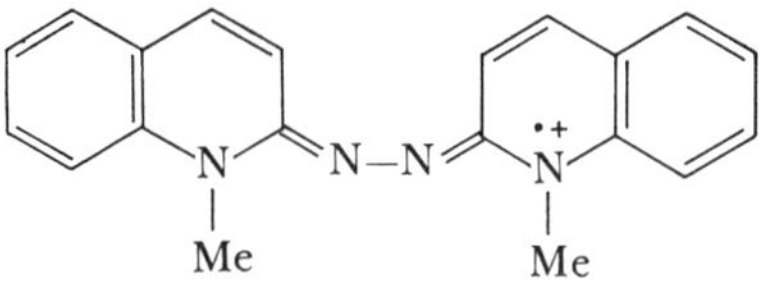

[13] $K_{comp} = 2 \cdot 14 \times 10^5$

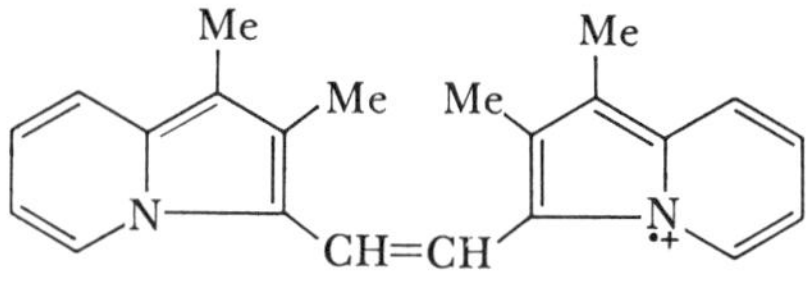

[14] $K_{comp} = 8 \times 10^2$

reverse reactions, and these for [13] were (in M^{-1} sec^{-1}) $1 \cdot 2 \times 10^9$ and $5 \cdot 7 \times 10^3$, while for [14] they were $6 \cdot 8 \times 10^8$ and $8 \cdot 1 \times 10^5$. Not all violene systems exist so far over on the cation radical side, however. Thus, K_{comp} for [15] is 16 in acetonitrile solution,

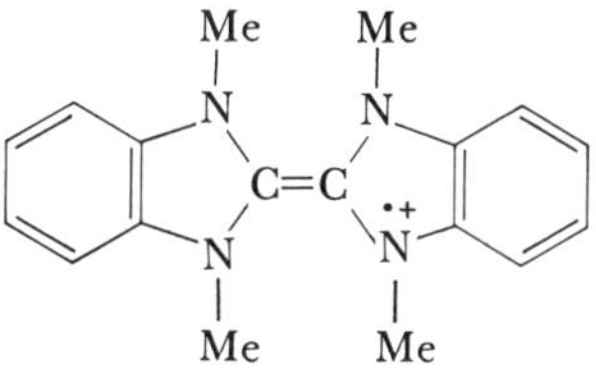

[15]

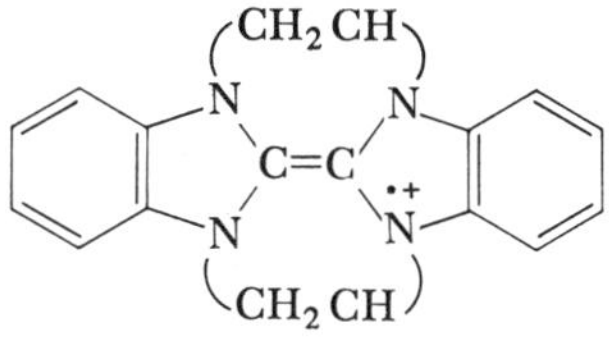

[16]

indicating that the equilibrium mixture contains 80% of the cation radical (see also, Bard and Phelps, 1970; Bard, 1971). In extraordinary contrast, K_{comp} for [16] could not be measured and was estimated to be about 10^{14} (Hünig et al., 1972, 1973).

Using optically transparent electrodes and a rapid scan spectrophotometer Gruver and Kuwana (1972) obtained $K_{comp} = 3 \times 10^{12}$ for the dimethyldihydrophenazine cation radical and 6×10^6 for the methylviologen cation radical [11].

Comproportionation has also been achieved with triphenylphosphonium methylide in benzonitrile (97) by Buck et al. (1972),

$$(C_6H_5)_2C = P(C_6H_5)_3 + (C_6H_5)_2\overset{+}{C}-\overset{+}{P}(C_6H_5)_3 \rightleftharpoons 2(C_6H_5)_2\overset{\cdot}{C}-\overset{+}{P}(C_6H_5)_3 \quad (97)$$

and with biacridyl in acidic ethanol (98) by Niizuma et al. (1972). In eqn (98), equilibrium is believed to lie much to the left; thus, disproportionation of the biacriden cation radical [17] is extensive.

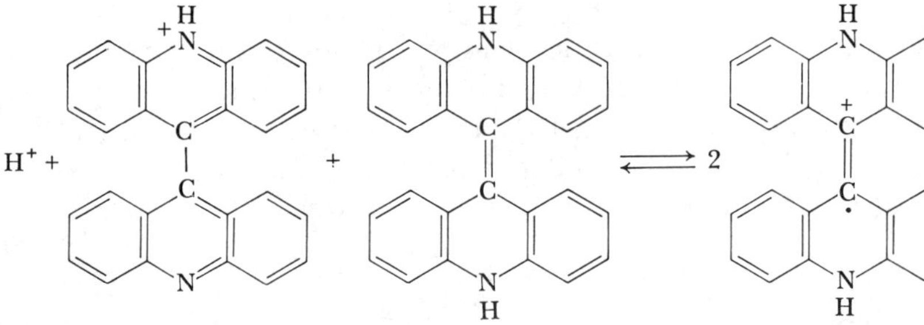

[Structures showing equation (98) with [17]]

Disproportionation is an important concept in cation radical reactions. Many reactions of cation radicals with nucleophiles have the stoichiometry of eqn (99) (or a variant of it). Without the aid of

$$2M^{\cdot +} + Nu^- \rightarrow M + M-Nu^+ \quad (99)$$

kinetics it is not possible to say what the mechanism of reaction is, and two important possibilities exist, namely, direct reaction with the cation radical [(100) and (101)] or reaction with the dication formed via disproportionation [(102) and (103)]. Each of these two

$$\begin{cases} M^{\cdot +} + Nu^- \rightarrow M-Nu^{\cdot} & (100) \\ M-Nu^{\cdot} + M^{\cdot +} \rightarrow M-Nu^+ + M & (101) \end{cases}$$

$$\begin{cases} 2M^{\cdot +} \rightleftharpoons M + M^{2+} & (102) \\ M^{2+} + Nu^- \rightarrow M-Nu^+ & (103) \end{cases}$$

sequences has been invoked in reactions of cation radicals with nucleophiles (such as water and pyridine) and a fair amount of controversy still remains. In particular it is difficult to distinguish electrochemically between the two sequences (which correspond with ECE and EEC processes respectively), and techniques of adapting electrochemical data to computer simulation of working curves for these processes are much in vogue (Feldberg, 1969; Cukman and Pravdic, 1974). More will be said of this in discussions of cation-radical chemistry.

The kinetic consequences of eqns (100) and (101) are that, making the reasonable assumption that (100) is rate-determining, the reaction of a nucleophile with a cation radical would be first order in nucleophile and cation radical. In contrast the consequences of eqns (102) and (103), assuming disproportionation to be a rapidly achieved equilibrium, are that reaction would be second order in cation radical, first order in nucleophile and inverse order in parent molecule. This mechanism was first proposed by Shine and Murata (1969) for the reaction of the thianthrene cation radical with water, and has since become a matter of considerable controversy (p. 229). Svanholm et al., (1974) have reported that the oxidative cyclization of tetraphenylethylene to diphenylphenanthrene is a clear-cut case of disproportionation.

Disproportionation constants, K_{disp}, have been measured from oxidation potentials in acetonitrile and other solvents for the thianthrene, 4,4'-dimethoxybiphenyl, and 9,10-di-p-anisylanthracene cation radicals (Hammerich and Parker, 1973). In acetonitrile they are, respectively, $2 \cdot 3 \times 10^{-9}$, $2 \cdot 7 \times 10^{-5}$ and $1 \cdot 9 \times 10^{-4}$, indicating that these cation radicals, like the violenes have a very small tendency toward disproportionation. K_{disp} for thianthrene cation radical (1×10^{-9}) and the tetrathioethylene cation radical (5×10^{-8}) have been measured in molten aluminum chloride-sodium chloride at 140° (Fung et al., 1973).

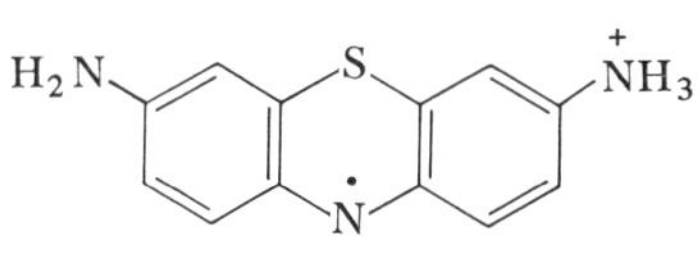

[18]

A recent study (Wildes et al., 1975) of the disproportionation of the radical cation semithionine ($TH_2^{+\cdot}$; [18]) is especially pertinent to the considerations noted above. The cation radical $TH_2^{+\cdot}$ undergoes disproportionation to thionine TH^+, and leucothionine TH_3^+ in acidic aqueous solutions (104). In 0·05 M aqueous sulfuric acid at 25° the

$$2TH_2^{\cdot +} \underset{}{\overset{k_d}{\rightleftharpoons}} TH_2^{2+} + TH_2$$

$$TH_2^{2+} \underset{}{\overset{-H^+}{\rightleftharpoons}} TH^+ \qquad (104)$$

$$TH_2 \underset{-H^+}{\overset{H^+}{\rightleftharpoons}} TH_3^+$$

value of k_d was $2·4 \times 10^9$ M^{-1} sec^{-1} and this value changes (by more than a factor of 10^3) for several aqueous organic solvents. Over the range investigated, log k_d showed a linear dependence on the solvent parameter Z indicating that the transition state for disproportionation of $TH_2^{\cdot +}$ has a substantial degree of electron transfer.

Ion-Pairing

The pairing of anion radicals with their counter cations is a wide-spread and now well documented phenomenon (Szwarc, 1972). In contrast, ion-pair phenomena in cation radical systems are not common, but see p. 222. Shifts in g-values caused by halide-ion interaction with the tetramethoxybenzene cation radical have been reported (Sullivan, 1973), and halide-ion splittings of metalloporphyrin cation radical esr spectra have been demonstrated (Fajer *et al.*, 1973).

4. ELECTRON TRANSFER REACTIONS OF CATION RADICALS

Among the principal reactions which cation radicals undergo are electron transfer, reaction with nucleophiles, reaction with aromatics and olefins, dimerization, and hydrogen abstraction (e.g. the Hoffmann–Loeffler reaction). Reactions in the gas phase are also being explored with increasing interest by mass spectrometry and ion-cyclotron resonance.

Most cation radicals are reduced by iodide ion. Since many compounds are oxidized to cation radicals by iodine, the two reactions are in equilibrium, and iodide-ion reduction occurs if an excess of iodide is used. This is frequently used for assaying the purity of isolated cation radical salts. Some cation radicals, such as that from perylene, are also reduced by chloride and bromide ion (Ristagno and Shine, 1971b).

Cation radicals are sometimes used to oxidize other molecules to their cation radicals by one-electron transfer (105). The success of

$$Ar_1H^{\cdot+} + Ar_2H \rightleftarrows Ar_1H + Ar_2H^{\cdot+} \tag{105}$$

the oxidation depends upon the relative oxidation potentials of Ar_1H and Ar_2H, and, if these are close, on being able to remove a salt of $Ar_2H^{\cdot+}$ from solution. Tris(p-bromophenyl)aminium hexachloroantimonate is used (see p. 194) and both thianthrene and dibenzodioxin cation radical perchlorates are also useful for one-electron oxidations, of zinc and magnesium tetraphenylporphyrins for example (Shine et al., 1975). Oxidation of magnesium tetraphenylporphyrin ($E_{1/2}$, 0·54 V) has also been achieved with the cation radical of zinc tetraphenylporphyrin ($E_{1/2}$, 0·71 V) by Fajer et al. (1973).

Equilibrium constants for a series of redox reactions (105) have been measured by Svanholm and Parker (1973), and are helpful in estimating how well one cation radical will serve in making another in the series.

Occasionally, electron transfer is a nuisance and competes with or overshadows a desired nucleophilic reaction. For example, reaction of thianthrene cation radical perchlorate with aniline leads to benzidine (among other products) instead of nucleophilic substitution in thianthrene (Shine and Kim, 1974).

Cation radicals may be oxidized further, and some of them become stable dications. This process is often characterized as the second wave in two successive anodic one-electron oxidations. The thianthrene dication is now well known (Shine and Piette, 1962; Hammerich and Parker, 1973), but has not been isolated. On the other hand, the diperchlorate of 2,3,7,8-tetramethoxythianthrene dication is a blue solid (Glass et al., 1973). Dications of some aromatics which are well-known for forming stable cation radicals (e.g. perylene) have been made in FSO_3H-SbF_5 solution (Brouwer and van Doorn, 1972). Irradiation of hexachlorobenzene cation radical in SbF_5-Cl_2 causes the formation of the dication (in this case, a triplet state) (Wasserman et al., 1974). Oxidation of metalloporphyrin cation radicals to stable (but reactive) dications is quite common (Dolphin et al., 1973).

Exchange Electron Transfer Reactions

The simplest electron transfer reactions of cation radicals are those involving electron transfer with parent molecules as shown in (106).

$$R^{\cdot+} + R \rightleftarrows R + R^{\cdot+} \qquad (106)$$

The rate constants for a number of cation radicals have been determined by magnetic resonance methods; the results are given in Table 2. Most rate constants have been determined by studying changes in the esr spectrum of the cation radical with increasing concentration of parent species. The first additions of parent causes a broadening in the individual hyperfine lines in the spectrum, and in this slow exchange region the rate constant may be obtained from eqn (107), where $\Delta\Delta H$ is the increase in peak-to-peak line width due

$$k = 1 \cdot 52 \times 10^7\, \Delta\Delta H/(1 - P_j)[R] \qquad (107)$$

to exchange (in gauss), [R] is the concentration of diamagnetic parent (in mole l^{-1}) and $(1 - P_j)$ is a correction which takes account of the hyperfine line used. With complicated spectra $\Delta\Delta H$ is difficult to determine directly, especially as hyperfine lines begin to merge with one another, and computer simulations are used to estimate the line broadening (e.g. Sorensen and Bruning, 1972; Hoefs and Bruning, 1972). At large concentrations of parent molecule the hyperfine components merge into a single line which narrows with increasing values of [R]. In this fast exchange region the rate constant can be determined using eqn (108) from a plot of line width

$$k = 2 \cdot 05 \times 10^7\, \nabla/\Delta H[R] \qquad (108)$$

(in gauss), ΔH, against $[R]^{-1}$. In eqn (108), ∇ is the second moment of the spectrum and is obtainable from the hyperfine coupling constants of the species. Corrections to eqn (108) for measurements made outside of the fast-exchange limit must frequently be applied, using methods described by Johnson and Holz (1969) (e.g. Kowert et al., 1972; Sorensen and Bruning, 1972).

These rate constants for exchange electron transfer are useful in predicting the rates of homogeneous electron transfer rates between different species using theoretical treatments proposed by Marcus (1968), (e.g. Shank and Dorfman, 1970) or the rates of heterogeneous electron transfer reactions involving cation radicals at electrodes. Differences in molecular geometry between parent and

Rate Constants For Homogeneous Electron Transfer Reactions

$$(R^{\cdot+} + R \rightleftharpoons R + R^{\cdot+})$$

Molecule	Solvent	Preparation of $R^{\cdot+}$ [b]	$k \times 10^{-9}$ (1 mole^{-1} sec^{-1})	Ref.[c]
Phenothiazine	CH_3CN	Elec. —0·1 M TEAP	6·7	1
	CH_3CN	Elec. —0·04 M TBAP	4·2	2
	CH_3CN	I_2	4·3	2
	5/1 $CHCl_3$–CH_3CN	I_2	1·8	2
10-Methylphenothiazine	CH_3CN	Elec. —0·1 M TEAP	2·2	1
Phenoxazine	CH_3CN	Elec. —0·1 M TEAP	4·5	1
Phenoxathiin	CH_3CN	Elec. —0·04 M TBAP	0·36	3
Dibenzo-p-dioxin	CH_3CN	Elec. —0·04 M TBAP	2·2	3
	3/1 $CHCl_3$–CH_3CN	Elec. —0·04 M TBAP	1·3	3
N,N,N',N'-tetramethyl-p-phenylenediamine	CH_3CN	Elec. —0·1 M TBAP	1·0	1
	CH_3CN	Air	0·20	6
	H_2O	Air	0·15	7
	H_2O	Air	0·21	7
N,N'-dimethyl-p-phenylenediamine	CH_3CN	Elec. —0·1 M TEAP	0·75	1
Tetramethylhydrazine	$CHCl_3$	I_2	~1	4
Tri-p-tolylamine	2/1 $CHCl_3$–CH_3CN	Elec. —0·04 M TBAP	0·66	2
	3/1 $CHCl_3$–CH_3CN	Elec. —0·04 M TBAP	0·55	2
	2/1 $CHCl_3$–CH_3CN	I_2	0·96	2
9,10-Diphenylanthracene	CH_2Cl_2	Elec. —0·04 M TBAP	0·34	5
	DCE	Elec. —0·04 M TBAP	0·37	5
	NE–DCE[a]	Elec. —0·04 M TBAP	0·40	5
	NE	Elec. —0·04 M TBAP	0·57	5

[a] DCE = dichloroethane. NE = nitroethane.
[b] Elec—electrolytic oxidation with supporting electrolytes of tetraethylammonium perchlorate (TEAP) or tetrabutylammonium perchlorate (TBAP). I_2-oxidation with iodine.
[c] 1. Kowert et al., 1972; 2. Sorensen and Bruning, 1973; 3. Sorensen and Bruning, 1972; 4. Romans et al., 1969; 5. Hoefs and Bruning, 1972; 6. Johnson, 1963; 7. Britt, 1964 (see also Bruce et al., 1956).

radical ion can affect the electron transfer rate. For example, Katz (1960) showed that electron transfer between the anion radical and dianion of cyclo-octatetraene (COT) is rapid, with a rate constant of about 10^9 l mole^{-1} s^{-1}, while that between parent and anion radical is much slower, the rate constant being 10^5 or less. These differences were attributed to the fact that the anion radical and dianion of COT are planar while the parent is nonplanar. The large rate constants observed for the heterocyclic molecules and their cation radicals, e.g. phenothiazine and phenoxazine, thus suggest very similar molecular conformations of parent and cation radical; this subject is discussed in more detail by Kowert et al. (1972). These authors also attempted to correlate the measured rate constants with molecular parameters, such as half-wave potentials, ionization potentials, HMO energies, and electron density distributions. While several correlations appear promising, there are insufficient data to draw any firm conclusions, especially when ion pairing effects may be present. Methods based on electrolytic generation of the cation radical are widely used and are usually carried out in the presence of a relatively high concentration of tetra-alkylammonium perchlorate. There is evidence that ion pair formation between the cation radical and perchlorate ion is possible, especially in solvents with low dielectric constants (Sorensen and Bruning, 1972). Thus electron transfer also involves rearrangement of the ion pair (i.e. transfer of a ClO_4^-) which tends to decrease the electron transfer rate compared to that of the free cation radical. This ion pair formation effect explains why the observed rate constants are lower in the lower dielectric constant CH_2Cl_2-CH_3CN mixtures than they are in CH_3CN alone.

Chemiluminescent and Electrogenerated Chemiluminescent Reactions of Cation Radicals

When the reaction between a cation radical ($D^{\cdot+}$) and an anion radical ($A^{\cdot-}$) involves the liberation of energy greater than that necessary to produce electronically excited states of A, D, or a molecular complex of these, the reaction (109) is often accompanied

$$A^{\cdot-} + D^{\cdot+} \rightarrow A^* + D \text{ or } A + D^* \tag{109}$$

by the emission of light. The excited states formed can be singlet or triplet, depending upon the energetics of the radical ion annihilation reaction and the rate of formation of the various states. These chemiluminescent (CL) reactions have been carried out by reaction of an anion radical (produced in an ethereal solvent by alkali metal ion reduction) with solid cation radical salts [typically the perchlorate salts of TMPD or tri-p-tolylamine (TPTA) cation radicals]. Alternatively, the ion radical species can be generated at an electrode immersed in a conducting solution prepared by adding A, D, and a non-electroactive salt (typically TBAP) to an aprotic solvent (e.g. CH_3CN, DMF, THF, CH_2Cl_2) with a wide available potential range and dielectric constant greater than ca 7. This mode of reaction, called electrogenerated chemiluminescence (ECL), occurs in the diffusion layer close to the electrode and allows a wide range of radical ions to be studied, even those that are not very stable, since they react in the rapid annihilation reaction a very short time after they are generated at the electrode. A number of reviews on ECL and CL have appeared (Kuwana, 1966; Bard et al., 1975; Zweig, 1968; Chandross, 1969; Hercules, 1969, 1971; Bard and Faulkner, 1975) so that only a brief outline of the basic concepts, and results will be presented here.

The energetics of the radical annihilation determine to a great extent the nature of the products formed. Generally, it is the enthalpy, ΔH°, for reaction (109) which is the useful parameter. This can be estimated from the electrochemical cyclic voltammetric peak potentials for the $A/A^{\cdot-}$ and $D^{\cdot+}/D$ couples $[E_p(A/A^{\cdot-})$ and $E_p(D^{\cdot+}/D)]$ by relation (110) (Weller and Zachariasse, 1967;

$$-\Delta H^\circ \text{ (in eV)} = E_p(D^{\cdot+}/D) - E_p(A/A^{\cdot-}) - 0 \cdot 16 \qquad (110)$$

Faulkner et al., 1972). Because the ion radical annihilation reaction is a very fast one, the Franck–Condon principle favors electronic excitation of a product molecule as a means of coping with the energy change of the reaction. Thus, if ΔH° is larger than the energy needed to form the excited singlet state of a product molecule, direct formation of $^1A^*$ or $^1D^*$ is possible. This occurs, for example, for the reaction of the radical cation and anion of 9,10-diphenylanthracene (DPA). When ΔH° is less than that required to produce the excited singlet states of A and D, but is large enough to produce a triplet state [e.g. that of A in (111)], then emission occurs only after a triplet-triplet annihilation step (112) can occur to form an

$$A^{\cdot-} + D^{\cdot+} \rightarrow {}^3A + D \qquad (111)$$

$$^3A + {}^3A \rightarrow A + {}^1A^* \qquad (112)$$

emitting excited singlet. An example of this sequence is the reaction of DPA anion radical and TMPD cation radical. Formation of the triplet state of the parent resulting from electron addition to the cation radical as shown in (113) and (114) is also possible in, for

$$A^{\cdot -} + D^{\cdot +} \to A + {}^3D \tag{113}$$

$$^3D + {}^3D \to D + {}^1D^* \tag{114}$$

example, the reaction of benzoquinone anion radical with rubrene cation radical. If $\Delta H°$ is less than that required to form either triplet state, it is still possible to form an emitting exciplex species directly in the annihilation reaction (115) (Weller and Zachariasse, 1971;

$$A^{\cdot -} + D^{\cdot +} \to (AD)^* \tag{115}$$

Zachariasse, 1975; Bard and Park, 1974). This occurs in the reaction of benzophenone anion radical with TPTA cation radical. Although this reaction is favored by low dielectric constant solvents, such as THF, it has also been observed in acetonitrile. Note that eqn (115) represents the reverse reaction of the formation of radical ions from dissociation of a photo-generated excited charge transfer complex. Finally, $\Delta H°$ may be so small that only ground state products are produced. When $\Delta G°$ for the annihilation reaction becomes positive, the reaction of A and D occurs spontaneously to form the radical ions. Representative CL and ECL reactions of radical ions, selected from the hundreds that have now been observed and investigated, are given in Table 3.

Some special advantages accrue from generating the radical ion reactants electrochemically. The electrochemical voltammetric investigations that precede the ECL experiment allow precise determination of peak potentials for formation of the ion radicals which are useful in estimating $\Delta H°$ for the reaction. Generation of the reactants electrochemically using closely controlled potentials prevents the production of undesired side products or the formation of more highly oxidized or reduced species. ECL can be generated either by a pulsing (transient) technique, alternately generating $A^{\cdot -}$ and $D^{\cdot +}$ at a single working electrode, or at a rotating ring-disk electrode (RRDE), where the reactant species are continuously generated at two closely spaced electrodes and mixed together by the hydrodynamic solution flow near the electrode surface thus producing steady-state emission. The theoretical treatments for ECL by both transient (Feldberg, 1966, 1966a; Bezman and Faulkner, 1972;

TABLE 3

Representative Chemiluminescent Radical Ion Redox Reactions

$$(A^{\cdot -} + D^{\cdot +} \rightarrow A^* + D \text{ or } D^* + A)^a$$

Cation radical $(D^{\cdot +})$	Anion radical $(A^{\cdot -})$	Solvent	$-\Delta H^\circ$ (eV)	Emitting species (energy, eV)	Mechanism[b] type	References[d]
DPA$^{\cdot +}$	DPA$^{\cdot -}$	DMF	3·08	^{1}DPA* (3·0)	S-route	1
TMPD$^{\cdot +}$	DPA$^{\cdot -}$	DMF	1·97	^{1}DPA* (3·0)	T-route	1
Rubrene$^{\cdot +}$	Rubrene$^{\cdot -}$	DMF	2·28	^{1}Rubrene* (2·3)	T-route	1, 2
Rubrene$^{\cdot +}$	Rubrene$^{\cdot -}$	THF	2·44	^{1}Rubrene* (2·3)	ST-route	2
Rubrene$^{\cdot +}$	BQ$^{\cdot -}$	DMF	1·33	^{1}Rubrene* (2·3)	T-route	1
TMPD$^{\cdot +}$	Rubrene$^{\cdot -}$	DMF	1·56	^{1}Rubrene* (2·3)	T-route	1
TH$^{\cdot +}$	PPD$^{\cdot -}$	CH$_3$CN	3·36	^{1}TH* (2·8) ^{1}PPD* (3·6)	S-route (TH) T-route (PPD)	3
10-MP$^{\cdot +}$	Fluoranthene$^{\cdot -}$	DMF	2·42	^{1}Fluoranthene* (3·0)	T-route	1, 4
TPP$^{\cdot +}$	TPP$^{\cdot -}$	DMF	2·92	^{1}TPP* (2·95)	S-route	1
DPBF$^{\cdot +}$	O$_2^{\cdot -}$	CH$_3$CN	2·2	O$_2^*$ (0·97)^c	S-route	5
TPhP$^{\cdot +}$	TPhP$^{\cdot -}$	CH$_2$Cl$_2$	2·15	^{1}TPhP* (1·90)	T-route (?)	6
TPTA$^{\cdot +}$	BP$^{\cdot -}$	THF	2·77	(BP$^-$TPTA$^+$)*	E-route	7, 8

[a] Abbreviations: DPA = 9,10-diphenylanthracene; TMPD = N,N,N',N'-tetraphenyl-p-phenylene diamine; BQ = p-benzoquinone; TH = thianthrene; PPD = 2,5-diphenyl-1,3,4-oxadiazole; 10-MP = 10-methylphenothiazine; DPBF = 1,3-diphenylisobenzofuran; TPP = 1,3,6,8-tetraphenylpyrene; TPhP = $\alpha,\beta,\gamma,\delta$-tetraphenylporphin; TPTA = tri-p-tolylamine; BP = benzophenone.
Solvents: DMF, N,N-dimethylformamide; THF, tetrahydrofuran.
[b] S-route: direct formation of emitting singlet state on radical ion reaction; T-route: formation of triplet on radical ion reaction, followed by triplet-triplet annihilation to form emitting species; ST-route: formation of emitting state by both S- and T-routes; E-route: direct formation of emitting exciplex on radical ion reaction.
[c] Emission not observed; the $^1\Delta_g$ state of O$_2$ is detected by trapping with DPBF.
[d] 1. See reviews, also Faulkner et al., 1972; 2. Tachikawa and Bard, 1974; 3. Keszthelyi et al., 1972; 4. Freed and Faulkner, 1971; 5. Mayeda and Bard, 1973; 6. Tokel et al., 1972; 7. Bard and Park, 1974; 8. Zachariasse, 1975.

Cruser and Bard, 1969) and RRDE (Maloy et al., 1971) methods have been presented, so that the emission intensity and electrolysis current can be used to obtain information about the efficiency and pathway of the process. Finally, the electrochemical cells make small scale experiments possible and are especially amenable to vacuum line and dry box techniques.

A number of techniques have been applied to probe the details of the ECL processes. Triplet intermediates resulting from the redox reaction have been identified by interception with *trans*-stilbene and determination of the extent of *trans*- to *cis*-isomerization (Freed and Faulkner, 1971) as well as by observing the effect of an external magnetic field on the emission intensity (Faulkner et al., 1972; Tachikawa and Bard, 1974). In the latter studies, it was shown that a magnetic field decreases the rate of the triplet-triplet annihilation reaction, (112) or (114), and also the rate of quenching of triplets by radical ions [eqn (116)]. Since the rate of the radical ion redox

$$^3A + D^{\cdot +} \rightarrow A + D^{\cdot +} \tag{116}$$

reaction is apparently unaffected, modulation of the ECL emission intensity by an external magnetic field provides evidence of triplet intermediates in the reaction mechanism. In connection with these magnetic field effect studies, cation radicals have been shown to be very efficient quenchers of excited states (Faulkner and Bard, 1969; Tachikawa and Bard, 1974a). The efficiency of production of excited states and photons in the ECL process has been determined in a number of investigations (Bard et al., 1973), and overall efficiencies (photons emitted per ion radical annihilation event) as high as 10–20% have been reported. Indeed, Freed and Faulkner (1972) reported near unit efficiency in the production of triplets in the reactions of fluoranthene anion radical and 10-phenylphenothiazine cation radical.

5. REACTIONS OF CATION RADICALS WITH NUCLEOPHILES

Cation radicals react with a variety of nucleophiles. Substitution often occurs, but there are many examples in which electron transfer occurs either entirely or in part. Eberson (1975) has provided a theoretical explanation for a number of reactions, based on the Dewar–Zimmerman rules.

Reaction with Water

This plays a very large role in cation radical chemistry. Because it is almost impossible to dry solvents completely, there is the likelihood of reaction with water in every anodic oxidation, and therefore particular concern with the water reaction appears in electrochemical literature.

Some tertiary amines become protonated at the anode during electrolysis; although current is used up, the overall reaction appears as if the amine were not oxidized. This is not the case, however, and the protonation has been represented by Russell (1963) as resulting from reduction of the amine cation radical by water in the solvent (117). In other cases, such as N-(p-dimethylaminophenyl)-

$$Et_3N^{\cdot +} + H_2O \rightarrow Et_3\overset{+}{N}H + \dot{O}H \qquad (117)$$

tetraphenylpyrrole (Cauquis and Genies, 1967), N,N-dimethyl-p-toluidine (Barbey et al., 1971), and tetramethylbenzidine (Delahaye et al., 1971), the water is represented as being oxidized to oxygen (118) although oxygen, in fact, has not been observed (Caullet,

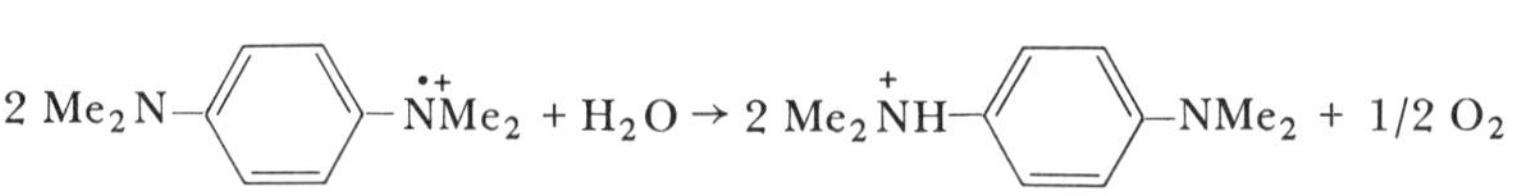

$$2\,Me_2N{-}\langle\bigcirc\rangle{-}\overset{\cdot +}{N}Me_2 + H_2O \rightarrow 2\,Me_2\overset{+}{N}H{-}\langle\bigcirc\rangle{-}NMe_2 + 1/2\,O_2 \qquad (118)$$

1973). The mechanism of this regeneration reaction, then, is not known.

The more common reaction with water is hydroxylation. Oxidation of 7,12-dimethylbenz[a]anthracene [19] by one-electron

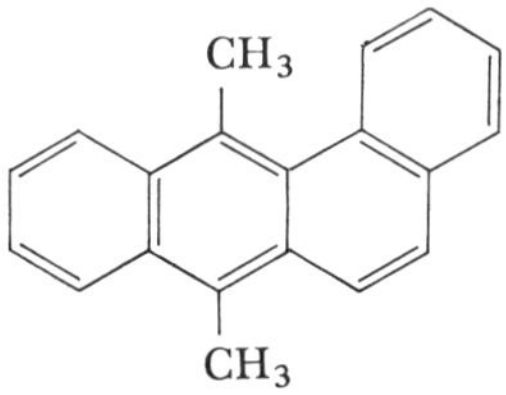

[19]

oxidants such as ceric ammonium nitrate leads among other products to 7-formyl-12-methyl- and 7-methyl-12-formyl-benzanthracene. Formation of these products is regarded as possibly beginning with attack of water at the 7- and 12-positions of the cation radical. Because these products are highly carcinogenic, it is also thought that the carcinogenicity of the hydrocarbon may be connected with the *in vivo* formation of its cation radical and reaction of the cation radical with water (Fried and Schumm, 1967).

Benzo[a]pyrene is hydroxylated to 6-hydroxybenzopyrene during anodic oxidation in acetonitrile and other solvents (Jeftic and Adams, 1970). Reaction of 9,10-diphenylanthracene cation radical [20] with water gives equal amounts of 9,10-dihydroxy-9,10-diphenylanthracene [23] and 9,10-diphenylanthracene. The reaction is first-order in cation radical and follows the sequence of eqns (119)-(121) (Sioda, 1968; Blount and Kuwana, 1970b). This

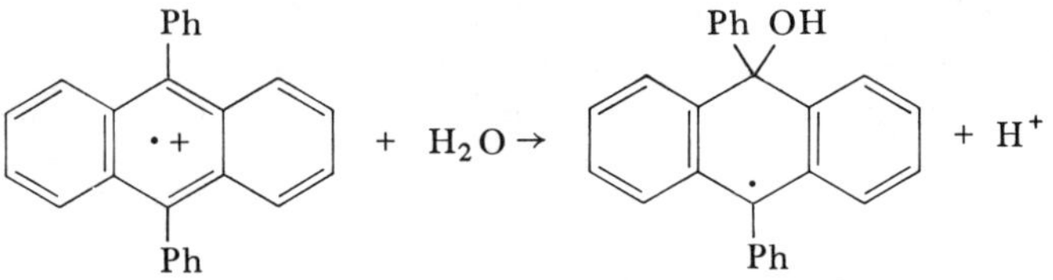

$$[20] + H_2O \rightarrow [21] + H^+ \quad (119)$$

$$[21] + [20] \rightarrow + [22] \quad (120)$$

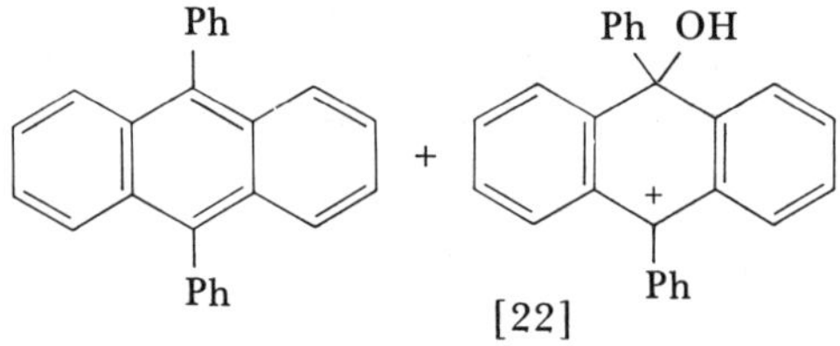

$$[22] + H_2O \rightarrow + H^+ \quad (121)$$

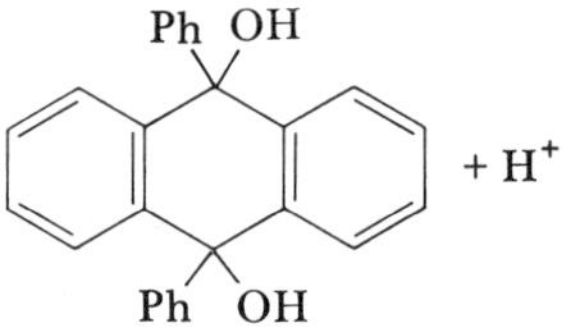

sequence [i.e. formation of [20] and eqns (119) and (120)] is analogous to an electrochemical ECE process, which is how the hydroxylation of benzo[a]pyrene is also viewed (Jeftic and Adams, 1970). An ECE process is also regarded as the route of hydroxylation of anthracene, 9,10-dihalogenoanthracenes, and 9-phenylanthracene (Parker, 1970, but see also Hammerich and Parker, 1974). Among these compounds, and in the case of benzo[a]pyrene, further oxidations occur at the anode, leading to quinones. Ristagno and Shine (1971a, b) found that perylene cation radical reacted with water to give perylene-3,10-quinone, in a sequence of reactions whose stoichiometry is summarized in (122). Perylene cation radical

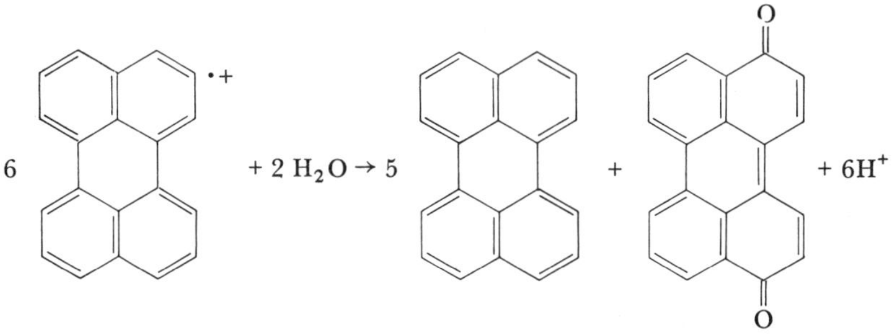

6 + 2 H$_2$O → 5 + + 6H$^+$ (122)

was one of the earliest to be characterized by esr spectroscopy. At that time (Aalbersberg et al., 1959a) reaction of a sulfuric acid solution of the cation radical with water gave only perylene, and it is now apparent that the quinone must have been overlooked in the sulfuric acid solution. Hydroxylation of dibenzodioxin cation radical gives the 2,3-quinone analogously (Cauquis and Maurey-Mey, 1972; Shine and Shade, 1974), and the French workers designate ECE sequences as the probable route, in preference to one that involves disproportionation of the cation radical.

The question of disproportionation is quite controversial in reactions of cation radicals with water. Murata and Shine (1969) found that the reaction of thianthrene cation radical (Th$^{\cdot +}$) with water gave quantitatively equal amounts of thianthrene (Th) and thianthrene 5-oxide (ThO) in a reaction that was second order in cation radical and inverse first order in thianthrene. They proposed therefore that the cation radical disproportionated in solution (123) and that the dication (Th^{2+}) reacted with water (124). The proposal has been

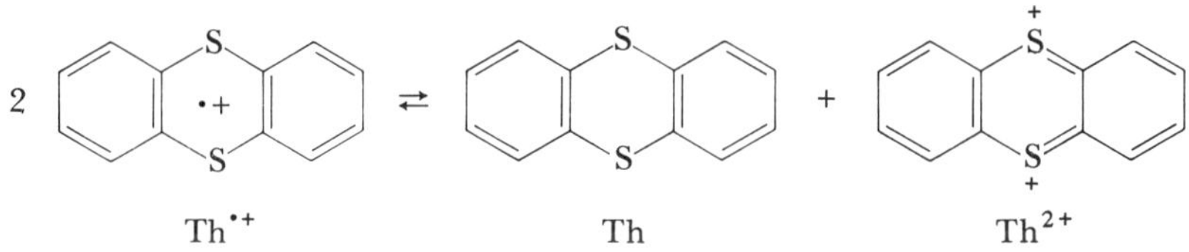

2 ⇌ + (123) 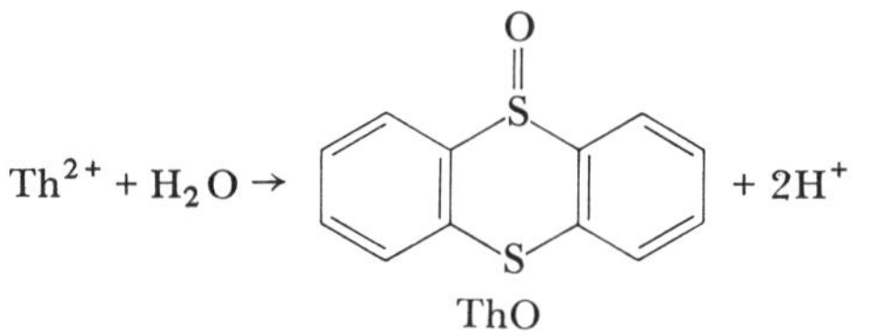
Th$^{\cdot +}$ Th Th^{2+}

Th^{2+} + H$_2$O → + 2H$^+$ (124)
ThO

criticized by Parker and co-workers on the grounds that the very small equilibrium constant for reaction (123) (approx. 10^{-9} M) would require an unacceptably large rate constant for (124) to fit the observed overall rate of reaction (Svanholm et al., 1975). At various times, therefore, it has been proposed that an ECE-type process prevails in which the cation radical is the kinetically active species (Parker and Eberson, 1970), or that the cation radical dimer reacts

$$2Th^{\cdot +} \rightleftarrows (Th^{\cdot +})_2 \qquad (125)$$

$$(Th^{\cdot +})_2 + H_2O \rightarrow Th + ThO + 2H^+ \qquad (126)$$

with water [eqns (125) and (126)] (Hammerich and Parker, 1972). A further alternative is that the reactive species is a water solvated complex of the radical ion, since the reaction is third order in water. A final solution to this hydroxylation reaction is apparently not yet at hand.

Hydroxylation involving disproportionation has also been proposed as the initial step in the anodic oxidation of 2,4,6-tri-t-butyl-aniline [24] to the mono-iminoquinone [25] (Cauquis et al., 1968), in the anodic oxidation of tetraphenylpyrrole to the hydroxy derivative [26] (Libert et al., 1973), and in the anodic oxidation of phenoxaselenine (Cauquis and Maurey-Mey, 1973). In contrast with the last result, the anodic oxidation of phenoxathiin to phenoxathiin 5-oxide is thought to be an ECE process in which the cation radical reacts with water (Barry et al., 1966). The reasoning behind this choice is that since Se is a larger atom and softer acid than S, reaction of the cation radical with water will be slow and disproportionation will control the reaction.

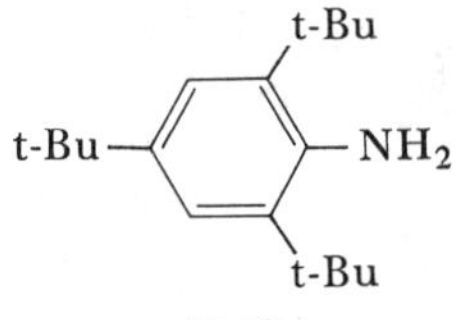

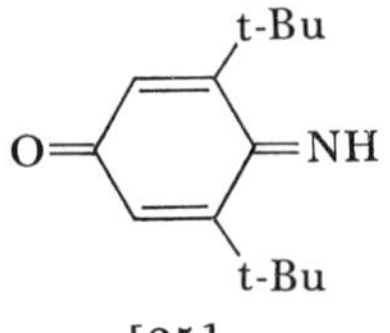

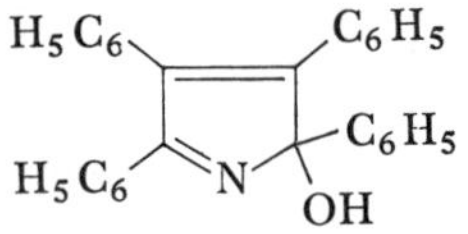

Reaction of water with organosulfur cation radicals and dications also plays a role in the exchange of sulfoxide oxygen with the oxygen atom of water in acid solutions (Oae, 1970; Oae et al., 1974).

Reaction with Alcohols

The most frequently encountered reactions are those in anodic oxidations in methanol solution (Fioshin et al., 1973). Reaction with solvent, called methoxylation, occurs and is believed to involve the cation radical of the substrate in either direct or indirect ways. A very common electrolyte solution is potassium hydroxide in methanol. N-Methylpyrrole gives a good yield of 2,2,5,5-tetramethoxy-N-methylpyrrole, and 2,6-dimethoxypyridine undergoes further methoxylation (Weinberg and Brown, 1966). N,N-Dimethylaniline is converted into N-methoxymethyl-N-methylaniline (Weinberg and Reddy, 1968), while ethyl vinyl ether is converted into 1,2-dimethoxy-1-ethoxyethane (Belleau and Au-Young, 1969). 9,10-Disubstituted anthracenes are dimethoxylated in good yields in methanol–sodium methoxide (127) (Parker et al., 1971; R = Et, Pr,

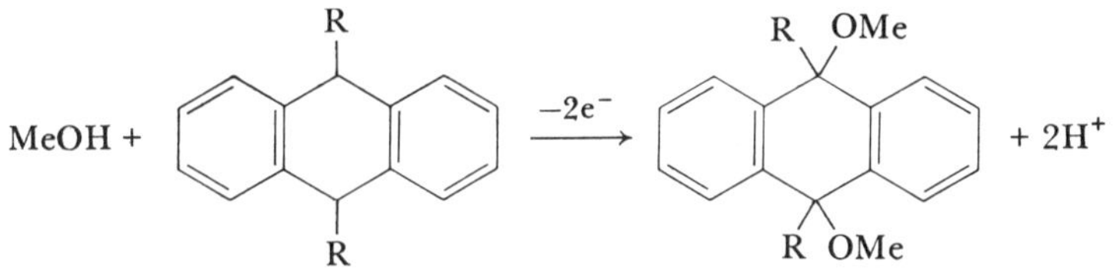

$$\text{MeOH} + \underset{R}{\overset{R}{\text{anthracene}}} \xrightarrow{-2e^-} \underset{R\ \text{OMe}}{\overset{R\ \text{OMe}}{\text{product}}} + 2H^+ \quad (127)$$

C_6H_5) and in methanol–potassium hydroxide (Weinberg and Belleau, 1973; R = OMe). Many similar examples can be found in the electrochemical literature. How cation radicals participate in methoxylations has not been defined precisely, but reasonable proposals are made in the literature.

It has been proposed (e.g. Weinberg and Reddy, 1968) that the cation radical is adsorbed on the anode surface, in which state it might react directly with methanol or undergo further oxidation to a dication in a step prior to or concerted with attack by methanol. Methoxylations of anthracenes (Parker et al., 1971) are regarded as ECE processes.

Tri-p-tolylaminium perchlorate is reduced by propanol and butanol in reactions which are first order in aminium ion and second order in alcohol (Belozerov et al., 1972), and it is proposed that a complex of cation radical and solvent breaks down as shown in (128). The alkoxy radical is presumed to end up as the corresponding aldehyde.

$$ClO_4^-\,(RO\cdots\overset{\bullet+}{Ar_3N}\cdots OR) \rightarrow Ar_3N + RO\cdot + ROH + HClO_4 \quad (128)$$
$$||$$
$$HH$$

Thianthrene cation radical perchlorate reacts slowly with alcohols. Reaction with the lower alkanols has not been defined but with

(—)-menthol and 4-t-butylcyclohexanol leads (129) to alkoxy-sulfonium perchlorates [27] (Shine et al., 1975b).

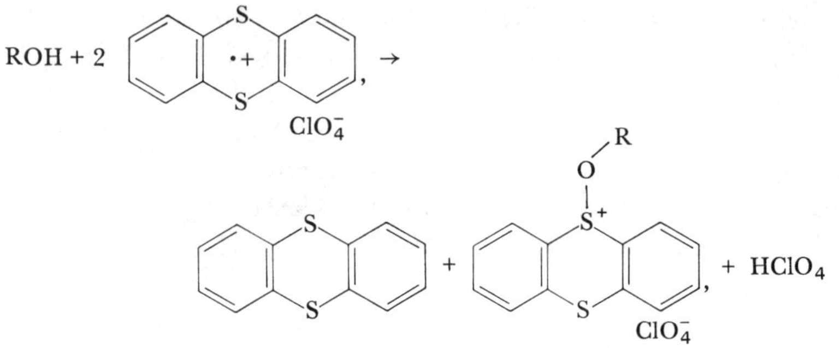

$$ROH + 2 \; \text{[thianthrene]}^{\cdot +} \; ClO_4^- \rightarrow \text{[thianthrene]} + \text{[thianthrene-OR]}^+ ClO_4^- + HClO_4 \quad (129)$$

[27]

Reaction with Cyanide Ion

This also appears frequently in the electrochemical literature as a process called cyanation. Anodic oxidation of aromatics and heterocyclic compounds in solutions containing cyanide ion gives nitriles. At one time, because of the low oxidation potential of cyanide ion itself (0·96 V vs SCE), it was thought that the cyanation of aromatics was a free radical reaction, involving the cyano radical formed at the anode (Koyama et al., 1965). This has since been shown not to be the case, in that the distribution of isomers from the cyanation of a monosubstituted aromatic was found to be quite unlike that of free-radical substitution (Susuki et al., 1968; Andreades and Zahnow, 1969). The most detailed comparisons have been made by Nilsson (1973). Aromatic nitriles were made anodically, and by the photolysis of ICN and the diazotization of cyanamide in the presence of the aromatics. The first method gave distributions of isomers quite different from the last two. A comparison of some of the results is given in Table 4. Photolysis of ICN and diazotization of cyanamide lead to reactions of the cyano radical. It is evident that anodic cyanation does not, and this is regarded as an ECE reaction involving the aromatic cation radical.

Anodic cyanation of diphenylamine gave 61% of 4-cyanodiphenylamine (Yoshida and Fueno, 1972) while similar cyanation of diphenylacetylene gave 60% of 4-cyanodiphenylacetylene (Yoshida and Fueno, 1973) in reactions also believed to involve the cation radical. Use of 2,5-dimethylthiophene (Yoshida et al., 1971) in

TABLE 4

Formation of Isomeric Aromatic Nitriles from Anodic Cyanation of C_6H_5X and from Photolysis of ICN in C_6H_5X (Nilsson, 1973)

X in C_6H_5X	Isomer distribution (%)[a]		
	o	m	p
OMe	53 (58)	0·1 (14)	47 (28)
Cl	50 (27)	0·5 (27)	50 (46)
C_6H_5	24 (44)	0·4 (28)	76 (28)
Me	40 (47)	8 (31)	52 (28)
NO_2	– (22)	– (63)	– (14)

[a] Numbers in parentheses are from ICN photolysis.

methanol solutions of cyanide ion produced cis- and trans-methoxycyano derivatives [28], which are also thought to be formed in typical ECE sequences.

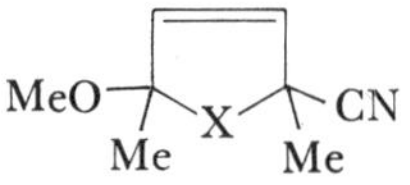

[28] X=S, O

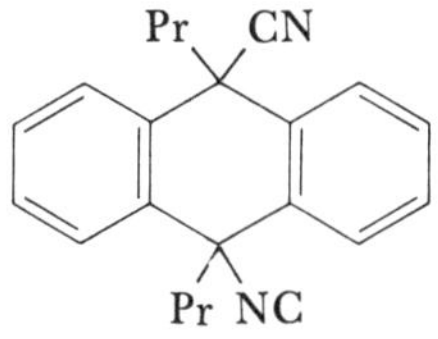

[29]

Because the cyanide ion is so easily oxidized its apparent ability to react with aromatic cation radicals instead of being oxidized by them reflects the competition so often encountered in cation radical chemistry between nucleophilicity and oxidizability of a nucleophile. The subject has not been treated analytically yet. In the present context, the tri-p-anisylaminium ion is reduced by cyanide ion (Papouchado et al., 1969) in a very fast overall second-order reaction (Blount et al., 1970). The cation radicals of thianthrene, phenothiazine, and phenoxathiin are also reduced by cyanide ion (Shine et al., 1974). In none of these cases, incidentally, is the fate known of the cyano radical presumed to be formed. Perylene cation radical perchlorate, on the other hand, reacts with cyanide ion in acetonitrile solution to give low (13%) yields of both 1- and 3-cyanoperylene (Shine and Ristagno, 1972).

Cyanide is a ambident ion. The only reported example of isonitrile formation in cation radical chemistry is in the anodic formation of [29] from 9,10-di-isopropylanthracene (Parker and Eberson, 1972).

Reaction with Halide Ions

To our knowledge no reaction of iodide ion as a nucleophile with a cation radical is known. Iodide ion reduces cation radicals very well and is frequently used for the iodimetric assay of cation-radical salts. Since the reduction is reversible and some compounds can be oxidized to the cation radical stage by iodine, an excess of iodide is used. Some cation radicals are also reduced by other halide ions; for example, that of 9,10-diphenylanthracene is reduced by bromide ion (Sioda, 1968), that of perylene by bromide and chloride ions (Ristagno and Shine, 1971b), and thianthrene radical cation to some extent by chloride ion (Murata and Shine, 1969). These reductions, particularly those by iodide ion, reflect again the competition between nucleophilicity and oxidizability of a nucleophile in reactions with cation radicals.

Fluoride ion reactions are somewhat puzzling. Oxidation of fluoride ion by a cation radical is out of the question and nucleophilic attack would appear to be certain. However, this reaction has failed completely with several cation radical perchlorates, such as those from perylene (Ristagno and Shine, 1971b), and phenothiazine (Shine *et al.*, 1972). It appears that fluoride ion is too weak a nucleophile to participate in such substitution reactions. On the other hand, several cases of anodic fluorination are known; oxidation of some aromatics, in solutions containing fluoride ion and at potentials lower than the oxidation potential of fluoride ion, has led to fluorination. This has occurred with naphthalene, which gave 4–5% of 1-fluoronaphthalene (Knunyants *et al.*, 1970), chloro-, bromo-, and fluorobenzene (Rozhkov *et al.*, 1971), trifluoromethylbenzene, which gave 60% of *m*-fluorination (Knunyants *et al.*, 1971), benzene, and some *p*-disubstituted benzenes (Rozhkov and Alyev, 1975). In the Russian work the electrolyte was Et_3NHF, or (mostly) $Et_4NF \cdot 2HF$ in acetonitrile. By using $Me_4NH \cdot 2HF$ in acetonitrile, it is possible to obtain 43% of 9,10-difluoro-9,10-diphenylanthracene from the aromatic (Ludman *et al.*, 1972). Each of these reactions is thought to involve the aromatic cation radical, but why cation radicals produced at an anode should react with fluoride ion while homogeneous reactions have failed is not known. Possibly, the electrolyte has particularly good nucleophilic characteristics. This idea has been entertained for the tetrafluoroborate ion (Koch *et al.*, 1973) because of the formation of *p*-toluyl fluoride in 65% yield from the anodic oxidation of diphenylmethyl *p*-tolyl ketone.

Apparently, the *p*-toluyl cation (not cation radical) is formed by oxidative scission of the ketone and is "trapped" by fluoride donation from the electrolyte $Me_4\overset{+}{N}\overset{-}{B}F_4$. In our context it is interesting that, in demonstrating the reversible one- and two-electron oxidation stages of thianthrene, Hammerich and Parker (1973) used the same electrolyte. Thus the BF_4^- ion is apparently not nucleophilic towards either the thianthrene cation radical or dication within the times of the cyclic voltammetry.

In support of an earlier proposal, Burdon and Parsons (1975) have used CNDO/2 and INDO calculations to show that the products of fluorination of aromatics by metal fluorides (AgF_2, CoF_3, MnF_3) can be correlated with the participation of aromatic cation-radical intermediates.

Fluorination of aromatics by XeF_2 in the presence of HF is thought, contrary to earlier belief, not to involve the aromatic cation radical. However, a cation radical *is* thought to accept fluoride ion in reactions of XeF_2 with easily oxidized aromatics (anisole) in the absence of HF (Anand et al., 1975). In this connection zinc tetraphenylporphyrin has been oxidized to its cation radical with xenon difluoride (Forman et al., 1971); although fluorine hyperfine splitting was found in the esr spectrum, reaction with fluoride ion did not occur.

Reaction of cation radicals with chloride and bromide ions is not too well defined in the literature. Phenothiazine cation radical perchlorate in acetonitrile solution containing halide ion gave 70–80% yields of 3-chloro- and 3-bromophenothiazine according to (130). Small amounts (4%) of 3,7-dihalogenophenothiazine were also

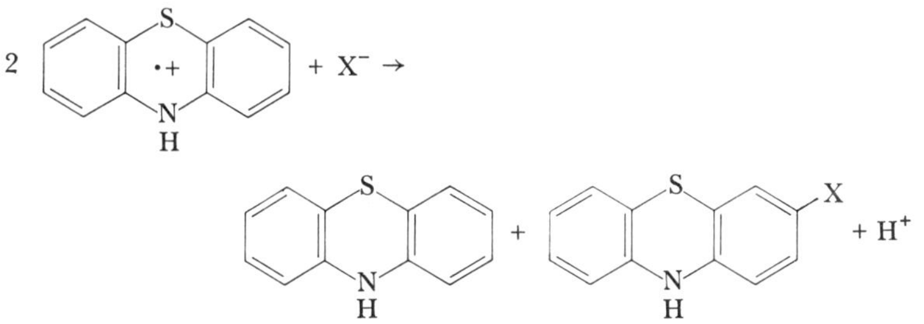

$$2 \; [\text{phenothiazine}]^{\cdot+} + X^- \rightarrow [\text{phenothiazine}] + [\text{3-X-phenothiazine}] + H^+ \quad (130)$$

formed, so that the reaction is not entirely clear (Shine et al., 1972). Alcais and Rau (1969) found that reaction of bromide with phenothiazine, and with 2-chloro- and 2-methoxyphenothiazine leads first, very rapidly, to the cation radical and next, by reaction with bromide ion, to bromination products.

Millington (1969) oxidized anthracene anodically in acetonitrile containing Et_4NBr, and, on the basis of the effect of the applied potential on yields of bromination products, attributed the formation of 9-bromo- and 9,10-dibromoanthracene to reaction of bromide ion with either the cation radical or dication of anthracene. The conclusion must be treated with caution, since the oxidation potential of bromide ion is much lower than that of anthracene.

Copper(II) chloride and bromide react with easily oxidized aromatics in refluxing benzene or chlorobenzene to give high yields of aryl halide. For example, 2-methoxy- and 2-ethoxynaphthalene gave >98% yields of 1-halogeno-2-alkoxynaphthalene; anisole gave 75% of p-chloro and 4% of o-chloroanisole (no m-chloroanisole was formed). Reaction is attributed to transfer of halogen atom to the cation radicals [eqn (130a)] (Bansal et al., 1973).

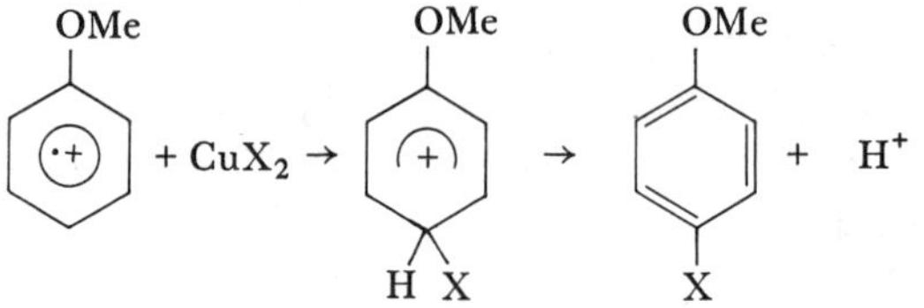

$$\text{[OMe-C}_6\text{H}_5]^{\cdot+} + CuX_2 \rightarrow \text{[OMe-C}_6\text{H}_5\text{(H)(X)}]^+ \rightarrow \text{OMe-C}_6\text{H}_4\text{-X} + H^+ \quad (130a)$$

Ledwith and Russell (1974b) have found that chlorination of benzene, toluene and other aromatic molecules is easily achieved in aqueous acetonitrile containing sodium peroxydisulfate and copper(II) chloride. Toluene, for example, gives no benzyl chloride but a mixture of chlorotoluenes (58% o-, 4% m-, and 38% p-) consistent with the spin distribution in the toluene cation radical. The amount of copper chloride used can be catalytic provided another source of chloride ion (LiCl) is added. Reaction is attributed to the very fast transfer of chlorine atom from copper(II) chloride to the cation radical (132); the metal halide is thus regarded as a trap for the aromatic cation radical. In the absence of copper(II) chloride, reactions of toluene with peroxydisulfate ion and chloride ion give

$$C_6H_5CH_3 + SO_4^{\cdot-} \rightarrow C_6H_5CH_3^{\cdot+} + SO_4^{2-} \quad (131)$$

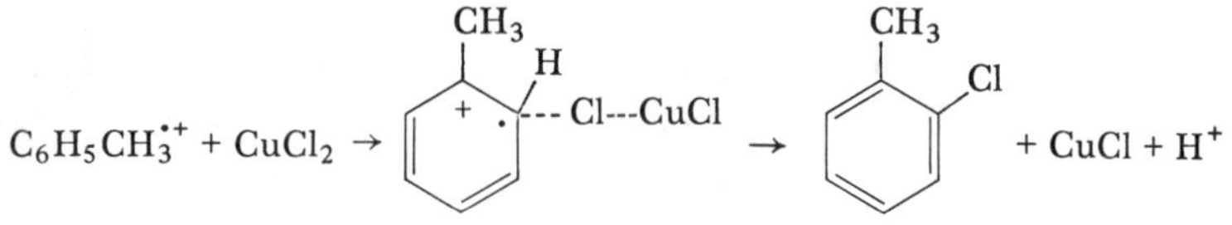

$$C_6H_5CH_3^{\cdot+} + CuCl_2 \rightarrow \text{[o-CH}_3\text{-C}_6\text{H}_5\text{(H)}\cdots\text{Cl}\cdots\text{CuCl]} \rightarrow o\text{-CH}_3\text{-C}_6\text{H}_4\text{-Cl} + CuCl + H^+ \quad (132)$$

$$CuCl + S_2O_8^{2-} + Cl^- \rightarrow CuCl_2 + SO_4^{2-} + SO_4^{\cdot-} \quad (133)$$

high yields of benzyl chloride, presumably by oxidative substitution reactions (Kochi, 1975) of intermediate benzyl radicals (134). It is

$$(PhCH_3)^{\cdot +} \rightarrow PhCH_2\cdot + H^+ \xrightarrow[Cl^-]{S_2O_8^{2-}} PhCH_2Cl + SO_4^{2-} + SO_4^{\cdot -} \quad (134)$$

worth noting that rather similar effects (i.e. side chain vs nuclear substitution) have been recently reported for reactions of aromatics with Fenton's Reagent (Walling and Johnson, 1975), and peroxydisulfate (Walling and Camaioni, 1975).

However, when the aromatic molecule is substituted in such a way as to stabilize intermediate cation radicals (relative to the corresponding benzyl radicals formed by deprotonation), direct nuclear chlorination may result without the intervention of copper(II) chloride. Thus, whereas toluene and p-xylene give no nuclear chlorinated products when allowed to react with peroxydisulfate and chloride ions in aqueous acetonitrile, m-xylene gives 10% ring chlorinated products and mesitylene gives 75% ring chlorination. Likewise it was found that aromatics forming even more stable cation radicals (anisole, phenol, naphthalene) gave high yields of nuclear chlorinated products when allowed to react with $S_2O_8^{2-}/Cl^-$ (135) (Ledwith and Russell, 1974c, 1975). Isomer distributions for

$$ArH + Cl^-/S_2O_8^{2-} \rightarrow ArCl + HSO_4^- + SO_4^{2-} \quad (135)$$

nuclear chlorinations by $S_2O_8^{2-}/Cl^-$ were identical to those obtained for comparable reactions catalysed by copper(II) chloride.

Exactly how the stabilized aromatic cation radical is converted into the nuclear chlorinated product, is not at present fully understood. As represented in eqn (135), nucleophilic substitution could arise from initial capture of the aromatic cation radical by chloride ion involving appropriate substituted cyclohexadienyl-type radicals ($\cdot$ArHCl), in which case the substitution pattern (at least the *ortho/para* ratio of products) might be expected to resemble more those from typical homolytic aromatic substitution processes rather than those from electrophilic substitutions, as observed experimentally. At present, there is a scarcity of significant mechanistic information relating to nucleophilic capture of aromatic cation radicals, although in every reported case (*vide infra*) the position of substitution corresponds with that arising from comparable electrophilic processes.

An alternative mechanism which cannot be eliminated on the basis of existing evidence and which has the attraction of yielding intermediate cyclohexadienyl derivatives similar to those produced by electrophilic chlorination, involves the production of chlorine anion radical (Cl_2^-) as the actual chlorinating species by reactions (136) and (137). Although direct oxidation of Cl^- by $S_2O_8^=$ is known

$$ArH^{\cdot+} + Cl^- \rightarrow ArH + Cl\cdot \qquad (136)$$

$$Cl\cdot + Cl^- \rightleftharpoons Cl_2^{\cdot-} \qquad (137)$$

to be very slow (presumably for kinetic reasons), electron transfer reaction between aromatic cation radicals and chloride ion would be favored by electrostatic forces, and it is worth noting that Standard Electrode Potentials (NHE) for the conversion $Cl_2 + 2e = 2Cl^-$ are 1.35 V (H_2O), 1·116 V (MeOH) and 0·58 V (CH_3CN) (Parsons, 1959). It follows therefore that the oxidation of Cl^- to $Cl\cdot$ becomes very much more likely in acetonitrile and (presumably) aqueous acetonitrile solvents. Certainly, oxidation of Cl^- by the aromatic cation radicals suggested as intermediates (136) cannot be ruled out on purely thermodynamic grounds. Such a process has, in fact, already been noted for halide ion reactions with perylene cation radical (*vide supra*). The equilibrium constant for formation of $Cl_2^{\cdot-}$ in reaction (137) is thought to be high ($> 10^5$ M^{-1}) and, with the relatively high chloride ion concentrations used in the work described, this would ensure that any $Cl\cdot$—generated by whatever mechanism—would exist as $Cl_2^{\cdot-}$.

It is important to note that atom transfer from $Cl_2^{\cdot-}$ to $ArH^{\cdot+}$ would obviously occur at the sites of highest spin density as proposed for the comparable atom transfer from $CuCl_2$. If properly substantiated, this would call into question the precise nature of chlorinations in general, especially in respect of the attendant concentrations of chloride ion and so-called halogen carriers such as Fe(III). For further discussion of the substitution patterns to be expected when cation radical intermediates are involved, attention is drawn to the stimulating review by Walling (1975).

Reaction with Amines

Pyridination

A number of aromatic hydrocarbons react with pyridine when oxidized either with iodine or anodically. N-Arylpyridinium ions are formed and the reaction is called pyridination. The reactive species is believed to be the aromatic cation radical. Benzo[a]pyrene has been widely studied because of its carcinogenicity and the belief in some circles that cation radicals may have a role in carcinogenic activity. Oxidation with iodine in the presence of pyridine gives the N-(6-benzo[a]pyrenyl)pyridinium ion [30] (Wilk et al., 1966;

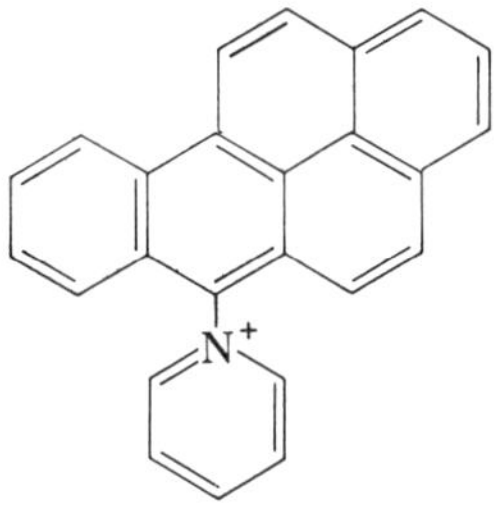

Rochlitz, 1967; Johnson and Calvin, 1973). Iodine and pyridine form a complex which is itself an iodinating agent, and this alone with benzo[a]pyrene gives high yields of 6-iodobenzo[a]pyrene. Pyridination occurs when an excess of pyridine is used. Anodic oxidation of benzo[a]pyrene in the presence of pyridine gave [30] in good yield (as the perchlorate); 4-picoline reacted similarly. On the other hand, with N-methylimidazole the product was 1-(1-benzo[a]pyrenyl)-3-methylimidazolium perchlorate [31]. The difference is thought by Blackburn and Will (1974) to have a possible connection with the carcinogenicity of benzo[a]pyrene in that, when its cation radical is generated in the vicinity of DNA, linkage to pyrimidine at the C-6 position could occur, whereas a different regiospecificity near a purine base might cause bonding to the imidazole ring and at the 1-position of the hydrocarbon.

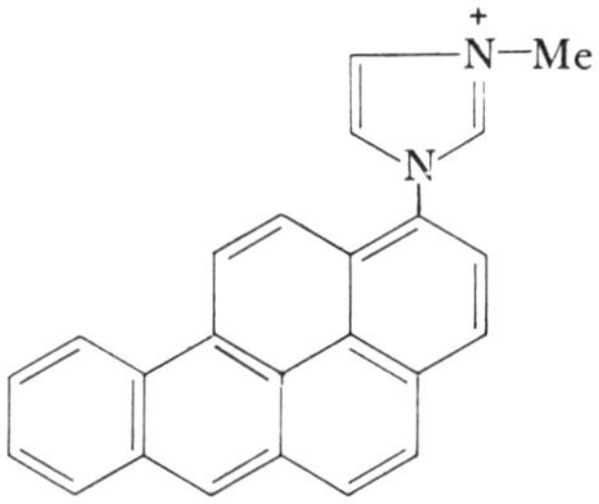

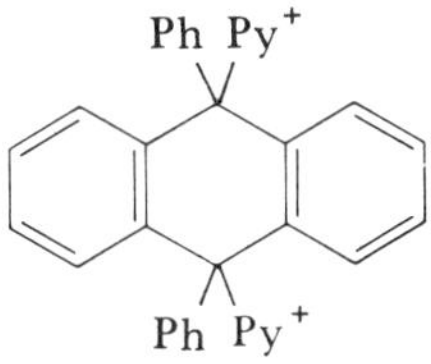

9,10-Diphenylanthracene (DPA) has also been an important substrate in pyridination reactions. Anodic reaction leads to a dipyridinium ion [32]. Marcoux (1971) found from the use of working curves that although the pyridination data could be described by both ECE and disproportionation processes, the data fitted disproportionation better. In contrast, Blount (1973), with transparent electrodes, and Svanholm and Parker (1973), with rotating disk electrodes, find that pyridination of DPA is an ECE reaction, following the pattern of Sioda's (1968) hydroxylation reaction. Anodic oxidation of DPA in the presence of 2,5-, 2,6-, and 3,5-

dimethylpyridine also gives the 9,10-dipyridinium products, whereas oxidation of 9,10-dimethylanthracene in similar circumstances leads to the hydrocarbon dimer [33]. Here, apparently, the lutidines

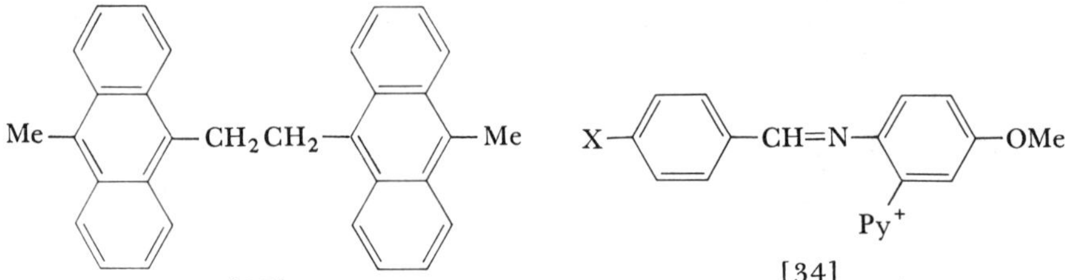

[33]

[34]

remove a proton from a methyl group of the hydrocarbon cation radical and the ensuing benzylic radical dimerizes (Parker and Eberson, 1970). Anodic oxidation of N-benzylidene-*p*-anisidines in the presence of pyridine gave *o*-substituted pyridinium perchlorates [34; X = H, Cl, OMe] in what are also believed to be ECE reactions (Masui and Ohmori, 1973).

Cation radicals of thianthrene (Shine *et al.*, 1972), phenoxathiin, N-methyl- and N-phenylphenothiazine (Shine *et al.*, 1975) react with pyridine as in (138), the pyridinium perchlorates being isolable in

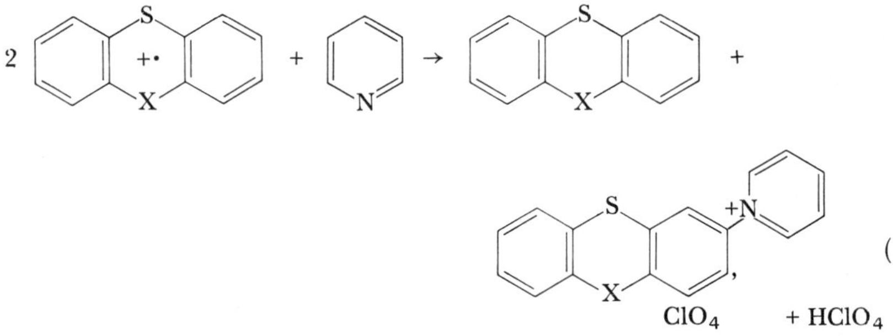

(138)

X = S, O, N-Me, N-C$_6$H$_5$

good yield. Reaction of the pyridine at ring carbon contrasts with reaction of alkylamines with these cation radicals as is discussed below. Whether or not the cation radical or the dication is the reactive intermediate in pyridination is not known as yet.

A rather special case of pyridination occurs during oxidation of pyridine in aqueous solution by peroxydisulfate ion (Ledwith and Russell, 1974a). Very high yields of isomeric pyridylpyridinium salts are obtained from reactions in neutral or acidic solutions whereas moderate yields of isomeric (neutral) bipyridyls are produced in alkaline solutions. Pyridine cation radical is thought to be the important reaction intermediate and, for acid or neutral solutions, the products arise by a chain process, as indicated in eqns (139)–(141).

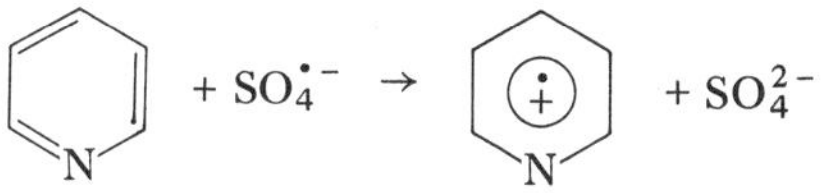

$$\text{C}_5\text{H}_5\text{N} + SO_4^{\cdot-} \rightarrow [\text{C}_5\text{H}_5\text{N}]^{\cdot+} + SO_4^{2-} \quad (139)$$

$$[\text{C}_5\text{H}_5\text{N}]^{\cdot+} + \text{C}_5\text{H}_5\text{N} \rightarrow [\text{pyridinyl-pyridinium}]^{\cdot+} \quad (140)$$

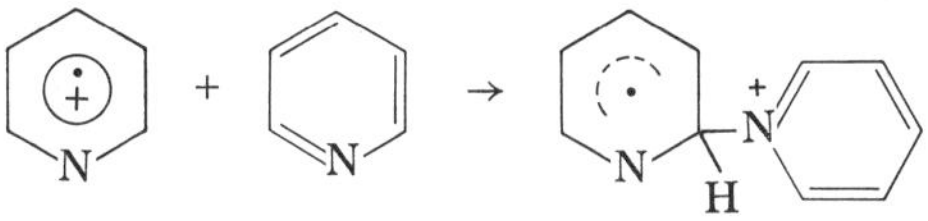

$$[\text{pyridinyl-pyridinium}]^{\cdot+} + S_2O_8^{2-} \rightarrow [\text{pyridyl-pyridinium}]^{+} + SO_4^{\cdot-} + HSO_4^{-} \quad (141)$$

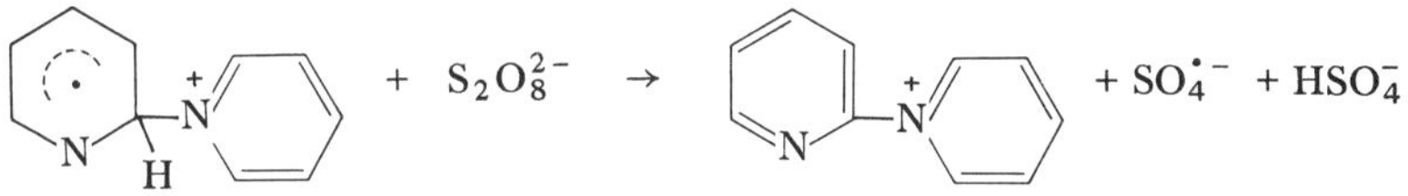

Formation of neutral bipyridyls requires overall abstraction of a hydrogen atom from each pyridine molecule and is much more difficult to explain. Possibilities include the deprotonation of pyridine cation radical to give, for example, the 2-pyridyl radical, or addition-elimination sequences of radicals such as $^{-}SO_3-N(C_5H_4)-H$. Pyridine cation radical was identified as a reaction intermediate by spin trapping with phenyl-N-t-butyl nitrone (142). The very stable

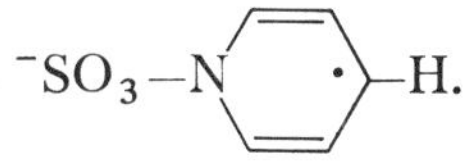

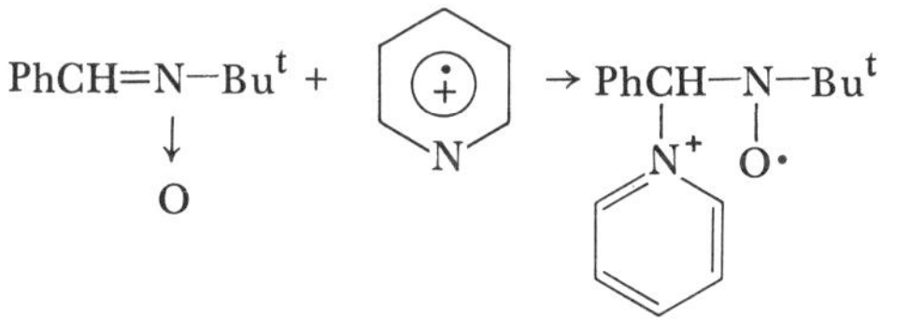

$$\text{PhCH=N-Bu}^t + [\text{C}_5\text{H}_5\text{N}]^{\cdot+} \rightarrow \text{PhCH-N-Bu}^t \quad (142)$$
$$\downarrow\qquad\qquad\qquad\qquad\qquad\qquad |\ \ \ |$$
$$\text{O}\qquad\qquad\qquad\qquad\qquad\qquad \text{N}^+\ \text{O}\cdot$$

(cationic) nitroxide radical produced gave an esr signal consisting of a 1:1:1 triplet of 1:2:2:1 quartets, analysed in terms of a nitroxide nitrogen triplet (a_N 1·37 mT) and an identical coupling of 0·29 mT for both the β-hydrogen and the pyridine nitrogen nuclei. It is interesting that this appears to be the only reported instance of cation radical characterization by spin trapping.

Alkylamines

Little is to be found in the literature on the reaction of cation radicals with alkylamines. Shang and Blount (1974) found that DPA$^{\cdot+}$ reacts with triethylamine in an ECE sequence to give the 9,10-diethylammonium ion [36], analogous to [32]. In fact, they find also that reaction of [32] with triethylamine in solution gives [36] in a two-step replacement sequence. Whether or not this means that in solution [32] dissociates into pyridine and a cation [35] which reacts with triethylamine (143) is not known.

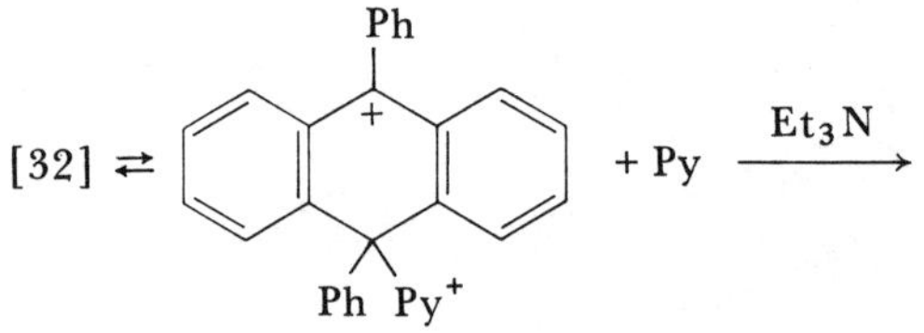

[32] ⇌ [structure with Ph, Ph, Py$^+$] + Py $\xrightarrow{Et_3N}$

[35]

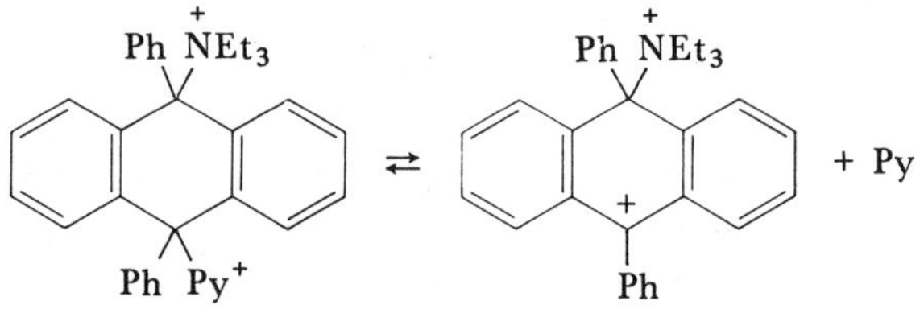

[structure with Ph NEt$_3^+$, Ph Py$^+$] ⇌ [structure with Ph NEt$_3^+$, Ph] + Py   (143)

[36]

Simple alkylamines react with heterocyclic sulfur-containing cation radicals to form sulfilimines [38] in reaction (144). Until

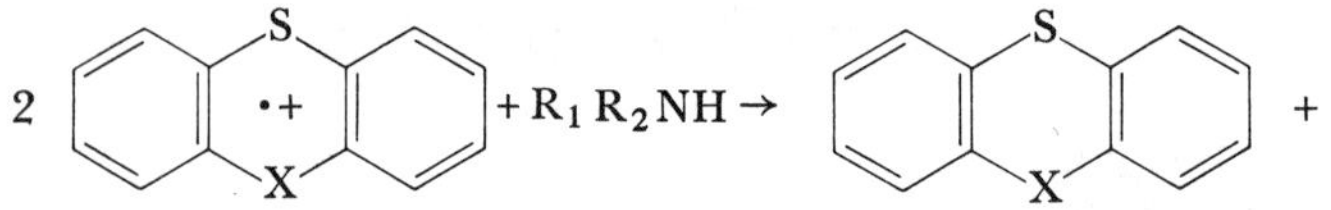

2 [structure of thianthrene-type cation radical with X] + R$_1$R$_2$NH → [structure with X] +

[37]

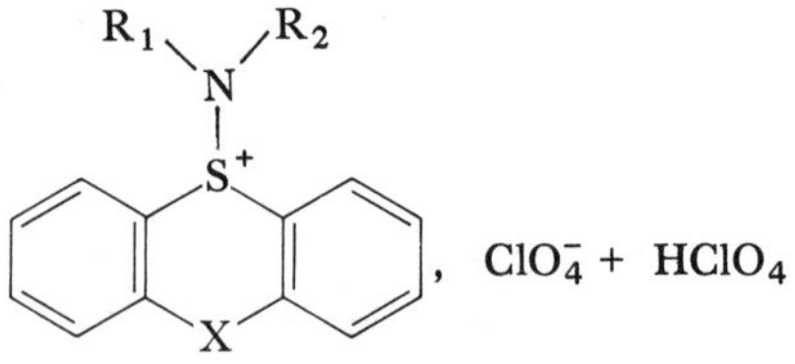

[structure [38] with R$_1$, R$_2$, N, S$^+$, X], ClO$_4^-$ + HClO$_4$ (144)

[38]

X = S, O, N-Me, N-C$_6$H$_5$; R$_1$ = H; R$_2$ = t-Bu
X = S, O, N-Me, N-C$_6$H$_5$; R$_1$ = Me; R$_2$ = Me
X = O, N-Me, N-C$_6$H$_5$; R$_1$ = H; R$_2$ = Me, Et, n-Pr, C$_6$H$_{11}$

recently it was thought that thianthrene radical cation [37; X = S], unlike its analogues [37; X = O, N-Me, N-Ph], gave [39] rather than [38] when treated with most primary amines (Kim and Shine, 1974). Unpublished work in the same laboratory has since shown that both

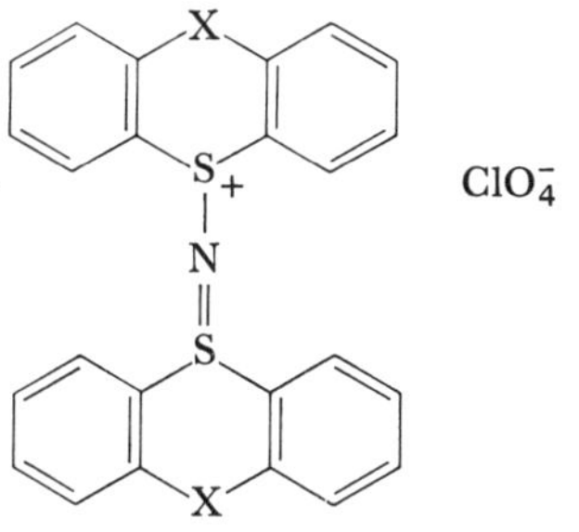

[39]

[38; R_1 = H, R_2 = alkyl] and [39] are in fact obtained, the latter in the smaller yields. How [39] is formed is not yet known. Product [39; X = N-Me, N-C_6H_5, O] is obtained when the appropriate cation radical reacts with ammonia (Bandlish et al., 1975; Mani and Shine, 1975).

The reactions of cation radicals described briefly above must be distinguished from amination reactions, in which amine cation radicals react with aromatics and olefins. These are described later.

Aromatic Amines

Aromatic amines have such low oxidation potentials that they are ordinarily oxidized when treated with a cation radical. The amine cation radical is then most likely to dimerize. The dimerization reaction may itself be looked upon as either the reaction of the amine cation radical with another molecule of amine or the dimerization of two cation radicals, as is illustrated with N,N-dimethylaniline in (145) and (146). The electrochemical literature is replete

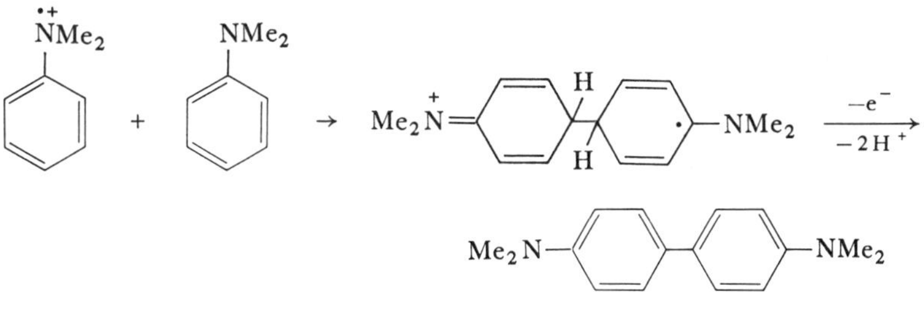

(145)

with examples of dimerizations in anodic oxidations of amines, while dimerizations in chemical and enzymatic (e.g. horseradish peroxidase) oxidations are also extensively described. These reactions are outside the scope of this chapter, however.

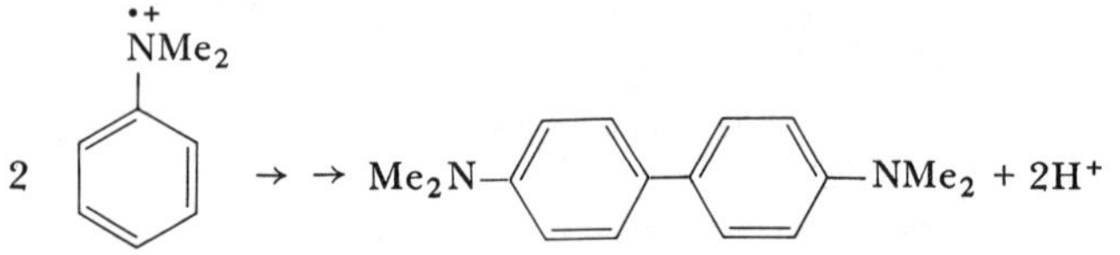

$$2 \; C_6H_5\overset{\cdot+}{NMe_2} \longrightarrow \longrightarrow Me_2N\text{-}C_6H_4\text{-}C_6H_4\text{-}NMe_2 + 2H^+ \quad (146)$$

It has been found (Kim et al., 1974) that the thianthrene cation radical reacts with acetanilide to give in good yield [40], which

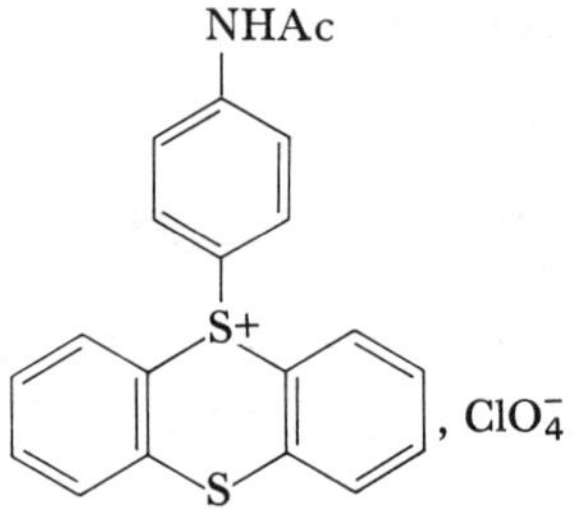

[40] (thianthrene-S⁺ with p-NHAc-phenyl substituent, ClO₄⁻)

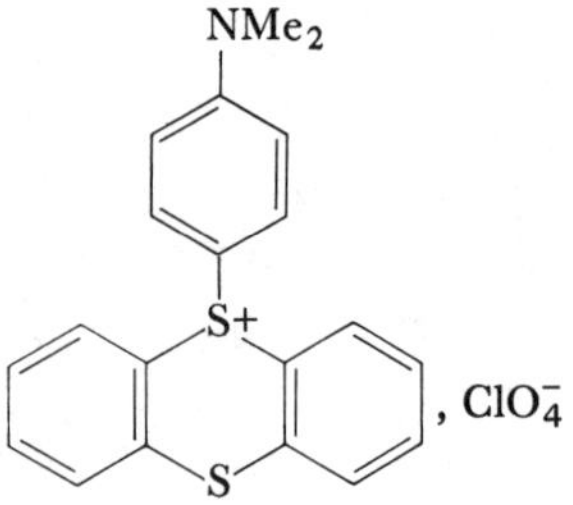

[41] (thianthrene-S⁺ with p-NMe₂-phenyl substituent, ClO₄⁻)

could be hydrolysed to give the amino compound. The latter could not be obtained by direct reaction of the cation radical with aniline. N,N-Dimethylaniline reacted with the thianthrene cation radical to give much tetramethylbenzidine but also some of the substitution product [41].

Amination and Related Reactions

Amination refers to the reaction of the aminium radical ($NH_3^{\cdot+}$) and its dialkyl derivatives ($R_2NH^{\cdot+}$) with organic substrates, particularly with olefins and aromatics. The aminium radical itself can be generated by the reduction of hydroxylamine O-sulfonic acid with ferrous ion (147) (Minisci and Galli, 1965). Dialkylaminium radicals are generated in several ways, the most common being the reduction of N-chlorodialkylamines by ferrous ion in acidic solution (148) and

$$R_2\overset{+}{N}HCl + Fe^{2+} \rightarrow R_2NH^{\cdot+} + Cl^- + Fe^{3+} \quad (148)$$

$$R_2NX \xrightarrow[H^+]{h\nu} R_2NH^{\cdot+} + X^{\cdot} \quad (149)$$

the photolysis of N-chloro- and N-nitrosodialkylamines in acidic solution [(149); X = Cl, NO]. Work with chloramines has been carried out mostly and this has been recently reviewed by Neale (1971), Sosnovsky and Rawlinson (1972), and Minisci (1975), while work with nitrosamines has been carried out and reviewed by Chow (1973). The reduction of an amine oxide by ferrous ion also gives an aminium radical (Lindsay Smith et al., 1973). Danen and Rickard (1972) have generated dialkylaminium radicals (R = Me, Et, n-Pr, iso-Pr) by reduction of N-chloramines by ferrous ion in 90% sulfuric and deuteriosulfuric acid and characterized them by esr spectroscopy. Ordinarily, the aminium ions are generated in acid solution in the presence of an organic substrate, with which reaction occurs in a chain sequence. With olefins, for example, a chloroalkylamine or a C-nitrosoalkylamine (or usually the isomeric oxime) is formed [(150)

$$R_2 NH^{\cdot+} + \diagup C{=}C\diagdown \rightarrow R_2 \overset{+}{N} H{-}\underset{|}{\overset{|}{C}}{-}\underset{|}{\overset{|}{C}}\cdot \qquad (150)$$

$$R_2 \overset{+}{N}H{-}\underset{|}{\overset{|}{C}}{-}\underset{|}{\overset{|}{C}}\cdot + R_2 \overset{+}{N}HX \rightarrow R_2 \overset{+}{N}H{-}\underset{|}{\overset{|}{C}}{-}\underset{|}{\overset{|}{C}}{-}X + R_2 NH^{\cdot+} \qquad (151)$$

and (151); X = Cl, NO]. Dialkylaminium radicals usually add in a 1,4-manner to 1,3-dienes giving [42] and in a 1,2-manner to allenes

$$R_2 \overset{+}{N}HCH_2 CH{=}CHCH_2 X \qquad\qquad R_2 \overset{+}{N}HCH_2 {-}\underset{X}{\overset{|}{C}}{=}CH_2$$

[42]   [43]

giving [43]. Addition to alkynes is usually followed by hydrolysis of the enamine function, and a carbonyl compound is formed (152).

$$R_2 \overset{+}{N}HCl + R'C{\equiv}CH \rightarrow R'C(Cl){=}CH\overset{+}{N}HR_2 \rightarrow RCHCl.CHO \qquad (152)$$

Reaction of N-chloramines with aromatics [(153)–(155)] has been illustrated most recently with alkylaromatics by Clerici et al. (1974). Amination of aromatics is an electrophilic rather than free-radical

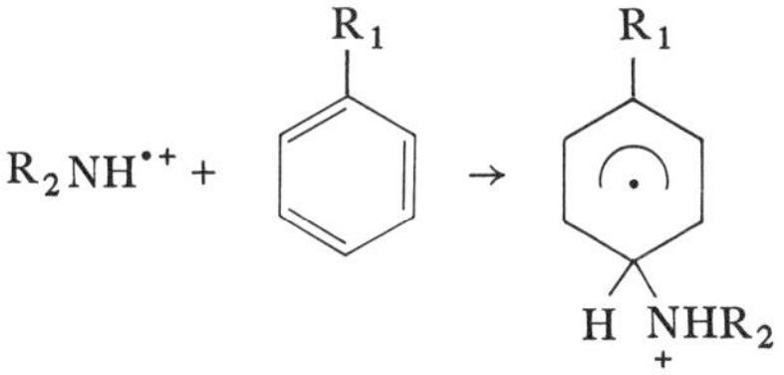

$$R_2 NH^{\cdot+} + \underset{}{\bigcirc}\!\!-\!R_1 \rightarrow \underset{H\;\;\overset{+}{N}HR_2}{\bigcirc}\!\!-\!R_1 \qquad (153)$$

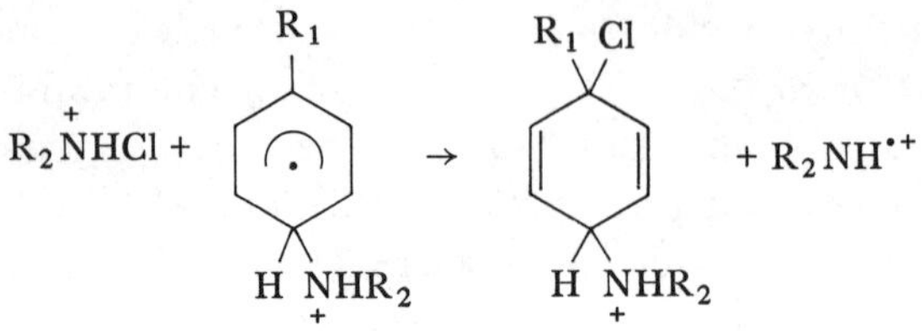

$$R_2\overset{+}{N}HCl + \underset{H\ \underset{+}{NHR_2}}{\overset{R_1}{\underset{\bullet}{\bigcirc}}} \rightarrow \underset{H\ \underset{+}{NHR_2}}{\overset{R_1\ Cl}{\bigcirc}} + R_2NH^{\bullet+} \quad (154)$$

$$\underset{H\ \underset{+}{NHR_2}}{\overset{R_1\curvearrowleft Cl}{\bigcirc}} \rightarrow H^+ + HCl + \underset{NR_2}{\overset{R_1}{\bigcirc}} \quad (155)$$

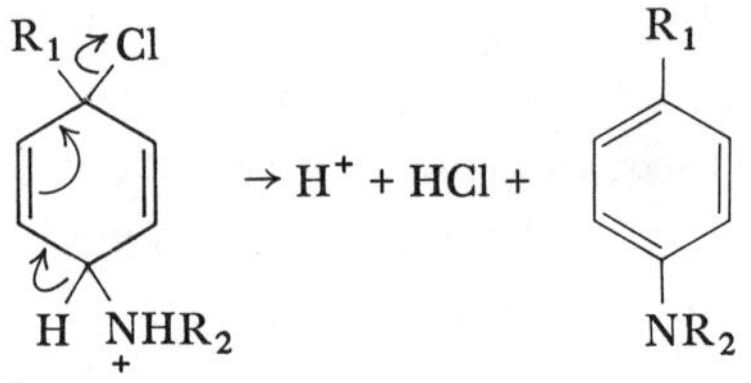

substitution reaction, although on the basis of partial rate factors it has been concluded that the transition state has some degree of charge transfer character [44].

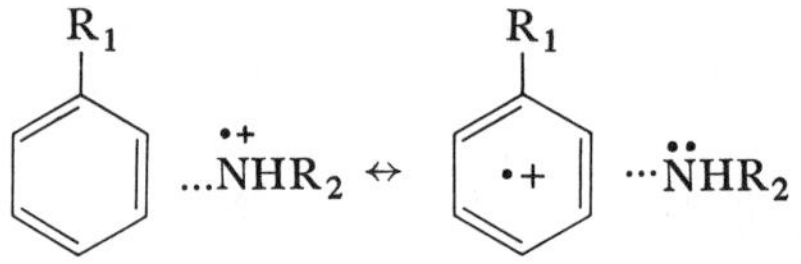

$$\underset{}{\overset{R_1}{\bigcirc}}\overset{\bullet+}{\cdots NHR_2} \leftrightarrow \underset{}{\overset{R_1}{\bigcirc}}^{\bullet+}\cdots\overset{\bullet\bullet}{N}HR_2$$

[44]

A long-known reaction, recognized only in recent years as involving aminium radicals, is the Hofmann–Loeffler rearrangement (Wolff, 1963). In this reaction, an alkylaminium radical, generated from an N-chloramine in strong sulfuric acid, undergoes an intramolecular hydrogen-atom transfer from C-4 or C-5 to nitrogen. Chlorination follows in a chain sequence [(156) and (157)]. The

$$CH_3CH_2(CH_2)_3\overset{\bullet+}{N}HR \rightarrow CH_3\overset{\bullet}{C}H(CH_2)_3\overset{+}{N}H_2R \quad (156)$$

$$CH_3\overset{\bullet}{C}H(CH_2)_3\overset{+}{N}H_2R + CH_3CH_2(CH_2)_3\overset{+}{N}HR \rightarrow$$
$$\phantom{CH_3\overset{\bullet}{C}H(CH_2)_3\overset{+}{N}H_2R + CH_3CH_2(CH_2)_3\overset{+}{N}HR \rightarrow}\underset{Cl}{|}$$

$$CH_3CH(CH_2)_3\overset{+}{N}H_2R + CH_3CH_2(CH_2)_3\overset{\bullet+}{N}HR \quad (157)$$
$$\underset{Cl}{|}$$

Hofmann–Loeffler rearrangement has a variety of forms, and a number of related reactions are known. Among these is the intermolecular chlorination of alkyl chlorides and esters shown in (158)

$$R_2NH^{\bullet+} + R'H \rightarrow R_2NH_2^+ + R'\bullet \quad (158)$$

$$R'\bullet + R_2\overset{+}{N}HCl \rightarrow R'Cl + R_2NH^{\bullet+} \quad (159)$$

and (159) (Minisci et al., 1970; Spanswick and Ingold, 1970). In these intermolecular reactions, all carbon atoms in a chain are monochlorinated, but the largest amount occurs at the penultimate ($\omega-1$) carbon atom. Predominant chlorination at the ($\omega-1$) carbon atom of alcohols, carboxylic acids, esters, and amides has also been achieved with the photodecomposition of N-chlorodiisopropylamine in concentrated sulfuric acid (Deno et al., 1971). Reduction of tributylamine N-oxide with ferrous ion in aqueous sulfuric acid leads to a mixture of 3-hydroxy- and 4-hydroxybutylamines [45] and

$$n\text{-Bu}_2\text{NCH}_2\text{CH}_2\overset{\text{OH}}{\underset{|}{\text{CH}}}\text{CH}_3 \qquad n\text{-Bu}_2\text{NCH}_2\text{CH}_2\text{CH}_2\text{OH}$$

$$[45] \qquad\qquad\qquad [46]$$

[46]. These compounds are produced formally by hydration of the cation formed by oxidation of the corresponding radical (160) which

$$(n\text{-Bu})_2\overset{+}{\text{N}}\text{HCH}_2\text{CH}_2\text{CH}_2\text{CH}_2\cdot + \text{Fe}^{3+} \rightarrow n\text{-Bu}_2\overset{+}{\text{N}}\text{HCH}_2\text{CH}_2\text{CH}_2\overset{+}{\text{CH}}_2 + \text{Fe}^{2+}$$

(160)

is obtained intramolecularly by a Hofmann–Loeffler rearrangement. The secondary cation is obtained from rearrangement of the primary cation and also from ferric-ion oxidation of the *intermolecularly-* formed secondary radical (Lindsay Smith et al., 1973).

Hydrogen-atom transfer reactions are also seen in the photodecomposition of N-nitrosamines in acidic methanol (161), and the

$$R_2\text{NH}^{\cdot+} + \text{CH}_3\text{OH} \rightarrow R_2\overset{+}{\text{NH}}_2 + {}^\cdot\text{CH}_2\text{OH} \qquad (161)$$

reason that acidic methanol can be used successfully as a solvent for amination of olefins is that the amination reaction is faster than the hydrogen-atom transfer (Chow et al., 1972).

Acetoxylation

Acetoxylation is commonly encountered in anodic oxidations in acetic acid or in solutions containing acetate ion (Eberson and Nyberg, 1976). Mechanistic studies have not been made, and explanations of anodic acetoxylation reactions in the literature are based on reasonable speculations and precedents from other anode reactions. Nuclear acetoxylation, for example, of anthracene and substituted anthracenes (Parker, 1970), is thought to follow the ECE pathway, while the formation of benzylic acetates from, say, methyl aromatics (Nyberg, 1970), appears to go through the sequence (162).

$$ArMe^{\cdot+} \rightarrow ArCH_2\cdot \rightarrow ArCH_2^+ \rightarrow ArCH_2OAc \qquad (162)$$

In this case, acetoxylation competes with dimerization of the benzylic radicals. Adsorption of the substrate and acetoxylation of the cation radical on the anode is thought to affect the distribution of isomers when aromatics are used (Eberson and Wilkinson, 1972).

Anodic acetoxylation of hindered phenols gave excellent yields (79%) of the cyclohexadienones [47; R = Me and t-Bu] (Ronlan and

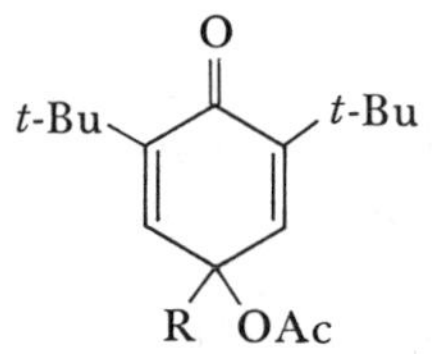

Parker, 1971), while even dimethylformamide, with an oxidation potential close to that of acetate ion, underwent acetoxylation by a sequence similar to that of (118), giving the product [48].

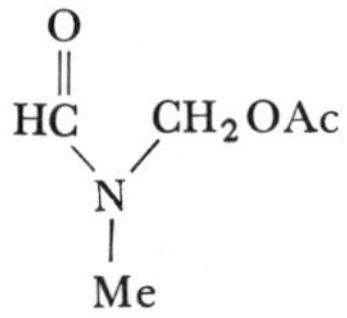

Acetoxylation during metal ion oxidations have already been described (p. 170). Trifluoroacetoxylation of aromatics occurs readily in oxidation by cobalt trifluoroacetate in trifluoroacetic acid (Kochi et al., 1973), and the acid itself is the nucleophile as shown in eqns (163)–(165). Comparisons of the nucleophilicity of TFA with

$$C_6H_6 + Co(III) \xrightleftharpoons{slow} C_6H_6^{\cdot+} + Co(II) \qquad (163)$$

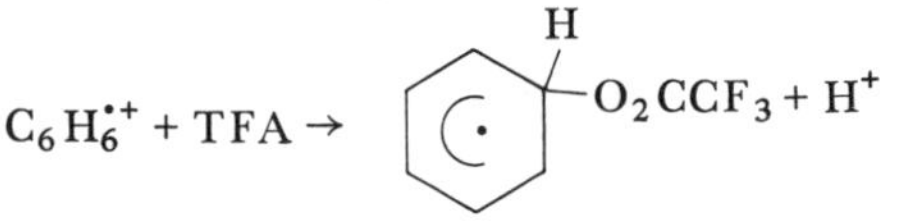

$$C_6H_6^{\cdot+} + TFA \rightarrow [\text{cyclohexadienyl radical with H and } O_2CCF_3] + H^+ \qquad (164)$$

$$[\text{cyclohexadienyl radical with H and } O_2CCF_3] + Co(III) \rightarrow C_6H_5O_2CCF_3 + H^+ + Co(II) \qquad (165)$$

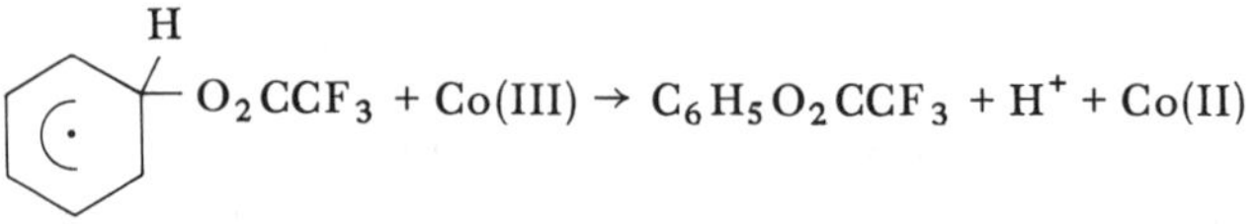

that of acetate and cyanide (also in anodic reactions) was sought in the o/p ratios for substitution in the chlorobenzene cation radical. From the data (cyanation 1·0, acetoxylation 0·64, trifluoroacetoxylation 0·52) it is concluded that the more nucleophilic cyanide ion is the less selective in reaction with the aryl cation radical. How much the selectivity is affected by the size of the nucleophile is not apparent, however.

Only a few reactions of acetate ion and other carboxylates with cation radicals are known other than in anodic oxidations. Thus, perylene cation radical gave the 3-acetate and benzoate (Ristagno and Shine, 1971b).

Reaction with Aromatics

We have already seen that aromatics can be aminated with amine cation radicals. Organosulfur cation radicals react with aromatics carrying an electron-donating group. The reaction appears to be electrophilic in nature (nitrobenzene does not react) and follows the stoichiometry shown in (166). The reactions are second order in

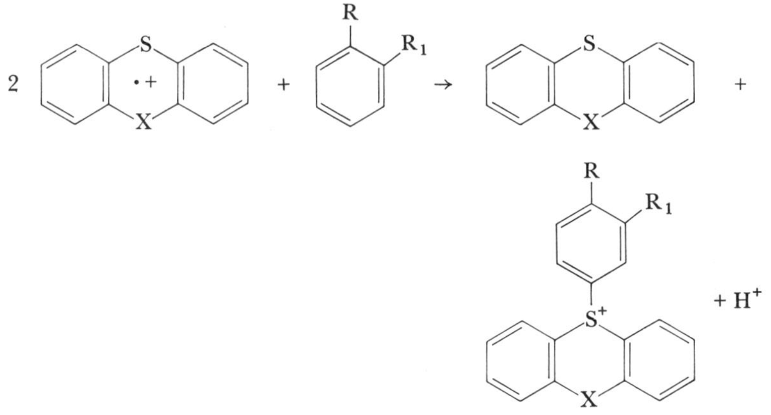

R	: OMe	NHAc	OH	NMe$_2$	OH	OH
R$_1$:	H	H	H	H	Cl	t-Bu

cation radical and inverse first order in the parent sulfide. They are thought therefore to involve disproportionation of the cation radical and reaction of the organosulfur dication with the aromatic (Silber and Shine, 1971; Kim et al., 1974). Analogous reactions have been observed in the anodic oxidation of phenyl sulfides (C$_6$H$_5$SR; R = phenyl and alkyl). The products were the sulfonium perchlorates [49] (Torii et al., 1973).

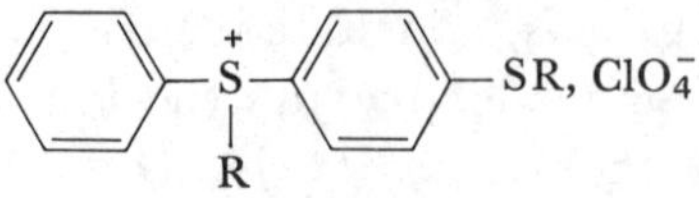

$$\text{[Ph]}-\underset{R}{\overset{+}{S}}-\text{[Ph]}-SR, ClO_4^-$$

[49]

Dimerization of aromatic compounds in anodic oxidations are very common, particularly among amines, phenols, and alkylaromatics (Eberson and Nyberg, 1973). Biaryls and their isomers are formed; for example, N,N-dimethylaniline gives tetramethylbenzidine. Some of these dimerizations may involve pairing of the cation radicals, while others (more commonly) involve reaction of the cation radical with its parent (an ECE reaction) or reaction of the dication (formed by two-electron oxidation or by disproportionation of the cation radical) with its parent. These reactions are too wide in scope to be dealt with here, but a recent study of the reactions of thianthrene cation radical (Th$^+$) with anisole (AnH) is particularly important in this general area and has a special bearing on the possible role of complicating disproportionation processes (Svanholm et al., 1975). It appears that the thianthrene cation radical reacts with anisole to give a complex (167) which reacts further by one of two pathways (168) and (169), depending on the concentration of

$$\text{Th}^{\cdot+} + \text{AnH} \rightleftharpoons (\text{Th–AnH})^{\cdot+} \qquad (167)$$

$$(\text{Th–AnH})^{\cdot+} + \text{Th}^{\cdot+} \rightleftharpoons (\text{Th–AnH})^{2+} + \text{Th} \qquad (168)$$

$$(\text{Th–AnH})^{2+} \rightarrow \text{Th-AnH}^+ + \text{H}^+ \qquad (169)$$

Th$^{\cdot+}$. Dibenzo-p-dioxin cation radical (DBO$^{\cdot+}$) was found to accelerate the rate of anisylation of Th$^{\cdot+}$ by a factor of 200 on account of the participation of DBO$^{\cdot+}$ in equilibrium (168); DBO$^{\cdot+}$ is a better oxidant than Th$^{\cdot+}$. A disproportionation mechanism was ruled out because the observed rate constants were found to be as much as 5×10^4 times higher than predicted for diffusion controlled reaction of Th^{2+}. The intervention of a molecular complex of the cation radical and the neutral aromatic could be important in many related oxidative coupling processes, especially those involving aromatic amines.

Reactions with Olefins

Aminium radicals add readily to olefins (p. 245). With other cation radicals electron transfer followed by dimerization (170) often occurs. In appropriate cases the dimer dication may be solvolysed or go on to initiate cationic polymerization of the olefin

$$2\ R_2C=CH_2^{\cdot+} \rightarrow R_2\overset{+}{C}-CH_2-CH_2-\overset{+}{C}R_2 \quad (170)$$

(Ledwith and Sherrington, 1972; Turcot *et al.*, 1974; Akbulut *et al.*, 1975). When the olefin is not susceptible to cationic homopolymerization, elimination to a 1,3-butadiene (171), or formation of

$$R_2\overset{+}{C}-CH_2-CH_2-\overset{+}{C}R_2 \rightarrow R_2C=CH-CH=CR_2 + 2H^+ \quad (171)$$

$$\underset{X}{\overset{R_1R_2N}{\diagdown}}C=CH_2 \xrightarrow{-2e^-} \underset{X}{\overset{R_1R_2N}{\diagdown}}\overset{+}{C}-CH_2CH_2\overset{+}{C}\underset{X}{\overset{NR_1R_2}{\diagup}} \quad (172)$$

stable dicationic salts results (172). Dimeric dications of this type have been made from 1,1-diphenylethylene by oxidation with $SbCl_5$ (Fleischfresser *et al.*, 1968; Bracke *et al.*, 1968), and from a series of electron-rich alkenes by oxidation with silver ion (Effenberger and Gerlach, 1970) or stable cation radicals (Bawn *et al.*, 1968; Bell *et al.*, 1974). Anodic coupling of unlike olefins has also been accomplished (Schäfer and Steckham, 1970).

A particularly interesting series of reactions has been characterized for the cation radical of the olefin N-vinyl carbazole (NVC [50]).

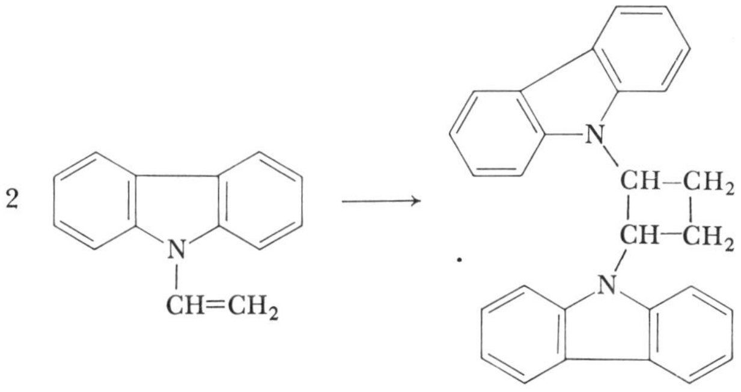

[50]   [51]   (173)

The wider scope of polymerization and co-polymerization processes involving [50$^{\cdot+}$] have been reviewed (Hyde and Ledwith, 1974), but special mention must be made of the very efficient cyclodimerization of NVC yielding *trans*-1,2-di-9-carbazylcyclobutane [51] (173). It is now clear that this novel cyclodimerization process involves a chain mechanism with propagating cation radical intermediates as shown in eqns (174) and (175). Initiation can be accomplished by a variety of

$$\text{RCH=CH}_2 \xrightarrow{-e^-} \text{RCH=CH}_2^{\cdot+} \xrightarrow{\text{RCH=CH}_2} \begin{bmatrix} \text{RCH--CH}_2 \\ | \quad | \\ \text{RCH--CH}_2 \end{bmatrix}^{\cdot+} \quad (174)$$

$$\begin{bmatrix} \text{RCH--CH}_2 \\ | \quad | \\ \text{RCH--CH}_2 \end{bmatrix}^{\cdot+} + \text{RCH=CH}_2 \rightarrow \begin{matrix} \text{RCH--CH}_2 \\ | \quad | \\ \text{RCH--CH}_2 \end{matrix} + \text{RCH=CH}_2^{\cdot+} \quad (175)$$

metal oxidants, photoexcited catalysts, anodic processes, and by radiolysis, and the chain lengths observed depend upon the experimental conditions (Ledwith, 1972; Kricka and Ledwith, 1974; Crellin and Ledwith, 1975). Although initially devised to explain the reactions of NVC, the cation radical chain process has now been applied to cyclodimerization of a wider variety of olefins including indene derivatives (Farid and Shealer, 1973; Farid et al., 1975), phenyl-vinyl ether (Evans et al., 1975), p-methoxystyrene (Yamamoto et al., 1975), and p-dimethylaminostyrene (Asanuma et al., 1975).

In an important extension of this type of photoinduced formation of cation radicals, presumably via exciplex intermediates, Neunteufel and Arnold (1973) have demonstrated that an oxidative dimer of 1,1-diphenylethylene, 1,1,4-triphenyl-1,2,3,4-tetrahydronaphthalene, is readily produced by photolysis of the olefin in the presence of methyl p-cyanobenzoate. Formation of the cation radical ($Ph_2C=CH_2$)$^{\cdot+}$ was confirmed by flash photolytic procedures. Similar photosensitized irradiations of $Ph_2C=CH_2$ in methanol yielded 2,2-diphenylethyl methyl ether rather than the oxidative dimer. Again it is thought that the ether product arises by methanolysis of the olefin cation radical. Rather similar anti-Markovnikov additions of methanol to 1-phenylcycloalkenes have been reported for photoreactions in the presence of methyl-p-cyanobenzoate (Shigemitsu and Arnold, 1975).

Reaction with Ketones

As far as we are aware, reactions of cation radicals with carbonyl compounds were until recently very rare. Cyclization of the 6-β-ketopropionic ester during oxidation of the metalloporphyrin [52] by iodine is thought by Cox et al., (1969) to occur in the metalloporphyrin cation radical. Kim and Shine (1974) found that cyanoacetamide reacted with thianthrene cation radical to give what now appears to be [53].

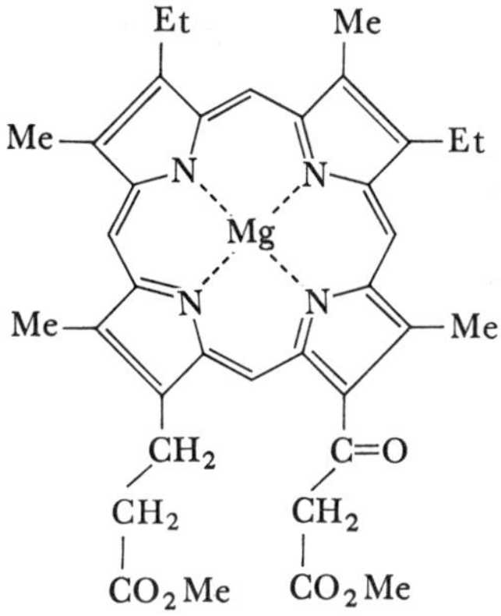

[52]

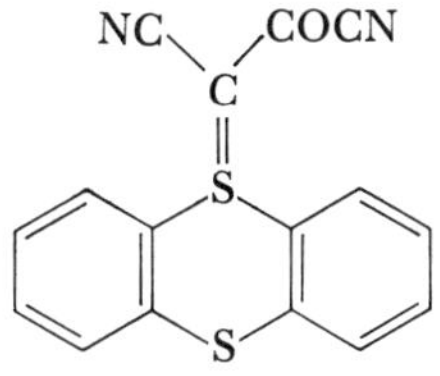

[53]

Analogously, a new series of reactions of organosulfur cation radicals has been discovered very recently (Shine *et al.*, 1975b). The products are either ketoalkylsulfonium perchlorates [eqn (176) for

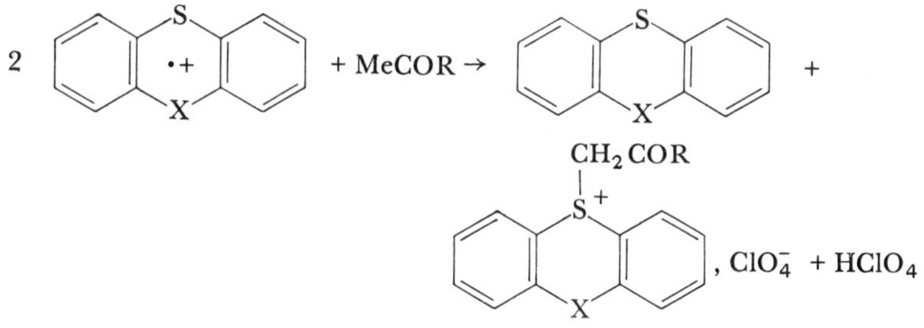

$$2 \text{ (thianthrene cation radical)} + \text{MeCOR} \rightarrow \text{(thianthrene)} + [54], ClO_4^- + HClO_4 \quad (176)$$

[54]

X = S, O, N-Me, N-C_6H_5; R = t-Bu, C_6H_5, 2-naphthyl

methyl ketones] or the corresponding ylides. A variety of ketones (e.g. acetone, 2-butanone, diisopropyl ketone, cyclopentanone, cyclohexanone) are usable. With 2-butanone reaction appears to occur at the 3-position. When ethyl benzoylacetate is used, an ylide is formed [55], the negative charge on the ylide carbon atom being

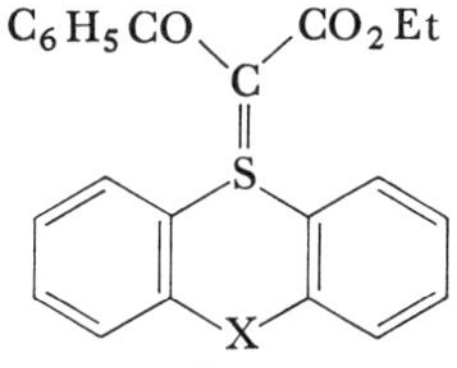

[55]

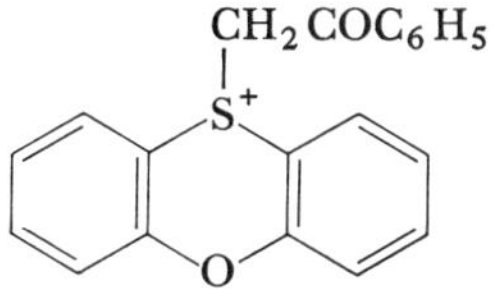

[56]

stabilized evidently by delocalization into the two carbonyl groups. Conversion of the sulfonium ion [56] into the corresponding ylide was achieved by reaction with triethylamine.

The mechanism of reaction of organosulfur cation radicals with ketones is not yet known, but may well involve the enolic form of the ketone. Reactions occur either in the neat ketone or in acetonitrile solution. Anodic oxidation of thianthrene in acetone also gave the anticipated product [54; X = S; R = Me].

The products [54] are useful organic intermediates and have a very large scope in reactions (Johnson, 1969). For example, reaction with nucleophiles gives the parent heterocycle and an α-substituted ketone.

6. CATION RADICALS FROM BIPYRIDYLIUM SALTS

Whereas cation radicals are formed from neutral molecules by one electron oxidation processes, bipyridylium salts readily give rise to stable cation radicals by one electron reduction processes. Several types of bipyridylium salts have been shown to have interesting herbicidal properties (Akhavein and Linscott, 1968) and the most important of these are exemplified in Scheme 3 by paraquat dichloride (PQ^{2+}) (1,1'-dimethyl-4,4'-bipyridylium dichloride) and diquat dibromide (PQ^{2+}; 6,7-dihydrodipyrido[1,2-a:2',1'-c]-pyrazine-di-ium dibromide). Herbicidal activity appears to depend, in part, on the ease of (reversible) one electron reduction of PQ^{2+} and

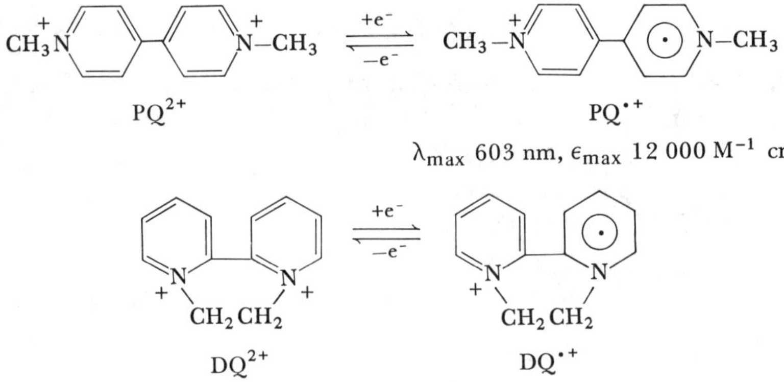

PQ^{2+} $PQ^{\cdot+}$
λ_{max} 603 nm, ϵ_{max} 12 000 M^{-1} cm^{-1}

DQ^{2+} $DQ^{\cdot+}$
λ_{max} 760 nm, ϵ_{max} 3109 M^{-1} cm^{-1}

Scheme 3

DQ^{2+} to form stable but air sensitive cation radicals $PQ^{·+}$ and $DQ^{·+}$ respectively (Dodge, 1971). One electron reduction of bipyridylium salts may be achieved electrochemically (Elofson and Edsberg, 1957), by chemical reducing agents (Kosower and Cotter, 1964; Yuen et al., 1967), and also by photolysis in the presence of primary and secondary alcohols (Hopkins et al., 1970; Johnson and Gutowsky, 1963). As in the case of stable cation radicals formed from neutral molecules, delocalization accounts for the great stability of $PQ^{·+}$ and $DQ^{·+}$; eighteen canonical forms may be written for each of these intensely colored cation radicals. Paraquat (PQ^{2+}) and diquat (DQ^{2+}) are useful oxidants in aqueous solution [$E_0 = -446$ and -349 mV (NHE) respectively] and it is of interest that their respective cation radicals are readily produced by reaction of the salts with aqueous alkali (Farrington et al., 1969).

Under conditions of high-vacuum the reaction between sodium hydroxide and paraquat dichloride in water is highly reproducible. With a 10-fold excess of alkali, the final yield of cation-radical (based on paraquat) increases with temperature from 28% at 30°C, to a limiting maximum of 66% at ca 85°C.

After destruction of the cation radical by acidification and admission of air to the system, the reaction products are paraquat dichloride, 1-methyl-4-(4-pyridyl)-pyridinium chloride [57] and formaldehyde [see (177)-(179)].

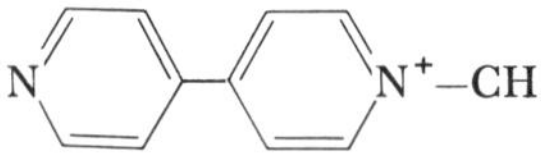

$$PQ^{2+} + OH^- \rightarrow N\text{◯-◯}N^+-CH_3 + CH_3OH \qquad (177)$$

[57]

$$CH_3OH + OH^- \rightarrow CH_3O^- + H_2O \qquad (178)$$
$$2PQ^{2+} + CH_3O^- \rightarrow CH_2O + 2PQ^{·+} + H^+ \qquad (179)$$

Methanol required for the reduction is provided by initial dequaternization of paraquat, and the maximum yield of cation-radical would be 66% as observed experimentally. The suggested mechanism is substantiated by addition of methanol to aqueous systems or by working in methanol solvent, where yields of $PQ^{·+}$ are quantitative. Other alcohols with α-hydrogen atoms will also reduce paraquat in basic media but solutions of potassium t-butoxide in t-butyl alcohol are inactive.

Oxidation of methoxide ion by the paraquat ion is essentially instantaneous in methanol or aqueous methanol when

[MeO⁻] ≥ 0·1 M. This behavior is in marked contrast to the more usual ring-substitution which occurs when pyridinium ions interact with bases (Bruck and Guttman, 1968; O'Leary and Stach, 1972).

Reaction rates for reduction of PQ^{2+} by CH_3O^- and CD_3O^- have been measured and shown to be almost identical, indicating the absence of any significant deuterium kinetic isotope effect, an important observation, since hydrogen or hydride abstraction would be a likely rate determining step in oxidation of methoxide ion. It seems reasonable therefore to assume that primary electron transfer between CH_3O^- and PQ^{2+} (180) is rate determining, steps (181) and (182) being much faster.

$$PQ^{2+} + {}^-OCH_3 \rightleftharpoons PQ^{\cdot+} + \dot{O}CH_3 \qquad (180)$$

$$CH_3O^\cdot + OH^- \rightleftharpoons CH_2O^{\cdot-} + H_2O \qquad (181)$$

$$CH_2O^{\cdot-} + PQ^{2+} \rightarrow PQ^{\cdot+} + CH_2O \qquad (182)$$

Formation of $PQ^{\cdot+}$ in alkaline solutions containing paraquat is not restricted to the simple examples noted above. Thus cyanohydrin anions are rapidly oxidized by paraquat in a sequence [(183) and (184)] which involves overall oxidation of aromatic aldehydes to

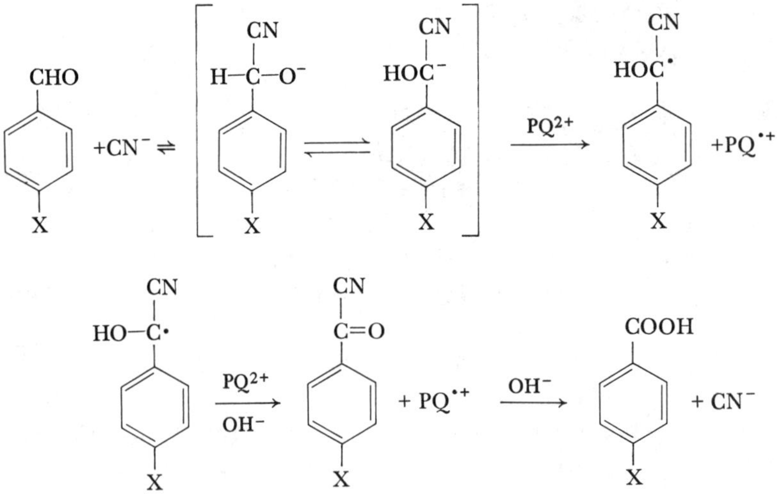

(X = CN, NO₂)

carboxylic acids (Kramer et al., 1967). A similar oxidation occurs when paraquat is allowed to react with the pseudo base formed in alkaline quinolinium ion solutions (Cooksey and Johnson, 1968).

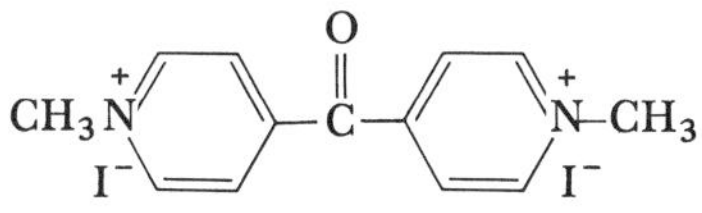

$$\text{CH}_3\overset{+}{\text{N}}\underset{\text{I}^-}{\diagup\!\!\!\bigcirc\!\!\!\diagdown}-\overset{\text{O}}{\underset{\|}{\text{C}}}-\underset{\text{I}^-}{\diagup\!\!\!\bigcirc\!\!\!\diagdown}\overset{+}{\text{N}}-\text{CH}_3$$

[58]

More difficult to explain, however, is the reported formation of $PQ^{•+}$ on treatment of the di-methiodide [58] with concentrated aqueous alkali (Minn et al., 1970; Geiger et al., 1971), although the reaction bears some resemblance to the well-established oxidative dimerization of N-methyl pyridinium salts by cyanide ion (Winters et al., 1970; Winters et al., 1967; Reuss and Winters, 1973).

Paraquat dichloride is reduced to the cation-radical $PQ^{•+}$ by uv irradiation of aqueous solutions containing primary or secondary alcohols (Hopkins et al., 1970). There is no reaction on irradiation in pure water or when t-butyl alcohol is added. Reaction rates are retarded by adding halide ion in the sequence or increasing quenching ability, $Cl^- < Br^- \ll I^-$. For all alcohols, stoichiometric amounts (by glc) of corresponding carbonyl compounds are produced (183). Reaction rates are proportional to the first power of

$$2PQ^{2+} + R^1R^2CHOH \xrightarrow{h\nu} 2PQ^{•+} + R^1R^2C=O + 2H^+ \qquad (183)$$

the light intensity and can be activated using 313 or 334 nm (but not 366 nm) irradiation, consistent with excitation via the long-wavelength tail of the absorption spectrum of paraquat. There is, however, experimental evidence for the intervention of paraquat–alcohol–counterion complexes in these processes (Brown et al., 1973).

A striking feature of the photoreduction of PQ^{2+} is the rapid decay in rate as reaction proceeds caused, apparently, by the cation radical product. $PQ^{•+}$ has a convenient "window" in its absorption spectrum at 310–340 nm, and the rate retardation has been shown to involve quenching of intermediates.

Fluorescence quenching experiments (with diquat dichloride), deuterium kinetic isotope effects, and the effects of triplet state quenchers have suggested a mechanism involving primary electron transfer to singlet excited paraquat [(184) and (185)].

An important step in the suggested mechanism is the oxidation of alcohols to alkoxy radicals by singlet excited paraquat. However, only when the alkoxy radical has an α-hydrogen atom is the product

$$PQ^{2+} \xrightarrow{h\nu} [PQ^{2+}]^* \xrightarrow{RCH_2OH} [PQ^{\cdot +}, RCH_2OH^{\cdot +}] \quad (184)$$

$$\downarrow RCH_2OH$$

$$2PQ^{\cdot +} + RCHO + H^+ \xleftarrow{PQ^{2+}} [PQ^{\cdot +}, RCH_2O^{\cdot}] + RCH_2OH_2^+ \quad (185)$$

$PQ^{\cdot +}$ observed experimentally. Otherwise the intermediates revert back to ground-state reactants, with apparent fluorescence quenching in the case of diquat. According to this mechanism t-butyl alcohol should be effective as a quencher for diquat fluorescence, as observed experimentally, but cannot produce $PQ^{\cdot +}$. Unambiguous confirmation of the intermediacy of alkoxy radicals has been provided by spin-trapping experiments with phenyl N-t-butyl nitrone (186).

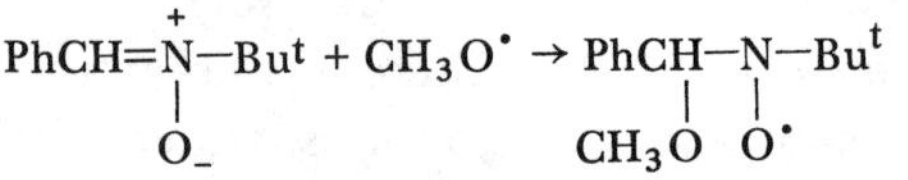

$$\underset{O_-}{PhCH=\overset{+}{N}-Bu^t} + CH_3O^{\cdot} \rightarrow \underset{CH_3O\ \ O^{\cdot}}{PhCH-N-Bu^t} \quad (186)$$

It is interesting to note that the photoinduced reactions of closely related compounds such as the dimethiodide of *trans*-1,2-bis(4-pyridyl)ethylene [59], while appearing to involve initial electron transfer processes, give high yields of products arising from addition of water, alcohols, ethers etc. at the double bond as in (189) for

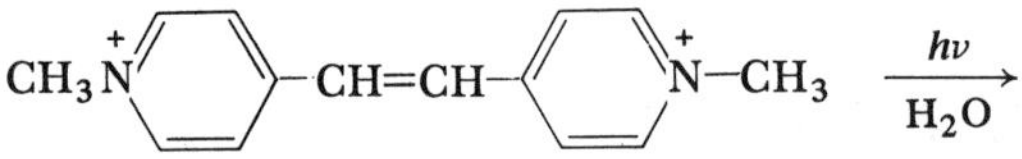

$$CH_3\overset{+}{N}\underset{[59]}{\diagdown}-CH=CH-\diagdown\overset{+}{N}-CH_3 \xrightarrow[H_2O]{h\nu}$$

$$CH_3\overset{+}{N}\diagdown-\underset{\underset{}{OH}}{CH}-CH_2-\diagdown\overset{+}{N}-CH_3 \quad (189)$$

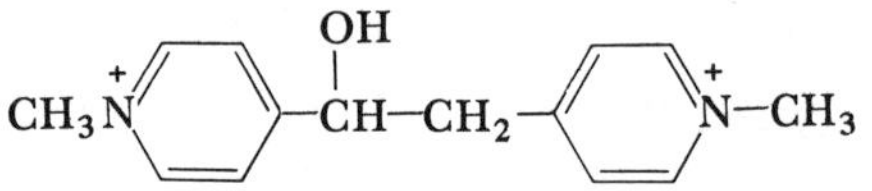

example (Happ *et al.*, 1971). The cation radicals of *cis*- and *trans*-forms of [59] undergo rapid interconversion to a stable planar *trans*-structure (Happ *et al.*, 1972).

Electron-Transfer Scavenging of Organic Radicals

Examination of the reactions outlined above shows that alkoxy radicals ($RCH_2O^{\cdot}$) should be produced by photochemical oxidation of alcohols and a necessary corollary of these processes is that tertiary alkoxy radicals, generated in the presence of paraquat cation radical, should immediately undergo electron transfer reduction forming t-alkoxide ion and paraquat dication. This has been confirmed unambiguously for t-butoxy radicals and paraquat cation radical, the former being generated photochemically from di-t-butyl peroxide and thermally from di-t-butyl hyponitrite (187) (Hopkins and Ledwith, 1971).

$$^tBuOOBu^t \xrightarrow[25°]{h\nu}$$
$$2^tBuO\cdot \xrightarrow{2PQ^{\cdot+}} 2^tBuO^- + 2PQ^{2+} \quad (187)$$
$$^tBuO-N=N-OBu^t \xrightarrow[50° (-N_2)]{\Delta}$$

Rate coefficients for thermal decomposition of di-t-butyl hyponitrite at 50° compared very well with those extrapolated from data in the literature, obtained by very different techniques, and reaction products contained one mole of base for each mole of $PQ^{\cdot+}$ consumed, strongly supporting the proposed electron transfer mechanism. It follows that, in 1:1 t-butyl alcohol-H_2O, t-butoxy radical must possess oxidizing power in excess of that of PQ^{2+}, for which $E_o = -446$ mV (NHE), and may, therefore, function as a primary one-electron oxidant in reactions at present interpreted in other ways.

In related work (Hyde and Ledwith, 1975), it has been shown that $PQ^{\cdot+}$ is a highly efficient scavenger for 2-cyano-isopropyl radicals produced by thermal decomposition of azo-bis-isobutyronitrile (AIBN). Representative kinetic data are shown in Table 5 and

TABLE 5

Decomposition of AIBN in the Presence of Paraquat Cation Radical. (Solvent, methanol; Temperature 70·0°C.)

10^3 [AIBN] (M)	10^3 [$PQ^{\cdot+}$] (M)	$10^7 \dfrac{-d[PQ^{\cdot+}]}{dt}$ (Ms^{-1})	$10^5\ k_d$ (apparent) (s^{-1})
1·305	1·29	0·71	2·7
2·67	1·29	1·56	2·9
4·02	1·29	2·30	2·8
4·74	1·29	2·77	2·9
5·00	1·29	2·83	2·8
6·25	1·29	3·44	2·8
5·00	0·65	2·85	2·8

indicate that the rate of disappearance of $PQ^{•+}$ is independent of $[PQ^{•+}]$ and essentially first order in [AIBN]. The average value of the rate coefficient for decomposition of AIBN ($\sim 2 \cdot 8 \times 10^{-5}$ s^{-1}) and the apparent activation energy (33 kcal mole^{-1}) are in excellent agreement with existing values in the literature (Van Hook and Tobolsky, 1958) and further demonstrate the reliability and convenience of determining rates of homolysis by radical scavenging with $PQ^{•+}$. It should be noted however, that in this case the scavenging does not involve simple electron transfer as in the related case of scavenging of t-butoxy radicals. Products of the reactions between $(CH_3)_2\dot{C}-CN$ and $PQ^{•+}$ are polymeric pyridinium salts, not amenable to the usual forms of characterization.

Formation of the highly coloured $PQ^{•+}$ may also be used as a convenient probe for formation of more reactive organic radicals.

An additional feature of the mechanism proposed for photochemical oxidation of alcohols by paraquat is that radicals of the type $RCH_2O•$ or, more likely, the rapidly formed isomeric radicals $R\dot{C}HOH$ must be scavenged by electron transfer reactions with ground state PQ^{2+}. Independent evidence for the latter process has been obtained by studying the thermal decomposition of benzpinacol in methanol containing paraquat dichloride (188). Paraquat

$$HO-\underset{\underset{C_6H_5}{|}}{\overset{\overset{C_6H_5}{|}}{C}}-\underset{\underset{C_6H_5}{|}}{\overset{\overset{C_6H_5}{|}}{C}}-OH \xrightarrow[70°]{\Delta} 2\ HO-\underset{\underset{C_6H_5}{|}}{\overset{\overset{C_6H_5}{|}}{C}}• \xrightarrow{2PQ^{2+}}$$

$$2PQ^{•+} + 2(C_6H_5)_2C=O + 2H^+ \quad (188)$$

oxidation of semipinacol radicals ($Ph_2\dot{C}-OH$) formed by unimolecular thermolysis of benzpinacol is apparently very efficient. Reaction rates, monitored by observing the appearance of $PQ^{•+}$, indicate an apparent activation energy of 30 kcal mole^{-1} for the decomposition of benzpinacol ($k \sim 6 \times 10^{-6}$ sec^{-1} at 71°C), figures in good agreement with data extrapolated from the literature for thermal decomposition of benzpinacol in the presence of oxygen (Neckars and Collenbrander, 1968).

A most significant observation is that apparent rates of homolysis for the various pinacols, determined by scavenging with PQ^{2+}, and really representing rates of escape of $Ar_2\dot{C}OH$ radicals from the solvent cage, correlate well with flash photometric data recently reported (Hammond et al., 1971) for rates of cage recombination of the same radicals.

The semipinacol radicals referred to are well established as intermediates in the photoreduction of aryl ketones and may be similarly oxidized by PQ^{2+} in typical photochemical reactions (Hyde and Ledwith, 1974b).

In marked contrast to the direct photoreduction of PQ^{2+} by aqueous isopropyl alcohol, where a limiting yield of $PQ^{\cdot+}$ is formed, the same process "sensitized" by benzophenone proceeds efficiently with complete conversion of PQ^{2+} to $PQ^{\cdot+}$. Chemical analysis shows that very little benzophenone is consumed during this part of the reaction, the quantum yield for formation of $PQ^{\cdot+}$ is 0·6, and the oxidation product is acetone. Under identical conditions the quantum yield for photoreduction of benzophenone in the absence of PQ^{2+} was 1·4.

It seems clear that benzophenone is here functioning as a "chemical sensitizer" for photoreduction of paraquat, the steps in the mechanism being set out in (189)–(191).

$$PH_2C=O \xrightarrow{h\nu} {}^1Ph_2C=O^* \rightarrow {}^3Ph_2C=O^* \quad (189)$$

$$^3Ph_2C=O^* + (CH_3)_2CHOH \rightarrow Ph_2\dot{C}-OH + (CH_3)_2\dot{C}-OH \quad (190)$$

$$Ph_2\dot{C}-OH + (CH_3)_2\dot{C}-OH + 2PQ^{2+} \rightarrow Ph_2C=O + (CH_3)_2C=O + 2PQ^{\cdot+} + 2H^+ \quad (191)$$

Experiments with other aryl ketones gave similar results and a common, important feature of the reactions is that continued photolysis, after complete reduction of PQ^{2+}, causes efficient bleaching of the radical, again without significant consumption of benzophenone. This photobleaching is caused by a chain reaction of $PQ^{\cdot+}$, initiated by transfer of a hydrogen atom from $Ar_2\dot{C}-OH$, and produces polymeric bipyridyls.

Electron transfer to PQ^{2+} is an important step in the photoinduced decarboxylation of several reducing organic acid anions. Thus formate, oxalate and benzilate anions have been shown to be effective photoreductants, in aqueous solution, for paraquat and diquat (Barnett et al., 1973).

Quantum yield determinations for the production of paraquat cation-radical were respectively 1·2 and 0·26 for benzilate and oxalate anions; formate ion was somewhat less reactive and the reactions are assumed to proceed as indicated (192) for oxalate and

$$PQ^{2+} + RCOO^- \xrightarrow{h\nu} PQ^{\cdot+} + RCOO^{\cdot} \xrightarrow{PQ^{2+}} PQ^{\cdot+} + CO_2 + (H^+ \text{ or } CO_2) \quad (192)$$

formate although the oxidation product (CO_2) was not identified. In the photoreduction by benzilate anion the oxidation product was shown to be benzophenone, formed according to eqn (193).

$$PQ^{2+} + Ph_2\underset{\underset{OH}{|}}{C}-COO^- \xrightarrow{h\nu} PQ^{\cdot+} + Ph_2\underset{\underset{OH}{|}}{\overset{\cdot}{C}}-COO \xrightarrow{-CO_2}$$

$$2\,Ph_2\underset{\underset{OH}{|}}{\overset{\cdot}{C}} \xrightarrow{PQ^{2+}} PQ^{\cdot+} + H^+ + Ph_2C=O \quad (193)$$

Paraquat does not absorb light of wavelength greater than about 345 nm but photoreduction of paraquat by oxalate and benzilate anions occurs with equal facility at 313 and 366 nm; for paraquat-benzilate anion combinations, photoactivity was observed at wavelengths greater than 400 nm. The absorption spectra of paraquat solutions containing these anions confirmed that absorption occurs at longer wavelengths than that from either the anion or paraquat alone.

Pyridinium and bipyridylium ions are known to give rise to charge-transfer spectra in combination with a variety of neutral and anionic donor species. It seems reasonable to conclude therefore that photoactivity arises from charge-transfer interaction between the bipyridylium cation and formate, oxalate, and benzilate ions (194).

$$PQ^{2+} + RCOO^- \rightleftharpoons [PQ^{2+}, {}^-OOCR \leftrightarrow PQ^{\cdot+}, \overset{\cdot}{O}OCR] \xrightarrow{h\nu} PQ^{\cdot+} + \overset{\cdot}{O}OCR \quad (194)$$

Represented in this way, the interaction between bipyridylium ions and carboxylate ions is strictly analogous to that between pyridinium ions and iodide ion, and photoactivity may result from absorption of a quantum of radiation by the ground state of the charge-transfer pair (intimate or contact-ion pair) or, following absorption, by a contact-pair formed on random collisional encounter.

The variety of electron transfer reactions described above for PQ^{2+} and $PQ^{\cdot+}$ obviously provide some (model compound) information related to the possible mechanism of action of PQ^{2+} in herbicidal applications. An even bigger and more applicable range of electron transfer processes involving PQ^{2+} and $PQ^{\cdot+}$ is provided by studies of the action of paraquat as a mediator (i.e. electron-transfer bridge), or terminal electron acceptor, in redox processes involving biologically active compounds (see for example Steckham and Kuwana, 1974; Krasnovsky, 1972). Whilst such studies properly reside outside the scope of this review, there are two aspects of bipyridylium ion chemistry equally relevant to model compound systems, biological systems, and commercial applications (Schoot et al., 1973). These are, respectively, the questions of disproportionation of $PQ^{\cdot+}$ and its rapid reaction with atmospheric oxygen.

We have already noted that although disproportionation of bipyridylium cation radicals is a well established phenomenon, the positions of equilibrium for paraquat and diquat lie overwhelmingly to the left hand side of eqn (197) (Hünig et al., 1973). It is also well

$$2 PQ^{\cdot+} \rightleftharpoons PQ^{2+} + PQ \tag{197}$$

$$(PQ \equiv CH_3-N\!\!\left\langle\!\!\!\begin{array}{c}\end{array}\!\!\!\right\rangle\!\!=\!\!\left\langle\!\!\!\begin{array}{c}\end{array}\!\!\!\right\rangle\!\!N-CH_3)$$

established (Elofson and Edsberg, 1957; Volke, 1968) that the first one-electron reduction stage of PQ^{2+} is independent of pH and that the second stage (i.e. $PQ^{\cdot+} + e^- \rightarrow PQ$) shows only a small systematic dependence on pH. These pH effects were estimated by polarographic reduction where there is always a substantial excess of unreduced PQ^{2+}.

Hopkins (1972) has shown however, that the cation radical of paraquat is rapidly destroyed by protic acids in water, provided that there is no free PQ^{2+} present. It appears that, although the position of equilibrium in (197) lies very much in favor of $PQ^{\cdot+}$, protic acid will rapidly shift the position by protonation of the reactive enamine PQ (198, 199). Independent studies of the reactions of PQ with

$$2PQ^{\cdot+} \rightleftharpoons PQ + PQ^{2+} \xrightarrow{H_3O^+} CH_3-\overset{+}{N}\!\!\left\langle\!\!\!\begin{array}{c}\end{array}\!\!\!\right\rangle\!\!\!-\!\!\!\left\langle\!\!\!\begin{array}{c}HH\\\end{array}\!\!\!\right\rangle\!\!N-CH_3 \tag{198}$$

[60]

$$[60] \rightarrow \text{polymeric products} \tag{199}$$

protic agents (Hopkins, 1972) show that conversion to $PQ^{\cdot+}$ occurs on reaction with methanol in acetone or cyclohexane solutions. Whilst the actual reaction mechanisms for these oxidations remain obscure, their importance in redox processes of PQ^{2+} and $PQ^{\cdot+}$ cannot be overlooked, especially for reactions of $PQ^{\cdot+}$, preformed to the exclusion of PQ^{2+}. For example, relative rates of destruction of $PQ^{\cdot+}$ by 10^{-3} M HCl in water at 50°C were found to increase in the order 3, 13, 62, 564 for standard reactions where the molar ratio

$PQ^{2+}/PQ^{\cdot+}$ was 69, 9, 1, 0 respectively. For all these processes at least 50% of the initial $PQ^{\cdot+}$ is recovered as PQ^{2+} and the remainder appears as a polymeric pyridinium salt, very similar to that formed by reactions of $Ar_2\dot{C}OH$ with $PQ^{\cdot+}$ (Hyde and Ledwith, 1974b). Interestingly the proposed intermediate vinylpyridinium salts [60] may arise equally well from addition of (H^+) to PQ, or by addition of $(H\cdot)$ to $PQ^{\cdot+}$.

Paraquat and related salts function as highly efficient herbicides in a catalytic manner requiring the presence of oxygen and of chloroplast photo-activation. It is widely assumed (Dodge, 1971) that the paraquat functions mainly as a catalytic co-factor in the overall photoreduction of molecular oxygen to peroxides which then act to destroy green plant. Reaction of $PQ^{\cdot+}$ with oxygen is extremely rapid and a recent kinetic study has provided important mechanistic details. Thus pulse radiolysis studies (Farrington et al., 1973) show that $PQ^{\cdot+}$ reacts with O_2 and $O_2^{\cdot-}$ with specific rates of $7 \cdot 7 \times 10^8$ and $6 \cdot 5 \times 10^8$ M s^{-1} respectively. The similarity in these rate constants is in marked contrast to the difference in redox potentials of O_2 and O_2^- ($-0 \cdot 59$ V and $+1 \cdot 12$ V respectively) [(200) and (201)]. These

$$PQ^{\cdot+} + O_2 \rightarrow PQ^{2+} + O_2^{\cdot-} \qquad (200)$$

$$PQ^{\cdot+} + O_2^{\cdot-} \rightarrow PQ^{2+} + O_2^{2-} \qquad (201)$$

results do not yet distinguish between $O_2^{\cdot-}$ or O_2^{2-} (or more probably their protonated adducts $HO_2^{\cdot}$, H_2O_2) as the important phytotoxic product produced by paraquat. Nevertheless, the high rates of reaction of $PQ^{\cdot+}$ with O_2 require care in the design and interpretation of electron transfer reactions of paraquat and its cation radical.

7. CONCLUDING REMARKS

Over a period of not more than one decade, interest in cation radicals has dramatically expanded so that it is now commonplace to suggest a role for such intermediates in a great variety of well established organic processes. The growth of interest can be ascribed to parallel developments in the feasibility of anodic oxidation processes, and in the recognition of overall electron transfer as an important mechanistic pathway for excited state deactivation in organic photochemistry. In contrast to related anion radical systems, it is much more difficult to generate cation radicals in an unambiguous way from purely thermal processes but major developments in this area can be expected to arise from the stimulus provided by other techniques.

Recognition of the importance of cation radicals seems likely to accelerate in the future and should be especially important for the interpretation of the diverse oxidations and reductions which dominate many biological processes.

ACKNOWLEDGEMENTS

The authors are extremely grateful to NATO, whose financial support made possible the writing of this review and several other collaborative projects. Individually we are also indebted to the National Science Foundation for Grants Nos. GP.31414X (AJB), and GP.25989X (HJS) and to the Science Research Council (AL) who have provided further support for studies of cation radical chemistry.

REFERENCES

Aalbersberg, W. I., and Mackor, E. L. (1960). *Trans. Faraday Soc.* **56**, 1351.
Aalbersberg, W. I., Hoijtink, G. J., Mackor, E. L., and Weijland, W. P. (1959a). *J. Chem. Soc.* 3049.
Aalbersberg, W. I., Hoijtink, G. J., Mackor, E. L., and Weijland, W. P. (1959b). *J. Chem. Soc.* 3055.
Aalbersberg, W. I., Gaaf, J., and Mackor, E. L. (1961). *J. Chem. Soc.* 905.
Adams, J. Q., and Nicksie, S. W. (1962). *J. Amer. Chem. Soc.* **84**, 4355.
Adams, R. N. (1969). "Electrochemistry at Solid Electrodes", Marcel Dekker, Inc. New York, N.Y.
Akbulut, U., Fernandez, J. E., and Birke, R. L. (1975). *J. Polymer Sci., Polym. Chemistry Edn.* **13**, 133.
Akhavein, A. A., and Linscott, D. L. (1968). *Residue Rev.* **23**, 97.
Albery, W. J., and Hitchman, M. L. (1971). "Ring-Disc Electrodes", Clarendon Press, Oxford.
Albrecht, A. C. (1970). *Accts. Chem. Res.* **3**, 328.
Alcacer, L., and Maki, A. H. (1974). *J. Phys. Chem.* **78**, 215.
Alcais, P., and Rau, M.-C. (1969). *Bull. Soc. Chim. France*, 3390.
Alcock, N. W., and Waddington, T. C. (1962). *J. Chem. Soc.* 2510.
Ambrose, J. F., and Nelson, R. F. (1968). *J. Electrochem. Soc.* **115**, 1159.
Anand, S. P., Quarterman, L. A., Hyman, H. H., Migliorese, K. G., and Filler, R. (1975). *J. Org. Chem.* **40**, 807.
Andreades, S., and Zahnow, E. W. (1969). *J. Amer. Chem. Soc.* **91**, 4181.
Andrieux, C. P., and Saveant, J. M. (1974). *J. Electroanal. Chem.* **57**, 27.

Andrieux, C. P., Nadjo, L., and Saveant, J. M. (1970). *J. Electroanal. Chem.* **26**, 147.
Andrieux, C. P., Nadjo, L., and Saveant, J. M. (1973). *J. Electroanal. Chem.* **42**, 223.
Andrulis, Jr., P. J., and Dewar, M. J. S. (1966). *J. Amer. Chem. Soc.* **88**, 5483.
Andrulis, Jr., P. J., Dewar, M. J. S., Dietz, R., and Hunt, R. L. (1966). *J. Amer. Chem. Soc.* **88**, 5473.
Asanuma, T., Yamamoto, M., and Nishijima, Y. (1975). *Chem. Comm.* 56.
Atkinson, T. V., and Bard, A. J. (1971). *J. Phys. Chem.* **75**, 2043.
Badger, B., and Brockelhurst, B. (1970). *Trans. Faraday Soc.* **66**, 2939.
Baizer, M. M., (ed.) (1973). "Organic Electrochemistry", Marcel Dekker, New York.
Bandlish, B. K., Padilla, A. G., and Shine, H. J. (1975). *J. Org. Chem.* **40**, 2590.
Banks, R. E., Farmell, L. F., Haszeldine, R. N., Preston, P. N., and Sutcliffe, L. H. (1964). *Proc. Chem. Soc.* 396.
Bansal, S. R., Nonhebel, D. C., and Mancilla, J. M. (1973). *Tetrahedron*, **29**, 993.
Barbey, G., Delahaye, D., and Caullet, C. (1971). *Bull. Soc. Chim. France*, 3377.
Bard, A. J. (1971). *Pure Appl. Chem.* **25**, 379.
Bard, A. J., and Faulkner, L. R. (1976). In "Electroanalytical Chemistry" (A. J. Bard, ed.), Vol. 10, Marcel Dekker, New York.
Bard, A. J., and Park, S.-M. (1975). "The Exciplex" (M. Gordon and W. R. Ware, eds.), Academic Press, New York, p. 305.
Bard, A. J., and Phelps, J. (1970). *J. Electroanal. Chem. App.* 2.
Bard, A. J., and Santhanam, K. S. V. (1970). In "Electroanalytical Chemistry" (A. J. Bard, ed.), Vol. 5, Chapter 3, Marcel Dekker, New York.
Bard, A. J., Keszthelyi, C. P., Tachikawa, H., and Tokel, N. E. (1973). In "Chemiluminescence and Bioluminescence" (D. M. Hercules, J. Lee, and M. Cormier, eds.), p. 193. Plenum Press, New York.
Bard, A. J., Santhanam, K. S. V., Cruser, S. A., and Faulkner, L. R. (1975). In "Fluorescence" (G. G. Guilbault, ed.), Chapter 14, Marcel Dekker, N.Y.
Barnett, J. R., Hopkins, A. S., and Ledwith, A. (1973). *J.C.S. Perkin II.* 80.
Barry, C., Cauquis, G., and Maurey, M. (1966). *Bull. Soc. Chim. France*, 2510.
Barton, D. H. R., Leclerc, G., Magnus, P. D., and Menzies, I. D. (1972). *Chem. Comm.* 447.
Bauer, D., Beck, J. P., and Texier, P. (1971). *Coll. Czech. Chem. Commun.* **36**, 940.
Bawn, C. E. H., and Sharp, J. A. (1957). *J. Chem. Soc.* 1854.
Bawn, C. E. H., Bell, F. A., and Ledwith, A. (1968). *Chem. Comm.* 599.
Bell, F. A., Ledwith, A., and Sherrington, D. C. (1969). *J. Chem. Soc. (B)* 2719.
Bell, F. A., Beresford, P., Kricka, L. J., and Ledwith, A. (1974). *J.C.S. Perkin I*, 1788.
Belleau, B., and Au-Young, Y. K. (1960). *Can. J. Chem.* **47**, 2117.
Belozerov, A. I., Tanaseichuk, B. S., Shegal, I. L., and Sherzhaknova, L. M. (1972). *J. Org. Chem. USSR* **8**, 1969.
Bennema, P., Hoijtink, G. J., Lupinski, J. H., and van Woorst, J. D. W. (1959). *Mol. Phys.* **2**, 431.
Bennion, B. C., Auborn, J. J., and Eyring, E. M. (1972). *J. Phys. Chem.* **76**, 701.
Beresford, P., Iles, D. H., Krika, L. J., and Ledwith, A. (1974). *J.C.S. Perkin I*, 276.
Beresford, P., and Ledwith, A. (1970). *Chem. Comm.* 15.
Beresford, P., Lambert, M. C., and Ledwith, A. (1970). *J. Chem. Soc., (C)* 2508.
Bernasconi, C. F., Bergstrom, R. G., and Hünig, S. (1971). *Chem. Comm.* 1485.
Bezman, R., and Faulkner, L. R. (1972). *J. Amer. Chem. Soc.* **94**, 3699.

Billon, J.-P. (1961). *Bull. Soc. Chim. France,* 1923.
Blackburn, G. M., and Will, J. P. (1974). *Chem. Comm.* 67.
Blanchi, J.-P., and Watkins, A. R. (1974). *Chem. Comm.* 265.
Blomgren, G. E., and Kommandeur, J. (1961). *J. Chem. Phys.* **35,** 1636.
Blount, H. N. (1973). *J. Electroanal. Chem.* **42,** 271.
Blount, H. N., and Kuwana, T. (1970a). *J. Amer. Chem. Soc.* **92,** 5773.
Blount, H. N., and Kuwana, T. (1970b). *J. Electroanal. Chem.* **27,** 464.
Blount, H. N., Winograd, N., and Kuwana, T. (1970). *J. Phys. Chem.* **74,** 3231.
Bolhuis, P. A., Akkerman, O. S., and Los, J. M. (1974). *Chem. Comm.* 870.
Bolton, J. R., and Carrington, A. (1961). *Proc. Chem. Soc.* 385.
Bolton, J. R., Carrington, A., and dos Santas Viega, J. (1962). *Mol. Phys.* **5,** 615.
Bracke, W., Cheng, W. J., Pearson, J. M., and Szwarc, M. (1969). *J. Amer. Chem. Soc.* **91,** 203.
Brass, K., and Fanta, K. (1936). *Ber.* **69,** 1.
Brass, K., and Tengler, E. (1931). *Ber.* **64,** 1650.
Britt, A. D. (1964). *J. Chem. Phys.* **41,** 3069.
Brivati, J. A., Hulme, R., and Symons, M. C. R. (1961). *Proc. Chem. Soc.* 384.
Brown, H. C., and Grayson, M. (1953). *J. Amer. Chem. Soc.* **75,** 6285.
Brown, N. M. D., Cowley, D. J., and Murphy, W. J. (1973). *Chem. Comm.* 592.
Brouwer, D. M. (1962). *J. Catalysis* **1,** 372.
Brouwer, D. M., and van Doorn, J. A. (1972). *Rec. Trav. Chim.* **91,** 1110.
Bruce, C. R., Norberg, R. E., and Weissman, S. I. (1956). *Rec. Trav. Chim.* **24,** 473.
Bruck, D., and Guttman, D. E. (1968). *J. Amer. Chem. Soc.* **90,** 4964.
Bruni, P., Colonna, M., and Greci, L. (1971). *Tetrahedron,* **27,** 5893.
Bruning, W. H., Nelson, R. F., Marcoux, L. S., and Adams, R. N. (1967). *J. Phys. Chem.* **71,** 3055.
Bryce-Smith, D., Clarke, M. T., Gilbert, A., Klunklin, G., and Manning, C. (1971). *Chem. Comm.* 916.
Buck, H. M., Bloemhoff, W., and Oosterhoff, L. J. (1960). *Tetrahedron Lett.* 5.
Buck, H. M., Huizer, A. H., Oldenburg, S. J., and Schipper, P. (1972). *Rec. Trav. Chim.* **89,** 1085.
Burdon, J., and Parsons, I. W. (1975). *Tetrahedron,* 31, 2401.
Burrows, H. D., Greatorex, D., and Kemp, T. J. (1972). *J. Phys. Chem.* **76,** 20.
Burrows, H. D., Kemp, T. J., and Welbourn, M. J. (1973). *J. Chem. Soc. Dalton* 969.
Butler, M. A., Ferraris, J. P., Bloch, A. N., and Cowan, D. O. (1974). *Chem. Phys. Lett.* **24,** 600.
Cadogan, K. D., and Albrecht, A. C. (1965). *J. Chem. Phys.* **43,** 2550.
Carrington, A., Dravnieks, F., and Symons, M. C. R. (1959). *J. Chem. Soc.* 947.
Carter, M. K., and Vincow, G. (1967a). *J. Chem. Phys.* **47,** 292.
Carter, M. K., and Vincow, G. (1967b). *J. Chem. Phys.* **47,** 302.
Caullet, C. (1973). Personal communication.
Cauquis, G., and Cros, J.-L. (1971). *Bull. Soc. Chim. France,* 3760.
Cauquis, G., and Genies, M. (1967). *Bull. Soc. Chim. France,* 3220.
Cauquis, G., and Maurey-Mey, M. (1972). *Bull. Soc. Chim. France,* 3588.
Cauquis, G., and Maurey-Mey, M. (1973). *Bull. Soc. Chim. France,* 291
Cauquis, G., and Parker, V. D. (1973). In "Organic Electrochemistry" (M. M. Baizer, ed.), Marcel Dekker, Inc., New York, N.Y.
Cauquis, G., Fauvelot, G., and Rigaudy, J. (1968). *Bull. Soc. Chim. France,* 4928.
Cauquis, G., Cros, J.-L., and Genies, M. (1971). *Bull. Soc. Chim. France,* 3765.
Chandross, E. A. (1969). *Trans. N.Y. Acad. Sci. Ser. 2* **31,** 571.

Chester, A. W. (1970). *J. Org. Chem.* **35**, 1797.
Chiang, T. C., and Reddoch, A. H. (1970). *J. Chem. Phys.* **52**, 1371.
Chow, Y. L. (1973). *Accounts Chem. Res.* **6**, 354.
Chow, Y. L., Lau, M. P., Perry, R. A., and Tam, J. N. S. (1972). *Can. J. Chem.* **50**, 1044.
Clark, D. B., Fleischmann, M., and Pletcher, D. (1973). *J. Perkin II*, 1578.
Clerici, A., Minisci, F., Perchinunno, M., and Porta, O. (1974). *J.C.S. Perkin II* 416.
Coffen, D. L., Chambers, J. Q., Williams, D. R., Garrett, P. E., and Canfield, N. D. (1971). *J. Amer. Chem. Soc.* **93**, 2258.
Cohen, S. G., Parola, A., and Parsons, G. H. (1973). *Chem. Rev.* **73**, 141.
Cooksey, D., and Johnson, M. D. (1968). *J. Chem. Soc.* 1191.
Cooper, J. T., and Forbes, W. F. (1968). *Can. J. Chem.* **46**, 1158.
Corio, P. L., and Shih, S. (1970). *J. Catalysis* **18**, 126.
Cowell, G. W., Ledwith, A., White, A. C., and Woods, H. J. (1970). *J. Chem. Soc. (B)* 227.
Cox, M. T., Howarth, T. T., Jackson, A. H., and Kenner, G. W. (1969). *J. Amer. Chem. Soc.* **91**, 1232.
Creason, S. C., Wheeler, J., and Nelson, R. F. (1972). *J. Org. Chem.* **37**, 4440.
Crellin, R. A., and Ledwith, A. (1975). *Macromolecules* **8**, 93.
Cruser, S. A., and Bard, A. J. (1969). *J. Amer. Chem. Soc.* **91**, 267.
Cukman, D., and Pravdic, V. (1974). *J. Electroanal. Chem.* **49**, 415.
Dalahaye, D., Barbey, G., and Caullet, C. (1971). *Bull. Soc. Chim. France*, 3082.
Dallinga, G., Mackor, E. L., and Verrijn Stuart, A. A. (1958). *Mol. Phys.* **1**, 123.
Danen, W. C., and Rickard, R. C. (1972). *J. Amer. Chem. Soc.* **94**, 3254.
Dannenberg, J. J. (1975). *Angew. Chem. Int. Ed.* **14**, 641.
Danyluk, S. S., and Schneider, W. G. (1962). *Can. J. Chem.* **40**, 1884.
Davidson, R. S. (1969). *Chem. Comm.* 1450.
de Boer, E., and Praat, A. P. (1964). *Mol. Phys.* **8**, 291.
Deno, N. C., Billings, W. E., Fishbein, R., Pierson, C., Whalen, R., and Wyckoff, J. C. (1971). *J. Amer. Chem. Soc.* **93**, 438.
de Sorgo, M., Wasserman, B., and Szwarc, M. (1972). *J. Phys. Chem.* **76**, 3468.
Dessau, R. M., Shih, S., and Heiba, E. I. (1970). *J. Amer. Chem. Soc.* **92**, 412.
Dewar, M. J. S., and Nakaya, T. (1968). *J. Amer. Chem. Soc.* **90**, 7134.
Dietz, R., and Larcombe, B. E. (1970). *J. Chem. Soc., B* 1369.
Dodge, A. D. (1971). *Endeavour* **30**, 130.
Dollish, F. R., and Hall, W. K. (1965). *J. Phys. Chem.* **69**, 4402.
Dollish, F. R., and Hall, W. K. (1967). *J. Phys. Chem.* **71**, 1005.
Dolphin, D., and Felton, R. H. (1974). *Accounts. Chem. Res.* **7**, 26.
Dolphin, D., Muljiani, Z., Rousseau, K., Borg, D. C., Fajer, J., and Felton, R. H. (1973). *Annal. Acad. Sci. New York* **206**, 177.
Duffey, Jr., W. (1962). *J. Chem. Phys.* **36**, 490.
Eastman, J. W., Engelsma, G., and Calvin, M. (1962). *J. Amer. Chem. Soc.* **84**, 1339.
Eberson, L. (1975). *Chem. Comm.*, 826.
Eberson, L., and Nyberg, K. (1976). *Adv. Phys. Org. Chem.* **12**, 1.
Eberson, L., and Wilkinson, R. G. (1972). *Acta Chem. Scand.* **26**, 1671.
Edlund, O., Kinell, P.-O., Lund, A., and Shimizu, A. (1967). *J. Chem. Phys.* **46**, 3679.
Effenberger, F., and Gerlach, O. (1970). *Tetrahedron Lett.* 1669.
Elofson, R. M., and Edsberg, R. L. (1957). *Canad. J. Chem.* **35**, 646.
Elson, I. H., and Kochi, J. K. (1973). *J. Amer. Chem. Soc.* **95**, 5060.
Engler, E. M., and Patel, V. V. (1975). *Chem. Comm.* 671.

Evans, T. R., Wake, R. W., and Jaenicke, O. (1975). In "The Exciplex" (M. Gordon and W. R. Ware, eds.), Academic Press, New York.
Fajer, J., Borg, D. C., Forman, A., Dolphin, D., and Felton, R. H. (1970). *J. Amer. Chem. Soc.* **92**, 3451.
Fajer, J., Borg, D. C., Forman, A., Felton, R. H., Vegh, L., and Dolphin, D. (1973). *Ann. Acad. Sci. New York* **206**, 349.
Fajer, J., Borg, D. C., Forman, A., Adler, A. D., and Varadi, V. (1974). *J. Amer. Chem. Soc.* **96**, 1238.
Farid, S., Hartman, S. E., and Evans, T. R. (1975). In "The Exciplex" (M. Gordon and W. R. Ware, eds.), Academic Press, New York.
Farid, S., and Shealer, S. E. (1973). *Chem. Comm.* 677.
Farrington, J. A., Ledwith, A., and Stam, M. F. (1969). *Chem. Comm.* 259.
Farrington, J. A., Ebert, M., Land, E. J., and Fletcher, K. (1973). *Biochem. Biophys. Acta* **314**, 372.
Faulkner, L. R., and Bard, A. J. (1969). *J. Amer. Chem. Soc.* **91**, 6497.
Faulkner, L. R., Tachikawa, H., and Bard, A. J. (1972). *J. Amer. Chem. Soc.* **94**, 691.
Fava, A., Sogo, P. B., and Calvin, M. (1957). *J. Amer. Chem. Soc.* **79**, 1078.
Feitelson, J., and Hayon, E. (1973). *J. Phys. Chem.* **77**, 10.
Feitelson, J., Hayon, E., and Treinin, A. (1973). *J. Amer. Chem. Soc.* **95**, 1025.
Feldberg, S. W. (1966). *J. Amer. Chem. Soc.* **88**, 390.
Feldberg, S. W. (1966a). *J. Phys. Chem.* **70**, 3928.
Feldberg, S. (1969). *J. Phys. Chem.* **73**, 1238.
Ferraris, J. P., Pocchler, T. O., Bloch, A. N., and Cowan, D. O. (1973). *Tetrahedron Lett.* 2553.
Fioshin, M. Ya., Mirkind, L. A., and Zhurinov, M. Zh. (1973). *Russ. Chem. Rev.* **42**, 293.
Fitzgerald, Jr. E. A., Wuelfing, Jr. P., and Richtol, H. H. (1971). *J. Phys. Chem.* **75**, 2737.
Fleischfresser, B. E., Cheng, W. J., Pearson, J. M., and Szwarc, M. (1968). *J. Amer. Chem. Soc.* **90**, 2172.
Flockhart, B. D., Scott, J. A. N., and Pink, R. C. (1966). *Trans. Far. Soc.* **62**, 730.
Forbes, W. F., and Sullivan, P. D. (1966). *J. Amer. Chem. Soc.* **88**, 2862.
Forman, A., Borg, D. C., Felton, R. H., and Fajer, J. (1971). *J. Amer. Chem. Soc.* **93**, 2790.
Foster, R. (1969a). "Organic Charge-Transfer Complexes", pp. 267-273, Academic Press, New York.
Foster, R. (1969b). "Organic Charge-Transfer Complexes", p. 326, Academic Press, New York.
Foster, R. (1974). "Molecular Complexes", Vol. II, p. 251, Paul Elek, London.
Frank, S. N., Bard, A. J., and Ledwith, A. (1975). *J. Electrochem. Soc.* **122**, 898.
Freed, D. J., and Faulkner, L. R. (1971). *J. Amer. Chem. Soc.* **93**, 2097, 3565.
Freed, D. J., and Faulkner, L. R. (1972). *J. Amer. Chem. Soc.* **94**, 4790.
Fried, J., and Schumm, D. E. (1967). *J. Amer. Chem. Soc.* **89**, 5508.
Fritsch, J. M., Weingarten, H., and Wilson, J. D. (1970). *J. Amer. Chem. Soc.* **92**, 4038.
Fry, A. J. (1972). "Synthetic Organic Electrochemistry", Harper and Row, N.Y.
Fuhrhop, J.-H., and Mauzerall, D. (1969). *J. Amer. Chem. Soc.* **91**, 4174.
Fuhrhop, J.-H., Wasser, P., Riesner, D., and Mauzerall, D. (1972). *J. Amer. Chem. Soc.* **94**, 7996.

Fuhrhop, J.-H., Kadish, K. M., and Davis, D. G. (1973). *J. Amer. Chem. Soc.* **95**, 5140.
Fung, K. W., Chambers, J. Q., and Mamantov, G. (1973). *J. Electroanal. Chem.* **47**, 81.
Geiger, F. E., Trichilo, C. L., Minn, F. L., and Filipescu, N. (1971). *J. Org. Chem.* **36**, 357.
Gibbons, W. A., Porter, G., and Savadatti, M. I. (1965). *Nature* **206**, 1355.
Glass, R. S., Britt, W. J., Miller, W. N., and Wilson, G. S. (1973). *J. Amer. Chem. Soc.* **95**, 2375.
Godfrey, T. S., and Porter, G. (1966). *Trans. Faraday Soc.* **62**, 7.
Gordon, M., and Ware, W. R. (1975). "The Exciplex", Academic Press, N.Y.
Gough, T. A., and Peover, M. E. (1966). "Polarography—1964", p. 1017, Macmillan, London.
Grace, J. A., and Symons, M. C. R. (1959). *J. Chem. Soc.* 958.
Gradowski, M., and Latowski, T. (1974). *Tetrahedron* **30**, 767.
Gruver, G. A., and Kuwana, T. (1972). *J. Electroanal. Chem.* **36**, 85.
Hammerich, O., and Parker, V. D. (1972). Abstracts, 4th Organosulfur Symposium, Ronneby-Brunn, Sweden.
Hammerich, O., and Parker, V. D. (1972). *Chem. Comm.* 156.
Hammerich, O., and Parker, V. D. (1973). *Electrochim. Acta* **18**, 537.
Hammerich, O., and Parker, V. D. (1974). *Chem. Comm.* 245; *J. Amer. Chem. Soc.* **96**, 4289.
Hammerich, O., Moe, S., and Parker, V. D. (1972). *Chem. Comm.* 156.
Hammond, G. S., Hamilton, E. J., Weiner, S. A., Hefter, H. J., and Gupta, A. (1971). IUPAC, XXIII Inter. Congr. of Pure and Appl. Chem., Vol. 4, p. 257, Butterworths, London.
Hanotier, J., Hanotier-Bridoux, M., and de Radzitsky, P. (1973). *J.C.S. Perkin II*, 382.
Hanson, P., and Norman, R. O. C. (1973). *J.C.S. Perkin II*, 264.
Happ, J. W., McCall, M. T., and Whitten, D. G. (1971). *J. Amer. Chem. Soc.* **93**, 5496.
Happ, J. W., Ferguson, J. A., and Whitten, D. G. (1972). *J. Org. Chem.* **37**, 1485.
Hausser, K. (1956). *Z. Naturforsch.* **11A**, 20.
Hausser, K. H., and Murrell, J. N. (1957). *J. Chem. Phys.* **27**, 500.
Heiba, E. I., and Dessau, R. M. (1971). *J. Amer. Chem. Soc.* **93**, 995.
Heiba, E. I., Dessau, R. M., and Koehl, Jr., W. J. (1968a). *J. Amer. Chem. Soc.* **90**, 1082.
Heiba, E. I., Dessau, R. M., and Koehl, Jr., W. J. (1968b). *J. Amer. Chem. Soc.* **90**, 2707.
Heiba, E. I., Dessau, R. M., and Koehl, Jr., W. J. (1968c). *J. Amer. Chem. Soc.* **90**, 5905.
Heiba, E. I., Dessau, R. M., and Koehl, Jr., W. J. (1969a). *J. Amer. Chem. Soc.* **91**, 138.
Heiba, E. I., Dessau, R. M., and Koehl, Jr., W. J. (1969b). *J. Amer. Chem. Soc.* **91**, 6830.
Hercules, D. M. (1969). *Accounts Chem. Res.* **2**, 301.
Hercules, D. M. (1971). In "Physical Methods of Organic Chemistry", 4th Ed., Part II (A. Weissberger and B. Rossiter, eds.), Academic Press, N.Y.
Hilpert, S., and Wolf, L. (1913). *Ber.* **46**, 2215.
Hirao, K.-I., and Yonemitsu, O. (1972). *Chem. Comm.* 812.
Hirschler, A. E. (1966). *J. Catalysis* **5**, 196.
Hirschler, A. E., and Hudson, J. O. (1964). *J. Catalysis* **3**, 329.

Hirschler, A. E., Neikam, W. C., Barmby, D. S., and James, R. L. (1965). *J. Catalysis* **4**, 628.
Hirschon, J. M., Gardner, D. M., and Fraenkel, G. K. (1953). *J. Amer. Chem. Soc.* **75**, 4115.
Hodgson, R. L., and Raley, J. H. (1965). *J. Catalysis* **4**, 6.
Hoefs, E. V., and Bruning, W. H. (1972). Private communication.
Hoijtink, G. J. (1970). In "Advances in Electrochemistry and Electrochemical Engineering" (P. Delahay and C. Tobias, eds.), Vol. 7, p. 221, Wiley-Interscience, N.Y.
Hoijtink, G. J., and Weijland, W. P. (1957). *Rec. Trav. Chim.* **76**, 836.
Holroyd, R. A., and Russell, R. L. (1974). *J. Phys. Chem.* **78**, 2128.
Holtz, H. D. (1972). *J. Org. Chem.* **37**, 2069.
Hopkins, A. S. (1972). Ph.D. Thesis, University of Liverpool.
Hopkins, A. S., and Ledwith, A. (1971). *Chem. Comm.* 830.
Hopkins, A. S., Ledwith, A., and Stam, M. F. (1970). *Chem. Comm.* 494.
Howarth, O. W., and Fraenkel, G. K. (1970). *J. Chem. Phys.* **52**, 6258.
Hughes, F., Kirk, R. D., and Patten, F. W. (1964). *J. Chem. Phys.* **40**, 872.
Hulme, R., and Symons, M. C. R. (1965a). *Nature* **206**, 293.
Hulme, R., and Symons, M. C. R. (1965b). *J. Chem. Soc.* 1120.
Hünig, S. (1967). *Pure and Appld. Chem.* **15**, 109.
Hünig, S., Schlaf, H., Kiesslich, G., and Scheutzow, D. (1969). *Tetrahedron Lett.* 2271.
Hünig, S., Scheutzow, D., and Schlaf, H. (1972). *Justus Liebigs. Ann.* **765**, 126.
Hünig, S., Gross, J., and Schenk, W. (1973). *Annalen* **142**, 324.
Hünig, S., Kiesslich, G., Quast, H., and Scheutzow, D. (1973). *Justus Liebigs Ann.* **1**, 310.
Hyde, P., and Ledwith, A. (1974a). "Molecular Complexes" (R. Foster, ed.), Vol. II, p. 173, Paul Elek, London.
Hyde, P., and Ledwith, A. (1974b). *J.C.S. Perkin II*, 1768.
Hyde, P., and Ledwith, A. (1975). Unpublished results.
Ichikawa, T., and Ohta, N. (1973). *J. Amer. Chem. Soc.* **95**, 8175.
Iida, Y., (1971). *Bull. Chem. Soc. Japan* **44**, 663.
Ilten, D. B., and Calvin, M. (1966). *J. Chem. Phys.* **42**, 3760.
Imura, T., Yamamoto, N., Tsubomura, H., and Kawabe, K. (1971). *Bull. Chem. Soc. Japan* **44**, 3185.
Isaacs, N. S. (1966). *J. Chem. Soc. (B)* 1053.
Isenberg, I., and Baird, S. L. (1962). *J. Amer. Chem. Soc.* **84**, 3803.
Itoh, M. (1974). *J. Amer. Chem. Soc.* **96**, 7390.
Janata, J., and Williams, M. B. (1972). *J. Phys. Chem.* **76**, 1178.
Jensen, B. S., and Parker, V. D. (1973). *Electrochim. Acta* **18**, 665.
Jeftic, L. J., and Adams, R. N. (1970). *J. Amer. Chem. Soc.* **92**, 1332.
Johnson, A. W. (1969). "Ylid Chemistry", Academic Press, New York.
Johnson, C. S., Jr., and Holz, J. B. (1959). *J. Chem. Phys.* **30**, 899.
Johnson, C. S. Jr. (1963). *J. Chem. Phys.* **39**, 2111.
Johnson, C. S. Jr., and Gutowsky, H. S. (1963). *J. Chem. Phys.* **39**, 58.
Johnson, M. D., and Calvin, M. (1973). *Nature* **241**, 271.
Johnson, P. V. (1971). *J. Chem. Soc. (A)* 2856.
Joschek, J.-I., and Grossweiner, L. I. (1966). *J. Amer. Chem. Soc.* **88**, 3261.
Kainer, H., and Hausser, K. H. (1953). *Ber.* **86**, 1563.
Kainer, H., and Wherle, A. (1955). *Ber.* **88**, 1147.
Katz, T. J. (1960). *J. Amer. Chem. Soc.* **82**, 3785.
Kawamori, A., Honda, A., Joo, N., Suzuki, K., and Ooshika, Y. (1966). *J. Chem. Phys.* **44**, 4363.

Kehrmann, F., and Diserens, L. (1915). *Ber.* **48**, 318.
Kehrmann, F., and Sandoz, M. (1917). *Ber.* **50**, 1673.
Kehrmann, F., Speitel, J., and Grandmougin, E. (1914). *Ber.* **47**, 2976.
Keszthelyi, C. P., Tachikawa, H., and Bard, A. J. (1972). *J. Amer. Chem. Soc.* **94**, 1522.
Kevan, L. (1973). In "Advances in Radiation Chemistry" (M. Burton and J. L. Magee, eds.), Vol. 4, Wiley Interscience, New York.
Khan, Z. H., and Khanna, B. N. (1974). *Can. J. Chem.* **52**, 827.
Kim, K., and Shine, H. J. (1974). *J. Org. Chem.* **39**, 2537.
Kim, K., Hull, V. J., and Shine, H. J. (1974). *J. Org. Chem.* in press.
Kimura, K., Yamada, H., and Tsubomura, H. (1968). *J. Chem. Phys.* **48**, 440.
Kimura, K., Yamazaki, T., and Katsumata, S. (1971). *J. Phys. Chem.* **75**, 1768.
Kimura, K., Achiba, Y., and Katsumata, S. (1973). *J. Phys. Chem.* **77**, 2520.
King, D. M., and Bard, A. J. (1965). *J. Amer. Chem. Soc.* **87**, 419.
Kira, A., Asai, S., and Imamura, M. (1972). *J. Phys. Chem.* **76**, 1119.
Knunyants, I. L., Rozhtov, I. N., Bukhtiarov, A. V., Gol'din, M. M., and Kudryavtsev, R. V. (1970). *Bull. Acad. Sci. USSR, Divn. Chem. Sci.* 1155.
Knunyants, I. L., Rozhtov, I. N., and Bukhtiarov, A. V. (1971). *Bull. Acad. Sci. USSR, Divn. Chem. Sci.* 1286.
Koch, V. R., Miller, L. L., Clark, D. B., Fleischmann, M., Joslin, T., and Pletcher, D. (1973). *J. Electroanal. Chem.* **43**, 318.
Kochi, J. K. (1975). *Tetrahedron Lett.* 41.
Kochi, J. K., Tang, R. T., and Bernath, T. (1973). *J. Amer. Chem. Soc.* **95**, 7114.
Komatsu, T., and Lund, A. (1972). *J. Phys. Chem.* **76**, 1727.
Komatsu, T., Lund, A., and Kinell, P.-O. (1972). *J. Phys. Chem.* **76**, 1721.
Kommandeur, J., and Hall, F. R. (1961). *J. Chem. Phys.* **34**, 129.
Kon, H., and Blois, M. S. (1958). *J. Chem. Phys.* **28**, 743.
Kosower, E. M. (1965). *Progr. Phys. Org. Chem.* **3**, 81.
Kosower, E. M. (1968). "Introduction to Physical Organic Chemistry", Wiley, New York.
Kosower, E. M., and Cotter, J. L. (1964). *J. Amer. Chem. Soc.* **86**, 5524.
Kowert, B. A., Marcoux, L., and Bard, A. J. (1972). *J. Amer. Chem. Soc.* **94**, 5538.
Koyama, K., Susuki, T., and Tsutsumi, S. (1965). *Tetrahedron Lett.* 627.
Kramer, D. N., Guilbault, G. G., and Miller, F. M. (1967). *J. Org. Chem.* **32**, 1163.
Krasnovsky, A. A. (1972). *Biophys. Journal* **12**, 749.
Kricka, L. J., and Ledwith, A. (1973). *J. Chem. Soc. Perkin I*, 294.
Kricka, L. J., and Ledwith, A. (1974). *Synthesis* 539.
Kudirka, P. J., and Nicholson, R. S. (1972). *Anal. Chem.* **44**, 1786.
Kurita, Y., Sonoda, T., and Sata, M. (1970). *J. Catalysis* **19**, 82.
Kuwana, T. (1966). In "Electroanalytical Chemistry" (A. J. Bard, ed.), Vol. 1, Chapter 3, Marcel Dekker, N.Y.
Land, E. J., and Porter, G. (1963). *Trans. Faraday Soc.* **59**, 2016, 2027.
Landsberg, R., and Müller, S. (1974). *Electrochim. Acta* **19**, 681.
Ledwith, A. (1972). *Accounts Chem. Res.* **5**, 133.
Ledwith, A. (1975). In "The Exciplex" (M. Gordon and W. R. Ware, eds.), Academic Press, New York, p. 209.
Ledwith, A., and Purbrick, M. B. (1973). *Polymer (Lond.)* **14**, 521.
Ledwith, A., and Russell, P. J. (1974a). *J.C.S. Perkin II*, 582.
Ledwith, A., and Russell, P. J. (1974b). *Chem. Comm.* 291.
Ledwith, A., and Russell, P. J. (1974c). *Chem. Comm.* 959.
Ledwith, A., and Russell, P. J. (1975). *J.C.S. Perkin II*, in press.

Ledwith, A., and Sherrington, D. C. (1972). *Macromolecular Syntheses* **4**, 183.
Lewis, G. N., and Bigeleisen, J. (1943). *J. Amer. Chem. Soc.* **65**, 2419.
Lewis, G. N., and Lipkin, D. (1942). *J. Amer. Chem. Soc.* **64**, 2801.
Lewis, I. C., and Singer, L. S. (1965). *J. Chem. Phys.* **43**, 2712.
Lewis, I. C., and Singer, L. S. (1966). *J. Chem. Phys.* **44**, 2082.
Lexa, D., and Reix, M. (1974). *J. Chim. Phys.* 511.
Libert, M., and Caullet, C. (1974). *Bull. Soc. Chim. France*, 800.
Libert, M., Caullet, C., and Huguet, J. (1972). *Bull. Soc. Chim. France*, 3639.
Libert, M., Caullet, C., and Barbey, G. (1973). *Bull. Soc. Chim. France*, 536.
Lindsay Smith, J. R., Norman, R. O. C., and Rowley, A. G. (1973). *J.C.S. Perkin I*, 566.
Liptay, W., Briegleb, G., and Schindler, K. (1962). *Z. Elektrochem.* **66**, 331.
Littler, J. S. (1971). In "Essays on Free Radical Chemistry" Special Publication No. 24, Ch. 15, Chemical Society (London).
Lucken, E. A. C. (1962). *J. Chem. Soc.* 4963.
Ludman, C. J., McCarron, E. M., and O'Malley, R. F. (1972). *J. Electrochem. Soc.* **119**, 874.
MacLean, C., and Van der Waals, J. H. (1957). *J. Chem. Phys.* **27**, 287.
Malachesky, P. A., Marcoux, L. S., and Adams, R. N. (1966). *J. Phys. Chem.* **70**, 2064.
Maloy, J. T., Prater, K. B., and Bard, A. J. (1971). *J. Amer. Chem. Soc.* **93**, 5959.
Mani, S. R., and Shine, H. J. (1975). *J. Org. Chem.* **40**, 2756.
Mann, C. K., and Barnes, K. K. (1970). "Electrochemical Reactions in Nonaqueous Systems, Marcel Dekker, New York.
Marcoux, L. S. (1971). *J. Amer. Chem. Soc.* **93**, 537.
Marcoux, L. S., and Adams, R. N. (1974). *J. Electroanal. Chem.* **49**, 111.
Marcoux, L. S., Fritsch, J. M., and Adams, R. N. (1967). *J. Amer. Chem. Soc.* **89**, 5766.
Marcus, R. A. (1968). *Electrochim. Acta* **13**, 995.
Masui, M., and Ohmori, H. (1973). *J.C.S. Perkin II* 1112.
Mataga, N. (1975). In "The Exciplex" (M. Gordon and W. R. Ware, eds.), Academic Press, New York, p. 113.
Matsunaga, Y. (1961). *Bull. Chem. Soc. Japan* **34**, 1293.
Matsunaga, Y., and Suzuki, Y. (1973). *Bull. Chem. Soc. Japan* **46**, 719.
Mayeda, E. A., and Bard, A. J. (1973). *J. Amer. Chem. Soc.* **95**, 6223.
McClelland, R. A., Norman, R. O. C., and Thomas, C. B. (1972). *J.C.S. Perkin I*, 562.
McConnell, H. M., and Lynden-Bell, R. M. (1962). *J. Chem. Phys.* **36**, 2393.
Medzhikov, A. A., Rozantsev, E. G., and Neiman, M. B. (1966). *Proc. Acad. Sci. USSR* **168**, 486.
Meites, L. (1960). In "Techniques of Organic Chemistry" (A. Weissberger, ed.), Chapter 49, Interscience, New York.
Meyer, K. (1910). *Ber.* **43**, 161.
Meyer, W. C. (1970). *J. Phys. Chem.* **74**, 2118, 2127.
Michaelis, L., and Granick, S. (1943). *J. Amer. Chem. Soc.* **65**, 1747.
Michaelis, L., Schubert, M. P., and Granick, S. (1939). *J. Amer. Chem. Soc.* **61**, 1981.
Michaelis, L., Granick, S., and Schubert, M. P. (1941). *J. Amer. Chem. Soc.* **63**, 351.
Miller, L. L., Nordbloom, G. D., and Mayeda, E. A. (1972). *J. Org. Chem.* **37**, 916.
Millington, J. P. (1969). *J. Chem. Soc. (B)* 982.
Minisci, F. (1975). *Accounts Chem. Res.* **8**, 165.

Minisci, F., and Galli, R. (1965). *Tetrahedron Lett.* 1679.
Minisci, F., Gardini, G. P., and Bertini, F. (1970). *Can. J. Chem.* **48**, 544.
Minn, F. L., Trichilo, C. L., Hunt, C. R., and Filipescu, N. (1970). *J. Amer. Chem. Soc.* **92**, 3600.
Möckel, H., Yuen, J., and Kevan, L. (1973). *J. Phys. Chem.* **77**, 3036.
Möckel, H., Bonifaviec, M., and Asmus, K.-D. (1974). *J. Phys. Chem.* **78**, 282.
Mulliken, R. S. (1952). *J. Amer. Chem. Soc.* **74**, 811; *J. Phys. Chem.* **56**, 801.
Mulliken, R. S., and Person, W. B. (1969). "Molecular Complexes", Chapter 16, Wiley, New York.
Murata, Y., and Shine, H. J. (1969). *J. Org. Chem.* **34**, 3368.
Nagai, S., Ohnishi, S.-I., and Nitta, I. (1971). *Bull. Chem. Soc. Japan* **44**, 230.
Neale, R. S. (1973). *Synthesis* **3**, 1.
Neckars, D. C., and Collenbrander, D. P. (1968). *Tetrahedron Lett.* 5045.
Nelson, R. F., and Feldberg, S. W. (1969). *J. Phys. Chem.* **73**, 2663.
Nelson, S. F., Weisman, G. R., Hintz, P. J., Olp, D., and Fahey, M. R. (1974). *J. Amer. Chem. Soc.* **96**, 2916.
Neunteufel, R. A., and Arnold, D. R. (1973). *J. Amer. Chem. Soc.* **95**, 4080.
Nicholson, R. S., and Shain, I. (1964). *Anal. Chem.* **36**, 706.
Nicholson, R. S., and Shain, I. (1965). *Anal. Chem.* **37**, 178.
Niizuma, S., Nakamaru, K., and Koizumi, M. (1972). *Chem. Lett. (Japan)* 59.
Nilsson, S. (1973). *Acta Chem. Scand.* **27**, 329.
Nishinaga, A., Ziemeck, P., and Matsuura, T. (1970). *J. Chem. Soc. (C)* 2613.
Norman, R. O. C., and Thomas, C. B. (1970). *J. Chem. Soc. (B)* 421.
Norman, R. O. C., Thomas, C. B., and Wilson, T. S. (1971). *J. Chem. Soc. (B)* 518.
Norman, R. O. C., Thomas, C. B., and Wilson, T. S. (1973a). *J.C.S. Perkin I*, 325.
Norman, R. O. C., Thomas, C. B., and Ward, P. J. (1973b). *J.C.S. Perkin I*, 2914.
Nyberg, K. (1970). *Acta Chem. Scand.* **24**, 473.
Oae, S. (1970). *Quart. Reports on Sulf. Chem.* **5**, 53.
Oae, S., Moriyama, M., Numata, T., and Kunieda, N. (1974). *Bull. Chem. Soc. Japan* **47**, 179, and references therein.
O'Leary, M. H., and Stach, R. W. (1972). *J. Org. Chem.* **37**, 1491.
Olmstead, M. L., Hamilton, R. G., and Nicholson, R. S. (1969). *Anal. Chem.* **41**, 260.
Onopchenko, A., Schulz, J. G. D., and Seekircher, R. (1972). *J. Org. Chem.* **37**, 2950.
Oohashi, Y., and Sakata, T. (1973). *Bull. Chem. Soc. Japan* **46**, 765.
Oster, G. K., and Yang, N.-L. (1973). *J. Phys. Chem.* **77**, 2159.
Ottolenghi, M. (1973). *Accounts Chem. Res.* **6**, 153.
Pac, C., Tosa, T., and Sakurai, H. (1972). *Bull. Chem. Soc. Japan* **45**, 1169.
Papouchado, L., Adams, R. N., and Feldberg, S. W. (1969). *J. Electroanal. Chem.* **21**, 408.
Parker, V. D. (1970). *Acta Chem. Scand.* **24**, 2768, 2775, 3151, 3171, 3455.
Parker, V. D., and Eberson, L. (1970a). *J. Amer. Chem. Soc.* **92**, 7488.
Parker, V. D., and Eberson, L. (1970b). *Acta Chem. Scand.* **24**, 3542.
Parker, V. D., and Eberson, L. (1972). *Chem. Comm.* 441.
Parker, V. D., Nyberg, K., and Eberson, L. (1969). *J. Electroanal. Chem.* **22**, 150.
Parker, V. D., Dirlan, J. P., and Eberson, L. (1971). *Acta Chem. Scand.* **25**, 341.
Parsons, R. (1959). In "Handbook of Electrochemical Constants" p. 73, Butterworth, London.
Paskovich, D. H., and Reddoch, A. H. (1972). *J. Amer. Chem.. Soc.* **94**, 6938.
Peover, M. E. (1967). In "Electroanalytical Chemistry" (A. J. Bard, ed.), Vol. 2, Chapter 1. Marcel Dekker Inc., N.Y.

Peover, M. E. (1968). *Electrochim. Acta* **13**, 1083.
Peover, M. E. (1971). In "Reactions of Molecules at Electrodes" (N. S. Hush, ed.), p. 259, Wiley Interscience, N.Y.
Peover, M. E., and White, B. S. (1967). *J. Electroanal. Chem.* **13**, 93.
Phelps, J., Santhanam, K. S. V., and Bard, A. J. (1967). *J. Amer. Chem. Soc.* **89**, 1752.
Phillips, T. E., Kistenmacher, T. J., Ferraris, J. P., and Cowan, D. O. (1973). *Chem. Comm.* 471.
Pittman, Jr., C. U., McManus, S. P., and Larsen, J. W. (1972). *Chem. Rev.* **72**, 357.
Pleskov, Yu. V., and Filinovsky, V. Yu. (1972). "Rotating Disk Electrodes", Moscow.
Popp, G. (1972). *J. Org. Chem.* **37**, 3058.
Porter, G. B., Simpson, J., and Baughan, E. C. (1970). *J. Chem. Soc. (A)* 2806.
Prater, K. B., and Bard, A. J. (1970). *J. Electrochem. Soc.* **117**, 335.
Prater, K. B., and Bard, A. J. (1970a). *J. Electrochem. Soc.* **117**, 1517.
Puglisi, V. J., and Bard, A. J. (1972). *J. Electrochem. Soc.* **119**, 833.
Pummerer, R., and Gassner, S. (1913). *Ber.* **46**, 2310.
Rao, V. R., and Ramakrishnan, V. (1971). *Chem. Comm.* 971.
Reddoch, A. H., Williams, D. F., and Chang, T. C. (1971). *J. Chem. Phys.* **54**, 2051.
Reuss, R. H., and Winters, L. J. (1973). *J. Org. Chem.* **38**, 3993.
Reynolds, R., Line, L. L., and Nelson, R. F. (1974). *J. Amer. Chem. Soc.* **96**, 1087.
Ristagno, C. V., and Shine, H. J. (1971a). *J. Amer. Chem. Soc.* **93**, 1811.
Ristagno, C. V., and Shine, H. J. (1971b). *J. Org. Chem.* **36**, 4050.
Roberts, R. M., Barter, C., and Stone, H. (1959). *J. Phys. Chem.* **63**, 2077.
Rochlitz, J. (1967). *Tetrahedron* **23**, 3043.
Romans, D., Bruning, W. H., and Michejda, C. J. (1969). *J. Amer. Chem. Soc.* **91**, 3859.
Ronlan, A., and Parker, V. D. (1970). *Chem. Comm.* 1567.
Ronlan, A., and Parker, V. D. (1971). *J. Chem. Soc. (C)* 3214.
Ronlan, A., and Parker, V. D. (1974). *Chem. Comm.* 33.
Ronlan, A., Coleman, J., Hammerich, O., and Parker, V. D. (1974). *J. Amer. Chem. Soc.* **96**, 845.
Rooney, J. J., and Pink, R. C. (1961). *Proc. Chem. Soc.* 142.
Rooney, J. J., and Pink, R. C. (1962). *Trans. Faraday Soc.* **58**, 1632.
Roth, H. D., and Lamola, A. A. (1974). *J. Amer. Chem. Soc.* **69**, 6270.
Rozhkov, I. N., Bukhtiarov, A. V., Gal'pern, E. G., and Knunyants, I. L. (1971). *Proc. Acad. Sci. USSR* **199**, 598.
Rozhkov, I. N. and Alyev, I. Y. (1975). *Tetrahedron* **31**, 977.
Rundel, W., and Scheffler, K. (1963). *Tetrahedron Lett.* 993.
Russell, C. D. (1963). *Anal. Chem.* **35**, 1291.
Sakata, T., and Nagakura, S. (1970). *Bull. Chem. Soc. Japan* **43**, 2414.
Sakata, K., Kamiya, Y., and Ohta, N. (1969). *Can. J. Chem.* **47**, 387.
Sakata, T., Onodera, A., Tsubomura, H., and Kawai, N. (1974). *J. Amer. Chem. Soc.* **96**, 3365.
Sato, Y., Kinoshita, M., Sano, M., and Akamatu, H. (1969). *Bull. Chem. Soc. Japan* **42**, 548, 3051.
Sato, H., and Aoyama, Y. (1973). *Bull. Chem. Soc. Japan* **46**, 631.
Saveant, J. M., and Vianello, E. (1963). *Electrochim. Acta* **8**, 905.
Saveant, J. M., and Vianello, E. (1965). *Electrochim. Acta* **10**, 905.
Saveant, J. M., and Vianello, E. (1967). *Electrochim. Acta* **12**, 1545.

Saveant, J. M., Andrieux, C. P., and Nadjo, L. (1973). *J. Electroanal. Chem.* **41**, 137.
Sawyer, D. T., and Roberts, Jr., J. L. (1974). "Experimental Electrochemistry for Chemists", Wiley-Interscience, New York, N.Y.
Schäfer, H., and Steckham, E. (1970). *Tetrahedron Lett.* 3835.
Schoot, C. J., Ponjee, J. J., van Dam, H. T., van Doorn, R. A., and Boleijn, P. T. (1973). *Appl. Phys. Lett.* **23**, 64.
Schwarz, F. P., and Albrecht, A. C. (1973). *J. Phys. Chem.* **77**, 2808.
Scott, J. A. N., Flockhart, B. D., and Pink, R. C. (1964). *Proc. Chem. Soc.* 134.
Scott, E. J. Y., and Chester, A. W. (1972). *J. Phys. Chem.* **76**, 1520.
Seo, E. T., Nelson, R. F., Fritsch, J. M., Marcoux, L. S., Leedy, D. W., and Adams, R. N. (1966). *J. Amer. Chem. Soc.* **88**, 3498.
Shang, D. T., and Blount, H. N. (1974). *J. Electroanal. Chem.* **54**, 305.
Shank, N. E., and Dorfman, L. M. (1970). *J. Chem. Phys.* **52**, 4441.
Shepard, F. E., Rooney, J. J., and Kemball, C. (1962). *J. Catalysis* **1**, 379.
Shigemitsu, Y., and Arnold, D. R. (1975). *Chem. Comm.* 407.
Shih, S., and Dessau, R. M. (1971). *J. Chem. Phys.* **55**, 3757.
Shimada, M., Masuhara, H., and Mataga, N. (1973). *Bull. Chem. Soc., Japan* **46**, 1903.
Shine, H. J., and Goodin, R. D. (1970). *J. Org. Chem.* **35**, 949.
Shine, H. J., Kim, K., and Mani, S. R. (1975). Unpublished work.
Shine, H. J., and Murata, Y. (1969). *J. Amer. Chem. Soc.* **91**, 1872.
Shine H. J., and Piette, L. (1962). *J. Amer. Chem. Soc.* **84**, 4798.
Shine, H. J., and Ristagno, C. V. (1972). *J. Org. Chem.* **37**, 3424.
Shine, H. J., and Shade, L. R. (1974). *J. Heterocycl. Chem.* **11**, 139.
Shine, H. J., and Silber, J. J. (1972). *J. Amer. Chem. Soc.* **94**, 1026.
Shine, H. J., and Sullivan, P. D. (1968). *J. Phys. Chem.* **72**, 1390.
Shine, H. J., Rahman, M., Seeger, H., and Wu, G. S. (1967). *J. Org. Chem.* **32**, 1901.
Shine, H. J., Silber, J. J., Bussey, R. J., and Okuyama, T. (1972). *J. Org. Chem.* **37**, 2691.
Sidgwick, N. V. (1966). "The Organic Chemistry of Nitrogen" (I. T. Millar, and H. D. Springall, eds.), Clarendon Press, Oxford.
Silber, J. J., and Shine, H. J. (1971). *J. Org. Chem.* **36**, 2623.
Singer, L. S., and Kommandeur, J. (1961). *J. Chem. Phys.* **34**, 133.
Singer, L. S., and Lewis, I. C. (1965). *J. Amer. Chem. Soc.* **87**, 4695.
Sioda, R. E. (1968). *J. Phys. Chem.* **72**, 2322.
Sorensen, S. P., and Bruning, W. H. (1972). *J. Amer. Chem. Soc.* **94**, 6352.
Sorensen, S. P., and Bruning, W. H. (1973). *J. Amer. Chem. Soc.* **94**, 2445.
Sosnovsky, G., and Rawlinson, D. J. (1972). In "Advances in Free Radical Chemistry" (G. H. Williams, ed.), Vol. IV, p. 203.
Spanswick, J., and Ingold, K. U. (1970). *Can. J. Chem.* **50**, 546, 554.
Stamires, D. M., and Turkevich, J. (1964). *J. Amer. Chem. Soc.* **86**, 749.
Stanienda, A., and Biebl, G. (1967). *Z. Physik. Chem.* (Neue Folge) **52**, 254.
Starkova, S. D., Chekhecheva, I. P., Nepgodina, O. I., Misina, V. P., and Medvdev, S. S. (1970). *Proc. Acad. Sci. USSR, Chem. Sectn.* 797.
Steckham, E., and Kuwana, T. (1974). *Berichte Bunsenges.* **78**, 253.
Stuart, J. D., and Ohnesorge, W. E. (1971). *J. Electroanal. Chem.* **30**, App. 11.
Stüwe, A., Weber-Schäfer, M., and Baumgärtel, H. (1974). *Ber. Bunsenges.* **78**, 309.
Sullivan, P. D. (1968). *J. Amer. Chem. Soc.* **90**, 3618.
Sullivan, P. D. (1973). *J. Amer. Chem. Soc.* **95**, 288.
Sullivan, P. D., and Norman, L. J. (1972). *Can. J. Chem.* **50**, 2141.
Susuki, T., Koyama, K., Omori, A., and Tsutsumi, S. (1968). *Bull. Chem. Soc. Japan* **41**, 2663.

Svanholm, U., and Parker, V. D. (1973a). *Acta Chem. Scand.* **27**, 1454.
Svanholm, U., and Parker, V. D. (1973b). *J.C.S. Perkin II*, 1594.
Svanholm, U., Ronlan, A., and Parker, V. D. (1974). *J. Amer. Chem. Soc.* **96**, 5108.
Svanholm, U., Hammerich, O., and Parker, V. D. (1975). *J. Amer. Chem. Soc.* **97**, 101.
Szwarc, M. (ed.) (1972). "Ions and Ion Pairs in Organic Reactions", Vol. 1, Wiley Interscience, N.Y.
Tachikawa, H., and Bard, A. J. (1974). *Chem. Phys. Letters* **26**, 246.
Tachikawa, H., and Bard, A. J. (1974a). *Chem. Phys. Letters* **26**, 10.
Takimoto, K., Nakayama, S., Suzuki, K., and Ooshika, Y. (1968). *Bull. Chem. Soc. Japan* **41**, 1974.
Takimoto, K., and Miura, M. (1972). *Bull. Chem. Soc. Japan* **45**, 2231.
Tanei, T. (1968). *Bull. Chem. Soc. Japan* **4**, 833.
Tokel, N. E., Keszthelyi, C. P., and Bard, A. J. (1972). *J. Amer. Chem. Soc.* **94**, 4872.
Tokunaga, H., Ono, Y., and Keii, T. (1971). *Bull. Chem. Soc. Japan* **46**, 3569.
Torii, S., Matsuyama, Y., Kawasaki, K., and Uneyama, K. (1973). *Bull. Chem. Soc. Japan* **46**, 2912.
Tosa, T., Pac, C., and Sakurai, H. (1969). *Tetrahedron Lett.* 3635.
Trahanovsky, W. S., and Brixius, D. W. (1973). *J. Amer. Chem. Soc.* **95**, 6778.
Tsujino, Y. (1968). *Tetrahedron Lett.* 2545, 4111.
Turcot, L., Glasel, A., and Funt, B. L. (1974). *J. Polym. Sci., Polym. Letters Edn.* **12**, 692.
van der Ploeg, R. E., de Korte, R. W., and Kooyman, E. C. (1968). *J. Catalysis* **10**, 52.
Van Hook, J. P., and Tobolsky, A. V. (1958). *J. Amer. Chem. Soc.* **80**, 779.
van Willigen, H., de Boer, E., Cooper, J. R., and Forbes, W. F. (1968). *J. Chem. Phys.* **49**, 1190.
Vincow, G. (1968). In "Radical Ions" (E. T. Kaiser and L. Kevan, eds.), Chapter 4, pp. 151-209, Wiley Interscience, New York.
Vincow, G., and Johnson, P. M. (1963). *J. Chem. Phys.* **39**, 1143.
Volke, J. (1968). *Coll. Czech. Chem. Comm.* **33**, 3044.
Walling, C. (1975). *Accounts Chem. Res.* **8**, 125.
Walling, C., and Camaioni, D. M. (1975). *J. Amer. Chem. Soc.* **97**, 1603.
Walling, C., and Johnson, R. A. (1975). *J. Amer. Chem. Soc.* **97**, 363.
Walter, R. I. (1955). *J. Amer. Chem. Soc.* **77**, 5999.
Walter, R. I. (1966). *J. Amer. Chem. Soc.* **88**, 1923.
Ward, R. L. (1963). *J. Chem. Phys.* **39**, 852.
Wasserman, E., Hutton, R. S., Kuck, V. J., and Chandross, E. A. (1974). *J. Amer. Chem. Soc.* **96**, 1965.
Weinberg, N. L., and Belleau, B. (1973). *Tetrahedron* **29**, 279.
Weinberg, N. L., and Brown, E. A. (1966). *J. Org. Chem.* **31**, 4054.
Weinberg, N. L., and Reddy, T. B. (1968). *J. Amer. Chem. Soc.* **90**, 91.
Weinstein, M., Kuszkat, K. A., and Dobkin, J. (1975). *Chem. Comm.* 68.
Weiss, J. (1946). *Trans. Faraday Soc.* **42**, 116.
Weissman, S. I., de Boer, E., and Conradi, J. J. (1957). *J. Chem. Phys.* **26**, 963.
Weitz, E., and Schwechten, H. W. (1926). *Ber.* **59B**, 2307.
Weitz, E., and Schwechten, H. W. (1927). *Ber.* **60B**, 545.
Weller, A. (1968). *Pure and Appl. Chem.* **16**, 115.
Weller, A., and Zachariasse, K. (1967). *J. Chem. Phys.* **46**, 4984.
Wheeler, L. O., Santhanam, K. S. V., and Bard, A. J. (1966). *J. Phys. Chem.* **70**, 404.
Wieland, H. (1907). *Ber.* **40**, 4260.

Wieland, H., and Wecker, E. (1910). *Ber.* **43**, 699.
Wildes, P. D., Lichtin, N. N. and Hoffman, M. Z. (1975). *J. Amer. Chem. Soc.* **97**, 2288.
Wilk, M., Bez, W., and Rochlitz, J. (1966). *Tetrahedron* **22**, 2599.
Winters, L. J., Borror, A. L., and Smith, N. G. (1967). *Tetrahedron Lett.* 2313.
Winters, L. J., Smith, N. G., and Cohen, M. I. (1970). *Chem. Comm.* 642.
Wolberg, A., and Manassen, J. (1970). *J. Amer. Chem. Soc.* **92**, 2982.
Wolff, M. E. (1963). *Chem. Rev.* **63**, 55.
Wu, C. Y., and Hall, W. K. (1967). *J. Catalysis* **8**, 394.
Wudl, F., and Southwick, E. W. (1974). *Chem. Comm.* 254.
Wurster, C., and Sendtner, R. (1879). *Ber.* **12**, 1803, 2071.
Yang, G. C., and Pohland, A. E. (1972). *J. Phys. Chem.* **76**, 1504.
Yang, G. C., and Pohland, A. E. (1973). In "Advances in Chemistry" (E. C. Blair, ed.), p. 33.
Yamada, H., and Kimura, K. (1969). *J. Chem. Phys.* **51**, 5733.
Yamamoto, M., Asanuma, T., and Nishijima, Y. (1975). *Chem. Comm.* 53.
Yamazaki, T., and Kimura, K. (1972). *J. Phys. Chem.* **76**, 1549.
Yokosawa, Y., and Miyashita, I. (1956). *J. Chem. Phys.* **25**, 796.
Yoshida, K., and Fueno, T. (1972). *J. Org. Chem.* **37**, 4145.
Yoshida, K., and Fueno, T. (1973). *J. Org. Chem.* **38**, 1045.
Yoshida, K., Saeki, T., and Fueno, T. (1971). *J. Org. Chem.* **36**, 3673.
Yoshino, A., Ohashi, M., and Yonezawa, T. (1971). *Chem. Comm.* 97.
Yuen, S. H., Bagner, J. E., and Myles, D. (1967). *Analyst* **92**, 375.
Zachariasse, K. (1975). "The Exciplex" (M. Gordon and W. R. Ware, eds.), Academic Press, New York, p. 275.
Zweig, A. (1968). *Adv. Photochem.* **6**, 425.

^{13}C Nmr Spectroscopy in Macromolecular Systems of Biochemical Interest

STEVEN N. ROSENTHAL and JANOS H. FENDLER

Department of Chemistry, Texas A & M University, College Station, Texas 77843, U.S.A.

1. Introduction	280
Principles of Natural Abundance Fourier Transfer ^{13}C Magnetic Resonance Spectroscopy	280
Chemical Shift Assignments	282
Spin-Lattice Relaxation	283
Aspects of ^{13}C Nmr of Biomacromolecules	286
2. Carbohydrates	287
Monosaccharides and their Derivatives	287
Disaccharides	304
Oligo and Polysaccharides	309
3. Nucleic Acids and Constituents	323
Nucleosides	324
Nucleotides	332
Nucleic Acids	346
4. Proteins and their Residues	350
Amino-acids and Peptides	350
Native and Denatured Proteins	371
5. Lipids	382
Phospholipids	382
Miscellaneous Lipids	388
6. Macromolecular Model Systems	390
Micellar Surfactants	390
Macrocyclic Compounds	394
References	406
Acknowledgements	406
Addendum	415

1. INTRODUCTION

Natural abundance carbon-13 nuclear magnetic resonance, ^{13}C nmr, spectroscopy is increasingly being utilized for the investigation of macromolecular systems. Information may be obtained directly for carbon atoms constituting the molecular backbone irrespective of the substituent atoms or groups. Recent developments in instrumentation have successfully overcome the inherently small sensitivity of ^{13}C nmr experiments on non-enriched samples which contain ca. 1·1% carbon-13. Briefly, the present day experimental technique takes advantage of stable spectrometers, wide-band proton decoupling and pulsed Fourier transform. The purpose of the present review is to guide the newcomer through the established methodology and to stimulate his interest in the application of this powerful technique. Principles underlying ^{13}C nmr spectroscopy are only treated briefly providing the most essential fundamentals. Background and details are available in authoritative recent books (Levy and Nelson, 1972; Stothers, 1972a; Breitmaier and Voelter, 1974) and review articles (Roberts, 1971; Breitmaier *et al.*, 1971a; Grutzner, 1972; Stothers, 1972b; Mochel, 1972; Gurd and Keim, 1973; Anet and Levy, 1973; Wilson and Stothers, 1973; Oster *et al.*, 1974). Emphasis will be placed on recent structural and conformational studies of relatively large molecules, particularly those of biochemical importance. The ^{13}C nmr of carbohydrates, nucleic acids and their constituents, proteins and their residues, lipids and macromolecular model systems will be discussed in detail and attempts will be made to cover the literature exhaustively. This somewhat arbitrary selection is dictated by the availability of recent summaries on ^{13}C spectroscopic measurements of steroids (Gray, 1973), alkaloids (Gray, 1973; Wenkert *et al.*, 1974) corrinoids and porphyrins (Gray, 1973) as well as on the utilization of ^{13}C in biosynthetic studies (Grutzner, 1972; Gray, 1973).

Principles of Natural Abundance Fourier Transform ^{13}C Magnetic Resonance Spectroscopy

A considerable advantage of ^{13}C nmr spectroscopy is the large range of chemical shifts. Resonance frequencies for most organic molecules are spread over 200 ppm. Additionally, with complete proton decoupling each magnetically non-equivalent carbon appears as a sharp single line and individual lines are resolvable if separated by only 0·1 ppm. Observation of over 30 individual ^{13}C resonance frequencies for a given macromolecule is now quite common. Such considerable resolving power clearly renders ^{13}C nmr a very effective tool for investigating subtle structural changes as well as providing a convenient method for the analysis of complex mixtures.

The price to pay for the high resolution of ^{13}C nmr is its low sensitivity. The smaller gyromagnetic ratio of ^{13}C coupled with its low natural abundance results in an approximately 6000-fold decrease in the sensitivity of a given sample compared to ^{1}H nmr experiments (Anet and Levy, 1973). Early ^{13}C nmr work was carried out, therefore, on solutions in the molar range. With present day instrumentation it is possible to observe ^{13}C nmr spectra for 10^{-2} M solutions of moderately small molecules (Levy and Nelson, 1972). Development of Fourier transform spectroscopy increased the sensitivity by an order of magnitude. This method relies on the time-averaged computer accumulation of data from repetitive equidistant rf pulsing of the entire spectrum (Farrar and Becker, 1971; Becker and Farrar, 1972). An added advantage of the Fourier transform spectroscopy is that spin-lattice relaxation times, T_1, for all the resolvable carbons in the molecule can be obtained and can be used to aid the assignment of ^{13}C resonance frequencies as well as to provide information on the forces controlling molecular motions (Levy, 1973). Determination of spin lattice relaxation times typically involves the consecutive application of 180° and 90° pulses at suitable intervals, t. The first pulse reverses the orientation of all the nuclear spins which immediately begin to relax (each at their own specific rate) but before all the spins can return to equilibrium the system is irradiated by the second pulse and sampled. Repeating the complete experiment using several appropriate intervals, t, values for T_1 can be determined (Levy, 1973). Values of T_1 can also be determined by progressive saturation using 90°, t, 90° pulse sequences (Freeman and Hill, 1971). Since this method is less time consuming, it is particularly useful for the analysis of larger molecules at low concentrations.

Wide band proton decoupling results not only in the appearance of individual carbon resonances as sharp singlets but also in an additional increase in sensitivity to maximum of 1·99. The increased sensitivity is the consequence of the collapse of proton decoupled multiplets and the nuclear Overhauser enhancement (Noggle and Schirmer, 1971).

Chemical Shift Assignments

Assignments of ^{13}C resonance frequencies to individual carbon atoms is, of course, the foremost task in the interpretation of the observed spectra. Several strategies have been applied to assist unequivocal assignments (Levy and Nelson, 1972; Stothers, 1972a, b). The initial appraisal is assisted by the established ranges for the resonance frequencies of carbon atoms bearing a variety of functional groups. ^{13}C Chemical shifts, relative to the methyl carbons of tetramethylsilane (TMS), of the most common functional groups are given in Table 1. Although benzene, carbon disulfide (CS_2), dimethyl sulfoxide (DMSO), dioxane, and methyl iodide were used as standards in earlier work, TMS is now generally accepted as the primary ^{13}C standard. All of the chemical shifts reported in this review have been converted to the TMS scale using the following chemical shift differences: benzene, +128·5 ppm; internal CS_2, +192·8 ppm; DMSO, +40·5 ppm; and dioxane, +67·4 ppm unless specified otherwise (see Stothers, 1972a). Available compilations of ^{13}C magnetic resonance frequencies (Johnson and Jankowski, 1972; Breitmaier et al., 1974) should also be consulted. The relatively large ranges of ^{13}C chemical shifts for given functional groups and their possible overlap necessitates the utilization of model compounds and chemical shift correlations. Such correlations are frequently additive and analogous to those observed for ^{1}H chemical shifts (Stothers, 1972a; Anet and Levy, 1973). Electron-withdrawing substituents attached to carbon atoms generally cause downfield shifts, the shifts being marked for aliphatic and aromatic systems (Stothers, 1972a, b). Comparisons of chemical shifts of model compounds with those to be assigned are simplified somewhat since in the absence of solute-solvent interactions, solvent and anisotropy effects are small (<1 ppm) and can therefore be neglected.

Determination of spin-lattice relaxation times also aids chemical shift assignments (*vide supra*). The more rapid reorientation of the methyl group in cholesteryl chloride with respect to the whole molecule, for example, results in a markedly longer relaxation time than that observed for the carbon atoms constituting the steroid skeleton (Allerhand et al., 1971b).

Selective enrichment of a given carbon atom in a molecule with ^{13}C will enhance its signal relative to the other carbons; thereby making its resonance more easily discerned. Similarly, deuteration of a carbon-bearing proton diminishes the resonance peak due to that carbon by virtue of losing the nuclear Overhauser enhancement (Koch and Perlin, 1970; Gorin, 1974; Breitmaier, 1974).

Lanthanide chemical shift reagents have also been employed to facilitate structural assignments, but they are less effective in ^{13}C experiments than in ^{1}H due to the incursion of both contact-type and pseudo-contact (i.e. anisotropy) shifts (Grutzner, 1972; Levy and Komoroski, 1974). For *non-shifting* paramagnetic spin-lattice relaxation reagents this complication does not arise, and this type of reagent has been profitably utilized to remove or suppress the nuclear Overhauser effect and to shorten long spin-lattice relaxation times (Levy and Komoroski, 1974).

In summary, strategies for chemical shift assignments include the careful consideration of available data on model compound correlations as well as examination of spin lattice relaxation times, off-resonance decoupling and isotopic substitution. In spite of all these considerations false assignments can easily be made. The reported ^{13}C resonance frequencies are often given in terms of tentative assignments. All the available data for ^{13}C resonance frequencies are presented, however, in the ensuing compilations. The cited reference of the most recent work should be consulted in the cases where there is more than one assignment.

Spin-Lattice Relaxation

Brief references have been made above to determinations of ^{13}C spin-lattice relaxation times. Utilization of these methods for obtaining information on molecular motions and interactions will be emphasized throughout the ensuing discussion. Standard books (Levy and Nelson, 1972; Stothers, 1972a), review articles (Anderson *et al.*, 1971; Becker *et al.*, 1971; Lyerla and Grant, 1972; Gray, 1973; Levy, 1973; Breitmaier *et al.*, 1975), and original publications should be consulted for details concerning the theory, technique and interpretation. Briefly, ^{13}C spin-lattice relaxation in ^{13}C[^{1}H] experiments originates from fluctuating magnetic fields with components at the ^{13}C Larmor frequency. These fields arise from dipole–dipole interaction among nuclei such as ^{13}C and ^{1}H or between nuclei and unpaired electrons. The spin-rotation mechanism is also often operative, particularly in the case of small molecules or flexible molecular segments. Other relaxation mechanisms such as scalar coupling or chemical shift anisotropy are much less important

TABLE 1
^{13}C Chemical Shift Ranges of Functional Groups[a]

General classification	Specific type of compound	400	300	200	100	0	−100	−200	−300
Carbonyls	Ketones			RCR′ (C=O)					
	Aldehydes			RCH (C=O)					
	Acids			RCOH					
	Esters			RCOR					
	Amides + imides			RCN					
	Anhydrides			−C(=O)−O−C(=O)−					
	Acid Halides			RCX (C=O)					
	Carbonates			−O−C(=O)−O−					
	α-Haloketones			−C(=O)−C−X					
	α-β Unsaturated ketones			−C(=O)−C=C−					
Oxygen-containing	Ethers				−C−O−				
	Alcohols				−C−O−				
Nitrogen-containing	Nitriles			R−C≡N					
	Isocyanides			R−N≡C					
	Oximes			C=N−OH					
				R−N=C=O					

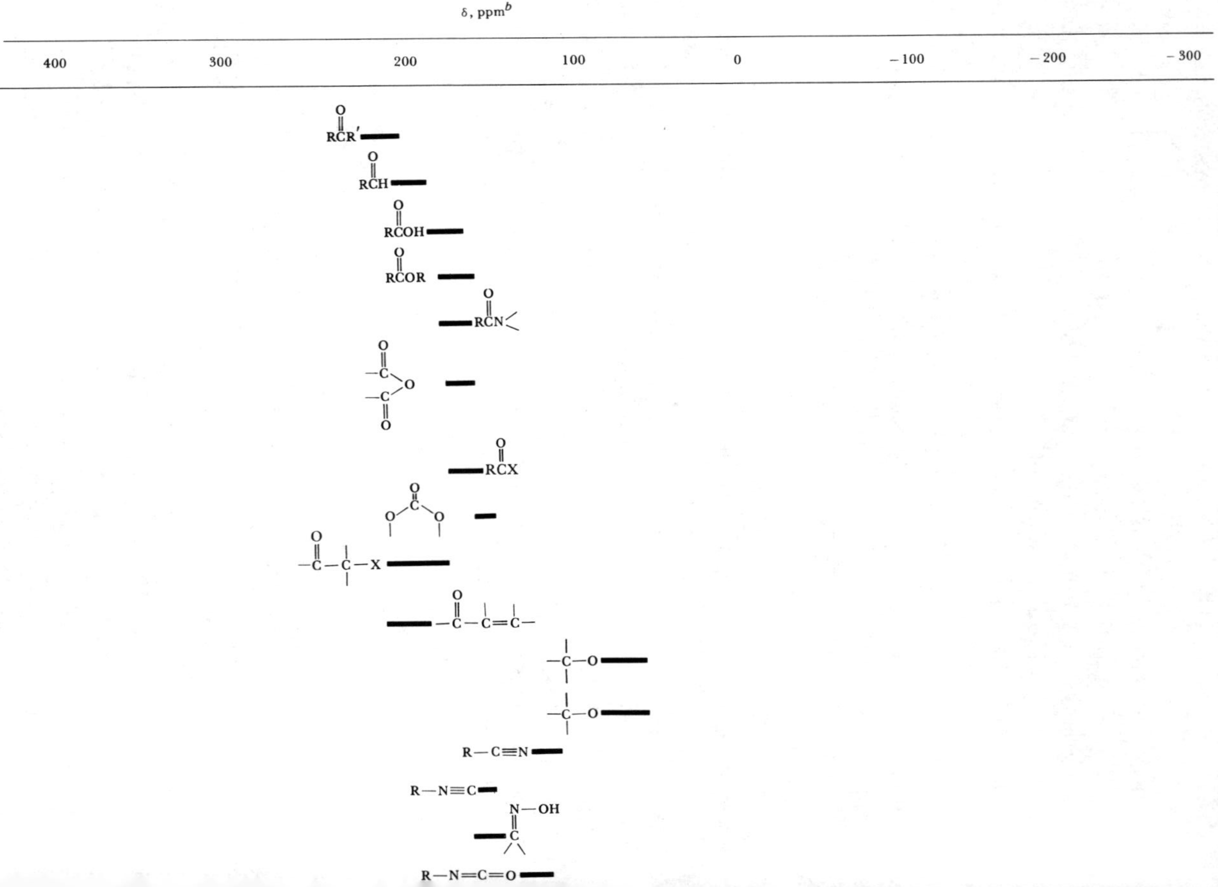

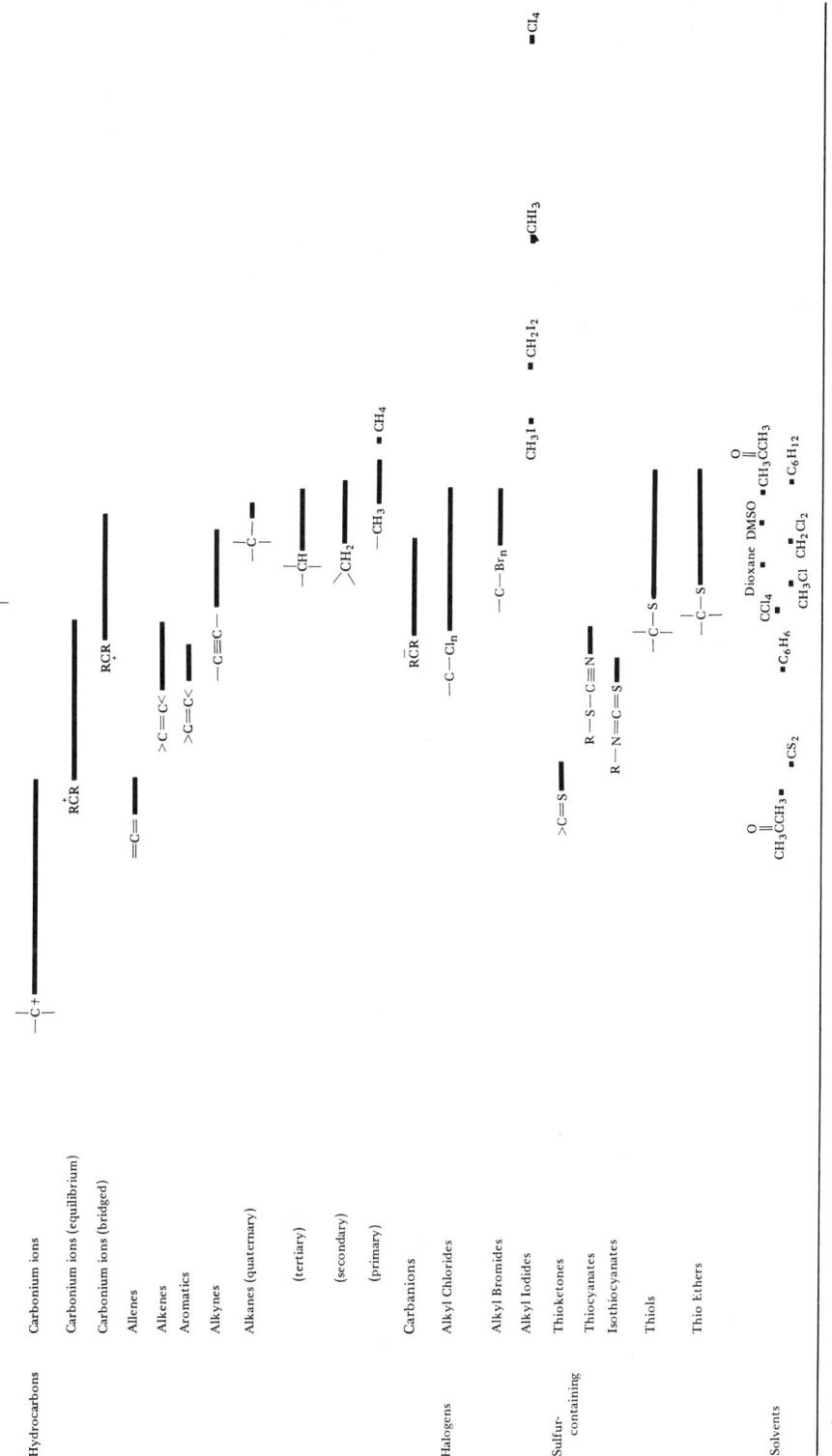

[a] Data taken from Breitmaier et al., 1971a and Levy and Nelson, 1972.
[b] Chemical shifts in parts per million relative to TMS.

in ^{13}C nmr. Substantial information may quite often be obtained from ^{13}C spin-lattice relaxation times in combination with nuclear Overhauser signal enhancement factors measured during ^{1}H decoupling by simple arithmetical manipulation (Levy, 1973). In addition to those specific applications to be cited subsequently, ^{13}C spin-lattice relaxation studies have provided information on internal rotation and tumbling of relatively small molecules, the extent and site of carbon protonation, intra- and intermolecular steric effects, hydrogen-bonding and ion-pairing. These applications have been summarized in detail by Levy (1973).

Aspects of ^{13}C nmr of Biomacromolecules

The high resolving power of ^{13}C FTnmr renders it an extremely sensitive tool for differentiating among magnetically discrete carbon atoms in biomacromolecules. Such sensitivity is achieved by wide-band proton decoupling and nuclear Overhauser enhancements. Since the magnitude of the nuclear Overhauser effects depends directly on the carbon relaxation mechanism, due care needs to be exercised prior to the interpretation of integrated ^{13}C resonance intensities (Allerhand et al., 1971b). An additional disadvantage of the noise decoupling is the loss of information on coupling constants. Off-resonance, single frequency and gated decoupling can, however, be used to obtain individual C—H coupling constants. Additional sensitivity is gained by using relatively large (up to 20 mm diameter) sample tubes (Allerhand et al., 1972). The combination of large sample tubes, proton noise decoupling and long time accumulation (overnight or longer) of Fourier transform spectra allows the determination of high resolution ^{13}C nmr spectra of 10^{-3} M macromolecules with sufficient sensitivity. Future developments in instrumentation and methodology will undoubtedly increase the detection sensitivity further. It should be recognized, however, that spectral resolution is limited by the appreciable number of chemically identical carbon atoms as well as by the frequently small differences in chemical shifts in the different environments. The presence of relatively few ^{13}C resonance frequencies in biopolymers is generally taken to indicate the occurrence of repeating structural units. While the gross features of the spectra are simplified, subtle conformational changes are not reflected. Additionally, substituents or groups whose relative concentration is small will have low signal-to-noise ratios and hence will not be resolvable.

Variation of the probe temperature can have a dramatic effect on the intensity and resolution of the resonances. An increase in temperature will generally allow more sample to be dissolved, and it can also be accompanied by a change in conformation of the sample, i.e. thermal denaturation. This conformational change usually results in a more resolved spectrum than that for the native molecule. Decreasing the temperature of solutions of small molecules leads to increased signal intensities due to a change in population of the energy levels. However, this is not effective for macromolecular systems because at low temperature there are inevitable viscosity increases which lead ultimately to the formation of gels and precipitates, resulting in line broadening.

It is evident that difficulties in the experimental techniques and data analysis will not hinder the explosive progress of ^{13}C nmr studies of biopolymers.

2. CARBOHYDRATES

Proton magnetic resonance studies have considerably assisted the elucidation of structures and conformations of mono-, di- and relatively simple oligosaccharides *in solution* (Hall, 1964; Inch, 1969, 1972; Kotowycz and Lemieux, 1973). While ^{13}C magnetic resonance spectroscopy profitably supplements the ^{1}H data, its real potential lies in the characterization of larger and more complex carbohydrates. The necessary foundations for such investigations are now well laid. Pioneering studies involved the assignment of the observed ^{13}C resonance frequencies for several monosaccharides to the specific carbon nuclei. Strategies employed for these assignments have been summarized (Stothers, 1972a) and will not, therefore, be reiterated here.

More recently reported chemical shift assignments and structural elucidation of the more complex carbohydrates as well as investigations of anomerization and conformation changes will form the ensuing discussion. Attention will be repeatedly drawn to the many advantages of ^{13}C spectroscopy, the potential of which has been as yet hardly explored.

Monosaccharides and their Derivatives

Table 2 summarizes the available ^{13}C chemical shift assignments for monosaccharides and their derivatives. The cited references and

TABLE 2

^{13}C Chemical Shifts of Monosaccharides, Derivatives, and Analogous Compounds
(Numbered formulae are on pp. 296–299)

Compound	δ, ppm[a]										
	C-1	C-2	C-3	C-4	C-5	C-6	C-7	Methyl	Phenyl	Other	Reference
α-D-Fructofuranose [1a][b]	62·1	105·3	83·0	77·0	82·2	62·1					Koerner et al., 1973
β-D-Fructofuranose [1b][b]	63·9	102·4	76·5	75·5	81·5	63·3					Koerner et al., 1973
α-D-Fructofuranose [1a][c]	63·8	105·5	82·9	77·0	82·2	61·9					Que and Gray, 1974
β-D-Fructofuranose [1b][c]	63·6	102·6	75·4	76·4	81·6	63·2					Que and Gray, 1974
α-D-Fructopyranose [1d][c]	64·1	99·1	70·5	68·4	70·0	64·7					Perlin et al., 1973
β-D-Fructopyranose [1d][d]	65·7	99·1	70·2	70·9	68·9	64·3					
α-D-Fructose-6-phosphate [2a][b]		105·34	82·59	76·93	81·36	64·47					Koerner et al., 1973
β-D-Fructose-6-phosphate [2b][b]	63·75	102·40	76·24	75·42	80·77	65·38					Koerner et al., 1973
α-D-Fructose-1,6-diphosphate [3a][b]		105·7	82·40	77·47	82·7						Koerner et al., 1973
β-D-Fructose-1,6-diphosphate [3b][b]	66·88	102·0	76·90	75·09	80·4	65·44					Koerner et al., 1973
α-D-Xylose [4a][e]	93·1	73·9	72·5	70·4	61·9						Stothers, 1972a
β-D-Xylose [4b][e]	97·6	75·1	76·8	70·2	64·0						Stothers, 1972a
Methyl-α-D-xylose [5a][e]	99·5	72·4	74·4	70·5	61·9			55·8			Stothers, 1972a
Methyl-β-D-xylose [5b][e]	105·0	73·9	76·9	70·5	65·9			57·8			Stothers, 1972a
Methyl-α-D-xylopyranoside [5a][f]	102·1	73·95	75·9	72·0	63·64			57·8			Breitmaier et al., 1971b
Methyl-β-D-xylopyranoside [5b][f]	106·55	75·5	78·4	71·8	67·7			59·7			Breitmaier et al., 1971b
α-D-Lyxose [6a][e]	95·0	71·5	71·1	68·5	63·9						Stothers, 1972a
β-D-Lyxose [6b][e]	95·0	70·8	73·7	67·6	64·9						Stothers, 1972a
α-D-Ribose [7a][e]	94·8	70·1	71·7	69·0	64·3						Stothers, 1972a
β-D-Ribose [7b][e]	95·3	72·6	72·6	70·0	64·3						Stothers, 1972a
Methyl-β-D-ribopyranoside [8a][f]	103·85	70·4	72·65	69·85	65·55			58·3			Breitmaier et al., 1971b
α-D-Arabinose [9a][e]	97·7	72·9	73·4	69·6	69·6						Stothers, 1972a
β-D-Arabinose [9b][e]	93·5	67·1	67·1	69·7	63·3						Stothers, 1972a
α-Methyl-D-arabinopyranoside [10a][f]	107·0	73·75	75·35	71·15	68·9			59·95			Breitmaier et al., 1971b
β-Methyl-D-arabinopyranoside [10b][f]	102·65	71·7	71·7	70·95	65·35			58·1			Breitmaier et al., 1971b
α-D-Glucose [11a][e]	93·0	72·5	73·8	70·6	72·3	61·8					Stothers, 1972a
β-D-Glucose [11b][e]	96·8	75·2	76·7	70·6	76·7	61·8					Stothers, 1972a
α-D-Glucopyranose [11a][g]	93·3	73·1	74·4	71·2	72·9	62·4					Usui et al., 1973b
β-D-Glucopyranose [11b][g]	97·1	75·6	77·3	71·2	77·3	62·4					Usui et al., 1973b
3-O-Methyl-α-D-glucopyranose [12a][g]	93·4	72·6	84·1	70·6	72·8	62·3		61·3			Usui et al., 1973b
					77·3	62·3		61·3			Usui et al., 1973b

289

Compound	C1	C2	C3	C4	C5	C6	OMe/other	Reference
3-O-Methyl-α-D-glucose [12a]e	93.1	72.1	83.7	70.1	72.3	61.7	60.9	Stothers, 1972a
3-O-Methyl-β-D-glucose [12b]e	97.0	74.7	86.2	69.9	76.7	61.7	60.6	Stothers, 1972a
2-Deoxy-α-D-glucose [13a]e	92.5	38.5						Stothers, 1972a
2-Deoxy-β-D-glucose [13b]e	94.6	40.8						Stothers, 1972a
Methyl-α-D-glucopyranoside [14a]e	100.1	72.5	74.2	70.7	72.3	61.7	55.8	Stothers, 1972a
Methyl-β-D glucopyranoside [14b]e	104.2	74.1	76.9	70.7	76.8	61.9	57.9	Stothers, 1972a
Methyl-α-D-glucopyranoside [14a]f	101.85	74.2	75.75	72.15	73.9	63.2	57.7	Breitmaier et al., 1971b
Methyl-β-D-glucopyranoside [14b]f	105.8	75.6	77.5	72.15	77.5	63.3	59.7	Breitmaier et al., 1971b
Methyl-α-D-glucopyranose [14a]g	100.5	73.1	74.8	71.4	72.8	62.3	56.8	Usui et al., 1973b
Methyl-β-D-glucopyranose [14b]g	104.5	74.6	77.3	71.2	77.3	62.4	58.8	Usui et al., 1973b
2-O-Methyl-α-D-glucopyranoseg	90.7	81.9	73.5	71.3	72.8	62.4	59.3	Usui et al., 1973b
2-O-Methyl-β-D-glucopyranoseg	97.1	85.2	76.8	71.3	77.3	62.4	61.7	Usui et al., 1973b
4-O-Methyl-α-D-glucopyranoseg	93.2	73.0	73.9	80.5	71.7	62.1	61.6	Usui et al., 1973b
4-O-Methyl-β-D-glucopyranoseg	97.1	75.8	76.7	80.5	76.1	62.1	61.6	Usui et al., 1973b
6-O-Methyl-α-D-glucopyranoseg	93.3	73.0	74.3	71.4	71.4	72.6	60.3	Usui et al., 1973b
6-O-Methyl-β-D-glucopyranoseg	97.3	75.8	77.2	71.4	75.8	72.6	60.3	Usui et al., 1973b
p-Nitrophenyl-α-D-glucopyranoside [15a]f	100.45	74.1	76.9	72.5	75.75	63.45		Breitmaier et al., 1971b
p-Nitrophenyl-β-D-glucopyranoside [15b]f	102.7	75.95	80.05	72.4	79.3	63.45	119.65, 128.25, 144.35, 164.85	Breitmaier et al., 1971b
m-Nitrophenyl-β-D-glucopyranoside [16a]f	103.55	76.05	79.95	72.6	79.2	63.55	119.45, 128.5, 144.55, 165.15 113.9, 119.55, 126.0, 133.45, 151.35, 160.55	Breitmaier et al., 1971b
o-Nitrophenyl-β-D-glucopyranoside [17a]f	103.25	75.95	80.05	72.4	79.5	63.45	119.85, 124.7, 121.5, 136.9, 143.05, 152.4	Breitmaier et al., 1971b
Phenyl-β-D-glucopyranoside [18a]f	103.05	75.8	79.5	72.4	79.3	63.55	119.0, 124.6, 132.15, 153.5	Breitmaier et al., 1971b
α-D-Mannose [19a]h	94.5	71.2	70.8	67.5	72.9	61.6		Gorin, 1973a
β-D-Mannose [19b]h	94.1	71.8	73.6	67.1	76.7	61.6		Gorin, 1973a
α-D-Mannose [19a]e	94.4	72.0	74.1	67.6	76.7	62.0		Stothers, 1972a
β-D-Mannose [19b]e	94.4	72.0	74.1	67.6	76.7	62.0		Stothers, 1972a
α-D-Mannose-6-phosphate [20a]i	94.8	71.3	70.6	67.1 or 66.7	72.6	63.7		Gorin, 1973a
β-D-Mannose-6-phosphate [20b]h	94.4	71.9	73.3	67.1 or 66.7	76.1	63.7		Gorin, 1973a
2,6-Di-O-methyl-α-D-mannose [21a]h	91.3	81.6	71.3 or 70.5	67.9	72.1	71.3 or 70.5	59.2, 59.0	Gorin, 1973a
2,6-Di-O-methyl-β-D-mannose [21b]h	94.5	81.9	73.9	67.6	75.3	71.3 or 70.5	59.2, 59.0	Gorin, 1973a

TABLE 2—continued

Compound	δ, ppm[a]							Methyl	Phenyl	Reference
	C-1	C-2	C-3	C-4	C-5	C-6	C-7			
Methyl-α-D-mannopyranoside [22a][e]	101.9	71.7	71.0	67.9	73.6	62.1		56.2		Stothers, 1972a
Methyl-β-D-mannopyranoside [22b][e]	102.2	71.5	74.2	68.0	77.5	62.3		57.8		Stothers, 1972a
Methyl-α-D-mannopyranoside [23a][f]	103.2	73.1	72.35	69.2	74.95	63.4		57.15		Breitmaier et al., 1971b
p-Nitrophenyl-α-D-mannopyranoside [23b][f]	101.5	73.4	72.5	69.45	78.0	63.75			119.55, 128.4, 144.45, 164.1	Breitmaier et al., 1971b
α-D-Galactose [24a][e]	93.2	70.2	69.4	70.2	71.2	62.1				Stothers, 1972a
β-D-Galactose [24b][e]	97.5	73.0	73.8	69.7	75.9	61.9				Stothers, 1972a
Methyl-α-D-galactopyranoside [25a][e]	100.7	70.8	71.1	69.8	72.1	62.7		56.5		Stothers, 1972a
Methyl-β-D-galactopyranoside [25b][e]	100.5	72.1	74.2	70.0	76.2	62.3		58.3		Stothers, 1972a
p-Nitrophenyl-α-D-galactopyranoside [26a][f]	100.75	75.5	71.2	70.45	72.05	63.0			119.55, 128.5, 144.15, 165.05	Breitmaier et al., 1971b
p-Nitrophenyl-β-D-galactopyranoside [26b][f]	103.35	73.15	76.05	71.1	78.55	63.45			119.45, 128.4, 144.45, 165.3	Breitmaier et al., 1971b
m-Nitrophenyl-β-D-galactopyranoside [27a][f]	104.2	73.25	76.05	71.1	78.65	63.45			113.9, 119.55, 120.2, 133.55, 173.05, 160.65	Breitmaier et al., 1971b
o-Nitrophenyl-β-D-galactopyranoside [28a][f]	104.1	73.15	76.25	71.1	78.65	63.45			119.95, 124.7, 127.5, 137.0, 152.55, 143.15	Breitmaier et al., 1971b
Phenyl-β-D-galactopyranoside [29a][f]	104.0	73.55	76.4	71.3	78.3	63.65			119.3, 124.9, 132.5, 160.4	Breitmaier et al., 1971b
α-L-Fucose [30a][e]	93.2	70.4	69.2	72.8	67.0	16.6				Stothers, 1972a
β-L-Fucose [30b][e]	97.3	72.8	74.0	72.4	71.5	16.6				Stothers, 1972a
α-L-Rhamnose [31a][e]	94.9	71.9	70.8	73.2	69.1	17.8				Stothers, 1972a
β-L-Rhamnose [31b][e]	94.5	72.3	73.9	72.8	72.8	17.8				Stothers, 1972a
Methyl-α-L-rhamnopyranoside [32a][f]	103.3	74.5	70.9	70.9	70.9	19.15		57.15		Breitmaier et al., 1971b
			72.0	72.0	72.0					
			72.8	72.8	72.8					
			or 72.8							
α-D-Allose [33a][e]	93.7	72.6	73.5	68.5	72.1	62.9				Stothers, 1972a
β-D-Allose [33b][e]	94.4	74.4	72.1	67.8	72.3	62.3				Stothers, 1972a
Methyl-α-D-isopyranoside [34][e]	120.4	71.8	72.7	71.2	71.7	61.1		56.7		Stothers, 1972a

Compound	C-1	C-2	C-3	C-4	C-5	C-6			Reference
Methyl-α-D-altropyranoside [35]e	102.0	70.9	65.7	70.9	62.2	56.3			Stothers, 1972a
Trans-fused methyl-4,6-O-benzylidene-α-D-allopyranoside [36a]i	100.2	67.9	68.8	78.1	56.9	68.8	101.5	55.7	Conway et al., 1972
Trans-fused methyl-4,6-O-benzylidene-α-D-altropyranoside [36b]i	101.6	69.6	68.8	76.0	57.8	68.8	101.8	55.0	Conway et al., 1972
Trans-fused methyl-4,6-O-benzylidene-β-D-altropyranoside [36c]i	99.4	70.8	68.6	76.5	62.8	68.6	101.8	56.4	Conway et al., 1972
Trans-fused methyl-4,6-O-benzylidene-α-D-glucopyranoside [36d]i	99.9	72.4	70.5	80.8	62.0	68.5	101.5	54.9	Conway et al., 1972
Trans-fused methyl-4,6-O-benzylidene-β-D-glucopyranoside [36e]i	104.2	74.2	72.9	80.3	65.9	68.3	101.5	56.8	Conway et al., 1972
Trans-fused methyl-4,6-O-benzylidene-α-D-mannopyranoside [36f]i	101.7	70.6	68.0	78.5	62.9	68.4	101.7	54.4	Conway et al., 1972
Trans-fused methyl-4,6-O-benzylidene-2,3-dideoxy-α-D-erythropyranoside [36g]i	98.1	29.0	23.5	77.9	64.5	69.1	101.6	54.2	Conway et al., 1972
Trans-fused methyl-4,6-O-benzylidene-3-deoxy-α-D-arabinopyranoside [36h]i	100.5	67.4	31.4	73.5	64.6	68.9	101.8	54.3	Conway et al., 1972
Trans-fused methyl-4,6-O-benzylidene-3-deoxy-α-D-ribopyranoside [36i]i	98.9	67.0	32.6	76.1	63.4	68.9	101.3	54.6	Conway et al., 1972
Trans-fused methyl-4,6-O-benzylidene-3-amino-α-D-glucopyranoside [36j]i	99.4	72.1	52.4	81.0	62.2	68.7	101.5	55.0	Conway et al., 1972
Trans-fused methyl-4,6-O-benzylidene-3-amino-α-D-altropyranoside [36k]i	101.6	69.4	52.0	75.9	57.6	68.8	101.8	54.8	Conway et al., 1972
Trans-fused methyl-4,6-O-benzylidene-3-amino-α-D-allopyranoside [36l]i	100.8	67.5	52.1	78.3	56.7	68.9	101.3	55.6	Conway et al., 1972
Trans-fused methyl-4,6-O-benzylidene-3-amino-α-D-mannopyranoside [36m]i	102.0	70.0	50.5	79.4	63.4	68.5	101.6	54.5	Conway et al., 1972
Trans-fused methyl-4,6-O-benzylidene-3-amino-β-D-glucopyranoside [36n]i	104.1	74.2	72.9	79.8	65.9	67.7	99.4	56.7	Conway et al., 1972

TABLE 2–continued

Compound	δ, ppm[a]							Methyl	Phenyl	Other	Reference
	C-1	C-2	C-3	C-4	C-5	C-6	C-7				
Methyl-4,6-O-benzylidene-3-deoxyl-3-nitro-4-glucopyranoside-2-ethylamine [37a][j]	98·25	60·08	87·70	78·41	62·06	68·89	101·35	55·48	C-1', 136·27 C-2', 6' 128·97 C-3', 5' 128·02 C-4', 125·87	CH_3CH_2NH 15·63 CH_3CH_2NH 41·59	Gurudata and Rajabalee, 1973
Methyl-4,6-O-benzylidene-3-deoxy-3-nitro-β-glucopyranoside-2-N-ethylamine [37b][j]	104·5	61·43	89·13	77·94	66·35	68·57	101·27	59·29	C-1', 136·27 C-2', 6' 129·05 C-3', 5' 128·10 C-4', 125·87	CH_3CH_2NH 15·56 CH_3CH_2NH 59·29	Gurudata and Rajabalee, 1973
Methyl-4,6-O-benzylidene-3-deoxy-3-nitro-α-glucopyranoside-2-pyrrole [38a][j]	99·52	61·98	84·52	79·05	61·98	68·97	101·43	54·76	136·35 128·97 128·02 125·95	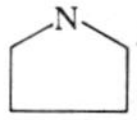 N⟨⟩ 48·33, 24·05	Gurudata and Rajabalee, 1973
Methyl-4,6-O-benzylidene-3-deoxy-3-nitro-β-glucopyranoside-2-pyrrole [38b][j]	102·62	62·30	87·14	78·33	66·27	68·65	101·11	56·51	136·27 128·97 128·02 125·87	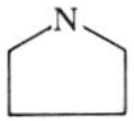 N⟨⟩ 48·17, 23·97	Gurudata and Rajabalee, 1973
Methyl-4,6-O-benzylidene-3-deoxy-3-nitro-α-glucopyranoside-2-piperidine [39a][j]	98·97	66·98	83·65	79·29	62·14	69·05	101·43	54·68	136·43 128·97 128·02 125·95	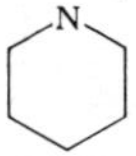 N⟨⟩ 50·87, 26·90, 24·68	Gurudata and Rajabalee, 1973
Methyl-4,6-O-benzylidene-3-deoxy-3-nitro-β-glucopyranoside piperidine [39b][j]	102·78	67·78	86·59	78·33	66·35	68·65	101·19	56·51	136·27 128·97 128·02 125·87	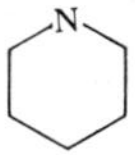 N⟨⟩ 57·03, 26·75	Gurudata and Rajabalee, 1973
Methyl-2,3,4,6-tetra-O-methyl-α-D-glucopyranoside[k]	98·16	82·58	84·28	80·61	70·98	72·41				24·52	Haverkamp et al., 1973
Methyl-2,3,4,6-tetra-O-methyl-β-D-glucopyranoside[k]	105·00	84·58	87·21	80·48	75·38	72·36					Haverkamp et al., 1973
2-Acetamido-2-deoxy-α-D-glucopyranoside[l]	92·1	55·3	72·0	71·4	72·8	61·9		23·3		C=O 175·7	Bundle et al., 1973

293

Compound	C-1	C-2	C-3	C-4	C-5	C-6				Other	Reference
2-Acetamido-2-deoxy-β-D-glucopyranoside[l]	96.2	58.0	75.2	71.2	77.2	62.0			23.5	C=O 175.9	Bundle et al., 1973
2-Acetamido-2-deoxy-α-D-mannopyranoside[l]	94.3	54.4	70.1	68.0	73.2	61.7			23.2	C=O 175.9	Bundle et al., 1973
2-Acetamido-2-deoxy-β-D-mannopyranoside[l]	94.3	55.3	73.2	67.8	77.5	61.7			23.2	C=O 176.8	Bundle et al., 1973
2-Acetamido-2-deoxy-α-D-galactopyranoside[l]	92.2	51.4	68.6	69.7	71.6	62.4			23.2	C=O 176.1	Bundle et al., 1973
2-Acetamido-2-deoxy-β-D-galactopyranoside[l]	96.5	54.9	72.3	69.0	76.3	62.2			23.4	C=O 175.8	Bundle et al., 1973
2-Acetamido-2-deoxy-α-D-glucopyranoside-1-phosphate[l]	93.9	55.3	72.6	71.2	73.2	61.9			23.2	C=O 175.8	Bundle et al., 1973
2-Acetamido-2-deoxy-β-D-glucopyranoside-1-phosphate[l]	96.7	57.7	75.3	71.2	77.4	62.3			23.6	C=O 176.2	Bundle et al., 1973
3-O-Acetyl-2-acetamido-2-deoxy-α-D-glucopyranoside[l]	92.2	53.4	75.1	69.1	72.6	61.7			23.2, 21.5	C=O 175.5, 175.1	Bundle et al., 1973
3-O-Acetyl-2-acetamido-2-deoxy-β-D-glucopyranoside[l]	95.6	56.3	77.0	69.1	77.0	61.8			23.0, 21.5	C=O 175.7, 175.1	Bundle et al., 1973
C-1',3'-dithianyl-4-t-butylcyclohexanol [40a][m]	73.80	35.95	23.65	46.60	23.65	34.95					Lukacs et al., 1972
[40b][m]	73.15	35.05	22.25	47.37	22.25	35.05					Lukacs et al., 1972
[40c][m]	99.80	72.05	70.80	72.95	57.30				55.45		Lukacs et al., 1972
[40d][m]	101.40	75.40	71.10	71.55	57.85				55.05		Lukacs et al., 1972
[40e][m]	98.50	35.80	72.85	78.00	59.25				55.25		Lukacs et al., 1972
[40f][m]	98.40	36.35	75.65	79.85	59.15	68.85	101.65			137.25, 128.80, 126.15, 127.95	
[40g][m]	98.40						100.90		54.70	137.00, 128.30, 125.70, 127.55	Lukacs et al., 1972
α-D-Glucose-3-sulfate[n]	104.45	83.20	80.50	80.50	73.05	67.95					Lukacs et al., 1972
β-D-Glucose-3-sulfate[n]	92.9	71.1	83.1	68.3	72.1	61.5					Honda et al., 1973
α-D-Glucose-6-sulfate[n]	96.5	73.8	85.2	68.3	76.3	61.5					Honda et al., 1973
β-D-Glucose-6-sulfate[n]	93.1	72.3	73.6	70.2	70.5	68.1					Honda et al., 1973
Tri-O-acetyl-1,2-O-(1-cyanoethylidene)-α-D-glucopyranosyl cyanide [41][o]	96.9	75.0	76.5	70.2	74.7	68.1			20.7		Honda et al., 1973
	97.9	74.4[q]	69.6[q]	68.2[q]	67.8[q]	63.2			24.5	C≡N 117.0, C=O 170.7, 169.7, 169.3	Coxon, 1973

TABLE 2—continued

Compound	δ, ppm[a]										
	C-1	C-2	C-3	C-4	C-5	C-6	C-7	Methyl	Phenyl	Other	Reference
Tetra-O-acetyl-β-D-glucopyranosyl cyanide [42][o]	66·8	69·4	73·3	67·8	77·3	61·8	20·6			C≡N 114·5 C=O 170·7, 170·3, 169·4, 169·0	Coxon, 1973
Tri-O-acetyl-β-D-xylopyranosyl cyanide [43][o]	68·4[q]	69·6	67·5[q]	66·0[q]	65·8		20·8 20·6 20·6			C≡N 115·0 C=O 169·9, 169·3	Coxon, 1973
Tetra-O-acetyl-β-D-galactopyranosyl cyanide [44][o]	66·6	71·2[q]	67·3[q]	67·1	75·8[q]	61·7				C≡N 114·9 C=O 170·7, 170·4, 170·1, 169·2	Coxon, 1973
+0·2 eq Pr(fod)$_3$[o]	64·6	68·8[q]	65·6[q]	65·6[q]	73·4[q]	58·8	19·5, 19·0, 18·8, 18·6			C≡N 113·5 C=O 167·3, 166·9, 164·9	Coxon, 1973
Tri-O-benzoyl-β-D-ribofuranosyl cyanide[o]	81·4[q]	74·9[q]	72·3[q]	69·8[q]	63·5				134·2, 134·0, 133·6, 130·1, 129·7, 128·8	C≡N 116·2 C=O 166·4, 165·4, 165·2	Coxon, 1973
Tri-O-benzyl-β-D-ribopyranosyl cyanide [45][o]	68·6[q]	67·3[q]	66·9[q]	65·6[q]	65·6				133·9, 133·7, 130·3, 130·0, 129·6, 128·9, 128·7	C≡N 115·7 C=O 165·6, 165·3	Coxon, 1973
Tri-O-acetyl-1,2-O-(1-cyanoethylidene)-α-D-galactopyranose [46][o]	99·4	73·2[q]	71·5[q]	70·2[q]	65·8	61·7	20·7			C≡N 117·1 C=O 170·6, 170·1	Coxon, 1973
1,5-Anhydro-D-glucitol [47][c]	61·4	81·2	70·6	78·3	70·3	69·7					Que and Gray, 1974
1,5-Anhydro-D-mannitol [48][c]	62·0	81·4	68·1	74·4	69·9	70·7					Que and Gray, 1974
1,5-Anhydro-D-altritol [49][c]	62·2	77·1	65·7	70·3	70·3	67·0					Que and Gray, 1974
1,5-Anhydro-D-talitol [50][c]	63·5	80·3	69·1	70·5	70·0	71·5					Que and Gray, 1974
1,5-Anhydro-L-galactitol [51][c]	62·2	80·3	70·0	75·0	67·3	70·0					Que and Gray, 1974
1,5-Anhydro-L-allitol [52][c]	61·1	76·9	67·0	69·9	66·7	66·0					Que and Gray, 1974
1,4-Anhydroerythritol [53][c]	72·2	71·7	71·7	72·2							Que and Gray, 1974
1,4-Anhydroribitol [54][c]	73·0	71·8	72·4	82·6	62·2						Que and Gray, 1974
1,4-Anhydroallitol [55][c]	72·7	72·3	72·8	82·8	72·1	63·1					Que and Gray, 1974
1,4-Anhydrolyxitol [56][c]	72·0	72·0	71·3	81·6	61·2						Que and Gray, 1974
1,4-Anhydromannitol [57][c]	71·1	72·2	71·8	81·0	70·2	63·9					Que and Gray, 1974
1,4-Anhydroxylitol [58][c]	73·5	77·5	77·1	81·7	60·7						Que and Gray, 1974
1,4-Anhydroglucitol [59][c]	74·2	77·2	76·7	80·7	68·8	64·4					Que and Gray, 1974
1,4-Anhydrothreitol [60][c]	73·7	77·1	77·1	73·7							Que and Gray, 1974
1,4-Anhydroarabinitol [61][c]	74·0	77·8	78·9	86·4	62·4						Que and Gray, 1974
2,5-Anhydroiditol [62][c]	60·9	81·2	77·5	77·5	81·2	60·9					Que and Gray, 1974
2,5-Anhydroglucitol [63][c]	61·0	81·8	77·8	78·9	85·6	62·5					Que and Gray, 1974
2,5-Anhydromannitol [64][c]	61·8	83·1	77·2	77·2	83·1	61·8					Que and Gray, 1974

Compound	C-1	C-2	C-3	C-4	C-5	C-6	Reference
α-L-Sorbopyranose [65a]n	65·4	98·8	74·8	71·3	70·3	62·6	Que and Gray, 1974
α-L-Sorbofuranose [65b]n	64·9	102·8	76·2	76·9	78·6	61·5	Que and Gray, 1974
α-L-Sorbopyranose [65a]n	65·8	99·4	72·6	76·0	71·4	63·7	Perlin et al., 1973
α-D-Tagatopyranose [66c]n	63·2	99·2	71·6	70·7	67·2	64·8	Que and Gray, 1974
α-D-Tagatofuranose [66d]n	64·4	99·3	70·1	64·6	70·3	61·1	Que and Gray, 1974
β-D-Tagatopyranose [66a]n		103·6	77·6	74·8	80·2	62·0	Que and Gray, 1974
β-D-Tagatofuranose [66b]n	64·1	98·7	71·3	73·1	81·0	62·8	Que and Gray, 1974
β-D-Tagatopyranose [66c]o	66·7	99·4	72·3	73·1	68·7	64·4	Perlin et al., 1973
β-D-Tagatofuranose [66d]o	66·1	100·2	66·3	66·4	71·0	61·4	Perlin et al., 1973
α-D-Psicopyranose [67c]n	85·9	98·7	71·2	66·4	66·7	58·8	Que and Gray, 1974
α-D-Psicofuranose [67d]n	64·9	99·5	71·2	64·0	69·9	65·0	Que and Gray, 1974
β-D-Psicopyranose [67a]n	64·1	104·4	71·2	72·6	83·7	62·2	Que and Gray, 1974
β-D-Psicofuranose [67b]n	63·6	106·7	75·6	71·9	83·7	63·3	Que and Gray, 1974
α-D-Psicofuranose [67a]p	64·17	104·01	71·24	72·56	84·3	64·17	Herve du Penhoat and Perlin, 1974
β-D-Psicofuranose [67b]p	63·31	106·27	75·62	71·87	63·65		Herve du Penhoat and Perlin, 1974
α-D-Psicopyranose [67c]p	64·98	99·15	66·44	65·94	62·24		Herve du Penhoat and Perlin, 1974
β-D-Psicopyranose [67d]p	64·98	98·45	71·24	66·72	58·86		Herve du Penhoat and Perlin, 1974
Methyl-α-D-psicofuranoseP	61·4	106·2	73·4	71·7	85·7	50·2	Herve du Penhoat and Perlin, 1974
Methyl-β-D-psicofuranoseP	58·2	110·2	75·6	72·8	84·6	49·6	Herve du Penhoat and Perlin, 1974
Methyl-β-D-psicopyranoseP	66·1	103·3	70·5	66·5	58·7	50·2 / 60·2	Herve du Penhoat and Perlin, 1974
Methyl-6-O-methyl-α-D-psicofuranoseP	61·6	106·2	73·3	71·9	83·6	50·5 / 60·5	Herve du Penhoat and Perlin, 1974
Methyl-6-O-methyl-β-D-psicofuranoseP	60·1	110·8	75·4	73·4	82·9	73·7 / 75·4	Herve du Penhoat and Perlin, 1974

Compound	O–C(CH$_3$)(CH$_3$)–O		C-2'	C-4'	C-5'	C-6'	C-2'–CH$_3$	C(CH$_3$)$_3$	C(CH$_3$)$_3$	Reference
[40a]m	108·20		53·75 / 60·95	29·90 / 30·55	25·15 / 25·70	29·90 / 30·55		31·85 / 32·15	27·20 / 27·40	Lukacs et al., 1972
[40b]m	108·20	25·45 / 25·55	54·70	29·80	24·80	30·30				Lukacs et al., 1972
[40c]m		25·55	55·80 / 53·10	26·55 / 30·10	24·05 / 25·80	26·95 / 30·65	25·15			Lukacs et al., 1972
[40d]m										Lukacs et al., 1972
[40e]m										Lukacs et al., 1972
[40f]m	109·60	25·35	58·50	26·00	24·15	26·00	24·90			Lukacs et al., 1972
[40g]m	112·20	26·35	50·50	29·10	26·75	29·10				Lukacs et al., 1972

(See over for footnotes to Table)

TABLE 2—continued (footnotes)

[a] Chemical shifts in parts per million downfield from tetramethylsilane, ($\delta = 0$ ppm); see cited reference for experimental details and for additional data.

[b] Obtained on a Varian XL-100 spectrometer at 23·5 MHz on 0·70 M solutions in D_2O; chemical shifts given relative to TMS.

[c] Obtained on a Varian XL-100-15 spectrometer with Fourier transform at 25·1 MHz on 1-2 M solutions in D_2O. Chemical shifts given relative to external CS_2 (chemical shifts converted to TMS scale using $\delta_{CS_2} = 193\cdot5$ ppm).

[d] Chemical shifts given relative to CS_2 standard (chemical shifts converted to TMS scale using $\delta_{CS_2} = 193\cdot7$ ppm).

[e] Data calculated in some cases from reference standards other than TMS; see Stothers (1972a) and references cited therein for experimental conditions and additional details.

[f] Obtained on a Brucker HFX-90-15 spectrometer with Fourier transform at 22·63 MHz and 30° on 0·5 g of methyl glucoside in 1·5 ml water and 0·5 g of aryl glucoside in 1·5 ml DMSO.

[g] Obtained on a JEOL-PS-100 spectrometer with Fourier transform at 25·1 MHz on 1–2 M aqueous solutions; chemical shifts given relative to internal TMS.

[h] Obtained on a Varian XL-100-15 spectrometer with Fourier transform at 27° relative to external TMS contained in a coaxial capillary; 0·5–1·0 g of compound in 4 ml D_2O.

[i] Obtained with Fourier transform at 15·08 MHz in $CDCl_3/CH_3OH$ (4/1, v/v); chemical shifts for the aromatic carbons are 136·8 (substituted), 128·7 (para), 127·8, and 125·8 ppm and for C-7 (CH_3) is 19·6 ppm.

[j] Obtained on a Varian XL-100 spectrometer with Fourier transform on 0·5 M solutions in $CDCl_3$ with an external deuterium lock (D_2O); chemical shifts given relative to internal TMS.

[k] Obtained on a Varian XL-100-15 spectrometer at 25·2 MHz on 5–20% solution in CD_3CN relative to internal TMS.

[l] Obtained on a Varian XL-100-15 spectrometer at 25·16 MHz on 100 mg of compound per ml of D_2O relative to external TMS.

[m] Obtained on a Brucker HFX-90 spectrometer with Fourier transform at 22·63 MHz in $CDCl_3$. Other assignments are given at the end of this table.

[n] Obtained on a JEOL JNM-PFT-100 spectrometer with Fourier transform at 25·1 MHz on 40% D_2O solution; chemical shifts given relative to CS_2 standard (chemical shifts converted to TMS scale using $\delta_{CS_2} = 192\cdot8$ ppm).

[o] Obtained on a Brucker HFX-11 spectrometer at 22·6 MHz on 0·2-1·1 M solutions in chloroform-d-hexafluorobenzene-tetramethylsilane (1 ml, 8·5 : 1 : 1 v/v).

[p] Obtained on a Varian HA-100 (CW mode) and Varian XL-100 (FT mode) spectrometers at 25·15 MHz; chemical shifts given relative to CS_2 standard (chemical shifts converted to TMS scale using $\delta_{CS_2} = 193\cdot7$ ppm).

[q] Not assigned.

Pentose Aldopyranoses

[1a] α-D-Fructofuranose; $R_1 = R_7 = CH_2OH$, $R_2 = R_3 = R_6 = OH$, $R_4 = R_5 = R_8 = H$
[1b] β-D-Fructofuranose; $R_2 = R_7 = CH_2OH$, $R_1 = R_3 = R_6 = OH$, $R_4 = R_5 = R_8 = H$
[2a] α-D-Fructose-6-phosphate; $R_1 = CH_2OH$, $R_7 = CH_2OPO_3^{2-}$, $R_2 = R_3 = R_6 = OH$, $R_4 = R_5 = R_8 = H$
[2b] β-D-Fructose-6-phosphate; $R_2 = CH_2OH$, $R_7 = CH_2OPO_3^{2-}$, $R_1 = R_3 = R_6 = OH$, $R_4 = R_5 = R_8 = H$
[3a] α-D-Fructose-1,6-diphosphate; $R_1 = R_7 = CH_2OPO_3^{2-}$, $R_2 = R_3 = R_6 = OH$, $R_4 = R_5 = R_8 = H$
[3b] **β-D-Fructose-1,6-diphosphate**; $R_2 = R_7 = CH_2OPO_3^{2-}$, $R_1 = R_3 = R_6 = OH$, $R_4 = R_5 = R_8 = H$

Hexose Aldopyranoses

[1c] α-D-Fructopyranose; $R_4 = R_5 = R_7 = R_9 = R_{10} = H$, $R_2 = R_3 = R_6 = R_8 = OH$, $R_1 = CH_2OH$
[1d] β-D-Fructopyranose; $R_4 = R_5 = R_7 = R_9 = R_{10} = H$, $R_1 = R_3 = R_6 = R_8 = OH$, $R_2 = CH_2OH$
[4a] α-D-Xylose; $R_1 = R_3 = R_6 = R_7 = R_9 = R_{10} = H$, $R_2 = R_4 = R_5 = R_8 = OH$
[4b] β-D-Xylose; $R_2 = R_3 = R_6 = R_7 = R_9 = R_{10} = H$, $R_1 = R_4 = R_5 = R_8 = OH$
[5a] Methyl-α-D-xylose; $R_1 = R_3 = R_6 = R_7 = R_9 = R_{10} = H$, $R_4 = R_5 = R_8 = OH$, $R_2 = OCH_3$
[5b] Methyl-β-D-xylose; $R_2 = R_3 = R_6 = R_7 = R_9 = R_{10} = H$, $R_4 = R_5 = R_8 = OH$, $R_1 = OCH_3$
[6a] α-D-Lyxose; $R_1 = R_4 = R_6 = R_7 = R_9 = R_{10} = H$, $R_2 = R_3 = R_5 = R_8 = OH$
[6b] β-D-Lyxose; $R_2 = R_4 = R_6 = R_7 = R_9 = R_{10} = H$, $R_1 = R_3 = R_5 = R_8 = OH$
[7a] α-D-Ribose; $R_1 = R_3 = R_5 = R_7 = R_9 = R_{10} = H$, $R_2 = R_4 = R_6 = R_8 = OH$
[7b] β-D-Ribose; $R_2 = R_3 = R_5 = R_7 = R_9 = R_{10} = H$, $R_1 = R_4 = R_6 = R_8 = OH$
[8a] Methyl-β-D-ribopyranoside; $R_2 = R_3 = R_5 = R_7 = R_9 = R_{10} = H$, $R_4 = R_6 = R_8 = OH$, $R_1 = OCH_3$
[9a] α-D-Arabinose; $R_1 = R_4 = R_5 = R_7 = R_9 = R_{10} = H$, $R_2 = R_3 = R_6 = R_8 = OH$
[9b] β-D-Arabinose; $R_2 = R_4 = R_5 = R_7 = R_9 = R_{10} = H$, $R_1 = R_3 = R_6 = R_8 = OH$
[10a] Methyl-α-D-arabinopyranoside; $R_1 = R_4 = R_5 = R_7 = R_9 = R_{10} = H$, $R_3 = R_6 = R_8 = OH$, $R_2 = OCH_3$
[10b] Methyl-β-D-arabinopyranoside; $R_2 = R_4 = R_5 = R_7 = R_9 = R_{10} = H$, $R_3 = R_6 = R_8 = OH$, $R_1 = OCH_3$
[11a] α-D-Glucopyranoside; $R_1 = R_3 = R_6 = R_7 = R_{10} = H$, $R_2 = R_4 = R_5 = R_8 = OH$, $R_9 = CH_2OH$
[11b] β-D-Glucopyranoside; $R_2 = R_3 = R_6 = R_7 = R_{10} = H$, $R_1 = R_4 = R_5 = R_8 = OH$, $R_9 = CH_2OH$
[12a] 3-O-Methyl-α-D-glucopyranoside; $R_1 = R_3 = R_6 = R_7 = R_{10} = H$, $R_2 = R_4 = R_8 = OH$, $R_5 = OCH_3$, $R_9 = CH_2OH$
[12b] 3-O-Methyl-β-D-glucopyranoside; $R_2 = R_3 = R_6 = R_7 = R_{10} = H$, $R_1 = R_4 = R_8 = OH$, $R_5 = OCH_3$, $R_9 = CH_2OH$
[13a] 2-Deoxy-α-D-glucose; $R_1 = R_3 = R_4 = R_6 = R_7 = R_{10} = H$, $R_2 = R_5 = R_8 = OH$, $R_9 = CH_2OH$
[13b] 2-Deoxy-β-D-glucose; $R_2 = R_3 = R_4 = R_6 = R_7 = R_{10} = H$, $R_1 = R_5 = R_8 = OH$, $R_9 = CH_2OH$
[14a] Methyl-α-D-glucopyranoside; $R_1 = R_3 = R_6 = R_7 = R_{10} = H$, $R_4 = R_5 = R_8 = OH$, $R_2 = OCH_3$, $R_9 = CH_2OH$
[14b] Methyl-β-D-glucopyranoside; $R_2 = R_3 = R_6 = R_7 = R_{10} = H$, $R_4 = R_5 = R_8 = OH$, $R_1 = OCH_3$, $R_9 = CH_2OH$
[15a] α-[p-Nitrophenyl]-D-glucopyranoside; $R_1 = R_3 = R_6 = R_7 = R_{10} = H$, $R_4 = R_5 = R_8 = OH$, $R_2 = OC_6H_4NO_2^-(p)$, $R_9 = CH_2OH$
[15b] β-[p-Nitrophenyl]-D-glucopyranoside; $R_2 = R_3 = R_6 = R_7 = R_{10} = H$, $R_4 = R_5 = R_8 = OH$, $R_1 = OC_6H_4NO_2^-(p)$, $R_9 = CH_2OH$
[16a] β-[m-Nitrophenyl]-D-glucopyranoside; $R_2 = R_3 = R_6 = R_7 = R_{10} = H$, $R_4 = R_5 = R_8 = OH$, $R_1 = OC_6H_4NO_2^-(m)$, $R_9 = CH_2OH$
[17a] β-[o-Nitrophenyl]-D-glucopyranoside; $R_2 = R_3 = R_6 = R_7 = R_{10} = H$, $R_4 = R_5 = R_8 = OH$, $R_1 = OC_6H_4NO_2^-(o)$, $R_9 = CH_2OH$
[18a] β-Phenyl-D-glucopyranoside; $R_2 = R_3 = R_6 = R_7 = R_{10} = H$, $R_4 = R_5 = R_8 = OH$, $R_1 = OC_6H_5$, $R_9 = CH_2OH$
[19a] α-D-Mannose; $R_2 = R_3 = R_5 = R_8 = OH$, $R_1 = R_4 = R_6 = R_7 = R_{10} = H$, $R_9 = CH_2OH$
[19b] β-D-Mannose; $R_1 = R_3 = R_5 = R_8 = OH$, $R_2 = R_4 = R_6 = R_7 = R_{10} = H$, $R_9 = CH_2OH$
[20a] α-D-Mannose-6-phosphate; $R_2 = R_3 = R_5 = R_8 = OH$, $R_1 = R_4 = R_6 = R_7 = R_{10} = H$, $R_9 = CH_2OPO_3^{2-}$
[20b] β-D-Mannose-6-phosphate; $R_1 = R_3 = R_5 = R_8 = OH$, $R_2 = R_4 = R_6 = R_7 = R_{10} = H$, $R_9 = CH_2OPO_3^{2-}$

TABLE 2—continued (footnotes)

[21a] 2,6-Di-O-methyl-α-D-mannose; $R_2 = R_5 = R_8 = OH$, $R_1 = R_4 = R_6 = R_7 = R_{10} = H$, $R_9 = CH_2OCH_3$
[21b] 2,6-Di-O-methyl-β-D-mannose; $R_1 = R_5 = R_8 = OH$, $R_2 = R_4 = R_6 = R_7 = R_{10} = H$, $R_9 = CH_2OCH_3$, $R_3 = OCH_3$
[22a] Methyl-α-D-mannopyranoside; $R_3 = R_5 = R_8 = OH$, $R_1 = R_4 = R_6 = R_7 = R_{10} = H$, $R_9 = CH_2OH$, $R_2 = OCH_3$
[22b] Methyl-β-D-mannopyranoside; $R_3 = R_5 = R_8 = OH$, $R_2 = R_4 = R_6 = R_7 = R_{10} = H$, $R_9 = CH_2OH$, $R_1 = OCH_3$
[23c] α-[p-Nitrophenyl]-D-mannopyranoside; $R_1 = R_4 = R_6 = R_7 = R_{10} = H$, $R_3 = R_5 = R_8 = OH$, $R_2 = OC_6H_4NO_2^-(p)$, $R_9 = CH_2OH$
[24a] α-D-Galactose; $R_1 = R_3 = R_6 = R_8 = R_{10} = H$, $R_2 = R_4 = R_5 = R_7 = OH$, $R_9 = CH_2OH$
[24b] β-D-Galactose; $R_2 = R_3 = R_6 = R_8 = R_{10} = H$, $R_1 = R_4 = R_5 = R_7 = OH$, $R_9 = CH_2OH$
[25a] Methyl-α-D-galactopyranoside; $R_1 = R_3 = R_6 = R_8 = R_{10} = H$, $R_4 = R_5 = R_7 = OH$, $R_2 = OCH_3$, $R_9 = CH_2OH$
[25b] Methyl-β-D-galactopyranoside; $R_2 = R_3 = R_6 = R_8 = R_{10} = H$, $R_4 = R_5 = R_7 = OH$, $R_1 = OCH_3$, $R_9 = CH_2OH$
[26a] α-[p-Nitrophenyl]-D-galactopyranoside; $R_1 = R_3 = R_6 = R_8 = R_{10} = H$, $R_4 = R_5 = R_7 = OH$, $R_2 = OC_6H_4NO_2^-(p)$, $R_9 = CH_2OH$
[26b] β-[p-Nitrophenyl]-D-galactopyranoside; $R_2 = R_3 = R_6 = R_8 = R_{10} = H$, $R_4 = R_5 = R_7 = OH$, $R_1 = OC_6H_4NO_2^-(p)$, $R_9 = CH_2OH$
[27a] β-[m-Nitrophenyl]-D-galactopyranoside; $R_2 = R_3 = R_6 = R_8 = R_{10} = H$, $R_4 = R_5 = R_7 = OH$, $R_1 = OC_6H_4NO_2^-(m)$, $R_9 = CH_2OH$
[28a] β-[o-Nitrophenyl]-D-galactopyranoside; $R_2 = R_3 = R_6 = R_8 = R_{10} = H$, $R_4 = R_5 = R_7 = OH$, $R_1 = OC_6H_4NO_2^-(o)$, $R_9 = CH_2OH$
[29a] β-Phenyl-D-galactopyranoside; $R_2 = R_3 = R_6 = R_8 = R_{10} = H$, $R_4 = R_5 = R_7 = OH$, $R_1 = OC_6H_5$, $R_9 = CH_2OH$
[30a] α-L-Fucose; $R_1 = R_4 = R_5 = R_7 = R_9 = H$, $R_2 = R_3 = R_6 = R_8 = OH$, $R_{10} = CH_3$
[30b] β-L-Fucose; $R_2 = R_4 = R_5 = R_7 = R_9 = H$, $R_1 = R_3 = R_6 = R_8 = OH$, $R_{10} = CH_3$
[31a] α-L-Rhamnose; $R_1 = R_3 = R_5 = R_8 = R_{10} = H$, $R_2 = R_4 = R_6 = R_7 = OH$, $R_9 = CH_3$
[31b] α-L-Rhamnose; $R_2 = R_3 = R_5 = R_8 = R_{10} = H$, $R_1 = R_4 = R_6 = R_7 = OH$, $R_9 = CH_3$
[32a] α-Methyl-L-rhamnopyranoside; $R_1 = R_3 = R_5 = R_8 = R_{10} = H$, $R_4 = R_6 = R_7 = OH$, $R_2 = OCH_3$, $R_9 = CH_3$
[33a] α-D-Allose; $R_1 = R_3 = R_5 = R_7 = R_{10} = H$, $R_2 = R_4 = R_6 = R_8 = OH$, $R_9 = CH_2OH$
[33b] β-D-Allose; $R_2 = R_3 = R_5 = R_7 = R_{10} = H$, $R_1 = R_4 = R_6 = R_8 = OH$, $R_9 = CH_2OH$
[34] Methyl-α-D-idopyranoside; $R_1 = R_4 = R_5 = R_8 = R_{10} = H$, $R_3 = R_6 = R_7 = OH$, $R_2 = OCH_3$, $R_9 = CH_2OH$
[35] Methyl-α-D-altropyranoside; $R_1 = R_4 = R_5 = R_7 = R_{10} = H$, $R_3 = R_6 = R_8 = OH$, $R_2 = OCH_3$, $R_9 = CH_2OH$
[41] Tri-O-acetyl-1,2-O-(1-cyanoethylidene)-α-D-glucopyranose; $R_1 = R_3 = R_6 = R_7 = R_{10} = H$, $R_2 = R_4 = 1,2-O-(1-cyanoethylidene)$, $R_5 = R_8 = AcO$, $R_9 = AcOCH_2$
[42] Tetra-O-acetyl-β-glucopyranosyl cyanide; $R_2 = R_3 = R_6 = R_7 = R_9 = R_{10} = H$, $R_1 = CN$, $R_4 = R_5 = R_8 = AcO$
[43] Tri-O-acetyl-β-D-xylopyranosyl cyanide; $R_2 = R_3 = R_6 = R_7 = R_9 = R_{10} = H$, $R_1 = CN$, $R_4 = R_5 = R_8 = AcO$
[44] Tetra-O-acetyl-β-D-galactopyranosyl cyanide; $R_2 = R_3 = R_6 = R_8 = R_{10} = H$, $R_1 = CN$, $R_4 = R_5 = R_7 = AcO$, $R_9 = AcOCH_2$
[45] Tri-O-benzoyl-β-D-ribopyranosyl cyanide; $R_2 = R_3 = R_5 = R_7 = R_9 = R_{10} = H$, $R_1 = CN$, $R_4 = R_6 = R_8 = BzO$
[46] Tri-O-acetyl-1,2-O-(1-cyanoethylidene)-α-L-galactopyranose; $R_1 = R_3 = R_6 = R_8 = R_{10} = H$, $R_2 = R_4 = 1,2-O-(1-cyanoethylidene)$, $R_5 = R_7 = AcO$, $R_9 = AcOCH_2$

Ketohexoses

[47] 1,5-Anhydro-D-glucitol; R=H
[65a] α-L-Sorbopyranose; R=OH

[48] 1,5-Anhydro-D-mannitol; R=H

299

[50] 1,5-Anhydro-D-talitol; R=H
[67d] β-D-Psicopyranose; R=OH

[52] 1,5-Anhydro-L-allitol; R=H
[67c] α-D-Psicopyranose; R=OH

[56] 1,4-Anhydrolyxitol; $R_1 = CH_2OH, R_1 = R_3 = H$
[57] 1,4-Anhydromannitol; $R_1 = CH(OH)CH_2OH, R_2 = R_3 = H$
[66a] α-D-Tagatofuranose; $R_1 = R_2 = CH_2OH, R_3 = OH$
[66b] β-D-Tagatofuranose; $R_1 = R_3 = CH_2OH, R_2 = OH$

[60] 1,4-Anhydrothreitol; $R_1 = R_2 = R_3 = R_4 = H$
[61] 1,4-Anhydroarabinitol; $R_1 = CH_2OH, R_2 = R_3 = R_4 = H$
[63] 2,4-Anhydroglucitol; $R_1 = R_2 = CH_2OH, R_3 = R_4 = H$
[64] 2,5-Anhydromannitol; $R_1 = R_3 = CH_2OH, R_2 = R_4 = H$
[65b] α-L-Sorbofuranose; $R_1 = H, R_2 = CH_2OH, R_3 = R_4 = CH_2OH$

[49] 1,5-Anhydro-D-altritol; R=H
[66d] β-D-Tagatopyranose; R=OH

[51] 1,5-Anhydro-L-galactitol; R=H
[66c] α-D-Tagatopyranose; R=OH

[53] 1,4-Anhydroerythritol; $R_1 = R_2 = R_3 = H$
[54] 1,4-Anhydroribitol; $R_1 = CH_2OH, R_2 = R_3 = H$
[55] 1,4-Anhydroallitol; $R_1 = CH(OH)CH_2OH, R_2 = R_3 = H$
[67a] α-D-Psicofuranose; $R_1 = R_2 = CH_2OH, R_3 = OH$
[67b] β-D-Psicofuranose; $R_1 = R_3 = CH_2OH, R_2 = OH$

[58] 1,4-Anhydroxylitol; $R_1 = CH_2OH, R_2 = H$
[59] 1,4-Anhydroglucitol; $R_1 = CH(OH)CH_2OH, R_2 = H$
[62] 2,5-Anhydroiditol; $R_1 = R_2 = CH_2OH$

Stothers' (1972a) book should be consulted for earlier work and additional details.

Current interest centers around establishing the percentages of different anomers present in equilibrium solutions (Lemieux and Stevens, 1966; Angyal, 1969; Doddrell and Allerhand, 1971b; Koerner et al., 1973; Herve du Penhoat and Perlin, 1974; Que and Gray, 1974). Although such investigations have utilized ^{1}H nmr spectroscopy (Lemieux and Stevens, 1966; Angyal, 1969) ^{13}C nmr allows the examination of carbohydrates which do not have proton bearing anomeric carbon atoms (fructose [1], for example). In addition, ^{13}C nmr spectroscopy should prove to be a powerful method for the determination of mutarotation rates. Determination of appropriate integrated resonance intensities is based upon the assumption that all carbons of the saccharides have identical nuclear Overhauser enhancements (Allerhand et al., 1971b). To date percentages of anomeric composition of D-fructose [1] (Doddrell and Allerhand, 1971b; Herve du Penhoat and Perlin, 1974; Que and Gray, 1974), D-fructose-6-phosphate [2] and D-fructose-1,6-diphosphate [3] (Koerner et al., 1973), D-psicose [67], L-sorbose [65] and D-tagatose [66] (Herve du Penhoat and Perlin, 1974; Que and Gray, 1974) have been determined (Table 3). The predominance of β anomer in fructose and fructose derivatives may well imply a similar predominance in the phosphofructokinase enzyme substrate complexes. The apparent discrepancies among these data originate from differences in spectral resolution. However, with appropriate care, ^{13}C FTnmr is capable of detecting anomers in concentrations of 1% or less. Determinations of the equilibrium composition of monosaccharides relies, of course, on the correct ^{13}C chemical shift assignments of the given anomers. For example, previous chemical shift assignments for [1c] and [1d] had to be modified by reversing the ^{13}C chemical shifts assigned to C-4 and C-5 (Table 2, Koerner et al., 1973). Other discrepancies of assignments exist, even for simple monosaccharides (Table 2), leading to the propagation of incorrect assignments of more complex molecules, such as nucleotides.

Differences in conformation are also manifested in the C—H coupling constants. Using ^{13}C enriched compounds, Perlin and Casu (1969) observed J(C-1 H-1)-values of 169 Hz and 160 Hz respectively for α- and β-D-glucopyranose. Using gated decoupling with a pulsed Fourier spectrometer, J(C-1 H-1) coupling constants for a variety of unenriched carbohydrates resulted in values of ca 10 Hz higher for

TABLE 3

Equilibrium Composition of Tautomeric Forms of Monosaccharides in Percentages
(Numbered formulae are on pp. 296-299)

Monosaccharide	Solution	α-pyranose	β-pyranose	α-furanose	β-furanose	Reference
D-Fructose [1]	Aqueous	3 ± 1	57 ± 6	9 ± 1	31 ± 3	Doddrell and Allerhand, 1971b
D-Fructose [1]	H_2O	<5	60	10	30	Herve du Penhoat and Perlin, 1974
	DMSO	<5	20	25	55	
D-Fructose [1]	H_2O	<2	72	5	23	Que and Gray, 1974
D-Fructose-6-phosphate [2]	D_2O			19 ± 2	81 ± 2	Koerner et al., 1973
D-Fructose-1,6-diphosphate [3]	D_2O			23 ± 4	77 ± 4	Koerner et al., 1973
D-Psicose [4]	H_2O	25	25	40	10	Herve du Penhoat and Perlin, 1974
	DMSO	20	15	50	15	
D-Psicose [4]	H_2O	26	21	38	15	Que and Gray, 1974
L-Sorbose [5]	H_2O	100				Herve du Penhoat and Perlin, 1974
	DMSO	95	5			
L-Sorbose [5]	H_2O	95		5		Que and Gray, 1974
D-Tagatose [6]	H_2O	90	10			Herve du Penhoat and Perlin, 1974
	DMSO	90	10			
D-Tagatose [6]	H_2O	71	15	5	9	Que and Gray, 1974

the equatorial than for the axial H-1 (Bock et al., 1973; Bock and Pedersen, 1974). Vicinal C–H coupling constants were found to depend on the dihedral angles in a manner analogous to that observed for H–H coupling constants (Schwarcz and Perlin, 1972). Combined data on H–H coupling constants and ^{13}C magnetic resonance frequencies of α- and β-D-idopyranosiduronic acids (structural components of heparin) indicated their conformation to be predominantly in the C-1(D) form, [68a] and [68b] respectively (Perlin et al., 1972a). Configurations of C-1,3-dithianyl branched chain carbohydrates [40a–40f] of the quaternary centres have been

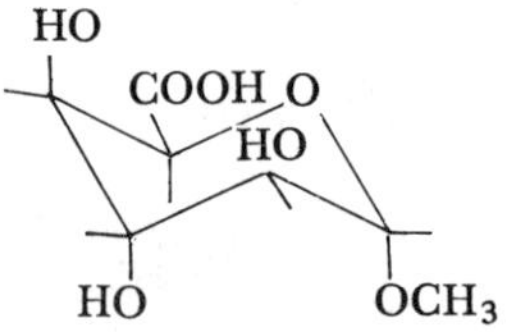

[68a]
Methyl-α-D-idopyranosiduronic acid

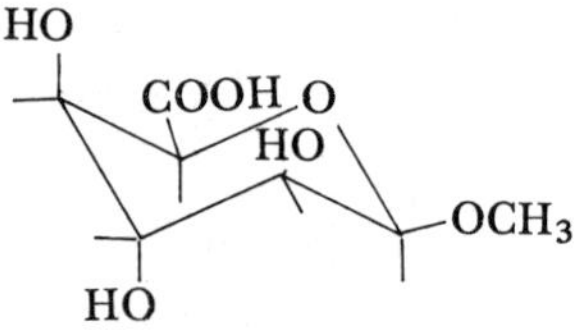

Methyl-β-D-idopyranosiduronic acid

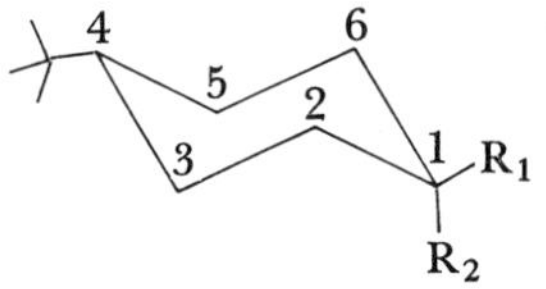

[40a] R_1 = OH
      R_2 = I
[40b] R_1 = I
      R_2 = OH

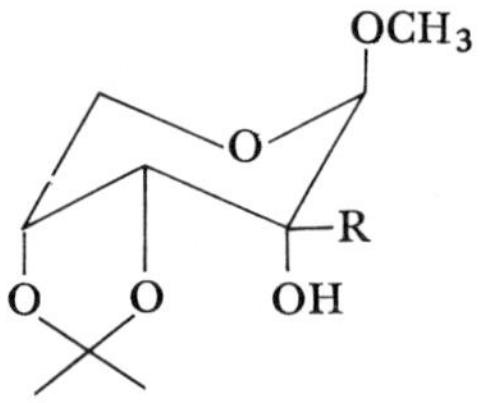

[40c] R_1 = I
[40d] R_1 = II

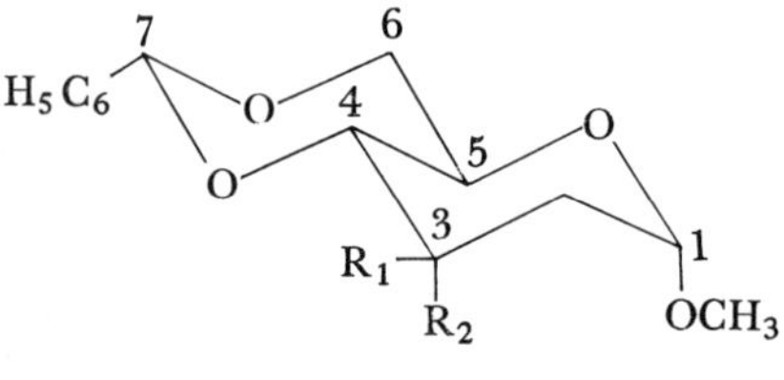

[40e] R_1 = I R_2 = OH
[40f] R_1 = OH or II R_2 = II or OH

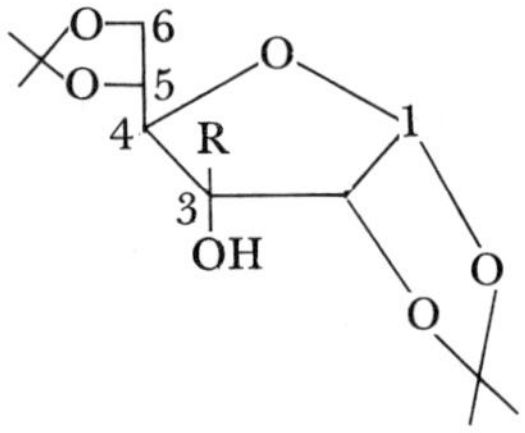

[40g] R = I

I =

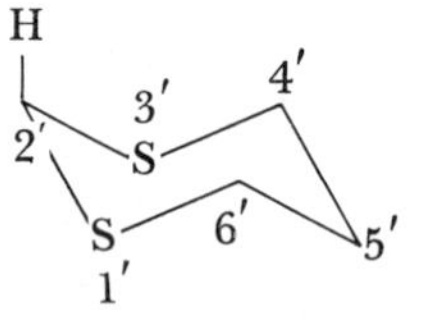

II =

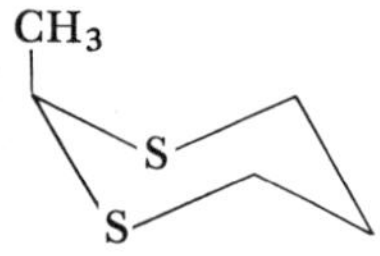

assigned with the aid of ^{13}C nmr spectroscopy (Table 2; Lukacs et al., 1972; Sepulchre et al., 1974).

The configurations of trans-fused bicyclobenzylidene derivatives [36a]-[36n] and [37a]-[39b] have been elucidated from their ^{13}C nmr spectra (Conway et al., 1972; Gurudata and Rajabalee, 1973).

for [36a-m], X = C$_6$H$_5$

for [36n], X = CH$_3$

Compound	Hexose Configuration	R$_1$	R$_2$	R$_3$	R$_4$	R$_5$	R$_6$
[36a]	α-allo	OMe	H	H	OH	OH	H
[36b]	α-altro	OMe	H	OH	H	OH	H
[36c]	β-altro	H	OMe	OH	H	OH	H
[36d]	α-gluco	OMe	H	H	OH	H	OH
[36e]	β-gluco	H	OMe	H	OH	H	OH
[36f]	α-manno	OMe	H	OH	H	H	OH
[36g]	2,3-dideoxy-α-erythro	OMe	H	H	H	H	H
[36h]	3-deoxy-α-arabino	OMe	H	OH	H	H	H
[36i]	3-deoxy-α-ribo	OMe	H	H	OH	H	H
[36j]	α-gluco	OMe	H	H	OH	H	NH$_2$
[36k]	α-altro	OMe	H	OH	H	NH$_2$	H
[36l]	α-allo	OMe	H	H	OH	NH$_2$	H
[36m]	α manno	OMe	H	OH	H	H	NH$_2$
[36n]	β-gluco	H	OMe	H	OH	H	OH

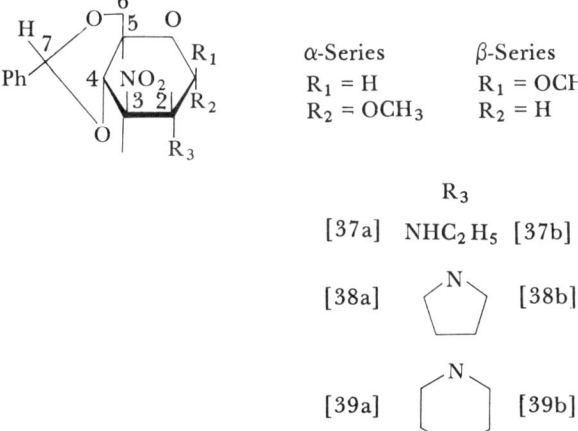

α-Series
R$_1$ = H
R$_2$ = OCH$_3$

β-Series
R$_1$ = OCH$_3$
R$_2$ = H

R$_3$

[37a] NHC$_2$H$_5$ [37b]

[38a] ⟨N⟩ [38b]

[39a] ⟨N⟩ [39b]

Disaccharides

Investigations of ^{13}C spectroscopy of sucrose [69] (Allerhand and Doddrell, 1971; Dorman and Roberts, 1971), lactose [70], cellobiose [71] (Dorman and Roberts, 1971) and maltose [72] (Dorman

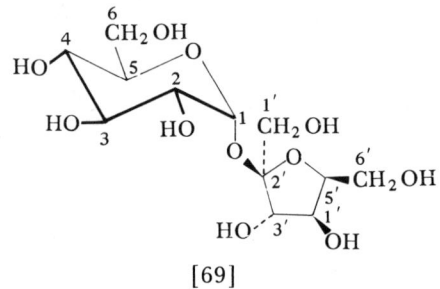

[69]

Sucrose (α-D-glucopyranosyl-β-D-fructofuranoside)

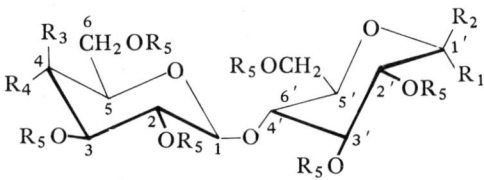

[70a] $R_1 = R_3 = OH$, $R_2 = R_4 = R_5 = H$
α-Lactose [4-O-(β-D-galactopyranosyl)-α-D-glucopyranose]
[70b] $R_2 = R_3 = OH$, $R_1 = R_4 = R_5 = H$
β-Lactose [4-O-(β-D-galactoryranosyl)-β-D-glucopyranose]
[71a] $R_1 = R_4 = OH$, $R_2 = R_3 = R_5 = H$
α-Cellobiose [4-O-(β-D-glucopyranosyl-α-D-glucopyranose]
[71b] $R_2 = R_4 = OH$, $R_1 = R_3 = R_5 = H$
β-Cellobiose [4-O-(β-D-glucopyranosyl-β-D-glucopyranose]

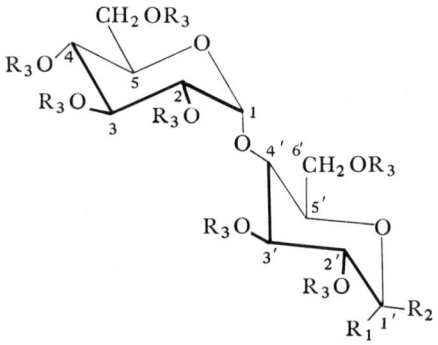

[72a] $R_1 = OH$, $R_2 = R_3 = H$
α-Maltose [4-O-(α-D-glucopyranosyl)-α-D-glucopyranose]
[72b] $R_2 = OH$, $R_1 = R_3 = H$
β-Maltose [4-O-(α-D-glucopyranosyl)-β-D-glucopyranose]

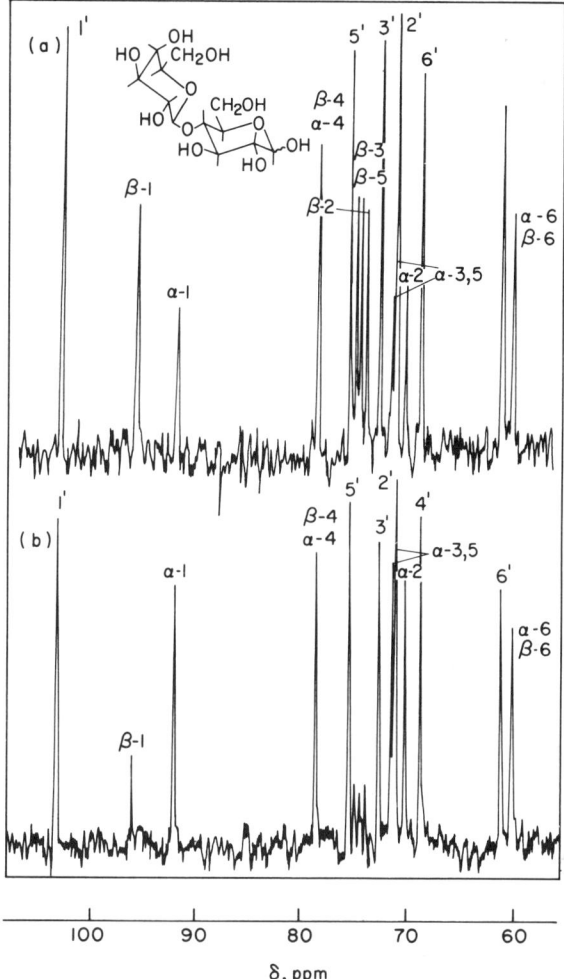

Figure 1. ^{13}C FTnmr spectra of a saturated lactose solution in water. (a), after dissolution; (b), after 3 days (taken from Voelter et al., 1973a).

[74] Epimaltose [4-O-α-D-glucopyranosyl-D-mannose]

TABLE 4
Chemical Shifts of Disaccharides and their Derivatives
(Numbered formulae are on pp. 304-5, 308)

Compound	1'	2'	3'	4'	5'	6'	1	2	3	4	5	6	Methyl	Reference
α-Cellobiose [71a][b]	103.5	74.3	76.9	70.7	76.9	61.8	92.9	72.5	74.3	79.9	72.5	61.3		Dorman and Roberts, 1971
α-Cellobiose [71a][c]	102.40	73.00	75.80	69.25	75.35	60.45	91.65	71.20	69.90	78.60	71.05	59.95		Voelter et al., 1971
α-Cellobiose [71a][d]	103.9	74.7	77.2	71.1	77.2	62.4	93.2	72.9	72.9	80.1	71.6	61.8		Usui et al., 1973b
β-Cellobiose [71b][b]	103.5	74.3	76.9	70.7	76.9	61.8	96.9	75.3	75.8	79.9	75.3	61.3		Dorman and Roberts, 1971
β-Cellobiose [71b][c]	102.40	73.00	75.80	69.25	75.35	60.45	95.55	73.75	74.10	78.50	74.60	59.75		Voelter et al., 1971
β-Cellobiose [71b][d]	103.9	74.7	77.2	71.1	77.2	62.4	97.1	75.7	76.1	80.1	75.7	61.8		Usui et al., 1973b
			77.4		77.4									
Methyl-β-Cellobioside[d]	103.7	74.3	77.0	70.7	76.9	61.9	104.2	74.0	75.7	80.1	75.5	61.4	58.2	Dorman and Roberts, 1971
Methyl-β-Cellobioside[d]	103.9	74.6	77.5	71.2	77.2	62.4	104.5	74.2	76.4	80.3	75.9	61.8	58.9	Usui et al., 1973b
α-Lactose [70a][b]	104.0	72.2	73.9	69.8	76.3	62.1	93.0	72.6	74.0	79.9	72.5	61.4		Dorman and Roberts, 1971
α-Lactose [70a][e]	103.00	71.05	72.60	68.60	75.40	61.10	91.85	70.20	71.20	78.40	71.20	60.15		Voelter et al., 1973a
											71.50			
β-Lactose [70b][b]	104.0	72.2	73.9	69.8	76.3	62.1	97.0	75.2	75.9	79.9	75.6	61.4		Dorman and Roberts, 1971
β-Lactose [70b][e]	103.00	71.05	72.60	68.60	75.40	61.10	95.80	73.90	74.45	78.40	74.45	60.15		Voelter et al., 1973a
											74.90			
Methyl-β-Lactose[b]	104.1	72.1	73.9	69.8	76.4	62.1	104.1	73.9	75.8	79.8	75.6	61.4	58.2	Dorman and Roberts, 1971
Methyl-β-Lactose[e]	104.15	72.20	73.80	69.80	76.55	62.55	104.15	74.00	76.00	79.75	76.00	61.35	58.50	Voelter et al., 1973a
	104.30						104.30		75.65		75.65			
α-Maltose [72a][b]	100.9	73.80	73.8	70.7	73.0	61.9	93.2	72.6	71.1	78.9	74.3	61.9		Dorman and Roberts, 1971
α-Maltose [72a][c]	99.35	73.85	72.70	69.15	72.50	60.35	91.65	71.50	69.80	76.85	71.15	60.45		Voelter et al., 1971
α-Maltose [72a][e]	99.60	72.70	72.70	69.40	71.70	60.55	91.95	69.95	69.95	76.95	69.95	60.55		Voelter et al., 1973a
		72.90	72.90			60.80		71.30	71.30		71.30	60.80		
								73.25	73.25		73.25			
α-Maltose [72b][d]	101.0	74.3	74.6	71.0	73.4	62.5	93.2	73.0	74.6	78.5	71.6	62.55		Usui et al., 1973b
β-Maltose [72b][c]	99.35	73.85	72.70	69.15	72.50	60.35	95.60	73.00	74.40	76.60	76.00	60.60		Voelter et al., 1971
β-Maltose [72b][e]	99.60	72.70	72.70	69.40	71.70	60.55	95.90	74.00	74.60	76.70	74.60	60.55		Voelter et al., 1973a
		72.90	72.90			60.80			76.25		76.25	60.80		
β-Maltose [72b][d]	101.0	74.3	74.6	71.0	73.4	62.5	97.1	75.7	77.8	78.5	76.1	62.5		Usui et al., 1973b
Methyl-β-Maltoside[b]	100.8	74.1	74.1	70.6	72.9	61.9	104.1	73.8	75.7	78.4	77.3	61.7	58.2	Dorman and Roberts, 1971
Methyl-β-Maltoside[d]	101.1	74.3	74.6	70.9	73.4	62.3	104.4	74.6	77.8	78.7	76.1	62.3	58.7	Usui et al., 1973b
Sucrose [69][b]	93.1	73.4	73.9	70.5	72.2	61.6	63.4	104.7	82.5	78.0	75.4	62.9		Dorman and Roberts, 1971
Sucrose [69][f]	93.0	73.3	73.7	70.3	72.1	61.3	63.4	104.6	77.6	82.3	75.1	62.6		Allerhand and Doddrell, 1971
Sucrose [69][g]	91.8	72.5	72.1	69.1	70.9	62.2	62.2	103.4	76.4 or	81.1	73.9			Binkley et al., 1972
			70.9		72.1	61.4	61.4		76.1					
						or 60.1	or 60.1							

Compound	C-1	C-2	C-3	C-4	C-5	C-6	C-1'	C-2'	C-3'	C-4'	C-5'	C-6'	Reference
α-Epilactose[e]	104·20	72·15	73·70	69·75	76·50	62·30	94·95	71·35	70·20	77·75	72·15	61·55	Voelter et al., 1973a
β-Epilactose[e]	104·20	72·15	73·70	69·75	76·50	62·30	94·85	71·80	74·90	77·35	76·10	61·55	Voelter et al., 1973a
α-Epimaltose [74a][e]	99·90	72·70	72·70	69·40	70·80	60·55	93·85	71·05	70·80	75·35	71·85	61·05	Voelter et al., 1973a
		72·95	72·95										
β-Epimaltose [74b][e]	99·90	72·70	72·70	69·40	70·80	60·55	93·65	71·25	73·65	74·75	74·75	61·05	Voelter et al., 1973a
		72·95	72·95							74·90	74·90		
α,α-Trehalose[d]	104·0	74·3	77·4	70·9	76·8	62·3	94·8	73·9	74·4	71·5	71·5	62·5	Usui et al., 1973b
		74·6							74·3				
β,β-Trehalose[d]							100·7	74·2	77·3	71·1	77·3	62·5	Usui et al., 1973b
α,β-Trehalose[d]							101·3	74·0	74·6	70·9	72·9	62·0	Usui et al., 1973b
α-Kojibiose[d]	95·7	73·1	74·4	71·1	73·1	62·0	90·8	77·1	73·1	71·1	73·1	62·0	Usui et al., 1973b
β-Kojibiose[d]	99·0	73·1	74·4	71·1	73·1	62·0	97·5	79·9	75·8	71·1	77·1	62·0	Usui et al., 1973b
Methyl-β-Kojibiose[d]	99·0	73·0	74·2	71·3	73·0	61·9	105·0	79·0	75·8	70·8	77·1	62·5	Usui et al., 1973b	58·9
α-Sophorose[d]	105·1	74·9	77·2	71·1	77·2	62·4	93·1	82·1	73·2	71·1	72·5	62·4	Usui et al., 1973b
β-Sophorose[d]	103·9	74·9	77·2	71·1	77·2	62·4	95·8	82·8	77·2	71·1	77·2	62·4	Usui et al., 1973b
Methyl-α-sophorose[d]	105·0	75·4	77·1	71·3	77·1	62·2	100·0	81·7	73·3	71·3	72·5	62·2	Usui et al., 1973b	56·2
α-Nigerose[d]	99·8	72·8	74·1	71·3	72·8	61·8	93·1	71·3	80·8	70·6	72·2	61·8	Usui et al., 1973b
β-Nigerose[d]	99·8	72·8	74·1	71·3	72·8	61·8	97·0	74·1	83·2	70·6	76·6	61·8	Usui et al., 1973b
α-Laminaribiose[d]	103·9	74·8	77·1	71·2	77·1	62·4	93·4	72·2	84·2	69·6	72·4	62·4	Usui et al., 1973b
β-Laminaribiose[d]	103·9	74·8	77·1	71·2	77·1	62·4	97·2	74·8	86·7	69·6	77·1	62·4	Usui et al., 1973b
α-Isomaltose[d]	99·4	73·3	75·0	71·3	73·8	62·5	93·8	73·3	75·0	71·3	71·3	67·4	Usui et al., 1973b
β-Isomaltose[d]	99·4	73·3	75·0	71·3	73·8	62·5	97·7	75·9	77·7	71·3	75·9	67·4	Usui et al., 1973b
α-Gentiobiose[d]	103·8	74·5	77·1	71·1	77·1	62·5	93·3	72·9	74·5	71·1	71·8	70·2	Usui et al., 1973b
β-Gentiobiose[d]	103·8	74·5	77·1	71·1	77·1	62·5	97·2	75·5	77·1	71·1	76·1	70·2	Usui et al., 1973b
Methyl-β-Gentiobiose[d]	104·0	74·0	77·2	71·0	77·2	62·5	104·5	74·0	71·0	71·2	76·1	70·0	Usui et al., 1973b	58·8
				71·2					71·2				

[a] Chemical shifts in parts per million downfield from tetramethylsilane, TMS (δ = 0 ppm); see cited reference for experimental details and for additional data.
[b] Obtained at 15·1 MHz using 1–2M aqueous solutions; original data measured relative to CS_2 and converted using $\delta_{CS_2} = 193.7$ ppm.
[c] Obtained with Fourier transform in D_2O relative to external tetramethylsilane (TMS).
[d] Obtained on a JEOL PS-100 at 25·1 MHz with Fourier transform (proton noise decoupled) relative to TMS.
[e] Obtained on a Bruker HFX-90 spectrometer at 22·628 MHz relative to external tetramethylsilane (TMS); concentrations of D-maltose and epimaltose were 300 mg and 102 mg per 1·5 ml D_2O, respectively.
[f] Obtained at 15·08 MHz and 36° using Fourier transform on 1M aqueous solutions; original data measured relative to CS_2 and converted using $\delta_{CS_2} = 193.7$ ppm.
[g] Obtained on a Varian XL-100-15 spectrometer with Fourier transform (proton noise decoupled) at 30° and on a modified Varian HA-100-12 at 25·15 MHz relative to internal tetramethylsilane (TMS).

and Roberts, 1971; Voelter et al., 1973a) in aqueous solution have been reported. Assignments of resonance signals to individual carbon atoms were based upon generalizations from the available chemical shifts of monosaccharides (Stothers, 1972a) assuming identical conformations of the pyranose ring in mono- and disaccharides. Typical of such generalizations are: (1) signals due to anomeric carbon atoms C-1 are observed at low fields; (2) those due to carbon bearing ether linkages (C-4′) are at ca. 78–80 ppm; (3) those ascribed to hydroxymethyl carbons (C-6 and C-6′) are at 59·7–61·9 ppm. Taking advantage of the recognized electronic and steric effects on ^{13}C resonance frequencies examination of the glycoside and acetate derivatives of these disaccharides have also aided the interpretation of the spectral data (Dorman and Roberts, 1971). Additionally, advantage has been taken of partially relaxed Fourier transform (PRFT) method (Allerhand and Doddrell, 1971). Table 4 summarizes the ^{13}C resonance frequencies assigned for disaccharides.

In aqueous solutions sucrose, lactose, maltose and cellobiose exist, of course, in anomeric equilibria of their α and β forms. Consequently more than twelve ^{13}C resonance signals are expected to be resolvable for each anomeric equilibrium pair. Indeed this is the case, although several anomeric signals overlap (Table 4; Allerhand and Doddrell, 1971; Dorman and Roberts, 1971; Voelter et al., 1973a). The significance of this observation is that rates and equilibria for anomerization of complex carbohydrates can be determined by ^{13}C

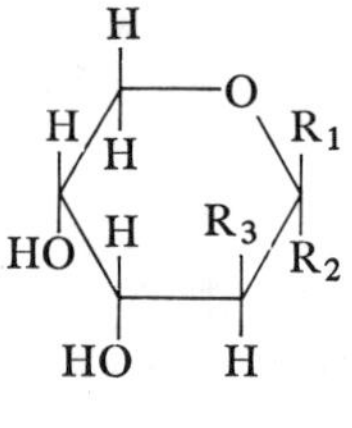

[73a] $R_1 = CH_2OH$, $R_2 = OH$
[73b] $R_1 = OH$, $R_2 = CH_2OH$

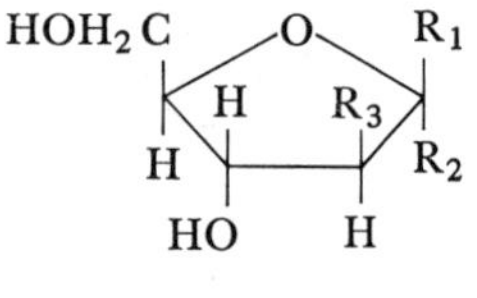

[73c] $R_1 = CH_2OH$, $R_2 = OH$
[73d] $R_1 = OH$, $R_2 = CH_2OH$

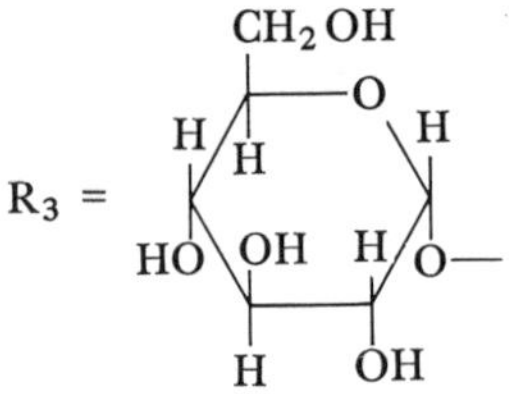

[73] Turanose [(α-D-glucopyranosyl)-D-fructose]

nmr spectroscopy. Figure 1 clearly indicates the time dependence of lactose mutarotation. Under favorable conditions, as in the case of cis-monosaccharides, measurements of integrated peak intensities provide directly the percentages of each anomer in the solution. Such measurements have established turanose [73] to be < 4% [73a], 39 ± 4% [73b], 20 ± 2% [73c] and 41 ± 4% [73d] (Doddrell and Allerhand, 1971b). The observed ^{13}C resonance signals in the equilibrium mixture of the anomeric forms of lactose [70] and cellobiose [71] were interpreted in terms of conformations which do not allow strong interactions between the two rings (Dorman and Roberts, 1971). Conversely, sucrose [69] and maltose [72], disaccharides joined through α-glycoside linkages, show conformational or steric interactions (Dorman and Roberts, 1971). Similar detailed conformational analyses of more complex carbohydrates are within the scope of ^{13}C magnetic resonance techniques and will undoubtedly be the subject of intense research scrutiny.

Oligo and Polysaccharides

The relative simplicity of proton-decoupled ^{13}C spectra is increasingly being utilized for identification and correlation of polysaccharides obtained from different natural and synthetic sources. The presence of signals with the same resonances are taken to indicate repeating structural units. The resonance frequencies of the antibiotic pseudo-trisaccharides gentamicin C_1 [75a], gentamicin C_{1a} [75b], gentamicin C_2 [75c], sisomicin [76], gentamine C_1 [84a], gentamine C_{1a} [84b], gentamine C_2 [84c] were assigned to the individual carbon atoms (Table 5) using the established strategies in correlating resonance frequencies of the appropriate carbon atoms of the structurally similar 2-deoxystreptamine [85] and methyl-β-garosaminide [83] (Fig. 2; Morton et al., 1973).

In addition to the gentamicin pseudotrisaccharides, ^{13}C nmr data on a relatively small number of structurally similar aminoglycoside antibiotics has appeared in the literature. The structural elucidation and characterization of hygromycin B [80], its degradation products hysoamine and hygromycin B_2 [87] and destomycin A [81], an isomer of hygromycin B, has been reported (Neuss et al., 1970). The ^{13}C nmr shifts of kanamycin A [78a] and B [78b] at two different pH-values have been used to confirm the presence of an additional amino-group in [78b] (Kotowycz and Lemieux, 1973). The ^{13}C nmr

TABLE 5

^{13}C Chemical Shifts of Oligosaccharides and Related Compounds

													δ, ppm^a		
					Ring A							Ring B			
Compound	Solvent	1	2	3	4	5	6	7	8	1	2	3	4	5	6
Pseudotrisaccharides															
Gentamicin C$_1$ [75a]	H$_2$O	101·4	20·3	67·5	73·3	68·7	22·9	38·1		51·8	36·8	50·9	88·6	75·4	87·9
Gentamicin C$_{1a}$ [75b]	D$_2$O	101·3	70·2	64·4	73·3	68·7	23·0	38·0		51·7	36·7	50·6	88·3	75·4	87·8
Gentamicin C$_2$ [75c]	D$_2$O	101·2	70·1	64·4	73·1	68·7	23·0	38·2		51·8	36·7	50·8	88·7	75·3	87·6
Sisomicin [76]	D$_2$O	101·5	70·0	64·3	73·0	68·5	22·9	37·9		51·7	36·4	50·3	85·2	75·3	87·7
Gentamicin CO$_2$ adduct	H$_2$O	101·2	70·2	64·4	73·3	68·7	22·4	37·8		51·6	36·7	50·5	87·8	75·4	87·8
Butirosin A [77a]	D$_2$O	101·0	57·8	76·7 76·5 75·3	73·6	76·7 76·5 75·3	43·7			52·3 51·6	36·7 36·2	52·3 51·6	86·8	84·1 83·9 83·0 82·7	76·7 76·5 75·3
(TFA)$_4$-butirosin A [77b]	CD$_3$OD	99·4	56·7	76·6 75·1 73·4 72·9	76·6 75·1 73·4 72·9	76·7 75·1 73·4 72·9	42·3				35·2 34·5		86·9	83·0 82·3 81·5	76·7 75·1 73·4 72·9
Butirosin A . 2H$_2$SO$_4$ [77c]	D$_2$O	96·9	56·0	71·3 70·5	73·3	71·3 70·5	42·7			51·4 51·0	33·1	51·4 51·0	84·9 82·9	84·9 82·9	76·6 76·0
(TFA)$_2$-butirosin A . H$_2$SO$_4$ [77d]	D$_2$O	97·5	55·8	72·0 70·3	72·6	72·0 70·3	41·9			51·3	33·5 32·9	51·3	84·2 82·7	84·2 82·7	76·5 75·7
Kanamycin A [78a]	pH 3·6	96·1	71·4	72·8	71·4	69·3	41·0			50·4	28·1	48·3	78·8	73·4	84·3
	pH 9·6	99·9	72·4	73·5	71·7	72·9	42·0			51·0	36·1	49·6	87·6	74·8	88·4
Kanamycin B [78b]	pH 5·5	86·3	54·2	68·9	71·2	69·6	40·7			50·2	28·9	48·9	78·7	74·7	84·2
	pH 10·6	100·5	55·6	73·8	71·7	73·0	41·9			50·7	35·8	49·6	86·8	74·7	88·2
	pH ≥ 11	100·1	55·3	73·4	71·2	72·8	41·6			50·2	35·4	49·2	86·4	74·2	87·7
Kanamycin B carbamate	pH ≥ 11	100·1	55·3	73·5	71·3	72·8	41·7			50·4	35·6	49·3	86·7	74·0	87·4
Tobramycin	pH ≥ 11	99·2	49·5	34·7	65·9	73·1	41·5			50·2	35·5	49·0	86·0	74·4	87·8
Tobramycin carbamate	pH ≥ 11	99·8	49·5	35·0	66·2	73·3	41·6			50·6	35·5	49·3	86·3	74·1	87·7
α-Streptomycin sulfate	D$_2$O	95·5	61·6	70·8	70·6	72·7	62·6	33·5		106·9	78·5	83·3	85·5	13·8	
α-Streptomycin sulfate	0·01 M NaOH	95·6	61·6	70·6	70·4	72·6	62·6	33·5		106·8	78·5	83·3	85·5	13·8	
α-Streptomycin sulfate	0·01 M HCl	95·4	61·6	70·6	70·3	72·6	62·6	33·4		106·8	78·5	83·3	85·5	13·8	
α-Streptomycin CaCl$_2$·3HCl	D$_2$O	95·3	61·4	70·4	70·1	72·3	62·3	33·3		106·4	78·2	82·9	85·5	13·5	
β-Streptomycin·3HCl	0·01 M HCl	95·5	61·4	69·9	69·2	70·3	61·4	32·0		107·6	76·9	81·7	86·2	12·0	
Dihydrostreptomycin sulfate	D$_2$O	94·5	61·4	70·3	70·3	72·5	62·0	33·0		106·7	78·7	81·7	84·9	13·8	
Ribostamycin [79a]	D$_2$O	99·8	56·4	74·1	72·3	73·9	42·7			51·2	36·7	51·2	83·0	85·0	78·4
	pD 9·5	98·7	56·0	71·6	72·3	73·1	42·0			51·3	35·6	50·7	81·6	85·5	77·7
Ribostamycin (H$_2$SO$_4$)$_{3/2}$ [79a]	D$_2$O	96·4	55·0	70·0	72·0	70·0	41·4			51·4	32·0	49·8	79·2	86·0	74·4
Tetra-N-acetyl-ribostamycin [79c]	D$_2$O	96·9	54·4	71·3	71·3	71·3	41·2			50·4	33·3	49·0	76·9	85·8	74·
Tetra-N-acetyl-2″,3″-O-isopropylidene-ribostamycin [79d]	D$_2$O	96·9	54·0	71·4	71·4	71·4	40·7			50·6	33·4	49·0	76·5	86·3	74·
SF-733D [79e]	D$_2$O	98·5	56·3	74·0	71·7	73·5	42·3			51·0	35·3	49·4	77·1	85·9	78·
SF-733X [79f]	D$_2$O	99·1	56·2	73·5	72·5	72·7	42·5			51·1	32·2	57·0	79·3	85·8	78·
	pD 4·5	97·4	54·4	71·6	69·3	69·8	41·5			51·1	26·9	54·4	77·6	84·7	73·
Hygromycin B [80]	aq.	59·1	33·1	51·6	78·2	86·6	74·7			100·6	76·1	75·6	74·3	73·2	62·
Destomycin A [81]	aq.	59·1	32·7	51·5		86·6				100·6	76·3	75·3	74·7	73·1	62·
Hexa-N-acetylneomycin B [82]	D$_2$O	97·0	54·2	71·4	71·4	71·4	40·8			50·4	33·2	48·9	76·7	86·0	74·
Pseudotrisaccharide residues															
Methyl-β-garosaminide [83]	D$_2$O	100·6	70·1	64·7	73·5	68·1	22·7	38·2	56·0						
Gentamine C$_1$ [84a]	D$_2$O									51·3	36·8	50·8	88·8	76·8	78·
Gentamine C$_{1a}$ [84b]	D$_2$O									51·3	36·8	50·5	88·1	76·8	78·
Gentamine C$_2$ [84c]	D$_2$O									51·3	36·8	50·7	88·7	76·8	78·
2-Deoxystreptamine [85]	D$_2$O									51·6	37·0	51·6	78·5	76·6	78·

(Numbered formulae are on pp. 315-319)

			Ring C							Other					Reference
1	2	3	4	5	6	7	8	1	2	3	4	5	6		
102·6	51·1	27·2	26·1	72·8	58·2	15·0	33·7								Morton et al., 1973
102·2	51·0	27·1	28·5	71·5	46·1										Morton et al., 1973
102·6	51·0	27·1	26·2	74·5	59·3	19·2									Morton et al., 1973
100·6	47·5	25·6	96·4	150·2	43·5										Morton et al., 1973
101·8	50·8	26·6	28·2	71·0	45·9										Morton et al., 1973

									Other						
									Hydroxyaminobutyric side chain						
								1	2	3	4				
112·2	84·1	84·1	84·1	62·7				178·1	72·0	36·2	38·9				Woo and Westland, 1973
	83·9	83·9	83·9												
	83·0	83·0	83·0												
	82·7	82·7	82·7										CF$_3$CO		
113·6	83·0	83·0	83·0	63·0				176·67	71·1	35·2	37·5	160·0			Woo and Westland, 1973
	82·3	82·3	82·3							34·5		158·8			
	81·5	81·5	81·5									158·3			
											CF$_3$	136·2			
												123·9			
												111·2			
113·9	88·2	88·2	88·2	62·7				177·64	71·8	33·1	38·8				Woo and Westland, 1973
	78·0	78·0	78·0												
	76·6	76·6	76·6												
	76·0	76·0	76·0												
113·7	88·1	88·1	88·1	62·3				177·42	71·7	33·5	38·7	160·0			Woo and Westland, 1973
	78·0	78·0	78·0							32·9					
	76·5	76·5	76·5												
	75·7	75·7	75·7												
100·7	68·7	55·7	66·2	72·8	60·7										Kotowycz and Lemieux, 1973
100·4	72·4	54·9	70·0	72·9	61·1										Kotowycz and Lemieux, 1973
100·8	68·5	55·5	66·1	73·4	60·5										Kotowycz and Lemieux, 1973
100·1	72·1	54·5	69·6	72·4	60·7										Kotowycz and Lemieux, 1973
99·5	71·7	54·2	69·1	71·9	60·2										Koch et al., 1974
99·5	71·6	54·2	69·2	70·0	63·5	158·6									Koch et al., 1974
99·1	71·6	54·2	69·2	71·9	60·2										Koch et al., 1974
99·2	71·6	54·3	69·2	70·1	63·5	158·9									Koch et al., 1974
90·5	79·1	59·6	74·4	60·0	74·0	71·8	159·1	158·5							Bock et al., 1974
90·6	79·0	59·3	74·4	60·0	73·9	71·7	159·0	158·5							Bock et al., 1974
90·6	79·0	59·3	74·4	60·0	73·9	71·8		159·1	158·5						Bock et al., 1974
90·3	78·7	59·1	74·1	59·6	73·7	71·7		158·6	158·2						Bock et al., 1974
96·1	77·9	57·5	73·6	58·5	73·6	81·1		157·7	157·7						Bock et al., 1974
64·3	78·7	59·1	74·3	59·8	73·7	71·5		158·8	158·3						Bock et al., 1974
109·1	75·7	70·5	83·4	62·6											Omoto et al., 1973
109·6	75·9	70·5	83·3	62·5											Omoto et al., 1973
110·0	76·1	70·0	83·3	62·1											Omoto et al., 1973
109·6	75·8	70·6	83·4	62·6				CH$_3$CO 23·0	CH$_3$CO 175·2, 174·9, 174·5, 174·1						Omoto et al., 1973
110·8	87·7	81·7	86·3	63·3				CH$_3$CO 23·2	CH$_3$CO 175·2, 174·7, 174						Omoto et al., 1973
								O–C(CH$_3$)(CH$_3$)–O 113·8			O–C(CH$_3$)(CH$_3$)–O 25·0 26·7				
109·6	75·9	70·7	83·3	62·7				–NHCOCH$_3$ 174·0	–NHCOCH$_3$ 23·5						Omoto et al., 1973
109·4	76·0	70·6	83·3	62·7				–NHCH$_2$COOH 50·4	–NHCH$_2$COOH 179·3						Omoto et al., 1973
111·8	76·4	70·9	83·1	61·3				–NHCH$_2$COOH 47·6	–NHCH$_2$COOH 171·9						Omoto et al., 1973
120·4	72·5	70·1	70·1	64·6	52·9	63·1		CH$_3$ 33·1							Neuss et al., 1970
120·3	72·5	70·1	70·1	64·6	53·0	63·0		CH$_3$ 32·7							Neuss et al., 1970
109·3	74·5	77·2	82·4	62·2				98·5	51·8	70·3	68·5	74·0	40·8		Rinehart et al., 1974

Morton et al., 1973

102·8	50·9	27·4	26·3	72·6	58·1	15·1	33·7								Morton et al., 1973
102·1	50·7	27·1	28·5	71·3	46·0										Morton et al., 1973
102·9	50·9	27·0	26·2	74·3	50·1	19·1									Morton et al., 1973
															Morton et al., 1973

TABLE 5—continued

Compound	Solvent	Ring A δ, ppm[a]								Ring B δ, ppm[a]					
		1	2	3	4	5	6	7	8	1	2	3	4	5	6
2-Deoxystreptamine [85]	D_2O									52·6	38·1	52·6	79·8	77·8	79·8
2-Deoxystreptamine [85]	D_2O									51·4	36·9	51·4	78·5	76·6	78·5
2-Deoxystreptamine [85]	pH ⩾ 11									50·7	36·3	50·7	77·7	75·8	77·7
Deoxystreptamine dihydrochloride	D_2O									52·6	30·5	52·6	74·6	77·0	74·6
Deoxystreptamine carbonate	D_2O									52·4	32·3	52·4	76·2	77·7	76·2
Di-N-acetyl-2-deoxystreptamine	D_2O									50·7	33·6	50·7	75·2	76·8	75·2
Mono-N-acetyl-2-deoxystreptamine	D_2O									51·1	34·6	50·6	77·3	76·5	75·2
Paromamine [86c]	D_2O	103·1	57·5	75·9 75·0	72·0	75·9 75·0	62·9			52·5 51·5	38·0	52·5 51·5	89·9	78·0	79·6
Paromamine [86c]	D_2O	101·9	52·3	74·6	70·8	73·8	61·5			51·2	36·7	50·4	88·7	76·7	78·2
Neamine [86a]	D_2O	102·4	57·2	75·4 74·6	73·5	75·4 74·6	43·3			52·5 51·3	37·3	52·5 51·3	88·3	78·0	79·1
Neamine [86a]	D_2O	101·5	56·2	74·4	72·4	73·4	42·6			51·4	36·5	50·3	87·7	76·9	78·1
Tetra-N-acetyl-neamine [86b]	D_2O	98·4	54·7	71·5	71·5	71·5	40·8			50·5	33·8	49·7	80·4	78·0	75·6
Hysomine	aq.	59·1	33·3	51·5	78·5	77·1	76·2								
Hygromycin B_2 [87]	aq.	59·1	33·0	51·6	78·2	86·4	74·7			102·3 51·5	72·0 34·2	69·8 57·7	69·5 76·6	77·1 76·6	62·3 78·5
Mono-N-carboxymethyl-2-deoxystreptamine	pD 11·0														
Mono-N-carboxymethyl-2-deoxystreptamine	pD 2·0									51·2	27·4	57·5	72·4	75·6	72·6
Butirosin P	D_2O	102·9	57·5	76·3 75·8 75·1	73·6	76·3 75·8 75·1	43·9			51·8 51·4	36·1 35·6	51·8 51·4	88·8	78·2	76·3 75·8 75·1
(s)-(−)-4-Amino-4-hydroxybutyric acid															
Oligosaccharides		Fructose								Glucose					
Raffinose [88]	0·6 M water	63·5 or 62·9	104·8	77·9	82·3	75·3	62·9 or 63·5			93·0	73·9 72·4 or 72·1	73·9 72·4 72·1	70·6 70·3	73·9 72·4 or 72·1	67·2
Stachyose [89]	0·5 M water	63·5 or 62·9	104·8	77·9	82·3	75·3	63·5 or 62·9			93·1	73·9 72·3 or 72·1	73·9 72·3 72·1	70·7 70·6 70·4	73·9 72·3 72·1	67·1
		Glucose								Fructose					
Nystose [90]	DMSO-d_6	92·0	74·0 or 71·6	72·8	69·8 or 71·6	72·8	61·9 61·7 61·2 61·1 61·0 or 60·5			61·9 61·7 61·2 61·1 61·0 or 60·5	103·7	82·1 or 77·7	77·7 82·1	74·6	61·9 61·7 61·2 61·1 61·0 or 60·5
1-Kestose [91]	DMSO-d_6	91·9	73·8	72·7 71·5	69·7	72·7 71·5	62·0 61·6 61·0 60·9 60·3			62·0 61·6 61·0 60·9 60·3	103·7	82·9 76·9	82·9 76·9	74·7	62·0 61·6 61·0 60·9 60·3
β-Maltotriose	1–2 M aq.	93·4	73·4	74·7	78·5	71·6	62·4			101·1	73·1	74·7	78·5	72·7	62·4
β-Maltotriose	1–2 M aq.	97·1	75·9	77·8	78·9	76·3	62·4			101·1	73·1	74·7	78·5	72·7	62·4
β-Panose	1–2 M aq.	93·3	73·1	76·6	78·5	71·6	62·4			101·1	74·3	76·6	71·2	71·2	67·4
β-Panose	1–2 M aq.	97·2	75·4	77·7	78·5	75·8	62·4			101·1	74·3	74·6	71·2	71·2	67·4
α-Isopanose	1–2 M aq.	93·8	73·3	74·6	71·0	71·3	67·5			99·3	73·3	74·6	78·8	71·6	62·3
β-Isopanose	1–2 M aq.	97·7	75·8	77·7	71·0	75·8	67·5			99·3	73·3	74·6	78·8	71·6	62·3
α-Gentiotriose	1–2 M aq.	93·4	73·1	74·3	71·2	71·2	70·0 or 70·2			103·7	74·6	77·1	71·2	76·1	70·0 or 70·2
β-Gentiotriose	1–2 M aq.	97·3	75·5	76·8	71·2	76·1	70·0 or 70·2			103·7	74·6	77·1	71·2	76·1	70·0 or 70·2
										Unassigned Chemical Shifts					
Stachyose [89]	DMSO-d_6	103·8	98·6	91·5	82·2	76·9	74·1	72·7	71·2	71·0	70·9	70·0	69·5	69·1	68·7
Raffinose [88]	DMSO	101·2	96·7	89·3	79·9	75·1	72·2	70·5	69·1	68·5	68·5	68·0	67·2	66·5	66·2
Planteose	DMSO	101·5	96·8	89·4	77·6	74·9	75·4	70·8	70·2	69·3	68·6	68·0	67·3	66·6	66·2

[a] Chemical shifts in parts per million downfield from tetramethylsilane (TMS); see cited reference for experimental conditions

			Ring C							Other					Reference
1	2	3	4	5	6	7	8	1	2	3	4	5	6		
															Woo and Westland, 1973
															Omoto et al., 1973
															Koch et al., 1974
															Woo and Westland, 1973
															Woo and Westland, 1973
								CH_3CO 23·2	CH_3CO 174·4						Omoto et al., 1973
								CH_3CO 25·0	CH_3CO 174·6						Omoto et al., 1973
															Woo and Westland, 1973
															Omoto et al., 1973
															Woo and Westland, 1973
															Omoto et al., 1973
															Omoto et al., 1973
								CH_3 32·9							Neuss et al., 1970
								CH_3 33·0							Neuss et al., 1970
								$NHCH_2COOH$ 50·6	$NHCH_2COOH$ 180·0						Omoto et al., 1973
								$NHCH_2COOH$ 47		$NHCH_2COOH$ 170·6					Omoto et al., 1973
								178·28	72·0	37·1	38·9				Woo and Westland, 1973
								182·17	72·4	33·4	39·1				Woo and Westland, 1973
		Galactose (terminal)								Galactose (internal)					
99·5	70·6	69·6	70·6	71·9	62·2										Allerhand and Doddrell, 1971
or 70·3		or 70·3													
99·1	70·7	69·6	70·7	72·0	62·2			99·4	70·7	69·8	69·8	70·7	67·5		Allerhand and Doddrell, 1971
	70·6		70·6						70·6	69·5	69·5	70·4			
	or 70·4		or 70·4						or 70·4						
			Fructose								Fructose				
61·9	103·2	81·9	81·9	72·8	61·9			61·9	102·8	81·9	81·9	74·6	61·9		Binkley et al., 1972
61·7		or 77·5	77·5		61·7			61·7		76·5	76·5		61·7		
61·2					61·2			61·2					61·2		
61·1					61·1			61·1					61·1		
61·0					61·0			61·0					61·0		
or 60·5					or 60·5			or 60·5					or 60·5		
62·0	103·2	82·9	82·9	72·8	62·0										Binkley et al., 1972
61·6		76·4	76·9		61·6										
61·0					61·0										
60·9					60·9										
60·3					60·3										
101·1	74·3	75·1	71·2	73·4	62·4										Usui et al., 1973b
101·1	74·3	75·1	71·2	73·4	62·4										Usui et al., 1973b
99·3	73·1	74·6	71·2	73·1	62·4										Usui et al., 1973b
99·3	73·1	74·6	71·2	73·1	62·4										Usui et al., 1973b
101·1	74·1	75·0	71·0	73·0	62·3										Usui et al., 1973b
101·1	74·1	75·0	71·0	73·0	62·3										Usui et al., 1973b
103·7	74·6	77·1	71·0	77·1	62·4										Usui et al., 1973b
103·7	74·6	77·1	71·0	77·1	62·4										Usui et al., 1973b
68·6	68·5	68·3	66·3	62·0	60·4										Binkley et al., 1972
64·5	60·0	59·6	58·3												Binkley et al., 1972
60·0	58·6	58·3													Binkley et al., 1972

and additional details.

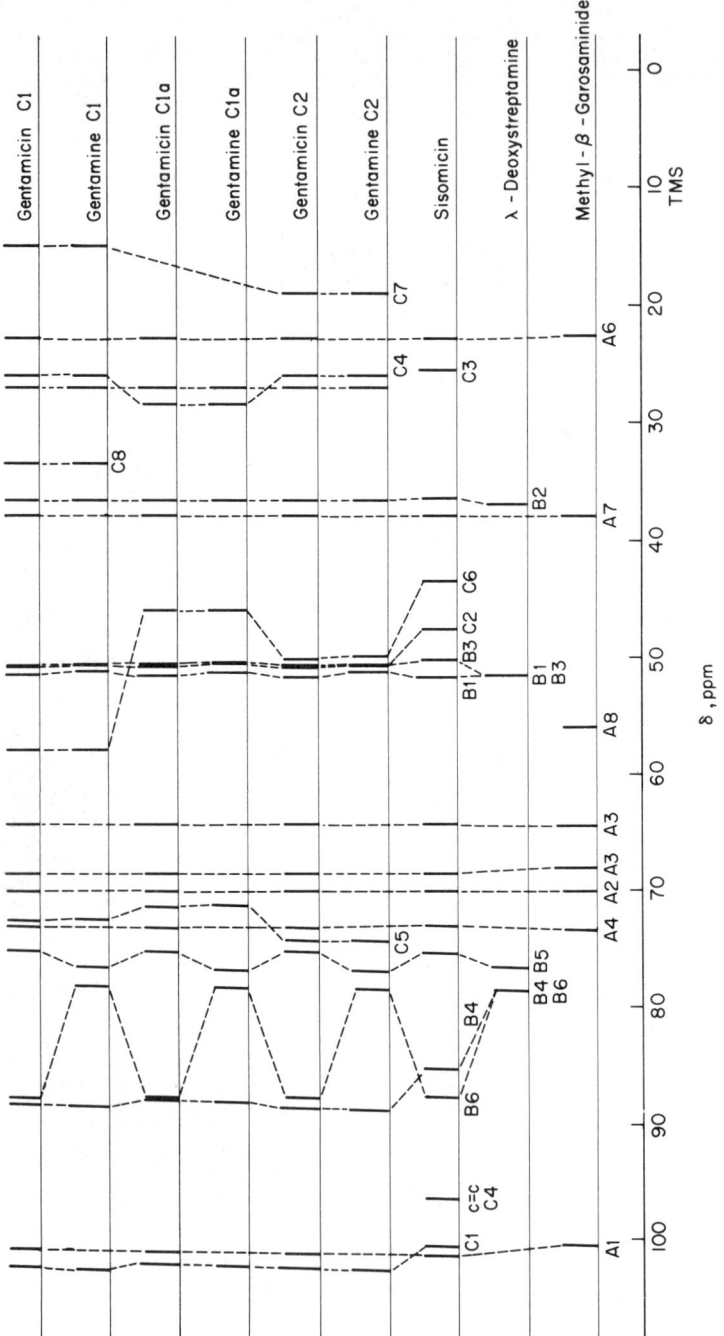

Figure 2. Correlation of the ^{13}C nmr chemical shifts for the pseudosaccharides and their fragments (taken from Morton et al., 1973).

spectra of ribostamycin [79a] and related compounds [79b-f, 86a-c] were recorded and an analysis of chemical shifts with structures was made (Omoto et al., 1973). ^{13}C spectral assignments of butirosin A [77a] and its related aminoglycosides [77b-d, 86a-c] has been determined from the individual structural components of butirosin (Woo and Westland, 1973). The utility of the ^{13}C nmr technique is clearly evident in the structural determinations of butirosins, which lack definite melting points, characteristic ultraviolet absorption and definitive infrared spectra. Mass spectrometry and ^{1}H nmr, although both are highly informative in structure studies, proved to be inconclusive for butirosins. The biosynthesis of the neomycin pseudotetrasaccharide antibiotics [82] has also been studied by means of ^{13}C nmr spectroscopy (Rinehart et al., 1974).

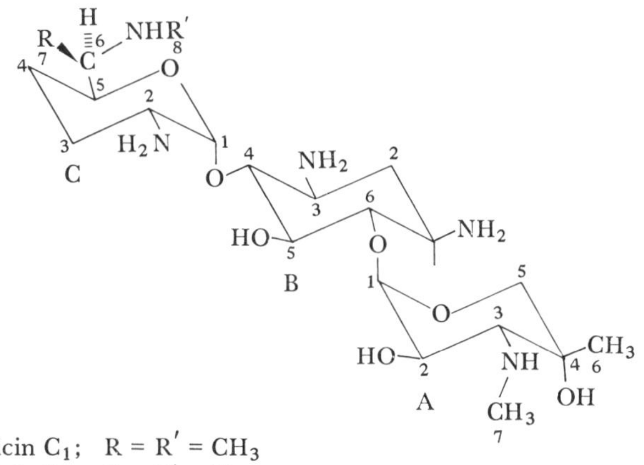

[75a] Gentamicin C_1; R = R' = CH_3
[75b] Gentamicin C_{1a}; R = R' = H
[75c] Gentamicin C_2; R = CH_3, R' = H

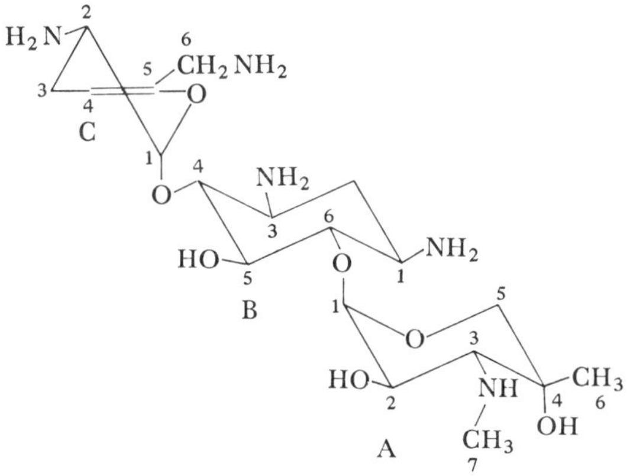

[76] Sisomicin

[Structure of butirosin shown]

[77a] Butirosin A; $R' = R'' = H$
[77b] $(TFA)_4$ butirosin, $R' = R'' = -COCF_3$
[77c] Butirosin A . $2H_2SO_4$; $R' = R'' = H . \frac{1}{2}H_2SO_4$
[77d] $(TFA)_2$ butirosin . H_2SO_4; $R' = H . \frac{1}{2}H_2SO_4$, $R'' = -COCF_3$

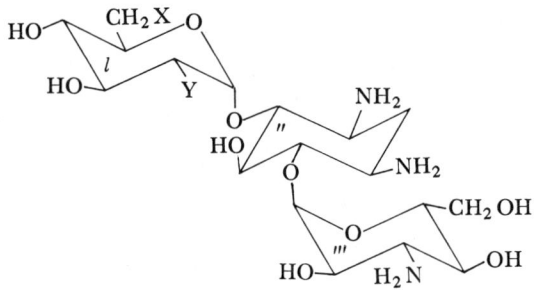

[78a] Kanamycin A; $X = NH_2$, $Y = OH$
[78b] Kanamycin B; $X = Y = NH_2$

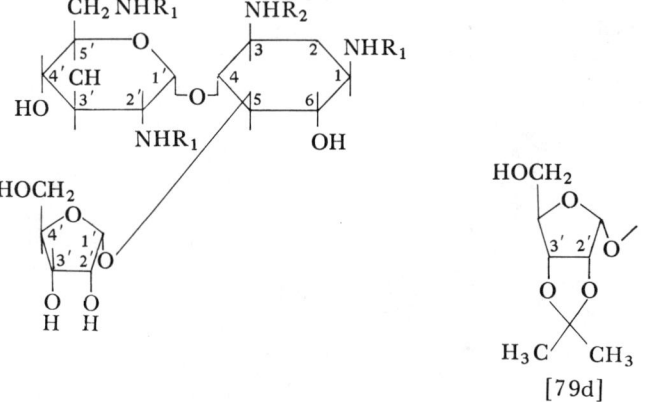

[79a] Ribostamycin; $R_1 = R_2 = H$
[79b] Ribostamycin sulfate; $R_1 = R_2 = H$, $(H_2SO_4)_{3/2}$
[79c] Tetra-N-acetylribostamycin; $R_1 = R_2 = COCH_3$
[79d] Tetra-N-acetyl-2′, 3′-isopropylidene ribostamycin; $R_1 = R_2 = COCH_3$
[79e] SF-733D; $R_1 = H$, $R_2 = COCH_3$
[79f] SF-733X; $R_1 = H$, $R_2 = CH_2COOH$

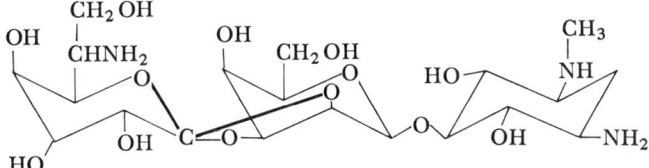

[80] Hygromycin B

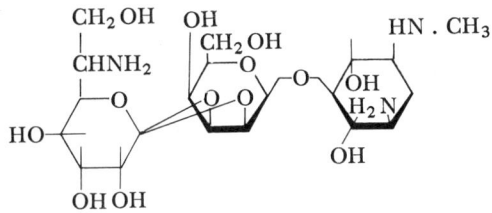

[81] Destomycin A

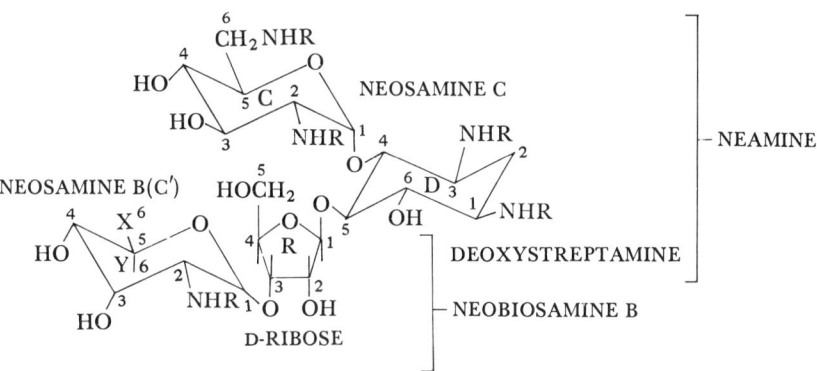

[82] Hexa-N-acetylneomycin; R = Ac; X = H; Y = CH$_2$NHAc

[83] Methyl-β-garosaminide

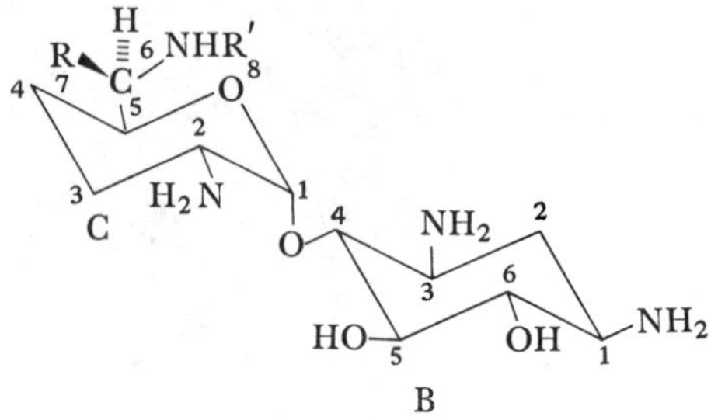

[84a] Gentamine C_1; R = R′ = CH_3
[84b] Gentamine C_{1a}; R = R′ = H
[84c] Gentamine C_2; R = CH_3, R′ = H

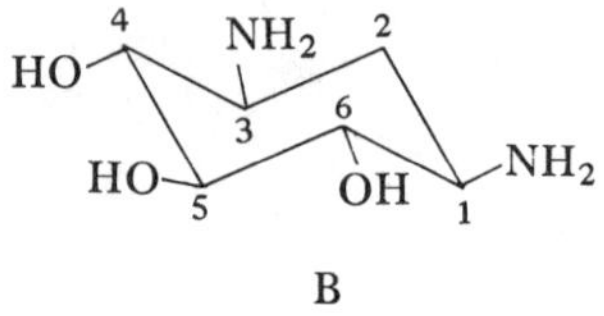

[85] 2-Deoxystreptamine

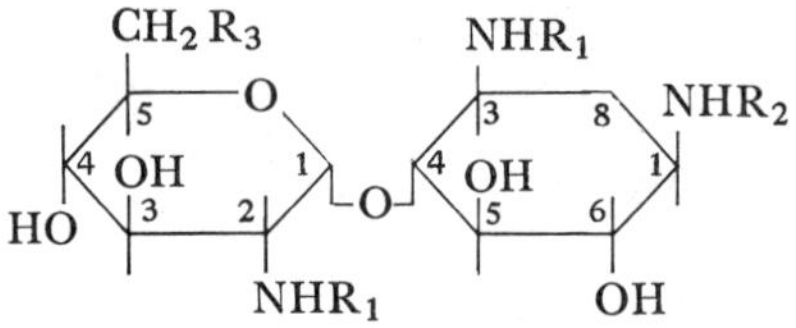

[86a] Neamine; $R_1 = R_2 = R_4 = H$; $R_3 = NH_2$
[86b] Tetra-N-acetylneamine; $R_1 = R_2 = R_4 = COCH_3$; $R_3 = NHCOCH_3$
[86c] Paramamine; $R_1 = R_2 = R_4 = H$; $R_3 = OH$

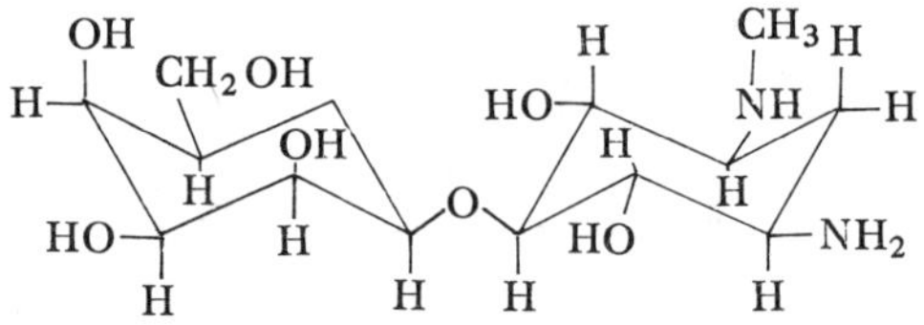

[87] Hygromycin B_2

Confidence in the accuracy of the given resonance frequencies was substantiated by the good agreement of the data obtained in three different laboratories, using different experimental techniques (Morton et al., 1973). ^{13}C Resonance frequencies of raffinose [88], melezitose [92] and stachyose [89] (Table 5) have been assigned to the individual carbon atoms analogously and by using partially relaxed Fourier transform nmr (Allerhand and Doddrell, 1971). Similarly, available assignments for α-D-glucopyranose, β-D-fructofuranose [1b] and sucrose [69] (Tables 2 and 4) have assisted the identification of the magnetically discrete carbon resonance frequencies of nystose [90] and 1-kestose [91] (Table 5; Binkley et al., 1972).

[88] Raffinose

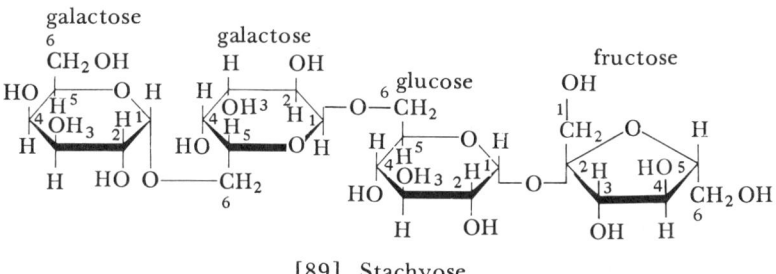

[89] Stachyose

[92] Melezitose

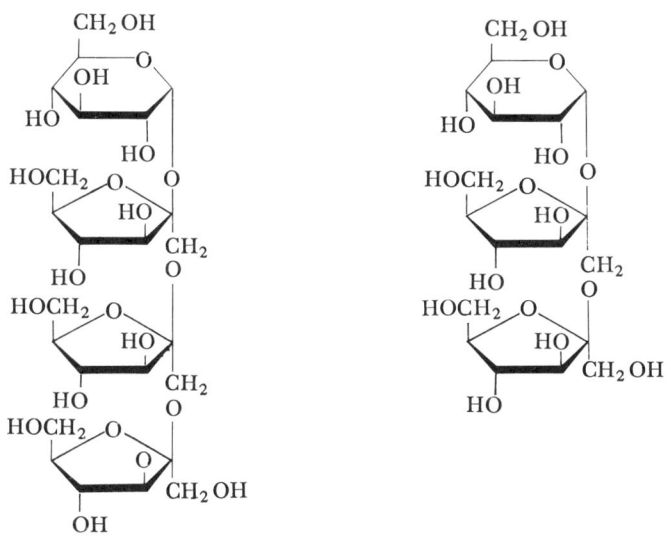

[90] Nystose   [91] 1-Kestose

TABLE 6

^{13}C Chemical Shifts of Polysaccharides

δ, ppm^a

Compound	Solvent	1	2	3	4	5	6	7	8	Other Carbons	Reference
Amylose (soluble starch)[b]	H$_2$O	100·7	78·2	74·3	72·5	72·2	61·6				Dorman and Roberts, 1971
	0·1 M NaOH	100·7	78·2	74·5	72·8	72·4	61·8				
	0·1 M NaOH	101·1	78·6	74·5	72·8	72·4	61·6				
	1 M NaOH	102·6	80·7	75·3	73·6	72·4	61·9				
	6·5 M LiBr	100·4	78·3	74·4	72·5	72·0	62·0				
Dextran[c]	D$_2$O	99·4	98·0	97·2	83·0	77·7	67·6	64·9	62·3		Usui et al., 1973a
Glucan (pD 7·0)[d]	D$_2$O	101·6	101·0	99·2	67·8	62·3	61·8				Jennings and Smith, 1973
Glucan (pD 14·1)[d]	D$_2$O	103·6	103·5	99·7	67·8	62·3	61·8				Jennings and Smith, 1973
Cellulose acetate[b]	CH$_2$Cl$_2$	101·5	77·0	73·4	62·8					CH$_3$ 21·7 C=O 171·3 170·9 170·5	Dorman and Roberts, 1971
Heparin[e]	D$_2$O	101·7	98·7	77·7	73·7–71·7	68·7	59·7			C=O 175·7	Perlin et al., 1972

		C-1	C-2	C-3	C-4	C-5	C-6	CH$_3$	C=O		
N. Meningitidis serogroup X[f]	D$_2$O	95·2	54·8	71·1	75·1	73·2	61·3	23·2	175·6		Bundle et al., 1974

Staphylococcus lactis 2102[f]	D₂O	95·2	55·0	71·5	70·3	73·0	65·3	23·0	175·6	Bundle et al., 1974
N. Meningitidis serogroup A[f]	D₂O	96·2 96·4	51·9 54·3	73·2 69·7	64·7 67·3	73·6	65·6	23·0 23·3 21·6	176·2 175·9 174·6	Bundle et al., 1974
N. Meningitidis serogroup A De-O-acetylated[f]	D₂O	96·4	54·3	69·7	67·1	73·5	65·6	23·3	175·8	Bundle et al., 1974

[a] Chemical shifts in parts per million downfield from tetramethylsilane (TMS); see cited reference for additional details.

[b] Obtained at 15·1 MHz with proton decoupling on 1–2 M aqueous solutions for the acetates reported relative to TMS and converted using $\delta CS_2 = 193 \cdot 7$. Peak 1 corresponds to C-1; Peak 2 to C-4; and Peak 6 to C-6. The other resonances have not been assigned to specific carbons.

[c] Obtained using a Jeol PFT-100 spectrometer and EC-6 computer system at 25·1 MHz with Fourier transform and proton noise decoupling on saturated solutions at room temperature (12 μsec, pulse width, 2·5 sec repetition time). The spectrum of dextran contains 15 distinct and distinguishable peaks, 8 of which have been tentatively assigned. The peaks between 77·7 and 67·6 ppm have been omitted from the table because they have not been assigned and their chemical shifts have not been reported. Peak assignments are as follows: Peak 1—anomeric carbon (α-1,6-linkage); Peak 2—anomeric carbon (α-1,6-linkage, with the α-1,2-interchain linkage); Peak 3—anomeric carbon (α-1,2-linkage); Peak 4—C-3 (α-1,3-linkage); Peak 5—C-2 (α-1,2-linkage); Peak 6—C-6 (α1,6-linear polymer); Peak 7–C-6 (bonded to anomeric oxygen of glucose unit substituted at the 2 position); and Peak 8—C-6 (non-substituted carbon).

[d] Resonances 1, 2 and 3 are for anomeric carbons (glycosidic) and 4, 5 and 6 are for exocyclic carbons.

[e] Obtained using a Varian XL-100-15 spectrometer with Fourier transform at 25·2 MHz (5000 Hz spectrum width, 75 μsec pulse width, 0·4 sec acquisition time) on 0·23 g of the sodium salt of heparin in 2·5 ml D₂O. Peak 1 is assigned to I-1; peak 2 to G-1; peak 3 to G-4, I-2, and I-4; peak 4 to G-3, G-5, I-3, and I-5; peak 5 to G-6; and peak 6 to G-2 where I refers to the α-L-idosyluronic acid and G to the 2-amino-2-deoxy-α-D-glucose residue.

[f] Obtained on a Varian XL-100-15 spectrometer at 25·16 MHz on 100 mg of compound per ml of D₂O relative to external TMS.

The observation of only six magnetically discrete ^{13}C signals in aqueous amylose (Table 6; Dorman and Roberts, 1971) have substantiated that its structure contains only repeating units of α-linked glucose molecules. Data obtained in combined ^{1}H and ^{13}C nmr investigations of dextran (Table 6) were rationalized in terms of a comb-like polymeric structure (Fig. 3; Usui et al., 1973a). Similarly, only twelve ^{13}C resonance frequencies were observed in the spectra of heparin (Table 6; Perlin et al., 1972b) consistent with the proposed structure of alternating 1 → 4 linked residues of α-L-iso-

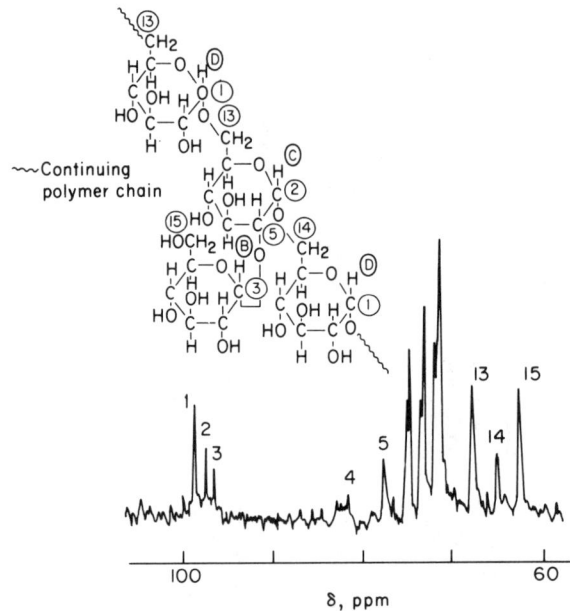

Figure 3. ^{13}C FTnmr spectrum of dextran at 25·1 MHz (taken from Usui et al., 1973a).

pyranosyluronic acid 3-sulfate and 2-deoxy-2-sulfamino-α-D-glucopyranosyl 6-sulfate. In another report, an additional broad resonance with a peak area of one carbon was found 102 ppm from TMS in the spectrum of heparin (Lasker and Chiu, 1973). This additional signal indicates a total of 13 carbon atoms and suggests that more than one type of uronic acid is present in heparin. Analysis of ^{13}C resonance frequencies of glucan has established the ratio of 1 → 4 to 1 → 6 linkages to be 2 : 1 in the homopolymer (Jennings and Smith, 1973). Linkages for several glucobioses were also determined in an earlier study (Yamaoka et al., 1971). The composition, sequence, anomeric configuration, position of linkages, location of substituents and conformation of group-specific antigens of *Neisseria Meningitidis* serogroups A and X have been determined using ^{13}C nmr techniques (Bundle *et al.*, 1974). Evidently ^{13}C resonance spectra of complex carbohydrates in solution is increasingly becoming a convenient and powerful method of establishing conformation, linkages and ratios of their constituents.

Polysaccharides isolated from yeasts of different origins were divided into groups according to their ^{13}C nmr spectra (Gorin and Spencer, 1972; Gorin 1973b). For example, polysaccharides isolated from the cell walls of six *Pichia* and *Hansenula* species were found to fall into three distinct groups: (1) a regular predominant structure consisting of $1 \rightarrow 2$ linked α-D-mannopyranose units interspersed with a single $1 \rightarrow 6$ unit; (2) three $1 \rightarrow 2$ linked α-D-mannopyranose units interspersed with a single $1 \rightarrow 6$ linked unit and (3) mixtures of (1) and (2) with three or four consecutively linked α-D-mannopyranose units joined to a $1 \rightarrow 6$ unit. Identification of the sugars formed on hydrolysis and investigation of the partially methylated glucosides by glpc have substantiated the validity of the ^{13}C nmr techniques (Gorin and Spencer, 1972).

3. NUCLEIC ACIDS AND CONSTITUENTS

The chemistry of nucleic acids, such as DNA and RNA, is extremely important and interesting to both chemists and biologists because of the multi-functional role these molecules perform in living cells. DNA, contained in the nucleus of almost all cells, is composed of four different nucleotides, namely, deoxyadenosine monophosphate (dAMP) [98], deoxythymidine monophosphate (dTMP), deoxyguanosine monophosphate (dGMP) [101] and deoxycytidine monophosphate (dCMP) [96]. It acts as a storage unit for the genetic information. This genetic information is encoded in DNA by a set of three nucleotides called a codon and each codon represents an amino acid (Frisch, 1966). In this way, genetic information stored in DNA, can be manifested by controlling the order of amino acids in biosynthetically produced proteins (Ingram, 1965). Messenger, transfer and ribosomal RNAs also play active parts in the biosynthesis of polypeptides.

In general, the composition of nucleic acids from less complex organic constituents involves the following overall scheme:

purine or pyrimidine + pentose sugar = nucleoside
nucleoside + phosphoric acid = nucleotide
n(nucleotides) = polynucleotide or nucleic acid

Within the last half decade a number of laboratories have undertaken investigations of the physical and chemical properties of nucleic acid constituents utilizing ^{13}C nmr techniques. The main areas of investigation were the structure and conformation of nucleosides and

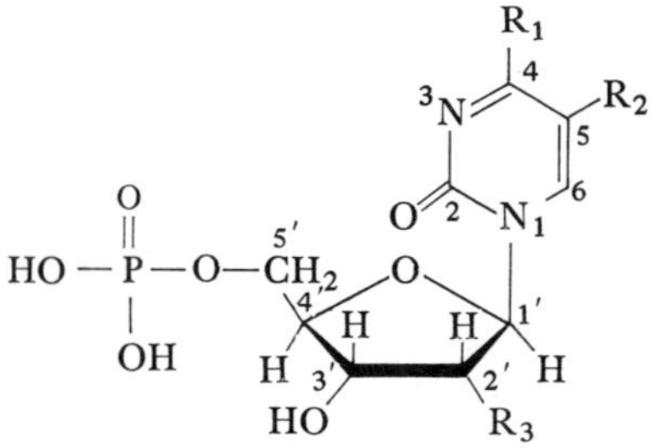

[93]–[96] 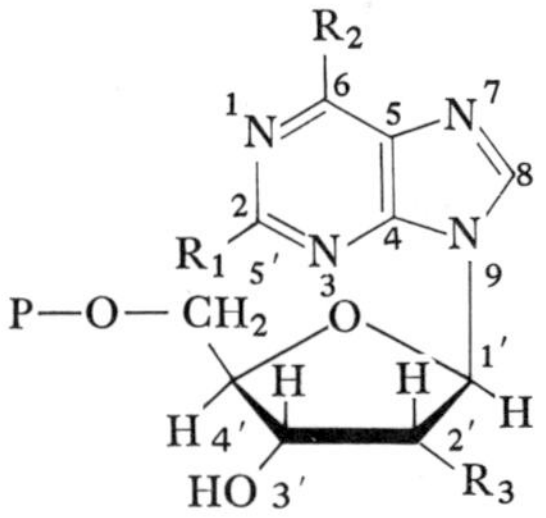 [97]–[102]

[93] Uridine monophosphate, UMP; $R_1 = R_3 = OH$, $R_2 = H$
[94] Thymidine monophosphate, TMP; $R_1 = OH$, $R_2 = CH_3$, $R_3 = H$
[95] Cytidine monophosphate, CMP; $R_1 = NH_2$, $R_2 = H$, $R_3 = OH$
[96] Deoxycytidine monophosphate, dCMP; $R_1 = NH_2$, $R_2 = R_3 = H$
[97] Adenosine monophosphate, AMP; $R_1 = H$, $R_2 = NH_2$, $R_3 = OH$, $P = PO_3H_2$
[98] Deoxyadenosine monophosphate, dAMP; $R_1 = R_3 = H$, $R_2 = NH_2$, $P = PO_3H$
[99] Adenosine triphosphate, ATP; $R_1 = H$, $R_2 = NH_2$, $R_3 = OH$, $P = P_3O_9H_4$
[100] Guanosine monophosphate, GMP; $R_1 = NH_2$, $R_2 = R_3 = OH$, $P = PO_3H_2$
[101] Deoxyguanosine monophosphate, dGMP; $R_1 = NH_2$, $R_2 = OH$, $R_3 = H$, $P = PO_3H_2$
[102] Inosine monophosphate, IMP; $R_1 = H$, $R_2 = R_3 = OH$, $P = PO_3H_2$

nucleotides, substituent effects, binding studies, base-stacking and models for polynucleotides and nucleic acids. Research in these areas prior to 1972 has been summarized (Levy and Nelson, 1972; Stothers, 1972a; Gray, 1973).

Nucleosides

Several relatively recent ^{13}C nmr studies of this class of compounds have been primarily concerned with nucleoside antibiotics (Kreishman et al., 1972; Krugh, 1973; Wenkert et al., 1973; Nouaille et al., 1974; Dea et al., 1974). The structure and conformation of pyrazomycin B [111b] have been elucidated by comparison of the chemical shifts of four related nucleosides [106ab, 108ab] and three C-nucleosides [107ab, 111a] (Wenkert et al., 1973). The ^{13}C nmr spectra of formycin A (7-amino-3-β-D-ribofuranosyl-1H-[4,3-d] pyrimidine) [109] and formycin B (1,6-dihydro-3-β-D-ribofuranosyl-7H-pyrazolo-[4,3-d]-pyrimidin-7-one) [110] exhibited unexpectedly broad resonances, 15 to 30 Hz at 25°, as opposed to the narrow peaks, less than 1·5 Hz wide, normally observed for other nucleosides (Krugh, 1973). These broadened peaks narrowed with increasing temperature so that at 90° all peaks exhibited line widths of less than 1·5 Hz. The peak widths were also relatively insensitive to variations in concentration. From the above observations and comparison with the ^{13}C nmr spectra of indoles, indazoles and pyrazoles under various experimental conditions (e.g., with the addition of H$_2$O or D$_2$O, acid or base), the peak broadening was

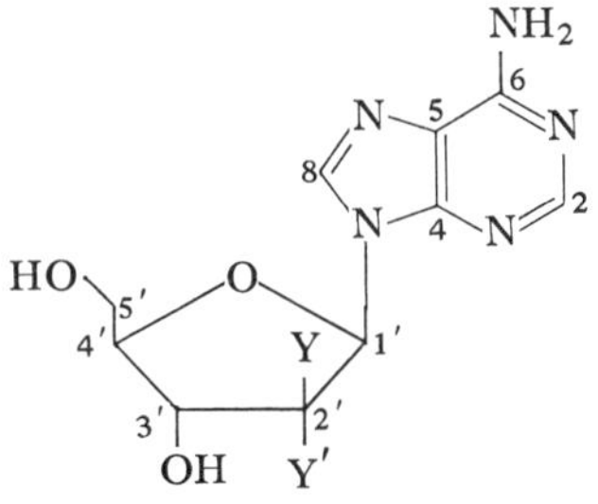

[108]

[108a] Adenosine
[108b] 9-(β-D-arabinofuranosyl) adenine

a, (β-ribo); Y = H, Y' = OH
b, (β-arabino); Y = OH, Y = H

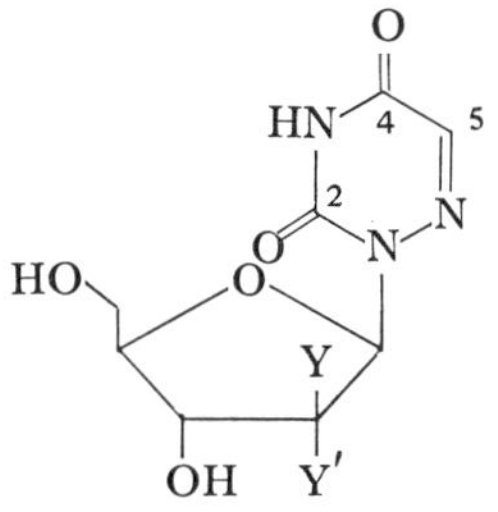

[106]

[106a] β-6-Azauridine
[106b] (β-D-arabinofuranosyl)-6-azauricid

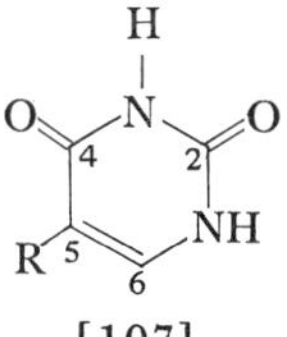

[107]

[107a] β-Pseudouridine
[107b] α-Pseudouridine

a, R = β-D-ribofuranosyl
b, R = α-D-ribofuranosyl

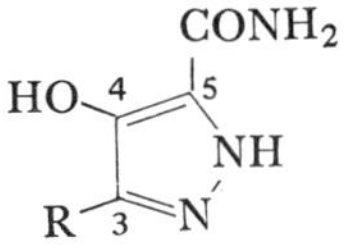

[111]

[111a] Pyrazomycin (β-D-ribofuranosyl-3-(4-hydroxy-5-carboxamido)-pyrazolyl
[111b] Pyrazomycin B (α-D-ribofuranosyl-3-(4-hydroxy-5-carboxamido)pyrazolyl)

TABLE 7
Chemical Shifts of Nucleosides
(Numbered formulae are on pp. 325, 330–31)

Compound	Solvent	δ, ppm[a]												Me	Reference
		C-2	C-4	C-5	C-6	C-8	C-9	1'	2'	3'	4'	5'			
Cytidine (103)[b]	DMSO	157·0	166·7	95·7	142·8			90·1	70·6	75·1	85·4	61·9		Stothers, 1972a
6-Methylcytidine (104)[b]	DMSO	157·4	166·2	96·6	155·4			92·9	71·2	71·9	86·2	63·3	20·9	Stothers, 1972a
5-Methylcytidine[c]	Water	159·9	167·9	104·2	141·4	12·6 (CH$_3$)								Komoroski and Allerhand, 1974
2-Thiocytidine[b]	DMSO	180·9	161·1	99·0	142·7			94·5	69·4	76·0	85·1	60·8		Stothers, 1972a
6-Hydroxycytidine[b]	DMSO	155·7	164·9	119·8	152·2			88·3	71·6	75·9	85·5	63·8		Stothers, 1972a
5-Azacytidine[b]	DMSO	154·6	166·8		157·5			90·6	70·2	75·1	85·3	61·3		Stothers, 1972a
Deoxycytidine[b]	DMF	157·0	167·0	95·9	142·5			86·6	40·6	71·8	88·6	62·7		Stothers, 1972a
Uridine [105][b]	DMSO	152·4	164·7	103·1	142·2			89·1	71·1	74·9	86·0	62·2		Stothers, 1972a
Uridine [105][d]	DMSO-d$_6$	151·2	163·8	102·2	141·6			88·3	74·0	70·3	85·2	61·3		Krugh, 1973
Uridine [105][e]	DMSO	152·3	164·6	103·3	142·3			89·3	75·1	71·4	86·3	62·4		Mantsch and Smith, 1973
Uridine [105][e]	Pyridine	152·3	164·6	102·5	141·2			90·5	76·0	71·2	86·3	61·9		Mantsch and Smith, 1973
Uridine [105][e]	D$_2$O	152·8	167·3	103·4	143·0			90·6	74·8	70·6	85·5	62·0		Mantsch and Smith, 1973
Uridine [105][e]	H$_2$O	152·9	167·4	103·5	143·1			90·7	75·0	70·8	85·5	62·0		Mantsch and Smith, 1973
5-Aminouridine[b]	DMSO	150·6	161·9	116·8	124·1			88·7	71·5	74·1	85·8	62·7		Stothers, 1972a
5-Hydroxyuridine[b]	DMSO	150·6	161·7	121·1	133·6			88·6	71·4	74·2	85·9	62·4		Stothers, 1972a
5-Bromouridine[b]	DMSO	150·9	160·0	96·8	141·4			89·7	70·3	75·1	85·8	61·3		Stothers, 1972a
5-Fluorouridine[b]	DMSO	150·3						89·5	70·6	74·9	86·0	61·5		Stothers, 1972a
6-Hydroxyuridine[b]	DMSO	167·4	168·1	41·3	152·5			89·5	71·4	73·1	85·8	63·8		Stothers, 1972a
[1-(β-D-ribofuranosyl)]barbituric acid][b]															
4-Thiouridine[b]	DMSO	149·1	191·1	113·7	136·9			89·7	70·6	75·0	86·1	61·7		Stothers, 1972a
2,4-Dithiouridine[b]	DMSO	173·8	187·0	118·7	135·6			94·3	69·6	75·7	85·7	60·6		Stothers, 1972a
6-Azauridine[b]	DMSO	149·5	157·6	137·4				90·7	71·4	73·6	85·7	63·0		Stothers, 1972a
6-Azauridine[c]	DMSO-d$_6$	148·7	156·8	136·6				89·7	72·6	70·6	84·9	62·3		Krugh, 1973
6-Hydroxy-5-azauridine[b]	DMSO	150·3	149·4		150·3			89·4	71·1	72·4	85·6	63·4		Stothers, 1972a
[1-(β-D-ribofuranosyl)cyanuric acid][b]															
5-Bromo-2'-deoxyuridine[b]	DMSO	150·7	160·1	96·6	141·2			86·0	~41	70·9	88·5	61·9		Stothers, 1972a
β-6-Azauridine [106a][f]	Water	148·7	157·3	136·6				89·5	72·7	70·0	83·8	61·4		Wenkert et al., 1973
(β-D-Arabinofuranosyl)-6-azauracil [106b][f]	Water	149·2	157·5	136·2				81·8	73·4	75·6	83·6	61·9		Wenkert et al., 1973
β-Pseudouridine [107a][f]	Water	152·3	164·8	110·0	141·0			78·6	72·9	71·3	82·9	61·1		Wenkert et al., 1973
α-Pseudouridine [107b][f]	Water	152·5	164·5	109·6	140·0			76·6	72·5	72·1	81·3	62·0		Wenkert et al., 1973
Pseudouridine[c]	Water	153·6	166·1	111·4	142·4									Komoroski and Allerhand, 1974

Compound	Solvent	C2	C4	C5	C6	C8/extra	C1'	C2'	C3'	C4'	C5'	CH3	Reference
Dihydrouridine[c]	Water	155.3	174.6	30.9	37.3		88.5	71.2	72.1	85.5	63.4		Komoroski and Allerhand, 1974
4-Deoxy-1-(β-D-ribofuranosyl)barbituric acid[b]	DMSO	152.1	142.4	101.2	164.4			71.2	72.6	85.5	63.4		Stothers, 1972a
4-Deoxy-1-(β-D-ribofuranosyl)-6-thionyl barbituric acid[b]	DMSO	149.3	136.8	114.4	194.0		94.0	70.8	72.6	85.5	63.2		Stothers, 1972a
2',4-Dideoxy-1-(β-D-ribofuranosyl)barbituric acid[b]	DMSO	152.0	142.0	101.3	164.2		81.6	36.9	71.9	86.9	62.5		Stothers, 1972a
4-Methylthio-1-(β-D-ribofuranosyl)barbituric acid[b]	DMSO	151.4	157.5	95.2	162.2		88.5	71.2	72.0	85.3	63.4		Stothers, 1972a
2'-Deoxy-4-methylthio-1-(β-D-ribofuranosyl)barbituric acid[b]	DMSO	151.6	157.3	95.4	162.6		82.1	37.9	72.2	87.3	62.8		Stothers, 1972a
2-Methoxy-2',3'-O-isopropylidene isocytidine[b]	DMSO	156.2	170.6	108.3	140.0		93.4	81.1	85.2	87.8	62.0	56.0	Stothers, 1972a
2',3'-O-Isopropylideneisocytidine[b]	DMSO	155.4	171.0	107.6	140.3		94.6	80.9	83.2	85.9	61.7		Stothers, 1972a
1-(β-D-arabinofuranosyl)cytosine[b]	DMSO	160.4	171.4	93.7	147.7		87.5	75.9	75.9	87.1	61.7		Stothers, 1972a
1-(β-D-arabinofuranosyl)uracil[b]	DMSO	151.5	143.5	101.0	164.6		86.1	76.5	76.5	85.7	61.7		Stothers, 1972a
3',5'-Anhydrothymidine[b]	DMSO	152.0	164.6	110.6	137.4		89.2	38.1	80.8	87.8	76.2		Stothers, 1972a
2,5'-Anhydro-2',3'-O-isopropylidene-uridine[b]	DMSO	157.7	171.2	109.7	143.8		97.6	85.2	84.7	82.0	75.1		Stothers, 1972a
2,2'-Anhydrouridine[b]	DMSO	160.8	172.6	109.5	137.8		89.9	89.9	75.8	89.9	61.8		Stothers, 1972a
5',6-Anhydrouridine-2',3'-O-isopropylidene-6-hydroxuridine[b]	DMSO	150.7	163.9	90.7	161.9		89.8	85.5	84.3	82.7	78.2		Stothers, 1972a
1-(5'-Deoxy-β-D-erythro-pent-4-enofuranosyl)uracil[b]	DMSO	151.5	164.0	103.5	141.7		90.5	70.3	73.0	162.9	86.4		Stothers, 1972a
1-(2',3'-Dideoxy-β-D-glycero-pent-2-enofuranosylthymine[b]	DMSO	152.1	165.2	110.3	138.0		88.8	136.1	127.2	90.2	63.7	13.4	Stothers, 1972a
Thymine[c]	DMSO-d6	151.6	165.0	107.8	137.8							11.9	Krugh, 1973
6-Azathymine[c]	DMSO-d6	149.7	157.5	142.7								15.9	Krugh, 1973
Thymidine[c]	DMSO-d6	150.5	163.8	109.5	136.2		83.9	39.5	70.5	87.3	61.4	12.3	Krugh, 1973
Thymidine[b]	DMSO	151.6	164.9	110.5	137.3		85.1	40.4	71.7	88.4	62.4	13.3	Stothers, 1972a
4-Thiothymidine[b]	DMSO	148.7	191.5	118.8	134.2	147.5	85.9	39.9	71.2	87.8	62.2	17.9	Stothers, 1972a
Purine[b]	DMSO	151.6	154.4	128.1	144.4	146.4							Stothers, 1972a
Nebularine[b]	DMSO	151.9	153.0	135.2	149.0	146.4	89.1	71.4	75.1	86.9	62.3		Stothers, 1972a
6-Chloro-9-(β-D-ribofuranosyl)purine[b]	DMSO	150.3	152.4	132.2	152.4	146.4	89.4	71.1	75.1	86.5	61.8		Stothers, 1972a
2-Amino-9-(β-D-ribofuranosyl)purine[b]	DMSO	160.6	154.9	124.8	150.8	142.6	88.4	71.4	74.8	86.6	62.3		Stothers, 1972a
2,6-Diamino-O-(β-D-ribofuranosyl)purine[b]	DMSO	161.3	152.6	83.3	157.6	115.0	89.0	72.3	75.0	87.4	63.4		Stothers, 1972a
Adenine[b]	DMSO	153.4	152.2	118.6	156.4	140.5							Stothers, 1972a
Adenine [108a][b]	DMSO	153.5	150.1	120.4	157.1	141.2	89.2	71.8	74.7	87.0	62.7		Stothers, 1972a
Adenosine [108a][b]	DMSO-d6	152.5	149.1	119.4	156.2	140.1	88.1	73.6	70.8	86.0	61.8		Stothers, 1972a
Adenosine [108a][f]	DMSO-d6	152.7	149.3	119.5	156.3	140.3	88.4	73.9	71.1	86.3	62.1		Wenkert et al., 1973
Adenosine [108a][g]	D2O	147.7	145.65	118.55	150.5	142.4	88.45	74.25	70.25	85.55	61.3		Breitmaier et al., 1972
6-N-Methyladenosine[b]	DMSO	153.6	149.3	120.9	156.3	140.8	89.4	72.0	75.0	87.1	63.0		Stothers, 1972a

TABLE 7—continued

Compound	Solvent	δ, ppm[a]												Me	Reference
		C-2	C-4	C-5	C-6	C-8	C-9	1'	2'	3'	4'	5'		
6-N-Methyladenosine[c]	Water	153·5	148·2	120·3	156·1	140·7	28·4 (CH₃)							Komoroski and Allerhand, 1974
1-Methyladenosine[c]	Water	148·6	147·6	120·2	151·9	144·0	38·4 (CH₃)							Komoroski and Allerhand, 1974
6-N, 6-N-Dimethyladenosine[c]	Water	152·2	149·0	119·7	154·4	139·0	39·6 (CH₃)							Komoroski and Allerhand, 1974
2-Chloroadenosine[b]	DMSO	154·2	151·3	119·4	157·6	141·2		88·9	71·5	74·9	86·8	62·6		Stothers, 1972a
8-Azaadenosine[c]	DMSO-d₆	156·1	148·8	124·1	156·7			89·7	72·9	70·7	86·2	61·8		Krugh, 1973
9-(β-D-Arabinofuranosyl)adenine [108b][f]	DMSO-d₆	152·8	149·7	118·4	154·3	140·8		84·1	75·1	76·0	84·1	61·0		Wenkert et al., 1973
Deoxyadenosine[b]		153·7	150·0	120·5	157·2	141·0		85·6	40·5	72·3	89·2	62·9		Stothers, 1972a
2-Deoxyadenosine[g]	D₂O	147·5	145·2	118·25	150·1	142·3		84·8	39·15	70·9	87·5	61·4		Breitmaier et al., 1972
9-α-D-Xylopyranosyladenine[f]	0·1 N HCl	147·8	145·1	117·8	150·3	143·5		80·7	69·25 or 69·6	67·95 or 67·55	69·6 or 69·25	67·55 or 67·95		Breitmaier and Voelter, 1973
9-β-D-Xylopyranosyladenine[f]	0·1 N HCl	148·35	144·9	118·7	150·2	143·05		84·25	71·65	76·4	68·85	68·1		Breitmaier and Voelter, 1973
9-β-D-Xylofuranosyladenine[f]	0·1 N HCl	149·2	147·6	118·35	153·1	141·45		89·65	75·2	80·15	82·95	60·0		Breitmaier and Voelter, 1973
9-α-D-Arabinopyranosyladenine[f]	0·1 N HCl	148·45	145·0	118·45	150·2	142·85		84·1	72·8	69·8 or 69·5	69·5	68·5		Breitmaier and Voelter, 1973
9-β-D-Arabinopyranosyladenine[f]	0·1 N HCl	147·7	144·7	117·95	150·1	143·3		80·4	69·8	65·6	69·4	63·35		Breitmaier and Voelter, 1973
9-α-D-Arabinofuranosyladenine[f]	0·1 N HCl	148·0	145·0	118·9	150·3	142·85		89·2	79·75	74·85	85·25	60·95		Breitmaier and Voelter, 1973
9-β-D-Ribopyranosyladenine[g]	D₂O	148·4	144·9	118·8	150·2	143·15		81·15	68·6	66·05	70·8	64·95		Breitmaier et al., 1972
9-α-D-Mannopyranosyladenine[i]	0·1 M HCl	148·45	145·10	118·70	150·50	143·05		81·90	71·00	67·85	67·20	78·35	59·90 (C6)	Voelter and Breitmaier, 1973
9-β-D-Mannopyranosyladenine[i]	0·1 M HCl	148·35	145·00	119·00	150·30	142·95		88·55	71·45	75·95	68·95	81·25	62·90 (C6)	Voelter and Breitmaier, 1973
9-α-D-Talopyranosyladenine[i]	0·1 M HCl	148·65	144·70	118·70	150·10	143·30		79·30	69·40	68·40	67·45	77·8	58·35 (C6)	Voelter and Breitmaier, 1973
Inosine[b]	DMSO	149·2	147·0	125·4	157·7	139·9		88·8	71·3	75·2	86·7	62·3		Stothers, 1972a

Compound	Solvent											Reference
Inosine[c]	DMSO-d6	148.2	146.0	124.5	156.7	138.9		87.9	74.3	70.4	85.8	61.4	Krugh, 1973
6-Thioinosine[b]	DMSO	146.1	144.7	136.3	176.7	142.0		88.5	70.9	75.2	86.5	62.0	Stothers, 1972a
Deoxyinosine[b]	DMF	148.5	146.3	125.2	157.4	139.4		84.4	40.2	71.4	88.8	62.2	Stothers, 1972a
Xanthosine[b]	DMSO	159.0	152.4	117.3	164.5	137.5		89.9	71.7	74.8	87.3	62.2	Stothers, 1972a
6-Thioxanthosine[b]	DMSO	158.5	149.1	127.3	183.0	137.5		88.9	71.7	75.1	87.1	62.2	Stothers, 1972a
Guanosine[b]	DMSO	154.6	152.4	117.6	157.8	136.9		87.3	71.6	74.9	86.4	62.2	Stothers, 1972a
6-Thioguanosine[b]	DMSO	154.3	149.0	129.4	176.2	139.5		87.9	71.4	76.0	86.5	62.4	Stothers, 1972a
Deoxyguanosine[b]	DMSO	154.7	152.0	117.7	158.2	136.7		83.9	40.4	71.9	88.7	62.9	Stothers, 1972a
1-Methylguanosine[c]	DMF	155.4	149.8	117.1	159.1	138.7	29.2 (CH3)						Komoroski and Allerhand, 1974
7-Methylguanosine[c]	Water	160.5 or 160.0	150.3	110.0	160.0 or 160.5	136.5	36.6						Komoroski and Allerhand, 1974
1-β-D-Ribofuranosyl-1,2,4-triazole-3-carboxamide (ribavirin)[j]	50% w/v D2O	157.6 (C-3)		147.4	164.0 (C=O)			92.8	76.3	71.6	86.3	62.8	Dea et al., 1974
1-β-D-Ribofuranosyl-1,2,4-triazole-3-carboxamide (ribavirin)[j]	32% w/v DMSO-d6	158.4 (C-3)		146.3	161.8 (C=O)			92.1	75.1	70.7	85.7	61.6	Dea et al., 1974
1-β-D-Ribofuranosyl-1,2,4-triazole-5-carboxamide (ribavirin)[j]	28% w/v DMSO-d6	151.5 (C-3)		148.8	160.0 (C=O)			90.3	74.5	71.3	85.5	62.5	Dea et al., 1974
Showdomycin[k]	DMSO-d6 Water	173.9	148.0	173.4	130.2 (C-3)			78.2	75.4	71.9	84.3	62.5	Elstner et al., 1973
Formycin A [109][c]	DMSO-d6	151.6	138.7	123.0	151.6	143.7		78.3	75.5	72.7	86.3	62.8	Krugh, 1973
Formycin B [110][c]	DMSO-d6	143.3	136.7	128.0	153.6	144.7		77.7	74.9	72.1	85.7	62.5	Krugh, 1973
Pyrazomycin [111a][e]	Water	132.7 (C-3)	139.3	127.3	163.7			75.4	73.8	70.8	84.1	61.1	Wenkert et al., 1973
Pyrazomycin B [111b][e]	Water	130.7 (C-3)	139.3	127.6	164.0			74.9	73.2	71.8	81.3	61.1	Wenkert et al., 1973

[a] Chemical shifts in parts per million relative to TMS.
[b] See original references for experimental details and additional data.
[c] Proton decoupled ^{13}C FTnmr spectra obtained on a "home-built" spectrometer at 15·18 MHz. ^{13}C chemical shifts corrected to TMS scale using δ_{CS_2} = 193·6.
[d] Spectra obtained on JEOL and Varian Spectrometers.
[e] Proton decoupled spectra obtained on a Varian XL-100-15 spectrometer with Fourier transform at 25·16 MHz on 0·3 M samples.
[f] Spectra obtained on a spectrometer with Fourier transform at 15·08 MHz.
[g] Proton decoupled ^{13}C FTnmr spectra obtained on a Bruker HFX-90 multi-nuclei spectrometer at 22·63 MHz on 0·25 M sample.
[h] Proton decoupled ^{13}C FTnmr spectra obtained on a Bruker HFX-90 multi-nuclei spectrometer at 22·628 MHz relative to external TMS.
[i] Proton decoupled ^{13}C FTnmr spectra obtained on a Bruker HFX-90 spectrometer at 22·63 MHz on 0·1 M samples.
[j] ^{13}C FTnmr spectra obtained on a Bruker HX-90 spectrometer at 22·6 MHz.
[k] Proton decoupled ^{13}C FTnmr spectra obtained on a "home-built" spectrometer at 15·18 MHz on 0·24 M sample. ^{13}C chemical shifts corrected to TMS scale using δ_{CS_2} = 193·7.
[l] Assignments of the 2' and 3' carbons for ribose moieties were incorrect in many earlier references, and should be reversed.

ascribed to prototropic tautomerization (Krugh, 1973). Similarly, the site of glucosylation in 3- and 5-substituted 1-β-ribofuranosyl-1,2,4-triazoles has been assigned from the ^{13}C chemical shifts of the triazole ring (Kreishman et al., 1972). The structure of a nucleoside antibiotic fluoride analog, 4-(3'-deoxy-3'-fluoro-1'-α-D-threofuranosyl)-N-2-phenyl-1,2,3-triazole, has been elucidated by comparison of the total proton decoupled ^{13}C nmr spectrum with the splittings in the partially proton decoupled spectrum (Nouaille et al., 1974). The requirements for the antiviral activity of 1-α-D-ribofuranosyl-1,2,4-triazole-3-carboxamide (ribavarin) and some related triazole nucleosides have been assessed by investigating their conformation and interactions in solution by ^{13}C and ^{1}H nmr (Dea et al., 1974). ^{13}C Nmr data for triazole nucleosides are given in Table 7.

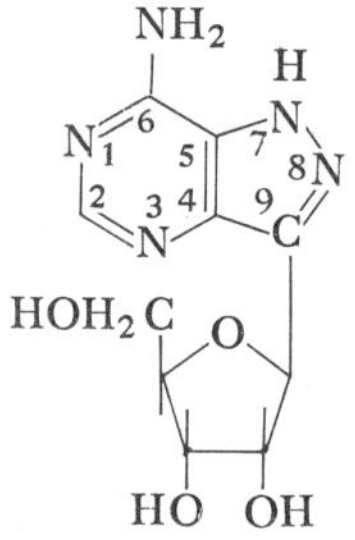

Formycin A
[109]

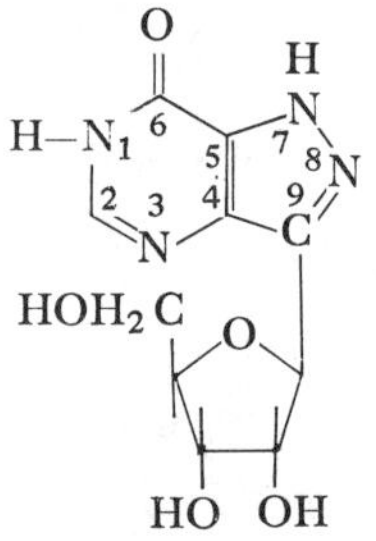

Formycin B
[110]

One of the biological consequences of nucleoside *syn/anti* conformational preference can be seen in the enzymatic synthesis of polynucleotides. Purine and pyrimidine nucleoside di- or triphosphate substrates with *syn* conformations will not react with the enzymes polynucleotide phosphorylase and RNA polymerase (Kapuler et al., 1970; Kapuler and Reich, 1971; Ikehara et al., 1969). Conformational analyses of nucleosides have also been carried out (Schweizer and Kreishman, 1973; Schweizer et al., 1973) by studying the torsional angle, χ, between the base and the sugar moieties about the glycosidic bond using the vicinal coupling constants of C-2 and H-1' determined from the natural abundance ^{13}C FTnmr spectra. This work confirmed the earlier ^{1}H nmr conclusions of Schweizer and coworkers (1971) that pyrimidine nucleosides have *anti*-conformations in solution while 6-substituted pyrimidine nucleosides exist as *syn*-conformers in solution due to steric hindrance. The relationship of vicinal coupling constants to the torsion angle has been studied using uridine, 2,5'-anhydro-1-β-D-ribofuranosyl uracil [112], and 2,2'-anhydro-1-β-D-arabino-furanosyl uracil [113], enriched with ^{13}C at the C-2 position (Lemieux *et al.*, 1972). Uridine was found to exist in the *anti*-conformation.

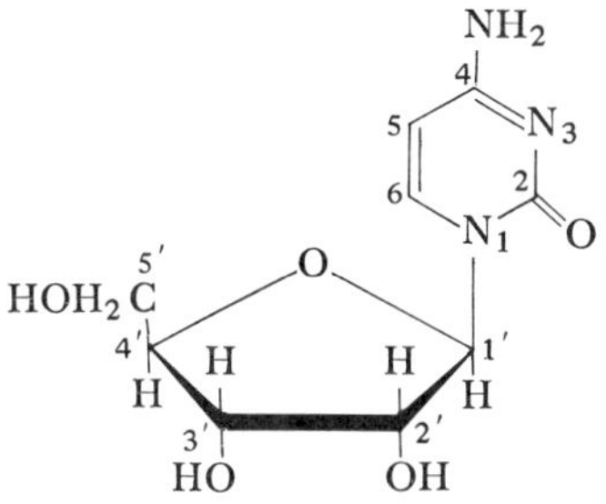

[103] Cytidine

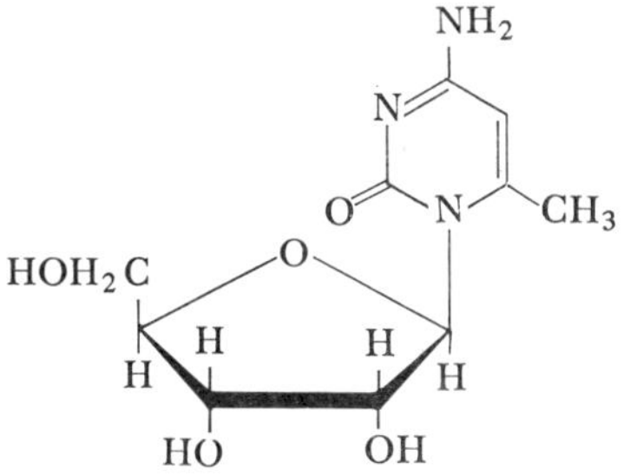

[104] Methyl cytidine

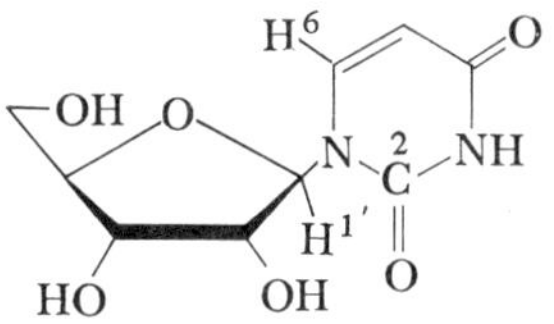

[105] Uridine

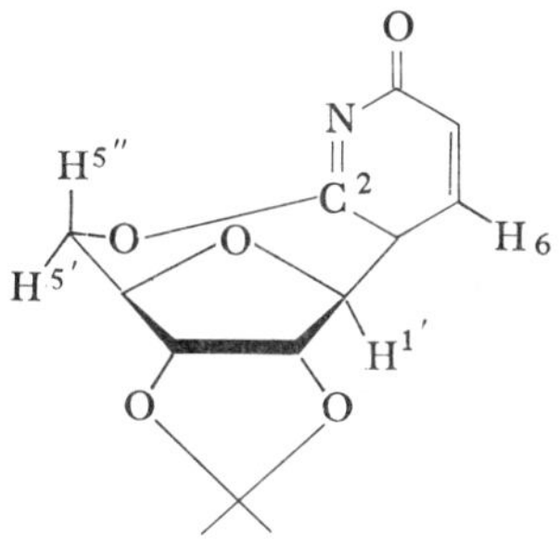

[112]

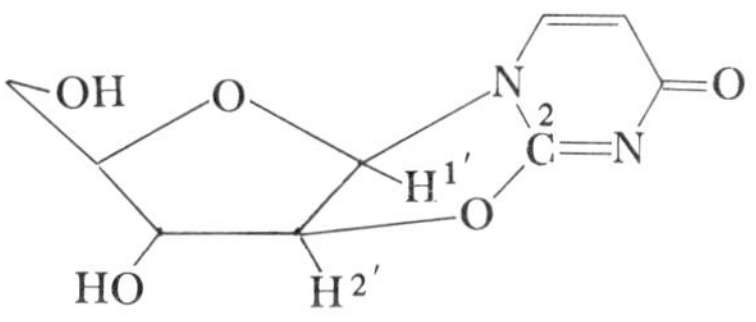

[113]

^{13}C Spectra of uridine [105] in dimethyl sulfoxide, pyridine, H_2O and D_2O have been recorded (Mansch and Smith, 1973). The chemical shifts of C-2', C-3' and C-5', i.e. the hydroxyl bearing carbons, shift upfield by 5 Hz in going from water to D_2O while the other resonances are unchanged within the experimental error of 2·5 Hz. In DMSO the carbonyl peaks are enhanced relative to water because proton exchange with the solvent at N-3 cannot occur, and the dipole-dipole interaction between the proton on the nitrogen and the carbonyl carbon atom therefore becomes important. In pyridine, the chemical shifts of uridine move both up and downfield, with respect to water, presumably as the consequence of a differential solvent interaction at the different carbons of uridine (Mantsch and Smith, 1973). The chemical shifts of nucleosides are summarized in Table 7.

Nucleotides

The first step in the investigations of nucleotide systems is the assignment of resonances to specific carbon atoms in the molecule. In addition to the techniques normally used (e.g. model compound

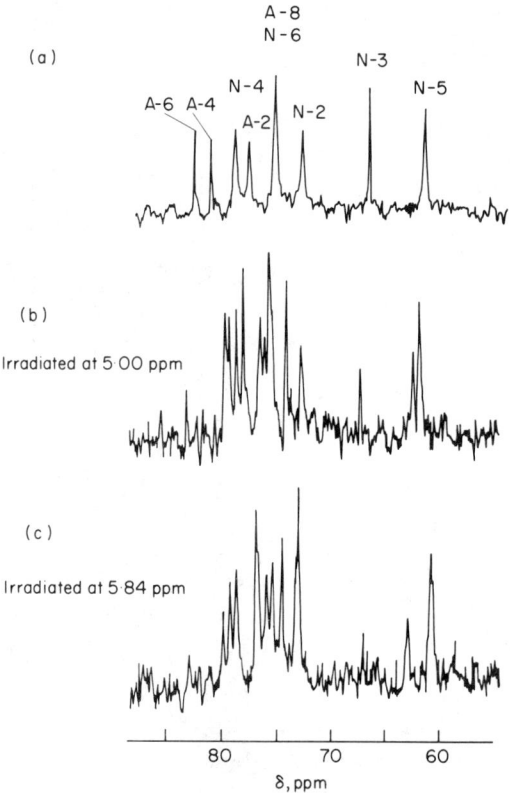

Figure 4. ^{13}C Spectrum of the aromatic region of NAD$^+$. The carbon atoms of the adenine ring are denoted by A and of the nicotinamide ring by N. (a), Noise decoupled spectrum; (b), ^{1}H off-resonance decoupled ^{13}C nmr spectrum of NAD, irradiated at 5·00 ppm downfield from dioxan in the ^{1}H spectrum; (c), ^{1}H off-resonance decoupled ^{13}C nmr spectrum of NAD$^+$, irradiated at 5·84 ppm downfield from dioxan in the ^{1}H spectrum. (Taken from Birdsall and Feeney, 1972.)

studies of chemical shifts and coupling constants, ionization studies, selective decoupling experiments and the use of paramagnetic shift reagents), the good correlations obtained between partially decoupled ^{13}C resonances vs proton irradiating frequencies con-

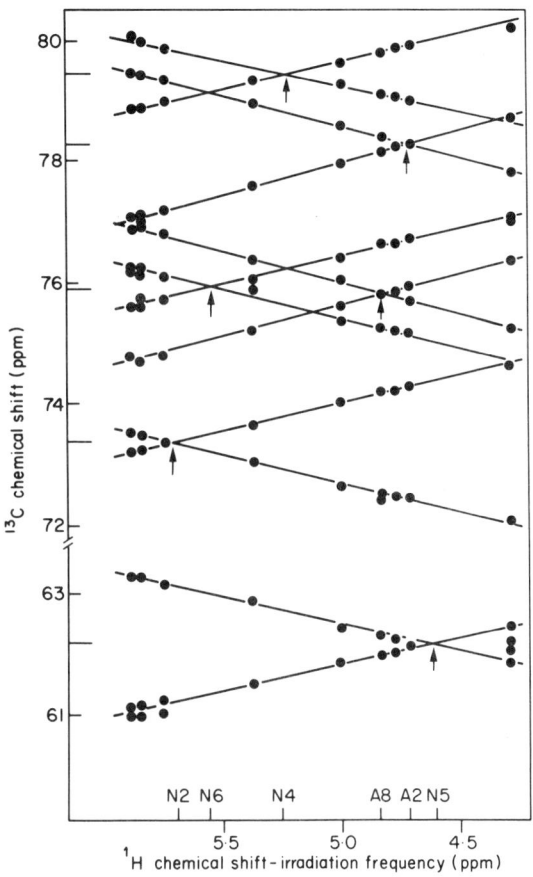

Figure 5. Plot of peak frequencies in the ^{1}H off-resonance selectively decoupled ^{13}C spectra of NAD$^+$ as a function of position of irradiation in the ^{1}H spectrum, expressed in ppm from internal dioxan. The positions of the peaks in the ^{1}H noise decoupled ^{13}C spectrum are shown by lines on the ordinate and the position of the proton peaks by lines on the abscissa. The arrows ↑ indicate the point of collapse of the ^{13}C doublet and the connection between a given ^{13}C peak and proton peak. The errors in the position of measurements are indicated by the size of the points except near the cross-over positions where the errors are larger (±0·15 ppm). (Taken from Birdsall et al., 1972a.)

firmed peak assignments graphically (Figs 4–6; Birdsall et al., 1972a; Birdsall and Feeney, 1972).

Data on ^{13}C chemical shifts of nucleotides are given in Table 8. Subsequent to the spectral assignments, conformational and

TABLE 8

^{13}C Chemical Shifts of Nucleotides
(Numbered formulae are on pp. 324 and 343)

Compound[b]	Conc., M	pH or pD	2	4	5	6	8	1'	2'	3'	4'	5'	Me	Reference
CMP [95][c]	1-2	neutral	158·4	167·1	97·4	142·7		90·0	70·6	75·2	84·2	64·2		Stothers, 1972a
UMP [93][c]	1-2	neutral	152·7	167·1	103·5	143·0		89·4	71·0	74·9	84·9	64·4		Stothers, 1972a
dCMP [96][c]	1-2	neutral	158·3	166·9	97·4	142·7		86·5	40·2	71·8	86·8	64·5		Stothers, 1972a
TMP [94][c]	1-2	neutral	152·4	167·1	112·3	138·3		85·7	39·4	72·1	86·8	64·9	12·6	Stothers, 1972a
AMP [97][c]	1-2	neutral	153·3	149·3	118·9	155·9	140·8	87·9	71·4	75·4	85·1	64·4		Stothers, 1972a
ATP [99][c]	1-2	neutral	153·2	149·2	118·9	155·7	140·5	87·8	71·0	75·1	84·7	66·1		Stothers, 1972a
IMP [102][c]	1-2	neutral	147·6	149·1	124·1	159·2	140·7	88·5	71·2	75·7	85·1	64·4		Stothers, 1972a
GMP [100][c]	1-2	neutral	154·5	151·9	116·6	159·2	138·2	87·9	71·5	75·2	85·0	64·8		Stothers, 1972a
dAMP [98][c]	1-2	neutral	153·1	149·0	119·0	155·9	140·7	84·4	40·0	72·3	87·0	64·8		Stothers, 1972a
dGMP [101][c]	1-2	neutral	154·7	151·8	116·6	159·4	138·0	84·1	40·0	72·2	87·0	64·9		Stothers, 1972a
3',5'-AMP[d]	0·15	7·2	153·82	149·05	119·52	156·20	140·78	92·69	73·47	72·91	78·48	68·54		Lapper et al., 1973
3',5'-UMP[d]	0·2	7·2	152·22	167·21	103·42	143·36		95·67	73·03	72·90	78·16	68·32		Lapper et al., 1973
3',5'-CMP[d]	0·2	7·2	158·18	167·53	97·38	143·08		96·18	73·37	72·91	78·28	68·50		Lapper et al., 1973
3',5'-GMP[d]	0·2	7·2	154·89	152·23	117·65	159·86	139·19	93·20	73·27	73·03	78·30	68·60		Lapper et al., 1973
3',5'-TMP[d]	0·2	7·2	152·51	167·33	112·96	138·91		86·53	35·71	76·01	77·48	68·40		Lapper et al., 1973
Dibutyryl-3',5'-AMP[d]			153·60	150·04	123·89	152·14	144·27	90·56	74·93	73·64	76·97	68·23		Lapper et al., 1973
2',3'-UMP[e]	0·15	7·2	152·42	167·41	103·46	144·79		93·74	81·74	78·26	86·25	62·12		Lapper and Smith, 1973
2',3'-CMP[e]	0·15	7·2	158·23	167·77	97·30	144·91		94·97	82·11	78·54	86·37	62·32		Lapper and Smith, 1973
2',3'-AMP[e]	0·15	7·2	153·70	149·24	119·83	156·40	141·69	90·60	81·56	78·78	86·43	62·40		Lapper and Smith, 1973
2',3'-GMP[e]	0·15	7·2	154·85	152·15	117·65	159·70	139·35	90·24	81·69	78·73	86·47	62·32		Lapper and Smith, 1973
5'-UMP[f]	0·2	7·8	153·22	167·61	103·74	143·16		89·63	75·04	71·10	85·11	64·37		Mantsch and Smith, 1973
3'-UMP[f]	0·2	7·8	152·90	167·41	103·46	143·00		90·34	74·65	73·14	84·86	61·88		Mantsch and Smith, 1973
2'-UMP[f]	0·2	7·8	153·22	167·73	103·66	143·96		90·32	76·51	70·96	85·33	62·32		Mantsch and Smith, 1973
Poly-U[f]	0·2	7·8	152·82	166·92	103·93	142·49		89·59	75·30	73·95	83·66	66·24		Mantsch and Smith, 1973
UpU[g]	0·4	7·2	152·7	167·1	103·5	142·8		90·5	74·4	74·2	84·8	61·7		Smith et al., 1973a
								90·3	75·0	70·7	84·0	66·2		
UpA (37°)[g](Up)	~0·1	7·2	152·6	167·0	103·4	142·6		90·3	74·2	74·2	84·7	61·7		Smith et al., 1973a
UpA (80°)[g](Up)	~0·1	7·2	153·7	166·2	104·1	143·3		91·0	75·1	74·6	85·3	62·4		Smith et al., 1973a
ApU[g](pU)	~0·1	7·2	153·7	166·2	103·7	141·5		90·6	75·3	70·5	85·7	65·9		Smith et al., 1973a
AMP [97][h]	0·1	2·5	146·4	149·7	119·8	151·6	143·7	89·3	75·9	71·5	85·4	65·6		Blumenstein and Raftery, 1973
AMP [97][h]	0·1	5·0	153·3	149·9	119·5	156·1	141·2	88·4	75·7	71·5	85·2	65·6		Blumenstein and Raftery, 1973
AMP [97][h]	0·1	8·5	154·0	150·1	119·6	156·6	141·4	88·2	75·7	71·8	85·9	64·6		Blumenstein and Raftery, 1973
AMP [97][i]	0·1	1·3	145·6	149·3	119·5	150·8	143·5	89·2	75·3	71·1	84·9	65·7		Birdsall and Feeney, 1972
AMP [97][i]	0·1	7·4	153·2	149·1	118·8	156·1	139·7	88·2	75·6	71·3	85·0	64·6		Birdsall and Feeney, 1972

TABLE 8—continued

Compound	Conc., M	pH or pD	Adenine Ring						Adenine Ribose					Reference
			A-2	A-4	A-5	A-6	A-8	A-1'	A-2'	A-3'	A-4'	A-5'	
5'-AMP[j]	0·6–0·7	7·0	153·2	149·6	119·2	155·8	141·0	88·0	75·4	71·4	85·3	64·8	Kotowycz and Hayamizu, 1973
3'-AMP[j]	0·6–0·7	7·0	153·0	148·9	120·7	155·8	141·2	89·3	74·7	74·2	86·1	62·5	Kotowycz and Hayamizu, 1973
2'-AMP[j]	0·6–0·7	7·0	149·9	149·2	119·6	155·9	141·6	88·4	77·0	71·4	86·2	62·4	Kotowycz and Hayamizu, 1973
ADP[k]	23 mg/ml		152·60	148·50	118·30	155·10	139·95	87·55	70·15	74·70	83·75	64·90	Breitmaier and Voelter, 1972
FAD[k]	75 mg/ml		152·40	148·15	117·75	154·65	139·20	87·45	70·25	75·15	83·85	65·40	Breitmaier and Voelter, 1972
UpA (37°)[g](pA)	~0·1	7·2	154·0	150·1		156·4	141·0	88·7	75·3	71·2	84·4	66·2	Smith et al., 1973a
UpA (80°)[g](pA)	~0·1	7·2	154·5			157·2	143·3	89·3	75·7	71·8	85·1	66·8	Smith et al., 1973a
ApU[g](Ap)	~0·1		153·8	149·3	120·8	156·7		90·0	75·4	74·3	83·6	62·3	Smith et al., 1973a
ApA, 3',5'[g]	~0·1	7·2	153·2	149·4	119·7	156·1		90·0	75·4	74·3	85·3	62·3	Smith et al., 1973a
			154·9	148·9	119·3	156·1		88·7	75·3	71·0	85·7	66·2	
ApA, 2',5' (37°)[g]	~0·1	7·2	153·9	149·2	119·0	156·4	141·7	88·8	78·3	71·6	86·7	62·5	Smith et al., 1973a
ApA, 2',5' (80°)[g]	~0·1	7·2	154·7	150·3		157·1	141·0	89·8	79·6	72·8	87·9	63·7	Smith et al., 1973a
			154·2	150·3				90·0	77·2	71·7	84·7	66·4	
NAD+ [114][h]	0·1	2·5	146·5	149·6	119·7	151·3	143·8	89·7	87·7	71·4	85·4	66·3	Blumenstein and Raftery, 1973
NAD+ [114][h]	0·1	7·5	154·1	150·2	119·7	156·6	141·1	88·1	75·2	71·6	85·1	66·5	Blumenstein and Raftery, 1973
NAD+ [114][h]	0·008	8·0	154·4				141·2	88·1	75·1	71·6	85·0	66·3	Blumenstein and Raftery, 1973
NAD+ [114][i]	0·15	1·0	154·7	149·2	119·3	150·7	143·5	89·1	75·5	71·1	84·8	66·3	Birdsall and Feeney, 1972
NAD+ [114][i]	0·3	7·05	153·5	149·5	119·0	155·8	140·7	87·9	74·9	71·4	84·5	65·8	Birdsall and Feeney, 1972
							140·5			71·2		66·3	
NAD+ [114][l]	0·2	1	146·0	149·5	119·6	150·9	143·7	89·6	75·9	71·5	85·2	66·7	Ellis et al., 1973
NAD+ [114][l]	0·2	2	146·2	149·5	119·6	151·0	143·7	89·5	76·0	71·5	85·3	66·4	Ellis et al., 1973
NAD+ [114][l]	0·2	3	146·8	149·5	119·6	151·4	143·1	89·3	76·0	71·6	85·3	66·4	Ellis et al., 1973
NAD+ [114][l]	0·2	4	148·5	149·6	119·5	152·5	142·9	89·1	75·8	71·6	85·2	66·3	Ellis et al., 1973
NAD+ [114][l]	0·2	5	152·5	149·8	119·4	135·3	141·3	88·2	75·5	71·7	85·1	66·3	Ellis et al., 1973
NAD+ [114][l]	0·2	6	153·6	149·9	119·4	156·1	141·1	88·1	75·4	71·8	85·0	66·2	Ellis et al., 1973
NAD+ [114][l]	0·2	7	153·7	149·9	119·3	156·2	140·8	88·1	75·4	71·7	85·0	66·2	Ellis et al., 1973
NADH[h]	0·1	8·0	154·1	150·1	119·7	156·5	141·0	88·5	71·4	75·7	84·7	66·3	Blumenstein and Raftery, 1973
NADH[i]	0·09	9·8	153·4	149·5	119·3	156·0	140·3	88·4	75·3	71·6	83·1	66·0	Birdsall and Feeney, 1972
										71·3	84·3	66·7	
										71·0			
α-NAD+[h]	0·05	2·5	146·6	149·7	119·7	151·5	143·7	89·2	71·5	76·0	85·4	66·5	Blumenstein and Raftery, 1973
α-NAD+[h]	0·05	7·5	154·9	150·0	119·5	156·7	141·2	80·0	71·5	75·8	85·3	66·7	Blumenstein and Raftery, 1973
NADP+ [115][h]	0·01	2·5	146·5	149·7	119·6	151·4	144·3	88·1	71·3	78·3	85·2	66·5	Blumenstein and Raftery, 1973
NADP+ [115][h]	0·1	5·1	153·2	150·0	119·3	155·8	151·6	87·2	71·2	77·6	84·7	66·5	Blumenstein and Raftery, 1973
NADP+ [115][h]	0·1	8·0	154·0	150·2	119·5	156·5	141·3	87·4	71·3	77·2	84·2	66·6	Blumenstein and Raftery, 1973

TABLE 8—continued

Compound	Conc., M	pH or pD	Nicotinamide Ring						Nicotinamide Ribose				Reference
			N-2	N-3	N-4	N-5	N-6	N-1'	N-2'	N-3'	N-4'	N-5'	
NADP+ [115]^i	0·1	2·0	145·5	149·1	119·9	159·5	143·6 / 143·4	87·8	78·0	71·0	84·9	66·1	Birdsall and Feeney, 1972
NADPH^h	0·1	8·1	154·0	150·4	120·0	156·8	141·2	88·1	71·4	77·5	83·9	66·6	Blumenstein and Raftery, 1973
NADPH^i	0·07	9·2	153·4	149·9	119·8	156·3	140·8	87·7	77·3	71·6 / 71·4 / 70·9	83·6 / 83·4	66·6 / 66·0	Birdsall and Feeney, 1972
NMN^h	0·05	4·0	141·3	135·4	147·5	133·8	143·9	101·3	79·0	72·2	88·5	65·3	Blumenstein and Raftery, 1973
NMN^h	0·05	8·0	141·2	135·4	147·7	133·8	144·6	101·7	79·0	72·6	89·5	64·2	Blumenstein and Raftery, 1973
NMN+^{i}	0·11	8·4	140·5		147·1	129·3	144·0	101·3	78·6	72·3	89·1	64·0	Birdsall and Feeney, 1972
NMN+^{i}	0·12	2·8	140·7	134·7	146·7	129·3	143·2	100·8	78·6	71·8	88·3	65·0	Birdsall and Feeney, 1972
NAD+ [114]^h	0·1	2·5	141·2	135·2	147·4	134·0	144·0	101·2	78·8	72·0	88·4	66·3	Blumenstein and Raftery, 1973
NAD+ [114]^h	0·1	7·5	141·2	135·0	147·1	134·0	143·8	101·3	78·8	71·8	88·7	66·1	Blumenstein and Raftery, 1973
NAD+ [114]^h	0·008	8·0	141·3		147·3		143·8	101·4	78·8	71·8	89·0	66·1	Blumenstein and Raftery, 1973
NAD+ [114]^i	0·15	1·0	140·9	134·9	147·0	129·7	143·5	100·9	78·5	71·6	87·8	66·3	Birdsall and Feeney, 1972
NAD+ [114]^i	0·3	7·05	140·7 / 140·5	134·4	146·7	129·5	143·2	100·9	78·5	71·4 / 71·2	87·9	65·8 / 66·3	Birdsall and Feeney, 1972
NAD+ [114]^i	0·2	1	141·1	135·1	147·3	130·0	143·7	101·1	78·8	71·9	88·1	66·7	Ellis et al., 1973
NAD+ [114]^i	0·2	2	141·1	135·1	147·3	130·0	143·7	101·2	78·8	71·9	88·3	66·4	Ellis et al., 1973
NAD+ [114]^i	0·2	3	141·1	135·1	147·3	130·0	143·8	101·2	78·9	71·9	88·3	66·6	Ellis et al., 1973
NAD+ [114]^i	0·2	4	141·1	135·1	147·2	130·0	143·7	101·2	78·9	72·0	88·3	66·6	Ellis et al., 1973
NAD+ [114]^i	0·2	5	141·1	134·9	147·1	129·9	143·6	101·2	78·9	71·9	88·2	66·8	Ellis et al., 1973
NAD+ [114]^i	0·2	6	141·1	134·8	147·0	129·9	143·5	101·2	78·8	71·8	88·1	66·7	Ellis et al., 1973
NAD+ [114]^i	0·2	7	141·1	134·8	146·9	129·8	143·5	101·2	78·8	71·7	88·1	66·7	Ellis et al., 1973
α-NAD+*^{g}	0·05	2·5	142·3	133·4	146·6	132·4	145·0	102·0	72·8	72·0	88·4	66·6	Blumenstein and Raftery, 1973
α-NAD+*^{g}	0·05	7·5	141·9	132·9	146·1	132·2	144·7	102·0	80·0	72·0	88·5	66·7	Blumenstein and Raftery, 1973
NADP+*^{g}	0·1	2·4	141·1	135·2	147·4	134·0	143·9	101·2	78·7	71·9	88·1	66·1	Blumenstein and Raftery, 1973
NADP+*^{g}	0·1	5·1	141·1	134·9	147·1	133·9	143·6	101·2	78·7	71·7	88·1	65·9	Blumenstein and Raftery, 1973
NADP+*^{g}	0·1	8·0	141·0	134·7	146·9	133·8	143·4	101·2	78·7	71·6	87·7	66·1	Blumenstein and Raftery, 1973
NADP+*^{g}	0·1	2·0	140·7	134·6	146·8	129·5	143·6 / 143·4	100·8	78·5	71·7	87·8	66·1	Birdsall and Feeney, 1972
Nicotinamide^i	1·0	6·78	148·3	129·7	136·9	124·9	152·4						Birdsall and Feeney, 1972
Nicotinamide^i	1·0	0	141·8	133·5	146·5	128·6	144·4						Birdsall and Feeney, 1972
N-Methyl nicotinamide^i	0·2	8·5	145·9	134·2	144·5	129·0	148·2						Birdsall and Feeney, 1972

		Dihydronicotinamide Ring						Dihydronicotinamide Ribose				Reference
NMNH$_2$[h]	0.05	5.2	139.5	101.8		106.7	126.2	96.2	71.8 72.2		83.8	66.0	Blumenstein and Raftery, 1973
NMNH$_2$[h]	0.05	8.5	139.6	101.9	23.0	106.7	126.1	96.3	71.9	71.7 72.0	84.4	65.1	Blumenstein and Raftery, 1973
NADH[h]	0.1	8.0	139.6	101.3	23.0	106.7	125.3	96.4			83.5	67.1	Blumenstein and Raftery, 1973
NADH[i]	0.09	9.8	139.0	102.0	22.8	106.0	124.7	95.9	71.6 71.3 71.0	71.6 71.3 71.0	83.1 84.3	66.0 66.7	Birdsall and Feeney, 1972
NADPH[h]	0.1	8.2	139.5	101.2		106.4	125.2	96.3		71.8 72.0	83.6	67.0	Blumenstein and Raftery, 1973
NADPH[i]	0.07	9.2	139.0	100.9	22.8	106.5	124.7	95.8	71.4 71.4 70.9	71.6 71.4 70.9	83.4	66.6 66.0	Birdsall and Feeney, 1972

		Isoalloxazine Ring										
FAD[k]	75 mg/ml	157.45 (R-2) 116.75 (R-9) 18.70 (R-7CH$_3$)	160.60 (R-4) 131.35 (R-11) 20.85 (R-8CH$_3$)	130.13 (R-6) 139.20 (R-12)	133.30 (R-7) 150.65 (R-13)	133.89 (R-8) 149.80 (R-14)		47.60	69.40	72.65	71.45	67.80	Breitmaier and Voelter, 1972

[a] Chemical shifts in parts per million relative to TMS; see cited reference for additional experimental details.

[b] Abbreviations used: NAD+ = nicotinamide adenine dinucleotide; NADH = reduced nicotinamide adenine dinucleotide; NADP+ = nicotinamide adenine dinucleotide phosphate; NADPH = reduced nicotinamide adenine dinucleotide phosphate; NMN, NMN+ = nicotinamide mononucleotide; NMNH$_2$ = reduced nicotinamide mononucleotide; α-NAD+ = α-nicotinamide adenine dinucleotide; AMP = 5′-adenylic acid; 3′,5′-AMP = adenosine 3′,5′-cyclic phosphate; 3′,5′-UMP = uridine 3′,5′-cyclic phosphate; 3′,5′-GMP = guanosine 3′,5′-cyclic phosphate; 2′,3′-cyclic cytidine 3′,5′-cyclic phosphate; 3′,5′-TMP = thymidine 3′,5′-cyclic phosphate; Dibutyryl-3′,5′-AMP = N^6,O$^{2'}$-dibutyryladenosine 3′,5′-cyclic phosphate; 2′,3′-UMP = uridine 2′,3′-cyclic monophosphate; 2′,3′-CMP = cytidine 2′,3′-cyclic monophosphate; 2′,3′-AMP = adenosine 2′,3′-cyclic monophosphate; 2′,3′-GMP = guanosine 2′,3′-cyclic monophosphate; 2′-UMP = uridine 2′-phosphate; 3′-UMP = uridine 3′-phosphate; 5′-UMP = uridine 5′-phosphate; Poly U = polyuridylic acid; ADP = adenosine 5′-diphosphate; FAD = flavin adenine dinucleotide; UpA, UpU, ApU and ApA = dinucleoside phosphates of uridine and/or adenine.

[c] See original references for experimental conditions and additional data.

[d] Proton decoupled ^{13}C FTnmr spectra obtained on a Varian XL-100-15 spectrometer at 25.2 MHz in D$_2$O; sodium salts of compounds except 3′,5′-UMP (barium salt). Chemical shifts relative to TMS in a concentric capillary tube.

[e] Proton decoupled ^{13}C FTnmr spectra obtained on a Varian XL-100-15 spectrometer at 25.2 MHz and 37° in D$_2$O; sodium salts of compounds except 2′,3′-UMP (barium salt) and 2′,3′-GMP (pyridinium salt). Chemical shifts relative to TMS in a coaxial tube.

[f] Proton decoupled ^{13}C FTnmr spectra obtained on a Varian XL-100-15 spectrometer at 37° in D$_2$O. Chemical shifts corrected to TMS scale using δC$_6$H$_6$ = 123.65 ppm.

[g] Proton-decoupled ^{13}C FTnmr spectra obtained on a Varian XL-100-15 spectrometer at 25.2 MHz and 37°. Chemical shifts relative to TMS in a concentric capillary tube.

[h] Proton-noise or single frequency decoupled ^{13}C FTnmr spectra obtained on a Varian XL-100-15D spectrometer at 25.1 MHz in D$_2$O on 0.1 M compound. Chemical shifts corrected to TMS scale using δCH$_3$CO$_2$H = 178.3 ppm.

[i] Proton-noise and single frequency decoupled ^{13}C FTnmr spectra obtained on a Varian XL-100 spectrometer at 25.2 MHz in D$_2$O. Chemical shifts corrected to TMS scale using δ$_\text{dioxane}$ = 67.4 ppm.

[j] Proton decoupled ^{13}C FTnmr spectra obtained on a Bruker HFX-90 spectrometer at 22.63 MHz and 27° in D$_2$O. Chemical shifts corrected to TMS scale using δ$_\text{dioxane}$ = 67.4 ppm.

[k] Proton-decoupled ^{13}C FTnmr spectra obtained on a Bruker HFX-90 spectrometer at 22.63 MHz. Chemical shifts given as susceptibility-corrected δ values relative to TMS.

[l] Proton-decoupled ^{13}C FTnmr spectra obtained on a Varian XL-100-15 spectrometer at 25.5 MHz. Chemical shifts corrected to TMS scale using δCS$_2$ = 193.7 ppm.

[m] Assignments of the 2′ and 3′ carbon atoms for the ribose moieties were incorrect in many of the earlier references, and should be reversed.

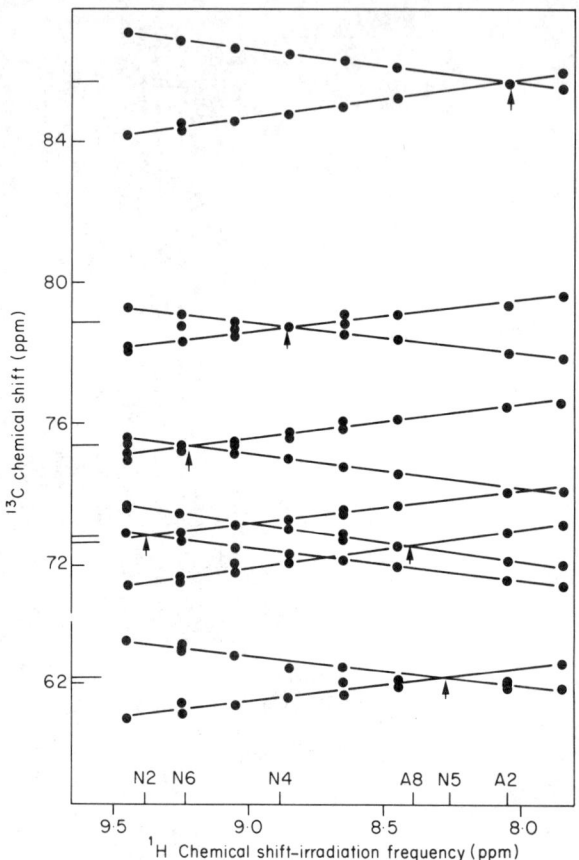

Figure 6. Plot of peak frequencies in the ^{1}H off-resonance selectively decoupled ^{13}C nmr spectra of NAD$^+$ at pH 7·05 as a function of position of irradiation in the ^{1}H spectrum, expressed in ppm downfield from internal dioxan. The positions of the peaks in the ^{1}H noise decoupled ^{13}C spectrum are shown by lines on the ordinate and the position of the proton peaks by lines on the abscissa. The arrows ↑ indicate the point of collapse of the ^{13}C doublet and the connection between a given ^{13}C peak and the assigned proton peak. The errors in the position of measurements are indicated by the size of the points except near the cross-over points where they are larger (±0·15 ppm). (Taken from Birdsall and Feeney, 1972.)

substituent effects, line broadening or relaxation times have been investigated. This information in turn can be applied to binding studies between nucleotides and small molecules. Nucleotides are also known to bind metallic ions (Izatt *et al.*, 1971; Phillips, 1966; Weser, 1968). The binding of manganese(II) ions to purine- and pyrimidine-nucleotides has been examined by ^{13}C nmr spectroscopy (Kotowycz and Hayamizu, 1973; Kotowycz and Suzuki, 1973a, Lam *et al.*, 1974). The resonances of C-5 and C-8 in AMP were selectively broadened on addition of 10^{-5} to 10^{-4} M Mn(II) ions (Figs 7-10). Information on the sensitivity of each position in the ring to added metal ions led to the conclusion that the metal ion was held near the N-7 position. The manganese(II)-ATP complex has also been studied by ^{13}C nmr techniques (Lam *et al.*, 1974). The results obtained indicated that the metal ion binds first to the phosphate groups of ATP and then to the N-7 nitrogen, in agreement with results for the purine monophosphate nucleotides. The effect of Cu(II) ions on proton-decoupled ^{13}C nmr spectra of purine nucleotides (Kotowycz and Suzuki, 1973b; Weser *et al.*, 1974) and pyrimidine nucleosides and nucleotides (Kotowycz, 1974) has also been determined. The location of copper(II) ion in pyrimidine nucleoside or nucleotide complexes has been shown to be near the N-3 position of the base (Kotowycz, 1974). Similarly, the preferential site of copper(II) ion in the AMP complex was shown to be the N-7 position of the purine ring (Weser *et al.*, 1974). Independent verification of this work has also been reported (Fritzsche *et al.*, 1973).

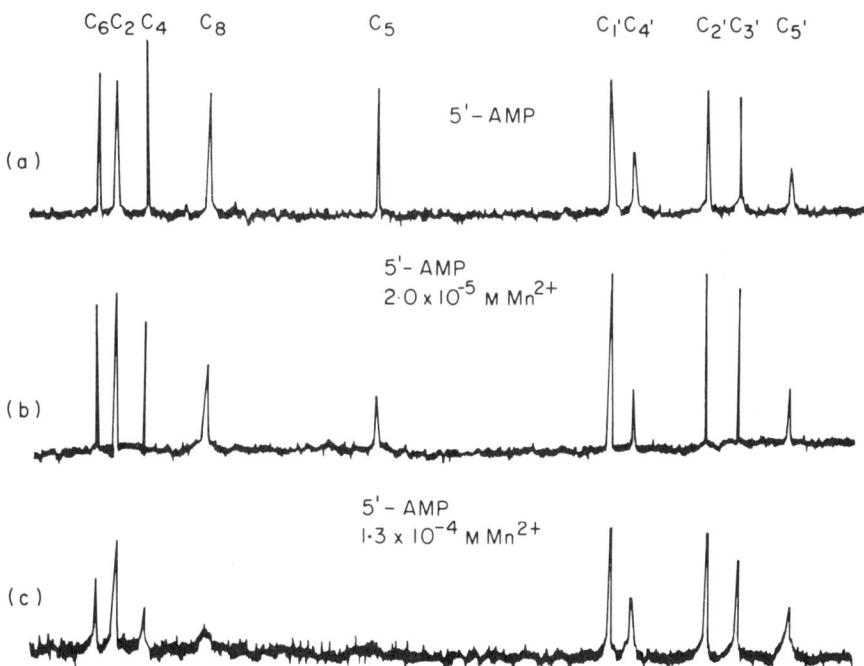

Figure 7. The effect of Mn^{2+} ions on the ^{13}C proton decoupled NMR spectra of 5'-AMP in D_2O (pD = 7.4) at 27°. Spectrum (a) is for the metal-free solution and the Mn^{2+} ion concentration is indicated for spectra (b) and (c). (Taken from Kotowycz and Hayamizu, 1973.)

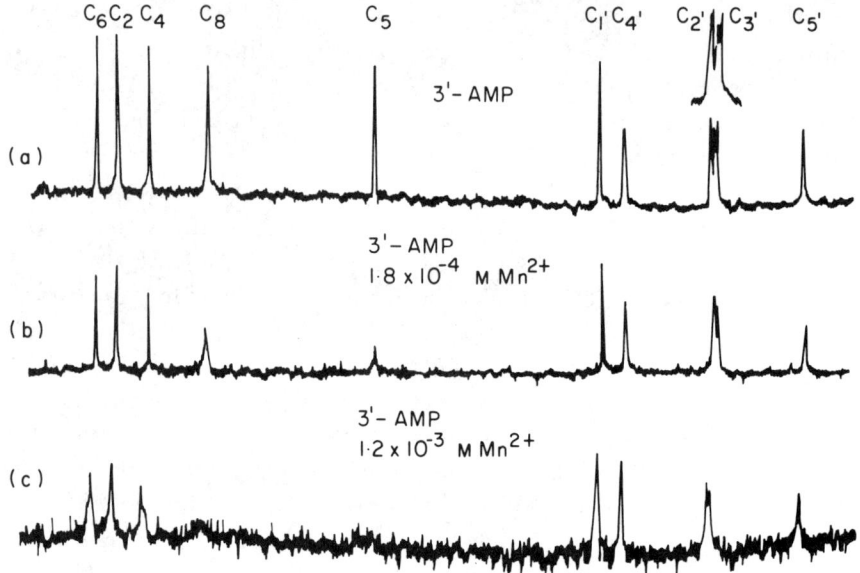

Figure 8. The effect of Mn^{2+} ions on the ^{13}C proton-decoupled nmr spectra of 3'-AMP in D_2O (pD = 7·4) at 27°. Spectrum (a) is for the metal-free solution and the Mn^{2+} ion concentration is indicated in spectra (b) and (c). (Taken from Kotowycz and Hayamizu, 1973.)

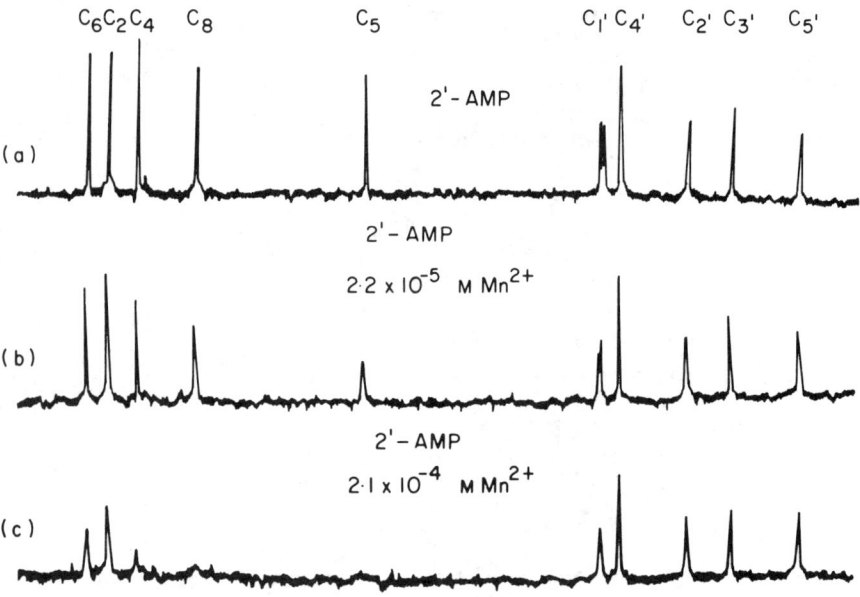

Figure 9. The effects of Mn^{2+} ions on the ^{13}C proton-decoupled nmr spectra of 2'-AMP in D_2O (pD = 7·4) at 27°. Spectrum (a) is for the metal free solution and the Mn^{2+} ion concentration is indicated in spectra (b) and (c). (Taken from Kotowycz and Hayamizu, 1973.)

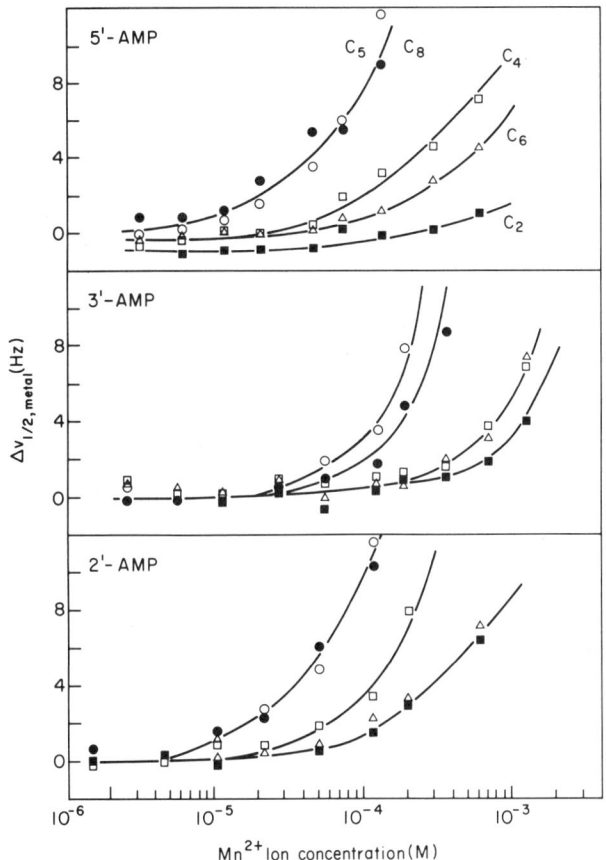

Figure 10. The dependence of $\Delta\nu_{1/2}$ (metal) for the five base carbon nuclei on the Mn^{2+} concentration. The different carbon nuclei are represented as follows: ■, C-2; □, C-4; ○, C-5; △, C-6; ●, C-8. (Taken from Kotowycz and Hayamizu, 1973.)

Another problem under current investigation by high resolution ^{13}C nmr is the interaction between and association of nucleotides and proteins. Line broadening effects and changes in the chemical shifts in the ^{13}C spectrum of ATP similar to the spectral changes noted for nucleotide metal ion complexation, were observed upon the addition of ferrocytochrome (Kayushin and Ajipa, 1973).

A number of ^{13}C nmr studies on the structure and conformation of nucleotides have been performed in the last few years (Birdsall and Feeney, 1972; Blumenstein and Raftery, 1972, 1973; Lapper et al., 1972, 1973; Lapper and Smith, 1973; Que et al., 1973; Smith et al., 1973a; Birdsall et al., 1973a). Guanosine tetraphosphate, a nucleotide which exhibits regulatory control in the interaction of protein synthesis and RNA accumulation in some bacteria, has been studied by ^{13}C nmr in order to determine the structure and sites of the phosphate groups (Que et al., 1973). Based on chemical shifts and line widths, the tetraphosphate was established as guanosine 5'-diphosphate, 3'-diphosphate [116]. Dihydronicotinamide adenine dinucleotide phosphate (NADPH), an important coenzyme in redox enzymatic reactions, and various allied nucleotides have been studied by ^{13}C nmr (Birdsall and Feeney, 1972; Blumenstein and Raftery, 1973; Ellis et al., 1973).

In solution these nucleotides will undergo a change in conformation from the folded to the unfolded form when (a) the pH is decreased below 4 (pK_a-value of nicotinamide protonation at N–1 = 4·0); (b) the temperature is raised above ambient and (c) alcohol is added to the aqueous solvent. The conformation of the ribose moieties in mono- and dinucleotides was proposed to be identical, based on the invariance, with the exception of the 5' carbon atoms, of their ^{13}C chemical shifts in these two environments. Changes in the chemical shifts of the C-5' carbon atoms reflect the difference between the mono- and pyrophosphate groups of the dinucleotide (Blumenstein and Raftery, 1973). The relaxation time of the carbonyl carbon of NAD$^+$ [114], T_1 = 14 ± 2 seconds, was independent of pH and hence of the conformation (Blumenstein and Raftery, 1973). In order to evaluate the effect of base stacking on the chemical shifts of NAD$^+$, the concentration was varied over the range of 0·008 to 0·1 M. However, the changes in chemical shift were slight, and the effects of base stacking were not readily detectable. The effect of base stacking on the ^{13}C chemical shifts of aqueous 5'-AMP manifests itself in a downfield shift of the base carbons as the concentration is decreased from 1·2 M to 0·04 M (Smith et al., 1973a). The ^{13}C and ^{1}H nmr data cannot distinguish between a straight and an alternate vertical stacking arrangement. However, an alternate stack seems to be the more reasonable configuration because it reduces the steric interactions between the ribosyl moieties. Preliminary studies on base-base stacking interactions and the conformation of the phosphodiester backbone in dinucleoside phosphates have also been reported (Smith et al., 1973a). The applicability to conformation of ^{13}C spin-lattice relaxation times of nicotinamide adenine dinucleotide and adenosine monophosphate in solution has been described (Hamill et al., 1974). Decreasing the concentration of AMP results in an increase in the T_1-values, a fact rationalized by assuming a decrease in size of the molecular

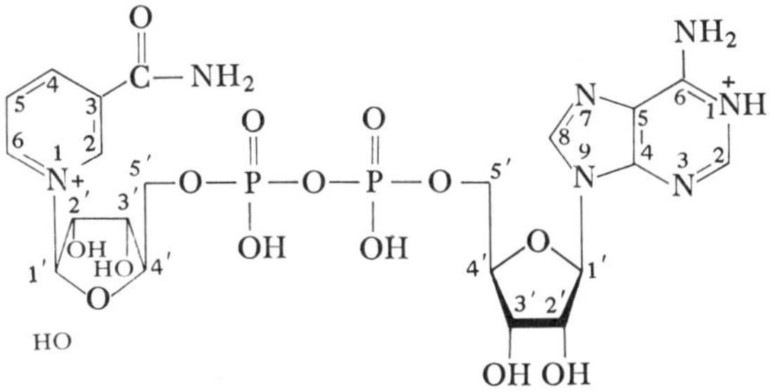

[114] NAD⁺

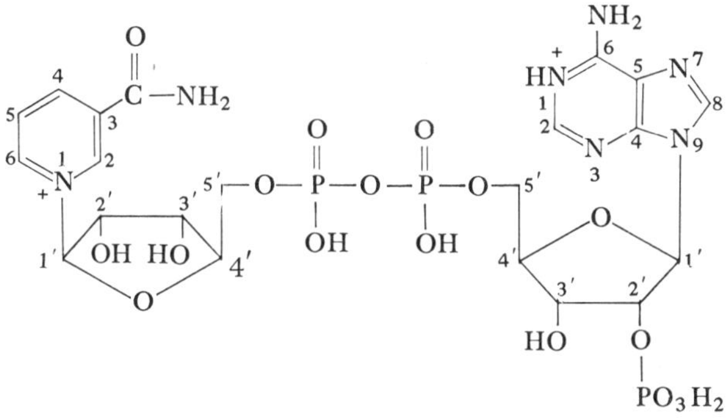

[115] NADP⁺

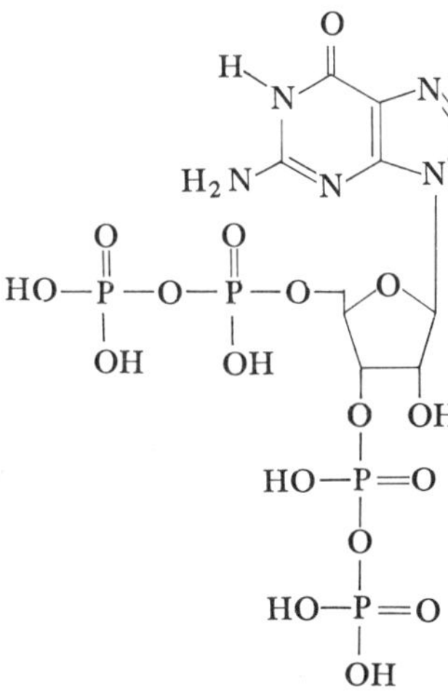

[116] Guanosine 5′-diphosphate, 3′-diphosphate

aggregates with the resultant faster tumbling rate (Smith et al., 1973a, Hamill et al., 1974). The ribose methine carbon atoms have slightly longer T_1-values than carbons A-2 and A-8 at 1·0 M and 0·4 M AMP, indicating that the ribose carbons have a greater degree of freedom and more mobility than the ring carbons. The proposed rationale for this behaviour is that, when AMP forms base stacks, the ribose portion of the molecule is positioned on the outside of aggregate (Schweizer et al., 1968), where it has considerably more mobility than the base carbons (Hamill et al., 1974). The T_1-values of NAD follow similar trends to those of AMP (Hamill et al., 1974) and support the base-base stacking model. For NAD the effect of pH on T_1-values is much smaller than the effect of dilution. The 1 : 2 complex formed in D_2O between actinomycin D and deoxyguanosine 5′-monophosphate [101] has also been investigated by ^1H (Arison and Hoogsteen, 1970; Danyluk and Victor, 1970; Krugh and Neely, 1973; Patel, 1974), ^{13}C (Patel, 1974) and ^{31}P nmr spectroscopy (Patel, 1974). However, the assignments of ^{13}C chemical shifts for actinomycin D were not available, and therefore no comparison was made for the free and complexed forms. The effect of ring current and stacking geometry on the chemical shifts were, however, discussed (Patel, 1974).

Another enzyme cofactor, flavine adenine dinucleotide (FAD) disodium salt, has been studied by ^{13}C nmr. Assignments of the resonances have been made by comparing the spectrum of FAD with its constituents, adenosine 5′-diphosphate trisodium salt and riboflavin 5′-monophosphate sodium salt (Breitmaier and Voelter, 1972).

Cyclic nucleotides have been investigated by Smith and coworkers (Lapper et al., 1972, 1973; Lapper and Smith, 1973; Smith et al., 1973a). Three-bond couplings between ^{13}C and ^{31}P, which vary from 1 to 10 Hz, have been observed in ^{13}C nmr spectra (Mantsch and Smith, 1972). The proton and ^{13}C nmr spectra of a number of 2′,3′- and 3′,5′-cyclic nucleotides were taken in order to examine the relationship between the coupling constants of ^{13}C and ^{31}P and the dihedral angles formed by the P–O–C and O–C–C planes. ^{31}P splittings for carbons 2′,3′,4′ and 5′ were observed in the spectra. The P–C-2′ coupling constants for six 3′,5′-nucleotides were 8·0 ± 0·3 Hz, indicating *trans* coupling (Lapper et al., 1972). The gauche coupling constant between C-4′ and P had an unusually large value of 4·5 Hz. This large coupling constant has been rationalized in the following manner: C-4′ had two means by which it could couple with the P and the larger coupling was composed of the sum of these two routes. Thus the gauche coupling constant was approximated as 2·3 Hz. An equation for calculating the coupling constant has been proposed (Lapper et al., 1973), where J_0 was taken to be 8·0 Hz and

$$^3J_{CP} = J_0 \cos^2 \gamma$$

γ was the CCOP dihedral angle. In addition, the coupling constants indicate that the conformations of the ribose and cyclic phosphate rings were similar for all 3′,5′-cyclic nucleotides (Table 9, Lapper et

TABLE 9
Carbon-13 Coupling Constants (Hz) for 3′,5′-Cyclic Nucleotides[a]

Nuclei	U	C	A	G	T	Dibutyryl-A
C-1′, P	<0·2	<0·2	<0·2	<0·2	<0·2	<0·2
C-2′, P	8·0	7·8	7·8	7·8 (8·3)	8·3	8·3
C-3′, P	4·5	4·3	4·3	3·8 (3·3)	4·8	4·5
C-4′, P	4·5	4·5	4·5	4·8	4·5	3·8
C-5′, P	7·0	7·0	7·0	7·3	7·0	7·3

[a]Taken from Lapper et al., 1973.
U, uridine; C, cytidine; A, adenosine; G, guanosine; T, thymidine.

al., 1973; Smith et al., 1973a). ^{13}C Nmr spectra and C–P coupling constants of some 2′,3′-cyclic nucleotides have also been examined (Table 10, Lapper and Smith, 1973; Smith et al., 1973a). Overall C–P couplings were smaller, less sensitive and not as reliable as their corresponding proton ^{31}P couplings for mononucleotides. In larger molecules, such as polynucleotides, where the ^{1}H spectra are very complex and difficult to interpret, C–P coupling constants will be of more service.

TABLE 10
Carbon-13 Coupling Constants (Hz) for 2′,3′-Cyclic Nucleotides[a]

Nuclei	U	C	A	G
1′P	6·8	6·3	3·8	4·8
2′P	2·5	2·5	1·8	2·3
3′P	0·7	0·5	1·0	1·0
4′P	2·5	2·2	4·3	3·5
5′P	0·2	<0·2	<0·2	<0·2

[a]Taken from Lapper and Smith, 1973.
U, uridine; C, cytidine; A, adenosine; G, guanosine.

Nucleic Acids

Until recently ^{13}C nmr studies of nucleic acids have not received the same attention as those of nucleosides and nucleotides. Polyuridylic acid (poly U), polyadenylic acid (Poly A), and unfractionated baker's yeast transfer ribonucleic acid (tRNA) have been investigated by ^{13}C nmr techniques (Table 11; Mantsch and Smith, 1972; Komoroski and Allerhand, 1972, 1974). Assignments for uridine monophosphates were based on changes in phosphorus–carbon splittings and on changes in the chemical shift of approximately 2·5 ppm on the phosphorylated carbon, which was caused by the deshielding effect of the phosphate group (Mantsch and Smith, 1972). Shifts in the resonances on going from the monomer to the polymer have been used to elucidate information on the nature of structural effects. Data on the dihedral angle have been obtained from the three-bond C–P coupling constants. The preferred molecular conformations of aqueous poly U (MW 130 000) were found to be the 4′-*trans* rotamer for the 5′-phosphate and the 2′-*trans* rotamer for the 3′-phosphate (Mantsch and Smith, 1972; Smith *et al.*, 1973a). Spectral changes in a variable-temperature ^{13}C nmr study of poly U in 0·5 M CsCl have been interpreted in terms of a helix-coil transition (Govil and Smith, 1973).

The first ^{13}C nmr study of nucleic acids was made on tRNAs because, for biopolymers, their molecular weights, 25 000 to 30 000, are low, and their chain lengths, 75 to 85 nucleotides, relatively short (Zachau, 1969). ^{13}C Nmr spectra of salmon sperm DNA, in D_2O at pD = 7·2 and 37°C and unfractionated tRNA have also been recorded (Smith *et al.*, 1973a); however, the resolution of these spectra is poor, especially for tRNA, and allows the assignment of only a few resonances. The high-resolution natural-abundance proton-decoupled ^{13}C FTnmr spectra at 15·18 MHz and the spin lattice relaxation times of aqueous unfractionated tRNA in the presence of magnesium(II) ions have been reported (Komoroski and Allerhand, 1972). Development of a probe for 20 mm O.D. sample tubes (Allerhand *et al.*, 1972; 1973c) coupled with the use of a special crystal filter (Allerhand *et al.*, 1973b) has led to an increase in sensitivity of a factor of 4 over commercially available FTnmr instruments. This increased sensitivity now makes it practical to observe the resonances from some minor bases in the natural abundance ^{13}C nmr spectrum of tRNA (Komoroski and Allerhand, 1974). The magnesium chloride : tRNA molar ratio was 8 in all cases

TABLE 11

^{13}C Chemical Shifts of Nucleic Acids and Related Compounds

Assignment	tRNA at 80°C[a]	tRNA at 52°C[a]	tRNA at 41°C[c]	Poly A at 59°C[a]	Poly U[b]	5′ NMP[a,c]
4H$_2$U			174·7			174·6
4C, 4U	167·1	166·6	166·6		166·93	167·3
6G	159·5	159·5	159·5			159·4
2C	158·1		156·8			158·6
6A	156·3		155·9	155·9		156·1
2G	154·6		154·4			154·7
2A	153·7			153·2		153·5
2U, 4G	152·4		152·3		152·82	152·9, 152·1
			150·7			
4A	149·7		147·2	148·9		149·5
6U, 6C	142·5		139·8		142·49	143·2, 142·9
8A	140·7		134·8	139·7		141·0
8G	138·3		134·9			138·4
5A	119·7	119·2	119·1	119·2		119·1
5G	117·5	117·0	117·0			116·8
5ψ, 5T			~111·6			111·4
5U	103·6	103·3	103·1		103·94	103·7
5C	97·2	96·4	96·2			97·5
1′C	90·4		~91·6			90·2
1′U	89·6	89·6	~91·6		89·59	89·6
1′A, 1′G	88·7		~91·6	88·5		88·1
4′	83·3	~81·6	~80·6	82·7	83·66	85·3–84·4
2′	74·5	73·6	~73·6	74·1	75·30	75·4–75·1
3′	74·5	73·6	~71·6	74·1	73·95	74·4–73·1
5′	65·9	~64·6	~64·6	65·7	66·24	65·0–64·4
OCH$_3$			56·0			
6H$_2$U, NCH$_3$			36·9			37·3
NCH$_3$			33·3			
5H$_2$U			31·1			30·9
NCH$_3$			28·8			
CCH$_3$			12·0			

[a] Chemical shifts in ppm converted to TMS scale using δ_{CS_2} = 193·6. Data obtained on a "home-built" spectrometer operating at 15·18 MHz. ~7·5 and 3 mM solutions of aqueous nucleic acids used. In tRNA at 52°C, 2C to 8G are poorly resolved. FT and decoupled mode – 5′NMP = 5′ Nucleotide Monophosphate (Komoroski and Allerhand, 1972).

[b] Chemical shifts in ppm relative to external TMS. Data obtained with a Varian XL-100-15 spectrometer FT and decoupled mode (Mantsch and Smith, 1972).

[c] Chemical shifts in ppm converted to TMS scale using δ_{CS_2} = 193·6. Data obtained on a "home-built" spectrometer operating at 15·18 MHz. 150 mg/ml solutions of aqueous nucleic acids used. FT and decoupled mode. ψ, pseudouridine; H$_2$U, dihydrouridine; T, ribosylthymine. (Komoroski and Allerhand, 1974.)

(Komoroski and Allerhand, 1972, 1974). The chemical shifts of tRNA are given in Table 11.

The effects of temperature and concentration on the spectra of nucleic acids were studied. As the temperature increased or the concentration decreased the line widths narrowed. At high concen-

TABLE 12

^{13}C Chemical Shifts of Amino Acids

Amino-Acid	C=O	α	β	γ	δ	ε	Other	Reference
Glycine (free)	173·2	42·2						Stothers, 1972a
(free)	173·1$_5$	42·5						Keim et al., 1973a
(peptide)	172·7$_5$	43·5						Keim et al., 1973a
Alanine (free)	176·5	51·3	17·0					Stothers, 1972a
(free)	176·6$_5$	51·6$_5$	17·2					Keim et al., 1973a
(peptide)	176·3	50·9	17·5					Keim et al., 1973a
Serine (free)	172·8	57·1	61·0					Stothers, 1972a
(free)	173·1	57·6	61·3					Keim et al., 1973b
(peptide)	173·1$_5$	56·6$_5$	62·1					Keim et al., 1973b
Cysteine	183·5	57·0	26·0					Stothers, 1972a
Phenylalanine	174·7	57·0	37·2				130·4-130·8 (o- and m-C's) 129·2 (p-C)	Stothers, 1972a
Histidine	174·6	55·0	28·7				136·9 (C-2) 117·9 (C-4)	Stothers, 1972a
Aspartic acid (free)	175·2	52·9	37·3	178·5				Stothers, 1972a
(free)	175·1	53·2$_5$	37·5	178·3$_5$				Keim et al., 1973b
(peptide)	174·8	52·5	39·4$_5$	178·5				Keim et al., 1973b
Asparagine (free)	175·7	52·6	35·9	175·4				Stothers, 1972a
(free)	175·7	52·6	36·1$_5$	175·7				Keim et al., 1973b
(peptide)	173·7	51·4	37·2$_5$	175·4				Keim et al., 1973b
Valine (free)	175·0	61·3	29·9	17·6 18·8				Stothers, 1972a
(free)	174·9	61·6	30·1	17·7 19·0				Keim et al., 1973a
(peptide)	174·8$_5$	60·7	30·8	18·4$_5$ 19·8				Keim et al., 1973a

Amino acid							Reference
Threonine (free)	173.7	61.2	66.8	20.2			Stothers, 1972a
(free)	173.6	61.4	66.9	20.4			Keim et al., 1973b
(peptide)	173.3	60.0	67.8$_5$	19.6			Keim et al., 1973b
Methionine	175.0	54.9	30.7	29.8		14.9	Stothers, 1972a
Glutamic acid (free)	175.3	55.4	27.8	34.2	182.0		Stothers, 1972a
(free)	175.6	55.9	28.0	34.6	182.1		Keim et al., 1973b
(peptide)	175.2	55.1	28.1$_5$	34.3	182.3		Keim et al., 1973b
Glutamine (free)	175.9	55.3	27.3	31.9	178.5		Stothers, 1972a
(free)	175.7	55.3	27.1	31.9	178.5		Keim et al., 1973b
(peptide)	174.6	54.3	27.5$_5$	32.1	178.7		Keim et al., 1973b
Leucine (free)	176.3	54.4	40.7	25.1	21.8		Stothers, 1972a
(free)	176.3	54.8	40.9	25.3	22.9		Keim et al., 1973a
					23.1		
(peptide)	175.9	53.6	40.5	25.1$_5$	22.1		Keim et al., 1973a
					23.1		
					21.6		
Isoleucine (free)	174.9	60.6	36.8	25.4	12.0	15.6	Stothers, 1972a
(free)	174.9	60.6	36.8	25.3$_5$	12.0	15.6	Keim et al., 1973a
(peptide)	174.8	59.6	36.9	25.3$_5$	11.3	15.6$_5$	Keim et al., 1973a
Arginine	174.9	54.8	28.2	24.6	41.2	157.2	Stothers, 1972a
Lysine	175.1	55.0	26.9	22.1	30.4	39.7	Stothers, 1972a
Proline		61.3	29.4	24.1	46.2		Stothers, 1972a
Tryptophan	175.2	56.1	27.4	108.4	127.5	137.3	Allerhand et al., 1973
					126.0	120.3	
					122.9	119.3	
						112.8 (ξ_2)	
						126.0 (ξ_3)	
						122.9 (ξ_3)	
						120.3 (η_2)	
						119.3 (η_2)	
Ornithine	174.7	55.1	28.4	23.9	40.1		Lyerla and Freedman, 1972

[a] Chemical shifts in parts per million downfield from TMS ($\delta = 0$ ppm); see cited reference for experimental details and for additional data.

tration some broadening of peaks were caused by tRNA aggregation. No C—P splittings were observable. The spectra of the folded and unfolded forms of tRNA were virtually identical with the exceptions of the resonance of the $4'$ carbon (shifted 1·5 ppm upfield) and the separation between $2'$ and $3'$ carbon resonances (collapsed in going from $52°$ to $80°C$). These changes were assumed to be indications of a change in conformation (Komoroski and Allerhand, 1972). However, the nature of these conformational changes are not well understood at present.

4. PROTEINS AND THEIR RESIDUES

Nuclear magnetic resonance studies of amino-acids, peptides and proteins have received a considerable amount of attention (Roberts and Jardetzky, 1970), but, owing to the large number of resonances, the broadness of the line widths and the relatively small variations in chemical shifts in proteins and other large biomolecules, the applicability of proton NMR to these systems has been severely limited. With the advent of commercially available fast Fourier transform proton noise decoupled nmr spectrometers, natural abundance ^{13}C FTnmr spectroscopy has become a practical alternative for examining proteins in solution. The preliminary investigations of amino-acids and simple peptides (Horsley and Sternlicht, 1968; Horsley et al., 1969, 1970; Sternlicht et al., 1971; Voelter et al., 1971; Gurd et al., 1971; Christl and Roberts, 1972) have been summarized (Stothers, 1972a; Levy and Nelson, 1972; Gray, 1973) and will not receive further discussion. However, in order to assist assignments, chemical shifts of the amino-acids and some simple peptides are given in Tables 12 and 13.

Amino-Acids and Peptides

Peptides containing three or more amino-acid residues have been demonstrated to be better models for protein studies than the amino-acids themselves. This is because the effect of peptide-bond formation on chemical shifts has been included, while other factors such as peptide conformation, solvent effects, ionization of terminal groups, and vicinal electrostatic and steric interactions have remained

TABLE 13

^{13}C Chemical Shifts of Amino-Acids in Protected Tetrapeptides[a]

Amino-Acid Residue	Solvent	δ, ppm[b]						
		C=O	α	β	γ	δ	ϵ	Other
Glycine	DMSO-d$_6$	168·7	42·1					
Alanine	DMSO-d$_6$	171·8	48·0	18·2				
	D$_2$O	175·8	50·8	17·7				
Valine	DMSO-D$_6$	170·4	57·0	30·8	19·0			
					17·0			
Leucine	DMSO-d$_6$	171·7	50·5	41·0	24·0	23·0		
						21·6		
Isoleucine	DMSO-d$_6$	170·6	56·2	37·1	24·2	11·0		
					15·1			
Proline (trans)	DMSO-d$_6$		59·1	29·0	24·1	45·8		
(cis)			58·7	31·7	22·0	46·8		
(trans)	D$_2$O	175·2	61·6	30·6	25·5	48·2		
(cis)			61·3	33·1	23·2	48·8		
Serine	DMSO-d$_6$	169·7	54·8	61·6				
	D$_2$O	172·6	56·6	62·3				
Threonine	DMSO-d$_6$	169·9	58·0	66·6	19·7			
	D$_2$O	172·7	60·2	68·3	20·0			
Aspartic Acid	DMSO-d$_6$	170·3	49·2	36·3	171·5			
	D$_2$O	174·2	52·7	39·8	178·4			
Asparagine	DMSO-d$_6$	170·9	49·3	37·1	171·2			
	D$_2$O	173·1	51·5	37·7	175·6			
Glutamic Acid	DMSO-d$_6$	170·9	51·4	27·6	30·0	173·9		
	D$_2$O	174·8	54·9	28·9	34·6	182·8		
Glutamine	DMSO-d$_6$	170·9	51·7	27·9	31·1	173·5		
	D$_2$O	174·0	54·1	28·1	32·2	179·0		
Lysine	DMSO-d$_6$	171·2	51·8	26·6	22·1	31·5	38·6	
	D$_2$O	174·7	54·4	27·5	23·1	31·8	40·5	
Arginine	DMSO-d$_6$	171·0	51·7	29·5	24·7	40·3		159·3
Cystine	DMSO-d$_6$	169·7	51·6	40·3				
Methionine	DMSO-d$_6$	170·9	51·5	32·0	29·3		14·6	
Phenylalanine	DMSO-d$_6$	170·9	53·6	37·7	137·8	127·9	129·0	126·1
Tyrosine	DMSO-d$_6$	171·0	53·9	36·9	127·9	130·1	114·9	155·5
Histidine	DMSO-d$_6$	170·0	51·8	27·6	130·3	116·8	133·6	
	D$_2$O	172·6	53·7	28·0	130·3	118·7	135·2	
Tryptophan	DMSO-d$_6$	171·2	53·1	27·9	109·7			ϵ_2 135·9
								δ_1 120·5
								ξ_3 123·5
								δ_2 127·2
								ξ_2 111·0
								ϵ_3 117·9
								η_2 118·2

[a] Data of Grathwohl and Wüthrich, 1974. Proton decoupled, natural abundance ^{13}C FTnmr spectra obtained on Varian XL-100 spectrometer at 25.16 MHz on TFA-Gly-Gly-L-X-Ala-OCH$_3$.

[b] Chemical shifts in parts per million relative to TMS.

essentially constant for the different central amino-acid residues (Gurd et al., 1971; Christl and Roberts, 1972; Horsley and Sternlicht, 1968). For these reasons, the ^{13}C chemical shifts, titration properties and relaxation times of pentapeptides, having the following general structure Gly-Gly-X-Gly-Gly, have been studied as models for protein systems (Gurd et al., 1972; Keim et al., 1973a, 1973b). For most aliphatic amino-acids, the difference in chemical shifts (in ppm) on formation of a peptide bond are as follows: carbonyl carbons, -0.2; C-α, -1.0; C-β, 0.2; C-γ, 0.2; C-δ, -0.4 (Keim et al., 1973a). The data are given in Table 12. In an analogous study, the ^{13}C chemical shifts of the 20 common amino-acids in the protected tetrapeptide with the structure TFA-Gly-Gly-L-X-L-Ala-OCH$_3$ have been determined (Table 13; Grathwohl and Wüthrich, 1974). The effect of different amino-acids on the shifts of the neighboring groups in the tetrapeptide was rather small except for proline.

The determination of the tautomeric equilibria for polypeptides containing histidyl residues was based on the observed predominance of the 1-H tautomer of the imidazole ring of L-histidine by ^{13}C nmr (Reynolds et al., 1973a).

Numerous investigations of ^{13}C chemical shifts of amino-acids as functions of pH have appeared in the literature (Stothers, 1972a; Levy and Nelson, 1972; Gray, 1973). In one study, computer analysis has been used to determine accurate pK-values and titration shifts (Quirt et al., 1974). A modified CNDO/2 program directly calculated the ^{13}C chemical shifts which were then used to rationalize the observed spectral changes on deprotonation of the amino-acids (Quirt et al., 1974). Carbon–carbon coupling constants have been determined using 85% ^{13}C-enriched amino-acids (Tran-Dinh et al., 1974). The influence of pH on the ^{13}C chemical shifts and coupling constants have been applied to conformational analysis of the amino-acids and the populations of different rotamers have been determined (Tran-Dinh et al., 1974). Values of T_1 for a few amino-acids in 2 M D$_2$O solutions at or near their isoelectric points have been reported. Additionally, the effect of pH on the T_1-values of the carboxyl carbon has been examined (Armitage et al., 1974). The pH-dependence of ^{13}C chemical shifts of the phenyl moiety in phenylglycine, phenylalanine and selected phenylalanyl dipeptides has been examined as part of investigation into the nature of π-inductive effects (Reynolds et al., 1973b).

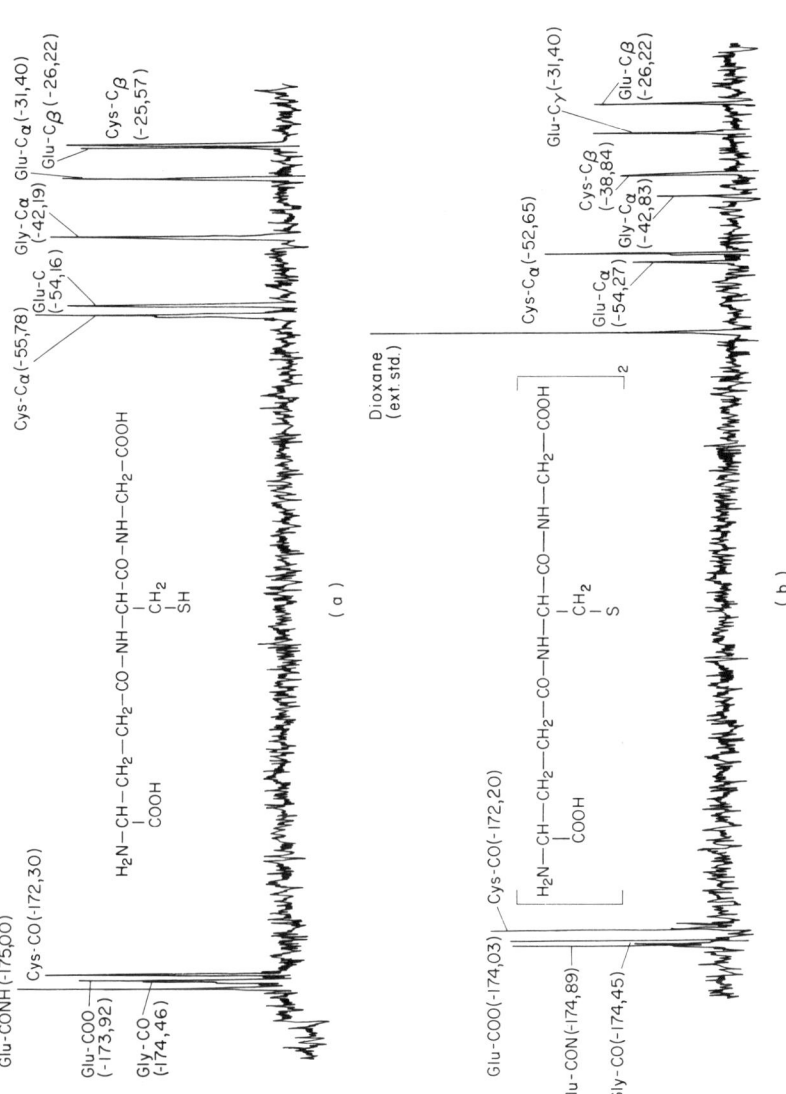

Figure 11. ^{13}C FTnmr spectra of glutathion, 0.3 M in water, pH = 3.52. (a), reduced form; (b), oxidized form. (Taken from Jung et al., 1972.)

TABLE 14

^{13}C Chemical Shifts of Proline Derivatives and Proline in Oligopeptides

Compound	Solvent	C=O	α	β	γ	δ	Other	Reference
Proline (zwitterion *trans*)	10% (w/v) aqueous	175·3	62·2	29·8	24·7	47·1		Dorman and Bovey, 1973
	D$_2$O, pH 1·0		(3·2)	(6·4)	(7·0)	(5·6)		Deslauriers et al., 1974b
	pH 6·4		(3·4)	(5·8)	(6·4)	(4·3)		
	pH 11·3		(2·9)	(5·0)	(5·6)	(4·4)		
N-Formylproline (*cis*)	10% (w/v) aqueous	177·0	60·5	30·2	23·3	45·1	CHO 165·3	Dorman and Bovey, 1973
(*trans*)	10% (w/v) aqueous	176·3	57·8	30·2	24·5	48·1	CHO 164·2	Dorman and Bovey, 1973
N-Acetylproline (*cis*)	10% (w/v) aqueous	177·2	61·8	31·7	23·4	47·5	CO 173·9 CH$_3$ 22·2	Dorman and Bovey, 1973
(*trans*)	10% (w/v) aqueous	177·1	59·9	30·3	25·1	49·3	CO 173·6 CH$_3$ 22·2	Dorman and Bovey, 1973
N-Acetylproline, methyl ester (*cis*)	10% (w/v) aqueous	177·1	61·5	31·8	23·2	47·5	CO 173·6 CH$_3$ 53·7	Dorman and Bovey, 1973
(*trans*)	10% (w/v) aqueous	177·1	59·8	30·0	25·2	49·2	CO 175·6 CH$_3$ 22·1 OCH$_3$ 53·7	Dorman and Bovey, 1973
N-Acetylproline amide (*cis*)	10% (w/v) aqueous	178·1	62·1	32·5	23·3	47·8	CO 173·8 CH$_3$ 22·3	Dorman and Bovey, 1973
(*trans*)	10% (w/v) aqueous	178·1	(1·9) 60·6 (1·8)	(3·4) 30·9 (·32)	(3·2) 24·9 (3·0)	(2·2) 49·4 (2·4)	CO 173·8 CH$_3$ 22·1	Dorman and Bovey, 1973
Glycylproline (*cis*)	10% (w/v) aqueous	179·1	62·5	32·4	23·2	48·1	Gly(CO) 166·4 Gly(α) 41·3	Dorman and Bovey, 1973

	Solvent							Assignments		Reference
(trans)	10% (w/v) aqueous	179·8	62·8	30·4	25·1	47·5		Gly(CO)	165·8	Dorman and Bovey, 1973
								Gly(α)	41·5	
t-Butoxycarbonyl-glycylproline (cis)		175·5	60·2	32·1	23·0	47·6		Gly(CO)	170·4	Torchia, 1974
								Gly(α)	43·6	
								BOC(CO)	158·1	
								C-quat.	80·5	
								CH$_3$	28·5	
(trans)		175·5	60·2	29·8	25·4	47·0		Gly(CO)		Torchia, 1974
								Gly(α)	43·6	
								BOC(CO)	158·1	
								C-quat.	80·5	
								CH$_3$	28·5	
t-Butyloxycarbonyl-glycylproline (cis)	CDCl$_3$	173·70	58·50	31·30	22·10	46·85		Gly(CO)	168·0	Voelter and Oster, 1973
								Gly(α)	42·85	
								BOC(CO)	156·45	
								C-quat.	79·85	
								CH$_3$	28·15	
(trans)	CDCl$_3$	174·15	59·15	28·70	24·50	46·05		Gly(CO)	168·40	Voelter and Oster, 1973
								Gly(α)	42·85	
								BOC(CO)	156·10	
								C-quat.	79·85	
								CH$_3$	28·15	
(cis)	DMSO	173·70	58·25	31·40	22·10	46·85		Gly(CO)	167·65	Voelter and Oster, 1973
								Gly(α)	42·60	
								BOC(CO)	156·00	
								C-quat.	78·20	
								CH$_3$	28·50	
(trans)	DMSO	173·70	58·90	29·00	24·70	45·75		Gly(CO)	167·65	Voelter and Oster, 1973
								Gly(α)	42·60	
								BOC(CO)	156·00	
								C-quat.	78·20	
								CH$_3$	28·50	

TABLE 14—continued

Compound	Solvent	C=O	α	β	γ	δ	Other		Reference
(cis)	H_2O, pH = 7	179·45	61·70	31·50	22·10	47·15	Gly(CO)	169·80	Voelter and Oster, 1973
							Gly(α)	42·40	
							BOC(CO)	157·85	
							C-quat.	81·15	
							CH_3	27·50	
(trans)	H_2O, pH = 7	178·65	61·70	29·25	24·15	46·50	Gly(CO)	169·05	Voelter and Oster, 1973
							Gly(α)	42·40	
							BOC(CO)	157·85	
							C-quat.	81·15	
							CH_3	27·50	
t-Butoxycarbonyl-alanyl proline (trans)	$CDCl_3$	174·15	58·90	28·15	24·70	46·95	Ala(CO)	172·60	Voelter and Oster, 1973
						47·70	Ala(α)	46·95	
								47·70	
							Ala(β)	17·70	
							BOC(CO)	155·25	
							C-quat.	79·75	
							CH_3	28·15	
(trans)	DMSO	173·60	58·70	28·80	24·80	46·50	Ala(CO)	171·35	Voelter and Oster, 1973
						47·70	Ala(α)	46·50	
								47·70	
							Ala(β)	16·95	
							BOC(CO)	155·25	
							C-quat.	78·20	
							CH_3	28·40	
(trans)	H_2O, pH = 7	179·45	61·95	29·35	24·50	47·35	Ala(CO)	173·05	Voelter and Oster, 1973
						48·25	Ala(α)	47·35	
								48·25	

Compound	Solvent						Reference
t-Butoxycarbonyl-proline	CDCl₃	178·35 176·60	58·80	30·75 29·35	24·15 23·50	46·70 46·20	Ala(β) 15·55 BOC(CO) 157·20 C-quat. 81·25 CH₃ 27·50	Voelter and Oster, 1972
	DMSO-d₆	174·55 174·15	58·90	30·55 29·65	24·15 23·40	46·40	BOC(CO) 155·35, 153·95, C-quat. 80·25 CH₃ 28·25 BOC(CO) 153·85 153·30 C-quat. 78·85 CH₃ 28·40	Voelter and Oster, 1972
	CD₃OD	174·70	58·60 58·05	29·90 29·15	23·30 22·65	45·55 45·20	BOC(CO) 153·95 C-quat. 79·50 CH₃ 26·75	Voelter and Oster, 1972
	H₂O	180·95	61·95	30·95	23·65	46·50	BOC(CO) 156·10 C-quat. 81·15 CH₃ 27·85	Voelter and Oster, 1972
Prolylphenylalanine (pH 0·9)	H₂O	170·7	61·2	31·1	25·1	48·2		Christl and Roberts, 1972
Phenylalanylproline (pH 0·9) (pH 5·5) (pH 10·5)	H₂O H₂O H₂O	176·3 179·2 180·1	61·1 63·2 63·2	30·0 32·4 32·4	25·9 23·4 23·6	48·8 47·8 48·0		Christl and Roberts, 1972 Christl and Roberts, 1972 Christl and Roberts, 1972
Phenylalanylprolyl-argine (pH ~ 7)	H₂O	174·0	62·2	27·8	24·4	49·0		Christl and Roberts, 1972
Polyproline I	H₂O	172·7	59·9	32·2	22·5	49·0		Dorman et al., 1973
Polyproline II	H₂O	172·3	59·3	28·7	25·4	48·4		Dorman et al., 1973
Polyproline	H₂O	172·6 (1·6)	59·4 (0·077)	28·8 (0·079)	25·4 (0·109)	48·5 (0·044)		Torchia and Lyerla, 1974
Polyhydroxyproline	H₂O	172·4 (1·5)	58·3 (0·069)	36·8 (0·058)	71·1 (0·113)	55·8 (0·041)		Torchia and Lyerla, 1974

TABLE 14—continued

Compound	Solvent	C=O	α	β	γ	δ	Other	Reference
Poly(Pro-Gly) (*trans*)	H_2O	175·8 (1·5)	61·8 (0·129)	30·8 (0·111)	25·4 (0·118)	48·3 (0·082)	Gly(CO) 170·4(1·5) Gly(α) 43·0(0·067)	Torchia and Lyerla, 1974
(*cis*)	H_2O	175·8 (1·5)	61·8 (0·129)	33·0 (0·111)	23·2 (0·118)	48·3 (0·082)	Gly(CO) 170·8(1·5) Gly(α) 43·0(0·067)	Torchia and Lyerla, 1974
Poly(Hyp-Gly) (*trans*)	H_2O	175·2 (1·5)	60·3 (0·189)	38·3 (0·092)	70·9 (0·177)	55·6 (0·076)	Gly(CO) 170·4(1·6) Gly(α) 43·0(0·069)	Torchia and Lyerla, 1974
(*cis*)	H_2O	175·2 (1·5)	60·3 (0·189)	40·7 (0·092)	69·1 (0·177)	55·6 (0·076)	Gly(CO) 171·0(1·6) Gly(α) 43·0(0·069)	Torchia and Lyerla, 1974
Poly(Gly$_1$-Gly$_2$-Pro$_3$-Gly$_4$) (*trans*)		176·2 (2·7)	62·0 (0·188)	30·4 (0·156)	25·4 (0·188)	48·1 (0·115)	Gly(CO) 170·5(3·4) Gly(α) 42·9(0·105) Gly$_1$(α) 43·7(0·105) Gly$_4$(CO) 173·0(2·2)	Torchia and Lyerla, 1974
		176·2 (2·7)	62·0 (0·188)	33·0 (0·156)	23·0 (0·188)	48·1 (0·115)	Gly(α) 42·9(0·105) Gly$_1$(α) 43·7(0·105) Gly$_4$(CO) 173·0(2·2)	
Glycyl-L-proline diketo-piperazine	DMSO	170·2	58·9	28·8	23·0	45·9		Patel, 1974
L-Prolyl-L-proline diketopiperazine	DMSO	166·8	60·6	28·1	23·8	45·5		Patel, 1974
Cyclo-(Pro-Gly)$_3$	CH_2Cl_2	171·2	60·4	27·7	26·5	47·2		Deber *et al.*, 1972
	DMSO	172·6 171·6 170·8	60·5 59·5	31·8 31·3 29·9	24·8 23·8 22·2	47·2 46·5		Deber *et al.*, 1972
Cyclo-(Pro-Gly)$_3$-Na	DMSO	173·3	59·7	29·3	24·8	46·4		Dorman *et al.*, 1973
Cyclo-(Pro-Gly-Ser)$_2$	H_2O	171·7	62·5	30·3	25·3	47·7		Dorman *et al.*, 1973
Gramicidin S	DMSO		61·0	30·2	25·1	47·5		Gibbons *et al.*, 1970
	CH_3OH		63·3	32·0	27·2	49·3		Gibbons *et al.*, 1970

Compound	Solvent					Reference
Antamanide	CD₃CN		62·1	32·6	26·0	49·3	Patel, 1973b
			61·9	32·0	26·0	48·4	
			60·1	29·9	23·2	47·8	
			59·8	29·6	22·4	47·7	
Antamanide-Na	CD₃CN		59·8	32·7	26·4	49·1	Patel, 1973c
			59·6	32·7	5·7	48·4	
			59·4	29·9	23·5	48·1	
			59·4	29·3	23·1	48·1	
Actinomycin D	1,4-Dioxane-d₈		59·9	32·2	24·0	48·4	Patel, 1974
	DMF		59·1	32·2	23·4	47·9	
Actinomycin D-deoxyguanosine	DMF		59·2	32·1	23·6	47·8	Patel, 1974
					23·4		
MSH-R-IF	D₂O	178·99	61·07	31·63	26·48	47·62	Deslauriers et al., 1973c
			(0·9)	(2·0)	(3·4)	(2·1)	
	DMSO-d₆	175·87	61·69	31·91	27·31	48·20	Deslauriers et al., 1973c
			(0·3)	(0·7)	(0·9)	(0·7)	
Dimethylamide MSH-R-IF	D₂O	178·33	61·12	31·66	26·12	47·63	Deslauriers et al., 1973c
			(0·7)	(1·2)	(1·6)	(1·4)	
	DMSO-d₆	175·39	61·62	32·03	27·31	48·24	Deslauriers et al., 1973c
			(0·8)	(1·0)	(1·4)	(1·1)	
TRH or TRF	pH = 4·0	177·81	61·56	33·08	25·73	49·13	Deslauriers et al., 1973b
				30·75	22·86	48·58	
Angiotensin II	pH = 4·0		61·93	30·35	25·63	49·15	Deslauriers et al., 1973b
					22·94		
	D₂O, pH = 4·5		0·08	0·20		0·11	
LRF	pH = 4·0		61·88	30·32	25·82	49·11	Deslauriers et al., 1973b
			(0·13)	(0·20)	(0·28)	(0·12)	

ᵃ Chemical shifts in parts per million relative to TMS. All chemical shifts corrected to TMS scale using $\delta_{CS_2} = 193·6$ ppm except those of Voelter and Oster (1972) and Deslauriers et al. (1973b, 1973c) which were measured relative to TMS. See cited reference for additional experimental details. Numbers in parentheses are T_1-values in seconds.

The sulfur-containing amino-acids, cysteine and methionine, a few of their derivatives and related compounds and a number of simple peptides containing these residues have been investigated by ^{13}C nmr (Jung et al., 1970, 1972a, b, 1974; Flohe et al., 1972). The ^{13}C nmr spectra of glutathione, a tripeptide composed of glutamic acid, cysteine and glycine, and its oxidized form have been recorded (Fig. 11, Jung et al., 1970, 1972a, 1974) and the ^{13}C chemical shifts have been reported. On going from cysteine to cystine in the oxidized form of glutathione large changes in the chemical shifts of C-α (−3 ppm) and C-β (13 ppm) occurred. The resonance peak of the —CH$_2$SSCH$_2$— moiety (41·6 ppm) has been observed in the ^{13}C nmr spectra of the peptide hormones oxytocin, vasopressin and insulin (Jung et al., 1970). Additionally, the effect of pH on the dissociation of amino-acids and peptides has been determined (Flohe et al., 1972; Jung et al., 1972a, b).

Numerous investigations of peptides containing proline have utilized ^{13}C nmr spectroscopy (Thomas and Williams, 1972; Voelter and Oster, 1972, 1973; Dorman and Bovey, 1973; Zimmer et al., 1972; Bovey, 1972; Deslauriers et al., 1973a, b, c; Gibbons et al., 1970; Pease et al., 1973; Deber et al., 1972; Smith et al., 1972; Lyerla and Freedman, 1972; Walter et al., 1973; Brewster et al., 1973; Wüthrich et al., 1972; Ovchinnikov et al., 1972; Patel, 1973b, 1973c; Voelter et al., 1973b; Torchia and Lyerla, 1974). The basic reason why oligopeptides which contain proline residues have been so extensively studied is that the amino-acid proline, through its pyrrolidine ring, which confers certain conformational restrictions on the system, plays an important role in the determination of the three dimensional structure of oligopeptides and proteins (Dorman and Bovey, 1973). ^{13}C chemical shifts of a number of proline derivatives and of proline residues in oligopeptides have been tabulated in Table 14. These data are useful in clarifying conformational details about the X-Pro bond in cyclo-(Pro-Gly)$_3$ (Deber et al., 1972), cyclo-(Pro-Ser-Gly)$_2$ (Dorman et al., 1974), poly-L-proline (Dorman et al., 1973), antamanide [131] (Patel, 1973b), actinomycin D (Patel, 1974) and oxytocin [118] (Brewster et al., 1973).

The conformational analysis of several linear proline-containing hormones has been considerably aided by ^{13}C nmr studies. The ratio of cis- and trans-isomers and the spin lattice relaxation times of thyrotropin-releasing factor or hormone (TRF or TRH), L-pyroglutamyl-L-histidyl-L-prolinamide, in deuterium oxide, deuterated dimethyl sulfoxide and deuterated pyridine have been determined

(Deslauriers et al., 1973a; Smith et al., 1973b). Luteinizing hormone-releasing factor (LRF), < Glu-His-Trp-Ser-Tyr-Gly-Leu-Arg-Pro-Gly-NH$_2$ (Deslauriers et al., 1973a; Smith et al., 1973b); [5-valine] antiotensin II β-amide, Asn-Arg-Val-Tyr-Val-His-Pro-Phe (Zimmer et al., 1972; Deslauriers et al., 1973b); and melanocyte-stimulating hormone releasing-inhibiting factor (MSH-R-IF or MIF), Pro-Leu-Gly-NH$_2$ (Deslauriers et al., 1973b; 1973c; Smith et al., 1973b) have been studied in a similar fashion. It should be noted that MSH-R-IF cannot exhibit *cis-trans* isomerism because the proline is located in the N-terminal position (Deslauriers et al., 1973b). Data are given in Table 15.

^{13}C Nmr spectroscopy has also been demonstrated to be an effective tool for the characterization of conformational effects in cyclic polypeptides (Bovey, 1972; Wüthrich et al., 1972; Deber et al., 1972; Smith et al., 1972; Deslauriers et al., 1972, 1974a, 1974b). In an attempt to elucidate the structure of the antibiotic viomycin the ^{1}H and ^{13}C nmr spectra were studied (Viglino et al., 1972). The ambiguity of ^{1}H nmr has led to the utilization of ^{13}C spectroscopy for the determination of the preferred conformation for cyclo (Gly-L-Pro-Gly)$_2$ [117a and b] (Pease et al., 1973). The data appear to support a conformer with a type II β-turn as opposed to one with a type II' β-turn (Pease et al., 1973). The effect of cyclization of the

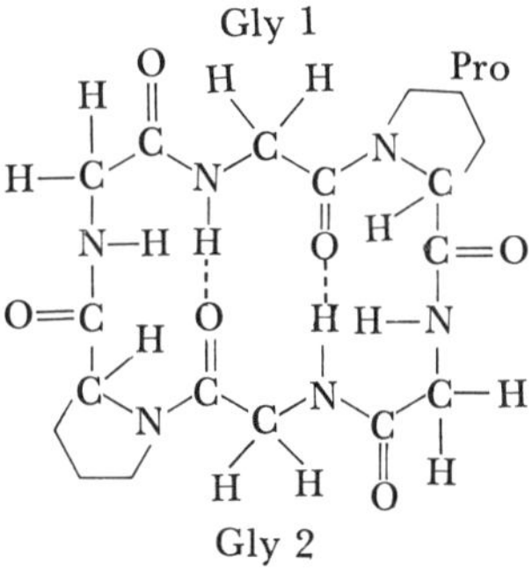

type II β turn
[117a]

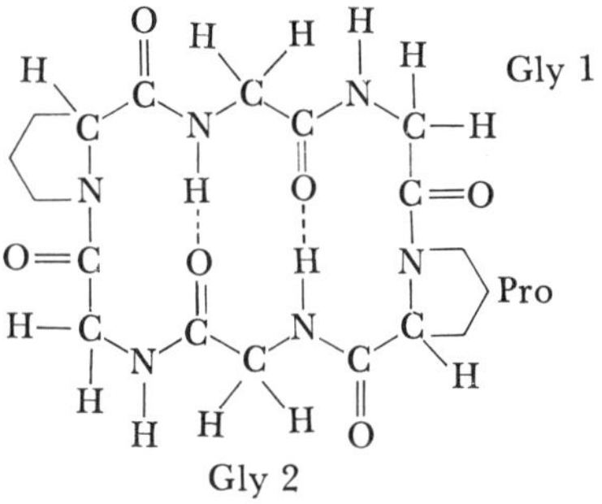

type II' β turn
[117b]

acyclic nonapeptide intermediate can be seen in the changes in the resonances of the α and carbonyl carbon atoms and in a downfield shift in the resonances of the ring portion of oxytocin (Deslauriers et al., 1972). In order to expand the scope of earlier investigations, comparisons have been made of the ^{13}C nmr spectra of oxytocin

TABLE 15

^{13}C Chemical Shifts of Selected Hormones and Antibiotics[a]

	Angiotensin II[b]	Antiotensin II[c]	Val-His-Pro-Phe[b]	LRF[c]	Gramicidin DMSO	SA[120][d] CH$_3$OH	Gramicidin S[e] [120]	Gramicidin DMSO	SA[120][f] CH$_3$OH
Asp C$_\alpha$	55·98								
C$_\beta$	36·13								
C=O	169·81								
	56·2								
	36·78								
	176·49								
Arg C$_\alpha$	54·47	51·65		52·17					
C$_\beta$	29·33	29·47		28·74					
C$_\gamma$		25·45		25·13					
C$_\delta$	41·69	41·73		40·98					
C$_\varepsilon$	157·78								
Val C$_\alpha$	60·14		59·33	61·05		58·0	61·8	60·30	58·0	61·8
C$_\beta$	31·38		31·17	32·89		30·9	32·5	32·04	32·4	33·5
C$_\gamma$	19·52		18·76	19·62		20·1	21·2	19·64	20·1	21·2
	18·92		17·9	17·9		19·2		19·47	19·2	
C=O			170·13	178·11					169·9	173·7
Tyr C$_\alpha$	57·28								
C$_\beta$	37·64								
C$_\gamma$	131·72								
C$_\delta$	129·02								
C$_\varepsilon$	116·45	119·85							
C$_\zeta$	155·56								
His C$_\alpha$	51·61	52·75	51·02	51·77	52·64	53·67			
C$_\beta$	26·96	28·79	27·13	26·85	29·44	28·28			
C$_\gamma$		133·66				129·86			
C$_\delta$	118·88	118·88	118·83	118·77	118·77	118·10			
C$_\varepsilon$	134·74	137·33	134·72	133·66		134·80			
C=O	170·88	172·5	170·76	170·78	172·93	172·64			

Pro C$_\alpha$	61·7	61·92	61·93	61·59	61·81	61·88	61·1	63·4	62·04	61·1	63·4
C$_\beta$	29·81	30·30	30·35	30·41	30·09	30·32	30·3	32·1		30·3	32·1
C$_\gamma$	25·45	25·45	25·63	25·67	25·45	25·82	25·2	27·3	24·53 or 24·70	24·2	26·2
C$_\delta$	49·08	49·08	49·15	49·19	48·97	49·11	47·6	49·4	47·93	47·6	49·4
C=O										169·0	174·6
Phe C$_\alpha$	55·23	57·28		55·44	57·28	51·4	54·1		51·64 or 52·72	55·1	57·3
C$_\beta$	38·07	38·61		37·53	38·5	36·0	38·8		37·39	36·9	38·8
C$_3$	137·65	138·63		137·65	138·52				136·93	135·9	138·1
C$_4$	130·59	130·59		130·53	130·53				130·40	129·1	131·6
C$_5$	129·89	129·67		127·62	129·62				129·66	128·0	131·0
C$_6$	128·27	127·94		128·27	127·94				128·43	126·6	128·8
C=O		178·65		175·96	128·65					170·2	174·6
Orn C$_\alpha$								57·3	55·97	52·3	54·1
C$_\beta$							32·4	33·5	30·62 or 30·93	30·9	32·5
C$_\gamma$							24·2	26·2	24·53 or 24·70	24·2	26·2
C$_\delta$								43·3	40·78 or 42·01	solvent	42·2
C=O										169·3	173·7
Leu C$_\alpha$							50·9	53·0	51·64 or 52·72	50·9	53·0
C$_\beta$							39·6	42·2	40·78 or 42·01	39·6	43·3
C$_\gamma$							24·2	26·2	25·71	25·2	27·3
C$_\delta$							23·7	24·8	23·25	23·7	24·8
C=O										170·8	174·6

[a] Chemical shifts in parts per million relative to TMS; see cited reference for additional details.
[b] Data of Zimmer et al., 1972. Proton decoupled, natural abundance ^{13}C FTnmr spectra obtained in a Bruker HFX 6 spectrometer at 22·628 MHz on 0·075 M angiotensin in D$_2$O and 0·17 M tetrapeptide. Chemical shifts reported at pHs ranging from 1·1 to 12·2.
[c] Data of Deslauriers et al., 1973b. Proton decoupled, natural abundance ^{13}C FTnmr spectra obtained on a Varian LX-100-15 spectrometer at 25·16 MHz on LRF (luteinizing hormone-releasing factor) at pH = 4·0. Chemical shifts obtained relative to TMS in a coaxial tube.
[d] Data of Gibbons et al., 1970. Proton decoupled, natural abundance ^{13}C FTnmr spectra obtained on a spectrometer at 25·15 MHz on samples containing 300–500 mg/2 ml. Chemical shifts corrected to TMS using δ_{CS_2} = 193·7 ppm.
[e] Data of Allerhand and Komoroski, 1973. Proton decoupled, natural abundance on ^{13}C FTnmr spectra obtained on a "home built" spectrometer at 15·18 MHz on sample containing 150 mg/ml CH$_3$OH.
[f] Data of Sogn et al., 1974. Proton decoupled ^{13}C FTnmr spectra on a Bruker HFX-90 spectrometer at 22·63 MHz. Chemical shifts corrected to TMS using δ_{CS_2} = 193·7 ppm.

[118] (Walter et al., 1973; Brewster et al., 1973; Lyerla and Freedman, 1972; Smith et al., 1973b), lysine vasopressin [119] (Walter et al., 1973; Lyerla and Freedman, 1972; Smith et al., 1973b), arginine vasopressin (Walter et al., 1973; Smith et al., 1973b), arginine vasotocin (Walter et al., 1973; Smith et al., 1973b), bacitracin (Lyerla and Freedman, 1972), [4-glycine]-oxytocin (Brewster et al., 1973), deamino-oxytocin (Brewster et al., 1973), and 7-D-proline-oxytocin (Brewster et al., 1973). Insight on the conformational flexibility of the neurohypophyseal hormones oxytocin [118] and lysine-vasopressin [119] has been gained from the T_1-values of their individual carbon atoms (Deslauriers et al., 1974a; Walter et al., 1974). Additionally, the effect of pH on the ^{13}C chemical shifts of oxytocin has been measured (Deslauriers et al., 1974c). The ^{13}C chemical shift data for these molecules are presented in Tables 16 and 17.

Gramicidin S [120], a cyclic decapeptide antibiotic, has been investigated by ^{13}C nmr (Gibbons et al., 1970; Johnson, 1971; Johnson and Jankowski, 1972; Allerhand and Komoroski, 1973; Sogn et al., 1974; Urry, 1974). Values of T_1 for gramicidin S, which

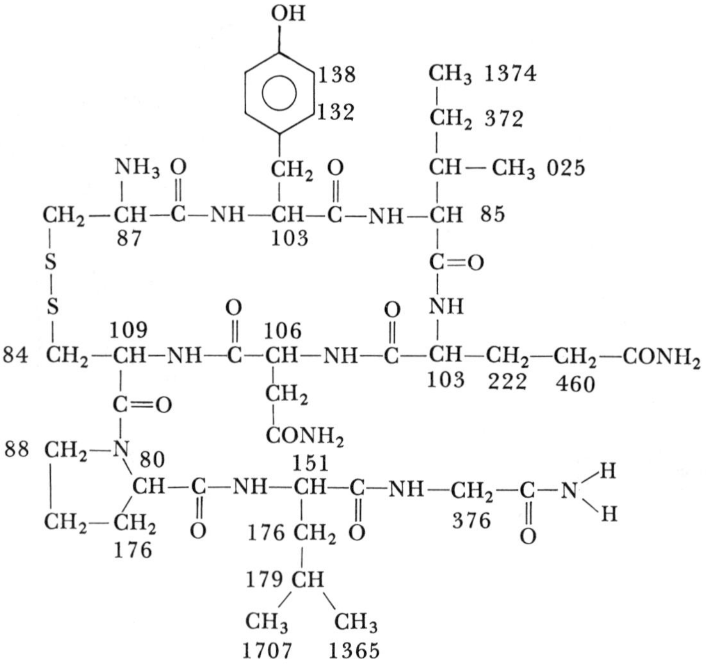

[118] Oxytocin with T_1-values (msec) in aqueous solution

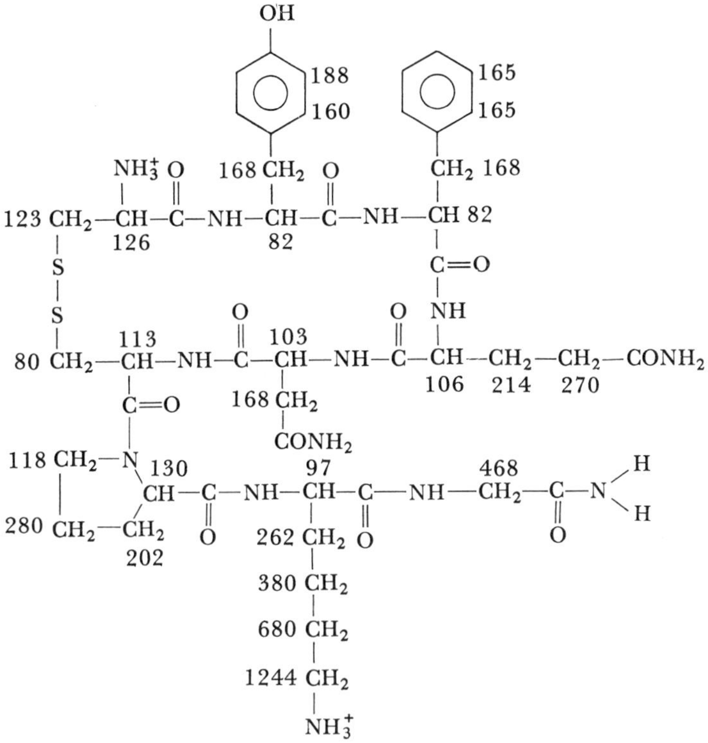

[119] Lysine Vasopressin with T_1-values (msec) in aqueous solution

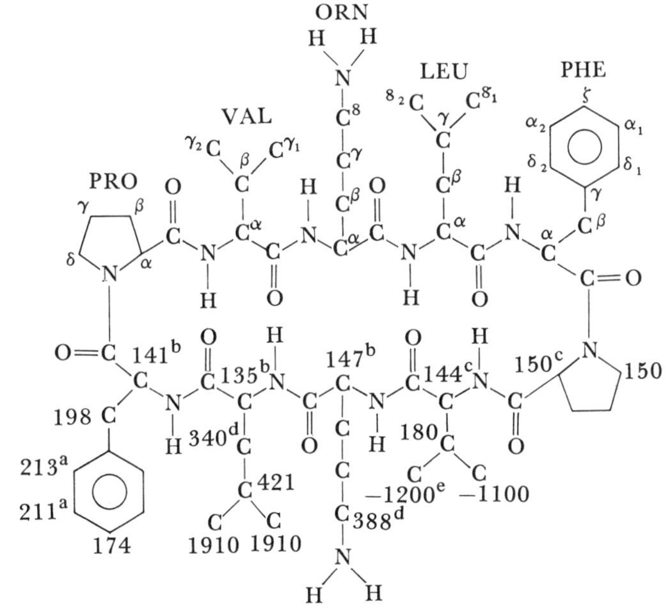

[120] Gramicidin S with T_1-values (msec) in methanol solution

TABLE 16

^{13}C Chemical Shifts of Oxytocin

Assignments		δ, ppm pD					DMSO-d$_6$[b]	pH 5·8[c]
		3·0[a]	4·0[a]	5·9[a]	7·0[a]	8·4[a]		
Gly	C$_\alpha$	43·29	43·30	43·31	43·31	43·34	43·62	43·2
	C=O	175·15	175·15	175·16	175·13			170·5
Leu	C$_\alpha$	53·81	53·81	53·81	53·80	53·79	53·06	53·6
	C$_\beta$	40·49	40·50	40·49	40·53	40·50	41·93	41·0
	C$_\gamma$	25·50	25·50	25·69	25·62	25·55	25·72	25·5, 25·3
	C$_\delta$	23·33	23·34	23·34	23·35	23·35	24·64	23·2
		21·88	21·89	21·89	21·91	21·89	23·13	21·7
	C=O	176·33	176·33	176·31	176·32		173·77	176·0
Pro	C$_\alpha$	61·83	61·84	61·85	61·85	61·85	61·69	61·6
	C$_\beta$	30·40	30·40	30·38	30·37	30·36	30·40	30·2
	C$_\gamma$	25·81	25·82	25·81	25·85	25·81	26·03	25·6
	C$_\delta$	49·07	49·08	49·07	49·06	49·04	48·49	48·9
	C=O	175·43	175·43	175·42	175·40			173·9, 173·6, 172·6
Asn	C$_\alpha$	51·52	51·52	51·54	51·62	51·66	52·07	51·4
	C$_\beta$	37·20	37·21	37·28	37·43	37·57	38·53	37·1, 36·8
	C$_\gamma$	175·69	175·69	175·72	175·77			175·4
	C=O	173·00	175·00	172·97	172·89			175·4
Gln	C$_\alpha$	56·58	56·59	56·28	56·17	56·11	55·64	52·4
	C$_\beta$	27·03	27·03	27·03	27·07	27·06	28·41	26·9

Cγ	32·24	32·24	32·27	32·29	32·31	33·19	32·1
Cδ	178·79	175·79	178·79	178·80			178·5, 175·1, 174·8
Ile Cα	61·11	61·11	60·95	60·66	60·41	60·61	60·8
Cβ	36·96	36·97	36·91	36·84	36·75	37·56	37·1
Cγ	25·50	25·50	25·50	25·53	25·55	26·01	25·5, 25·3
Cγ	16·10	16·16	16·12	16·19	16·24	17·09	15·9
Cδ	11·80	11·80	11·83	11·92	11·97	12·79	11·6
C=O	174·34	174·34	174·27	174·23			174·0
Tyr Cα	56·31	56·30	56·53	56·33	56·22	56·24	56·4, 56·1
Cβ	37·20	37·21	37·28	37·43	37·31	37·36	37·1, 36·8
C_1	129·02	129·02	129·11	129·27	129·37	129·45	128·7
C_2	131·67	131·67	131·68	131·71	131·73	131·64	131·3
C_3	116·84	116·84	116·81	116·77	116·76	116·53	116·6
C_4	155·80	155·84	155·79	155·75	155·80	157·39	155·5
C=O	173·78	173·78	173·93	174·89			175·1, 174·8
Cys-1 Cα	52·52	52·53	52·62	52·84	53·00	54·93	56·4, 56·1
Cβ	39·54	39·54	39·75	40·09	40·27	41·61	40·3, 39·4
C=O	170·89	179·89	170·91	170·99		169·40	169·7
Cys-1 Cα	53·45	53·45	53·81	54·54	55·14	45·69	53·6
Cβ	40·73	40·63	41·25	42·76	43·70	41·41	40·3, 39·4
C=O	168·86	168·86		174·23			173·9, 173·6, 172·6

[a] Data of Walter et al., 1973. Proton decoupled, natural abundance ^{13}C FTnmr spectra obtained on a Varian XL-100-15 spectrometer at 25·16 MHz and 37°. Chemical shifts are relative to external TMS.
[b] Data of Deslauriers et al., 1972. Proton decoupled, natural abundance ^{13}C FTnmr spectra obtained on a Varian XL-100-15 spectrometer at 25·16 MHz and 37°. Chemical shifts are relative to external TMS.
[c] Data of Lyerla and Freedman, 1972. Proton decoupled, natural abundance ^{13}C FTnmr spectra obtained on a Varian XL-100-15 spectrometer at 25·2 MHz and 31 ± 3°. Chemical shifts corrected to TMS scale using δCS_2 = 193·5 ppm.

TABLE 17

^{13}C Chemical shifts of Cyclic Peptides and Precursors[a]
(Numbered formulae are on pp. 365 and 370)

		[121][b]	[122][b]	[123][b]	[124][b]	[125][b]	[126][b]	Lys VP[c] [119]	Lys VP[d] [119]	Arg VP[c]	Arg Oxy-tocin[c]	Deamino ala Oxytocin[c]	Deamino lys VP[c]	Bacitracin[d]
Gly	C_α	43·60	43·60	43·60	43·60	43·72	43·60	43·23	43·1	43·20	43·22	43·31	43·22	
	C=O	172·46	172·46	172·46	172·46	172·50	172·50	175·07	170·4	175·17	175·02	175·15	175·06	
Leu	C_α	53·06	53·06	53·06	53·06	53·06	53·06					53·77		53·2, 53·7
	C_β	41·73	41·81	41·81	42·01	41·93	41·93					40·44		41·1
	C_γ	25·84	25·85	25·83	25·83	25·83	25·91					25·49		25·1
	C_δ	24·52	24·52	24·52	24·52	24·52	24·52					23·35		21·9
		23·13	23·13	23·13	23·13	23·13	23·13					21·83		23·1
	C=O	173·65	173·65	173·69	173·65	173·65	173·89					176·33		174·7
Pro	C_α	61·49	61·49	61·61	61·61	61·61	61·61	61·78	61·5	61·80	61·87	61·89	61·80	
	C_β	30·41	30·48	30·48	30·48	30·48	30·29	30·50	30·3	30·42	30·45	30·40	30·47	
	C_γ	26·03	26·03	26·03	26·03	26·03	25·91	25·86	25·6	25·85	25·83	25·76	25·87	
	C_δ	48·49	48·57	48·57	48·57	48·57	48·57	49·11	48·9	49·09	49·08	49·13	49·10	
	C=O	173·13	173·13	173·17	173·1	173·13	173·01	175·42	173·7, 173·5, 173·8, 173·1	175·36	175·43	175·44	175·47	
Cys-6	C_α	54·13	52·27	52·34	52·36	52·46	52·27	52·28	55·9	52·25	52·41	53·40	53·26	78·2
	C_β	37·36	37·24	37·24	37·24	37·36	37·24	39·64	40·4, 39·7	30·56	39·36	36·24	36·78	
	C=O	171·06	170·63	170·59	170·59	170·63	170·51	170·85	170·0	170·88	170·89	171·09	170·92	172·0, 172·9
Asn	C_α		53·35	51·27	51·27	51·35	51·35	51·34	51·2	51·34	51·64	51·84	51·40	52·0
	C_β		39·03	38·63	38·63	38·63	38·75	37·57	37·2, 36·9	37·58	37·16	37·85	38·13	37·4
	C_γ		175·85	173·01	172·85	173·01	172·50	175·62	175·3	175·04	175·67	175·89	175·94	175·6
	C=O		175·85	173·01	172·85	172·81	172·89	173·34	175·3, 173·1, 172·8, 173·7, 173·5	137·37	173·18	173·33	173·05	176·2, 175·6
Gln	C_α			56·12	53·74	53·74	53·85	56·14	52·2	56·18	56·43	56·24	56·01	
	C_β			29·29	29·61	29·61	29·61	26·93	26·8	26·89	26·97	26·93	27·03	
	C_γ			33·07	32·99	33·07	33·07	32·21	32·1	32·19	32·19	32·29	32·26	
	C=O			175·52	175·56	175·76	175·76	178·92	178·5	178·86	178·80	178·94	178·94	
	C=O			172·34	172·30	172·30	172·30	174·18	178·5, 175·1, 174·9	174·19	174·20	174·33	174·23	
Ile	C_α				60·89	58·50	58·67				61·32	59·89		58·4, 59·2, 59·4
	C_β				38·04	38·24	38·16				37·02	36·88		36·8
	C_γ				26·03	26·03	25·91				25·59	25·49		25·3
	C_δ				16·97	16·97	16·97				16·03	16·30		15·8, 15·5, 15·1
					12·53	12·60	12·60				11·77	12·07		12·0, 11·3, 11·1
	C=O				172·54	172·81	172·38				174·42	174·33		174·3, 174·0, 173·8, 173·5, 172·0, 172·9

α	57·99	55·52	56·48	56·4	56·40	56·43	55·74
C_β	38·24	38·16	37·57	37·2, 36·9	37·58	37·16	36·62
C4	158·59	157·39	155·78	155·4	155·77	155·80	155·56
C2	131·83	131·76	131·68	131·3	131·69	131·66	131·51
C1	128·97	129·17	128·99	128·6	128·97	129·02	129·46
C3	115·94	116·45	116·75	116·5	116·75	116·85	116·52
C=O	173·25	173·13	173·34	175·1, 174·9	137·37		174·00
Cys-1 C_α		50·27	53·54	53·8	53·48	53·31	
C_β		37·04	40·92	40·4, 39·7	41·02	40·93	
C=O		171·74	168·72	173·7, 173·5	168·69	168·86	
				173·1, 172·8			
Benzyl-oxycarbonyl-C_1	138·51	138·5	138·58	138·59			
C_2	129·52	129·97	129·97	129·97			
C_3	128·38	129·25	129·25	129·17			
C_4	128·18	128·36	128·38	128·38			
CH_2	67·25	67·05	66·85	67·17			
Benzyl CH_2	38·38	33·98	34·96	34·96	34·06		
Benzyl aromatic	128·18	128·38	128·38	128·38			
	139·98	140·10	140·10	140·10			
	130·44	130·56	130·56	130·56			
	129·96						
Arg C_α					54·66	54·64	
C_β					29·00	29·01	
C_γ					25·56	25·83	
C_δ					41·66	41·67	
C_ϵ							
C=O					175·17	175·24	
Lys C_α					54·91	54·7	53·2, 53·7
C_β					27·35	27·3	31·4
C_γ					23·21	23·0	22·9
C_δ					31·24	31·1	28·3
C_ϵ					40·43	41·5	40·1
C=O					175·52	173·7, 173·5,	174·3, 174·0, 173·8,
						172·8, 173·1	173·5
Phe C_α					56·78	56·3	56·0
C_β					37·11	37·5	37·8
C1					137·42	137·1	137·3
C2					130·13	129·9, 129·7	129·7, 130·0
C3					130·30	129·9, 129·7	129·7, 120·0
C4					128·48	128·1	128·1
C=O					173·98	174·7	174·3, 174·9, 173·8,
							173·5

TABLE 17—continued

		[121][b]	[122][b]	[123][b]	[124][b]	[125][b]	[126][b]	Lys VP[c] [119]	Lys VP[d] [119]	Arg VP[c]	Arg Oxytocin[c]	Deamino ala Oxytocin[c]	Deamino lys VP[c]	Bacitracin[d]
Ala	C_α											51·02		
	C_β											17·82		
	C=O											176·59		
MPr	C_α												39·88	
	C_β												34·46	
	C=O													
Glu	C_α											40·73		54·5
	C_β											35·16		27·3
	C_γ													31·9
	C=O											175·44		178·4, 174·5
Asp	C_α													52·3
	C_β													38·8
	C_γ													175·6
	C=O													175·6, 172·9, 172·0
Orn	C_α													53·2, 53·7
	C_β													28·3
	C_γ													24·4
	C_δ													39·8
	C=O													
His	C_α													55·0
	C_β													27·7
	C2													134·6
	C4													118·3
	C5													129·1
	C=O													

[a] Chemical shifts in parts per million relative to TMS; see cited reference for additional experimental details.
[b] Data of Deslauriers et al., 1972. Proton decoupled ^{13}C FTnmr spectra obtained on a Varian XL-100-15 spectrometer at 25·16 MHz and 37° in DMSO-d$_2$. [121] = benzyloxycarbonyl-Cys(benzyl)-Pro-Leu-Gly-NH$_2$; [122] = benzyloxycarbonyl-Asn-Cys(benzyl)-Pro-Leu-Gly-NH$_2$; [123] = benzyloxycarbonyl-Gln-Asn-Cys(benzyl)-Pro-Leu-Gly-NH$_2$; [124] = benzyloxycarbonyl-Ile-Gln-Asn-Cys(benzyl)-Pro-Leu-Gly-NH$_2$; [125] = benzyloxycarbonyl-Try(benzyl)-Ile-Gln-Asn-Cys(benzyl)-Pro-Leu-Gly-NH$_2$; [126] = benzyloxycarbonyl-Cys(benzyl)-Try-Ile-Gln-Asn-Cys(benzyl)-Pro-Leu-Gly-NH$_2$.
[c] Data of Walter et al., 1973. Proton-decoupled ^{13}C FTnmr spectra obtained on a Varian XL-100-15 spectrometer at 25·16 MHz and 37°. Conditions: Lys VP [lysine(8) vasopressin], pD 3·9, D$_2$O, 40 mg/ml; Arg VP [arginine(8) vasopressin], pD 3·6.D$_2$O; Arg oxytocin [arginine(8) oxytocin], pD 3·5, D$_2$O, 38 mg/ml; Deamino ala Oxytocin [deamino alanine(2) oxytocin], pD 3·5, D$_2$O; Deamino lys VP [deamino lysine(8) vasopressin], pD 3·5; D$_2$O.
[d] Data of Lyerla and Freedman, 1972. Proton-decoupled ^{13}C FTnmr spectra obtained on a Varian XL-100-15 spectrometer at 25·2 MHz and 31 ± 3°. Conditions: Lys-VP [lysine-vasopressin (acetate form)], 0·09M, pH 5·8; Bacitracin, 0·07M, pH 3·6. Chemical shifts corrected to TMS scale using δCS_2 = 193·5 ppm.

are comparable to those of lysine vasopressin and oxytocin, are useful in obtaining rates of internal rotation of the side chains (Allerhand and Komoroski, 1973). The complete and unambiguous assignment of all the carbon resonances of gramicidin S-A has been reported (Sogn et al., 1974), utilizing selective biosynthetic enrichment techniques. ^{13}C Chemical shift data for gramicidin S are given in Table 15.

Native and Denatured Proteins

^{13}C Nmr investigations of proteins and/or polypeptide systems can be classified into three general categories: (1) ^{13}C chemical shifts and spin lattice relaxation times of native and specifically labelled or ^{13}C-enriched proteins and subunits; (2) helix-coil transitions of polypeptides; and (3) binding studies, protein-substrate interactions and complex formation.

Natural abundance ^{13}C nmr spectra of hen egg-white lysozyme (Lauterbur, 1970; Chien and Brandts, 1971; Allerhand et al., 1973a; Freedman et al., 1973; Bradbury and Norton, 1973), bovine pancreatic ribonuclease A (Allerhand et al., 1970, 1971a; Glushko et al., 1972), a few RNase residues (Freedman et al., 1973), gelatin (Chien and Wise, 1973), oxidized and reduced horse-heart cytochrome C (Oldfield and Allerhand, 1973), elastin in unstretched calf ligamentum nuclae (Torchia and Piez, 1973), collagen in insoluble calf achilles tendon in equilibrium with 2·0 M KCNS (Torchia and Piez, 1973), horse hemoglobin (Conti and Paci, 1971), rabbit hemoglobin (Moon and Richards, 1972a), human hemoglobin (Conti and Paci, 1971; Moon and Richards, 1972a), sperm whale myoglobin (Conti and Paci, 1971) and carboxymethylated harbor seal apomyoglobin (Nigen et al. 1973a) have been recorded. The ^{13}C chemical shifts and the available spin lattice relaxation times have been given in Table 18. The ^{13}C spectra of enriched ferredoxin (Packer et al., 1972), specifically enriched tryptophan synthetase α subunit (Browne et al., 1973a, b), selectively enriched α-lytic protease (Hunkapiller et al., 1973a, b), enriched ^{13}C-containing analogue forms of semisynthetic ribonuclease-S (Chaiken et al., 1973), specifically labeled ribonuclease (1-15) peptide (Freedman et al., 1971; Chaiken et al., 1973), and myoglobins carboxymethylated with enriched bromoacetate (Nigen et al., 1972, 1973b) have also been studied.

The helix-coil transitions of poly(N-δ-carbobenzoxy-L-ornithine) (PCBO) (Boccalon et al., 1972), poly(γ-benzyl-L-glutamate) (PBLG) (Paolillo et al., 1972; Allerhand and Oldfield, 1973; Bradbury et al., 1973), poly-L-methionine (PLM) (Tadokoro et al., 1973) poly-L-aspartate polymers (Bradbury et al., 1973) in deuterochloroform-trifluoroacetic acid (TFA) systems and poly-L-glutamic acid (PGA) (Lyerla et al., 1973), poly-L-lysine (Saito and Smith, 1973) and copoly(L-glutamic acid, L-lysine, L-alanine) (Bradbury et al., 1973) in aqueous solution have been studied using ^{13}C nmr spectroscopy. Chemical shift data and spin lattice relaxation times are in Table 20. Additionally, using ^{13}C-enriched glycine, ^{13}C nmr has been used effectively for monitoring the change in conformation, i.e., helix formation, in going from the isolated and disordered semisynthetic ribonuclease-S-fragment to the folded semisynthetic complex (Chaiken, 1974). One of the earliest references in the literature on the applicability of ^{13}C spectroscopy for monitoring helix-coil transitions of polypeptides was the study of PCBO (Boccalon et al., 1972). The changes in ^{13}C chemical shifts of the carbons in PCBO relative to a solvent mixture containing 0·01 volume fraction TFA were plotted versus the percentage TFA in solution (Fig. 12). The δ and aromatic carbons of PCBO were rather insensitive to the amount of TFA in solution and therefore were not plotted. Changes in the ^{13}C chemical shifts for the helix-coil transition of PBLG and T_1-values of the helix have been determined (Paolillo et al., 1972). ^{13}C Spin lattice relaxation times, T_1, spin-spin relaxation times, T_2, and nuclear Overhauser enhancements, NOE, for the α-carbons of PBLG of various molecular weights have been used to study transitions from rigid to flexible forms of this polymer (Allerhand and Oldfield, 1973). Effective rotational correlation times, τ_{eff}, calculated from T_1- and NOE-values, for the α-carbons were 24–32 nanoseconds for the helical form and approximately 0·8 nanoseconds for the random coil (Allerhand and Oldfield, 1973). The transition from the α-helix to the random-coil of PLM causes the resonances of the α- and carbonyl carbons to move upfield 2·3 and 3·4 ppm respectively (Tadokoro et al., 1973), which is consistent with results obtained for PBLG and PCBO. Further work is required before the reasons for the chemical shift differences between the corresponding carbons in the helical and random-coil forms in deuterochloroform-TFA systems can be elucidated. Plots of chemical shifts and relaxation times vs. pH have been used to study the helix-coil transition of poly-L-lysine hydrochloride in aqueous solution (Saito and Smith,

1973). Additionally, T_1 data for smaller oligomers of lysine have also been measured (Saito and Smith, 1974). Similarly, titration curves have been applied to the study of helix-coil transitions in aqueous solutions of poly-L-glutamic acid (Lyerla et al., 1973).

A number of protein-substrate interactions and binding studies have also been examined using ^{13}C nmr spectroscopic techniques. Interactions of ^{13}C enriched pyruvates with Mn(II), Mn(II)-pyruvate carboxylase and Mn(II)-pyruvate kinase have been studied using the

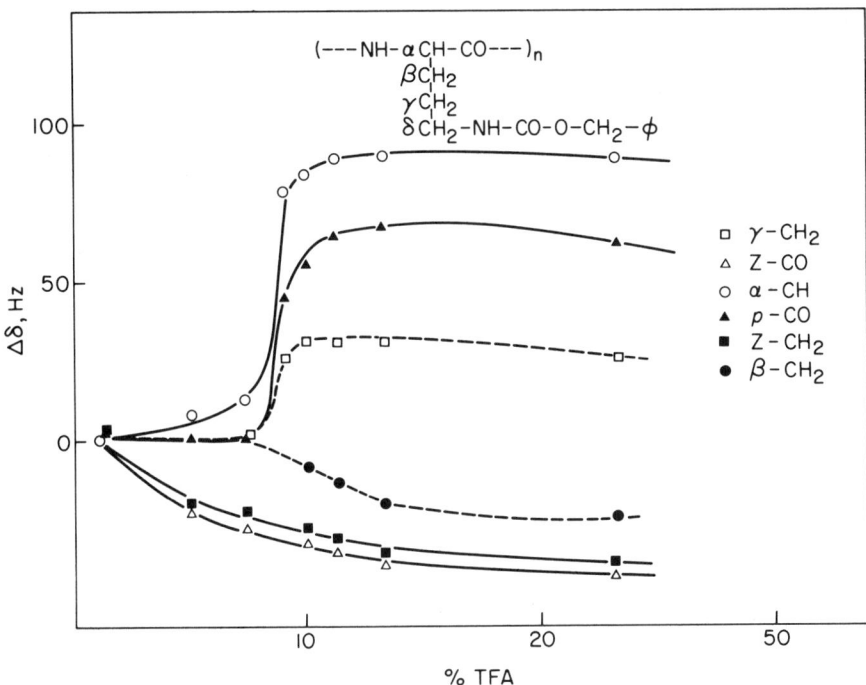

Figure 12. Relative shifts of the various carbon atoms of poly(N-δ-carbobenzoxyl-L-ornithine) in different mixtures of $CDCl_3$-TFA. The shifts are referred to their value in a mixture containing 0·01 volume fraction of TFA. (Taken from Paolillo et al., 1972.)

longitudinal ($1/T_1$) and transverse ($1/T_2$) relaxation rates of the enriched carbons (Fung et al., 1973; Mildvan et al., 1973). Distances of 7·1 to 8·5 Å between the Mn atom and the pyruvate carbon atoms in the Mn(II) metallo-enzyme-pyruvate complexes have been calculated using the estimated correlation times, ~5 x 10^{-9} sec, from the frequency dependence of the ^{13}C longitudinal relaxation rates (Fung et al., 1973; Mildvan et al., 1973). Interaction of pyruvate with transcarboxylate, a metallobiotin enzyme containing Co(II), Zn(II) and Cu(II) which catalyses the reversible transfer of the carboxyl group of (S)-methylmalonyl-CoA to pyruvate to form propionyl-CoA and oxalacetate, has been studied by ^{13}C spectroscopy (Fung et al., 1974). In a similar type of study, the binding of ^{13}C-enriched methyl-D-glucopyranosides to transition metal derivatives of concanavalin A has been investigated (Brewer et al., 1973a, b, c). The α-anomer was found to be bound to the protein in C-1 chair conformation with its 3- and 4-carbons closest to the metal ion at an average distance of 10 Å (Brewer et al., 1973a). The α- and β-anomers are bound in different orientations while retaining their C-1 chair conformations (Brewer et al., 1973b, c). This accounts for the difference in binding constants of the saccharides. Spin lattice relaxation times of α-methyl-D-glucopyranoside in the presence of transition metal derivatives of concanavalin A have been reported (Brewer et al., 1973c; Villafranca and Viola, 1974). The conformation of N-acetyl-L-tryptophan in its complex with chymotrypsin has also been studied using ^{13}C spectroscopy (Rodgers and Roberts, 1973).

^{13}C Spectroscopy has also been used to study the interaction of Mn(II) and bicarbonate with D-ribulose 1,5-diphosphate carboxylase (Miziorko and Mildvan, 1974). The binding of manganese by ribulose 1,5-diphosphate carboxylase is dependent upon the bicarbonate concentration. The effects of enzyme and Mn(II) on the longitudinal and transverse ^{13}C relaxation rates of $H^{13}CO_3^-$ were determined. A distance of 5·4 ± 0·1 Å for the Mn(II)-^{13}C separation was calculated using the appropriate form of the Solomon–Bloembergen equation, i.e.,

$$r = 512\, (T_{1M} f(\tau_c))^{1/6}$$

where $f(\tau_c)$ is the correlation function and T_{1M} is the relaxation time in the presence of metal.

Several ^{13}C nmr studies on the binding of enriched carbon monoxide to various hemoglobins (Hb) have been made (Conti and Paci, 1971; Moon and Richards, 1972a, b; Matwiyoff et al., 1973; Antonini et al., 1973; Vergamini et al., 1973; Shulman et al., 1973) in order to investigate the possibility of different interactions of the hemoglobin subunits with ligands. The nature of these interactions should be reflected in the binding affinities of the ligands at the active site and the conformations of the subunits. The ^{13}C chemical shift data for the HbCO complexes are given in Table 19. Two resonances, which are located near 206 ppm from TMS, were

TABLE 18
13C Chemical Shifts of Proteins[a]

Tentative Assignment		Gelatin[b]	Ribonuclease[c]	Myoglobin[d]	Lysozyme[e]	Denatured lysozyme[e]	Lysozyme[f]	α-Lactalbumin[f]	α-Chymotrypsin[f]
Glycine	C_α	43·2 (55)	43·7				43·3	41·0	43·5
Alanine	C_α	50·4 (120)	50·9				53·8-50·3	53·7-47·9	51·0-50·3
	C_β	17·2 (120)	17·2 (118)	17·5 (211)			16·5	16·9	17·4
Serine	C_α	55·9	56·4-53·9 (35)				56·1-56·0	55·5	56·1
	C_β	61·9	61·9-60·1				61·5	62·9-60·4	61·8
Cystine	C_α		56·4-53·9 (35)				53·8-50·3	53·7-47·9	51·0-50·3
	C_β		37·0				36·5-30·0	36·3	36·7-30·8
Phenylalanine	C_α	55·9	56·4-53·9 (35)				56·1-56·0	55·5	56·1
	C_β	37·6	37·0				36·5-30·0	37·3	36·7-30·8
	C_γ		137·2	137·2 (346)	138·7, 138·4, 138·0, 137·7, 137·4, 136·1	137·2	136·7	136·0	136·4
	C_{ϕ_1} (C_{ϵ_1}, C_{ϵ_2}, C_{δ_1}, C_{δ_2}, C_ζ)	136·7, 129·8, 129·1, 127·6	129·4 (93)	129·7 (101)	130·3, 129·8, 129·3, 128·7, 128·2, 127·5, 126·8, 126·7, 126·4	129·7	129·3	129·3	129·5
							128·9	128·9	129·1
			128·0 (93)	128·0 (78)			127·3	127·2	127·4
Histidine	C_α	54·2 (92)	56·4-53·9 (35)				53·8-50·3	53·7-47·9	54·7-53·1
	C_β	26·9	27·1	27·3 (127)			26·7	27·1-26·7	27·0
	C_γ		129·4	131·4 (412)					
	C_{δ_2}		118·2	121·4					
	C_{ϵ_1}		134·5	138·2 (86)				135·0	134·0
Aspartic acid	C_α	52·1	56·4-53·9 (35)				53·8-50·3	53·7-47·9	51·0-50·3
	C_β	37·6	37·0				36·5-30·0	36·3	36·7-30·8
	C_γ	177·8-177·4	177·2	178·2					
Asparagine	C_α	52·1	56·4-53·9 (35)				53·8-50·3	53·7-47·9	51·0-50·3
	C_β	37·6	37·0				36·5-30·0	36·3	36·7-30·8
	C_γ	177·8-177·4	177·2	177·9					
Valine	C_α	60·0-59·4 (84)	61·9-60·1 (35)				60·0-59·1	59·0	60·0
	C_β	30·8	32·0-30·9	31·2			36·5-30·0	31·2-29·7	36·7-30·8
	C_{γ_1}	19·0	19·6	19·7			18·7	18·7	19·1
	C_{γ_2}	18·2	19·6				18·0	18·1	18·6

TABLE 18—continued

Tentative Assignment		Gelatin[b]	Ribonuclease[c]	Myoglobin[d]	Lysozyme[e]	Denatured lysozyme[e]	Lysozyme[f]	α-Lactalbumin[f]	α-Chymotrypsin[f]
Threonine	C_α	60·0–59·4 (84)	61·9–60·1 (35)				60·0–59·1	59·0	60·0
	C_β	67·6	67·9 (38)	19·7			67·0	67·4	67·4
	C_γ	19·4	19·6 (148)				19·2	19·0	19·6
Methionine	C_α	54·2 (92)	56·4–53·9				53·8–50·3	53·7–47·9	54·7–53·1
	C_β	30·8	32·0–30·9				36·5–30·0	31·2–29·7	36·7–30·8
	C_γ	29·9	32·0–30·9				36·5–30·0	31·2–29·7	36·7–30·8
	C_ε	15·2	15·3				14·7	14·8	13·0
Glutamic acid	C_α	54·2 (92)	56·4–53·9 (35)	29·0 (127)			53·8–50·3	53·7–47·9	54·7–53·1
	C_β	28·6	28·7 (S)	34·5 (81)			28·3	27·9	27·0
	C_γ	34·1 (72)	32·0–30·9	181·2			36·5–30·0	33·7	36·7–30·8
	C_δ	180·9	178·2						
Glutamine	C_α	54·2 (92)	56·4–53·9 (35)				53·8–50·3	53·7–47·9	54·7–53·1
	C_β	26·9 (84)	27·1				26·7	27·1–26·7	27·0
	C_γ	30·8	32·0–30·9				36·5–30·0	31·2–29·7	36·7–30·8
	C_δ	177·8–177·4	177·2						
Leucine	C_α	54·2 (92)	56·4–53·9 (35)	40·3 (182)			53·8–50·3	53·7–47·9	54·7–53·1
	C_β	40·1 (84)	40·3	25·5 (110)			39·9	39·7	40·1
	C_γ	25·0	25·3	23·3 (171)			27·7	24·5	25·0
	C_{δ_1}	22·5	22·9	21·9 (286)			22·6	22·5	22·8
	C_{δ_2}	21·4	21·7 (S)				21·4	20·9	21·4
Isoleucine	C_α	60·0–59·4 (84)	61·9–60·1				60·0–59·1	59·0	60·0
	C_β	37·6	37·0	25·5			36·5–30·0	36·3	36·7–30·8
	C_{γ_1}	25·0		15·7			24·7	24·5	25·0
	C_{γ_2}	16·2	15·8				15·2	14·8	15·3
	C_{δ_1}	11·2	11·2				10·8	10·7	11·1
Arginine	C_α	54·2 (92)	56·4–53·9 (35)	29·0			53·8–50·3	53·7–47·9	54·7–53·1
	C_β	28·6	28·7				28·3	27·9	27·0
	C_γ	25·0	25·3				24·7	24·5	25·0
	C_δ	41·3 (84)	41·1	41·4	41·4		41·1	41·0	
	C_ζ	157·4 (1400)	157·7 (35)	157·7		157·7	157·3	156·8	
Lysine	C_α	54·2 (92)	56·4–53·9	31·2 (55)			53·8–50·3	53·7–47·9	54·7–53·1
	C_β	30·8 (150)	32·0–30·9 (66)	23·3 (171)			36·5–30·0	31·2–29·7	36·7–30·8
	C_γ	22·5 (115)	22·9 (100)	27·3 (127)			22·6	22·5	22·8
	C_δ	26·9	27·1 (188)	40·3 (182)			26·7	27·1–26·7	27·0
	C_ε	40·1 (280)	40·3 (278)				39·9	39·7	40·1

C_β	29.9 (80)	32.0–30.9				31.8
C_γ	25.0 (80)	25.3		36.5–30.0		36.7–30.8
C_δ	47.6 (92)			24.7	31.2–29.7	25.0
Tyrosine C_α	55.9	56.4–53.9 (35)		56.1–56.0	24.5	
C_β	37.6	37.0		36.5–30.0	55.5	56.1
C_γ		128.8 (S)	128.7	128.2	36.3	36.7–30.8
					127.9	127.4
C_{δ_1}, δ_2		131.2 (49)	130.3, 129.8,			
			129.3, 128.7,			
			128.2, 127.5,	130.6	130.5	130.8
$C_{\epsilon_1}, \epsilon_2$		116.4 (44)	126.8, 126.7,	116.0	115.6	115.9
			126.4			
C_ζ		155.5	156.1, 154.1	154.9	154.6	
			155.6			
Hydroxy proline C_α	60.0–59.4 (84)					
C_β	37.6					
C_γ	70.4 (84)					
C_δ	55.9					
Tryptophan C_γ			109.9			
			112.3, 111.6,			
			110.5, 109.9,			
			108.5			
C_δ			127.9			
			130.3, 129.8,			
			129.3, 128.7,			
			128.2, 127.5,			
			126.8, 126.7,			
			126.4			
C_δ			138.7, 138.4,			
			138.0, 137.7,			
			137.4, 136.1			
Carbonyls	175.7–168.8 (650)	178.7–170.7	175.8–172.7	175.6	181.1–171.8	177.7–170.4

[a] Chemical shifts in parts per million relative to TMS. T_1-values in milliseconds are given in parentheses.
[b] Data of Chien and Wise, 1973. Proton decoupled natural abundance ^{13}C FTnmr spectra obtained on a Bruker HFX-100 spectrometer at 22.63 MHz and 31° on 15% (w/v) gelatin in 0.2 M KCl at pH = 7.0. Chemical shifts corrected to TMS scale using δCS_2 = 193.7 ppm.
[c] Data of Glushko et al., 1972. Proton decoupled, natural abundance ^{13}C FTnmr spectra of ribonuclease A obtained on a "home built" spectrometer at 15.074 MHz and 45 ± 3° on 19 mM ribonuclease A at pH = 1.46. Chemical shifts corrected to TMS scale using δCS_2 = 193.7 ppm.
[d] Data of Nigen et al., 1973a. Proton decoupled, natural abundance ^{13}C FTnmr spectra of modified harbor seal apomyoglobin obtained on a "home built" spectrometer at 15.08 MHz and 33–35° on 8 mM denatured harbor seal apomyoglobin prepared by carboxymethylation in the presence of 8 M urea, pH = 6.89. Chemical shifts corrected to TMS scale using δCS_2 = 193.7 ppm.
[e] Data of Allerhand et al., 1973. Proton decoupled, natural abundance ^{13}C FTnmr spectra of native and denatured egg white lysozyme obtained on a "home built" spectrometer at 15.18 MHz on 25% (w/v) native lysozyme in 0.1 M NaCl at pH = 4.0 and 40° and on 25% (w/v) denatured lysozyme in 0.1 M NaCl and 6.2 M guanidinium chloride at pH = 4.3 and 45°.
[f] Data of Bradbury and Norton 1973. Proton decoupled, natural abundance ^{13}C FTnmr spectra of lysozyme, α-lactalbumin and α-chymotrypsin obtained on a Bruker HX-90 spectrometer at 22.63 MHz on 15–30″ lysozyme in water, 75° and pH 2.8; on 15–20% α-chymotrypsin, 6 M guanidine-HCl, 25°, and pH 2.7; and on 15–20% α-lactalbumin, 6 M guanidine HCl, 25° and pH 7.4. Chemical shifts corrected to TMS scale using δCS_2 = 192.8 ppm.

observed on formation of HbCO complexes and were assigned to binding with either α- or β-subunits (Moon and Richards, 1972b). These peaks were pH-independent over the ranges given in Table 19. Values of T_1 for the bound carbon monoxide of ~0·3 sec were recorded for a 3 mM solution of rabbit HbCO at pH 7·0 (Moon and Richards, 1972b). The intensity of the low-field signal of rabbit HbCO in the presence of oxygen decreased more rapidly than the high-field signal, indicating different binding affinities of the subunits. Kinetic studies using the stopped-flow method have demonstrated that CO will replace oxygen faster in the β-subunit than in the α-subunit in many hemoglobins (Olson et al., 1971). With this information the low field signal was tentatively assigned to CO bound to the α-subunit (Moon and Richards, 1972b). Data from ^{13}C nmr and ir spectroscopy have been used to demonstrate that three different sites for CO binding can exist in rabbit blood (Matwiyoff et al., 1973). The chemical shifts due to the binding of CO with α- and β-subunits of hemoglobin were also identified using isolated α- and β-chains (Antonini et al., 1973; Vergamini et al., 1973). In addition, Hb M_{IWATE}, a variant of human hemoglobin containing a ferric ion in the heme units of the α-chains which do not bind CO, was used to confirm the assignment for the CO–β-subunit complex (Vergamini et al., 1973). Similar studies of red blood cells suspensions treated with carbon monoxide, carbon dioxide or cyanide (Matwiyoff and Needham, 1972) and human hemoglobin treated with carbon dioxide and bicarbonate-carbonate (Morrow et al., 1973) have also been reported. More recently, the effect of pH, in the range 6·7–8·2, on the carbamylation of the N-terminal valine α-amino- and the lysine ε-amino-groups of human adult and sickle cell carboxyhemoglobins has been compared using ^{13}C spectroscopy (Moon et al., 1974). The carbamylation of α-amino-groups of N-terminal valine is strongly pH-dependent while that of the lysine ε-amino-group is rather insensitive to change in pH. The enzyme carbonic anhydrase has also been studied (Feeney et al., 1973; Koenig et al., 1973). The signal of the cyanide carbon in the inhibitor complex formed by adding cyanide to bovine carbonic anhydrase B was 74·4 ppm, which was centrally located between the shifts of hydrogen cyanide (45·3 ppm) and cyanide ion (98·3 ppm) (Feeney et al., 1973). Rate and equilibrium constants for the enzymatic reaction of human carbonic anhydrase B with enriched carbon dioxide have been determined from the line widths of carbon dioxide and bicarbonate ion resonances (Koenig et al., 1973).

TABLE 19

^{13}C Chemical Shifts of ^{13}CO in Carbonyl Hemoglobins

pH	Hemoglobin	δ, ppm[a] α	β	Reference
6·6–7·3	Isolated Chains (7 mM heme)[b]	206·98	206·60	Vergamini et al., 1973
6·6–7·3	Hemoglobin A (7 mM heme)[b]	207·04	206·61	Vergamini et al., 1973
6·6–7·3	Hemoglobin A (20 mM heme)[b]	207·15	206·77	Vergamini et al., 1973
6·6–7·3	Hemoglobin M$_{\text{IWATE}}$[b]		206·36	Vergamini et al., 1973
6·6–7·3	Hemoglobin A (1/2 in the form of met Hemoglobin)[b]	206·98	206·39	Vergamini et al., 1973
6·6–7·3	Hemoglobin A (7 mM heme and 8 M in Urea)[b]	207·18	206·69	Vergamini et al., 1973
6·79–7·49	Sperm whale Myoglobin[c]	207·69		Moon and Richards, 1972b
5·42	Sperm whale Myoglobin[c]	207·41		Moon and Richards, 1972b
6·35–7·90	Adult human Hemoglobin[c]	206·54	206·06	Moon and Richards, 1972b
6·90	Fetal human Hemoglobin[c]	206·54	206·10	Moon and Richards, 1972b
6·21	Mouse Hemoglobin[c]	206·38	205·94	Moon and Richards, 1972b
6·94–7·38	Rabbit Hemoglobin[c]	207·96	205·98	Moon and Richards, 1972b
7·0	Hemoglobin[d]	207·41	206·95	Antonini et al., 1973
7·0	Isolated chains[d]	207·50	206·72	Antonini et al., 1973

[a] Chemical Shifts in parts per million relative to TMS.
[b] Proton decoupled ^{13}C FTnmr spectra obtained on a Varian XL-100-15 spectrometer operating at 25·2 MHz. 90% ^{13}C-enriched CO was used to carboxylate the hemoglobin.
[c] Spectra recorded on a Varian XL-100-15 spectrometer operating at 25·15 MHz. Concentrations varied from 2 to 5 mM in 0·1 M NaCl. CO was 90% enriched. Chemical shifts corrected to TMS scale using δ CS$_2$ = 192·8 ppm.
[d] Proton decoupled ^{13}C FTnmr spectra obtained on a Varian XL-100 spectrometer operating at 25·15 MHz. Concentrations were 15 mM for all samples. CO was 90% enriched. Chemical shifts corrected to TMS scale using δ CS$_2$ = 193·7 ppm.

TABLE 20

13C Chemical Shifts for the Helix-Coil Transition of Polyamino Acids

Compound	Solvent	C_α	C_β	C_γ	C_δ	C_ϵ	C=O amide	Other	Reference
Poly-L-lysine hydrochloride (37°)[b]	D_2O	55·3	31·6	23·8	30·7	41·1			Saito and Smith, 1973
(60°)[b]	D_2O	55·1	32·0	23·7	30·5	41·2			Saito and Smith, 1973
(90°)[b]	D_2O	55·6	32·4	23·7	30·6	41·3			Saito and Smith, 1973
PBLG[c,d]	3% TFA/97% CDCl$_3$	56·3 (0·03 ± 0·01)	25·9 (0·03 ± 0·01)	30·7 (0·03 ± 0·01)	170·5 (1·93 ± 0·1)		173·1 (0·9 ± 0·1)	Bz 65·9 (0·1 ± 0·02) (0·264) $C_{2,6}$ 126·6 (0·81) 127·1 ± 0·05) C_1 133·6 (3·3 ± 0·1)	Paolillo et al., 1972
PBLG[c,d]	29% TFA/71% CDCl$_3$	53·2 (0·035) (0·052)	27·1 (0·065) (0·085)	30·6 (0·108) (0·139)	173·2		171·1	Bz 67·6 (0·419) $C_{2,6}$ 126·6 127·1 C_1 133·6	Paolillo et al., 1972
Poly-met[e]	10% TFA/90% CDCl$_3$	56·78 (0·11)	31·06 (0·09)	31·06 (0·09)	—	16·07 (1·01)	177·22 (0·95)		Tadokoro et al., 1973

Sample	Solvent							Reference
Poly-met[e]	35% TFA/65% CDCl$_3$	54.23 (0.14)	30.45 (0.13 ±0.02)	30.45 (0.13 ±0.02)	—	15.27 (1.06 ±0.01)	173.94 (0.91)	Tadokoro et al., 1973
		56.48 (0.11)	30.82 (0.09 ±0.03)	30.82 (0.09 ±0.03)	—	15.49 (0.84 ±0.03)	177.34 (1.18)	
			30.99 (0.12 ±0.02)	30.99 (0.12 ±0.02)				
Poly-met[e]	65% TFA/35% CDCl$_3$	54.11 (0.08)	30.88 (0.06)	30.15 (0.12)	—	14.80 (0.60)	173.94 (0.90)	Tadokoro et al., 1973
PGA[f]	pH = 8.0	54.66	28.95				174.55	Lyerla et al., 1973
PGA[f]	pH = 4.7	56.47	26.94				176.34	Lyerla et al., 1973

[a] All chemical shifts in parts per million relative to TMS. T_1 values in seconds are given in parentheses.
[b] Spectra were obtained on a Varian XL-100-15 spectrometer with Fourier transform and decoupling. Chemical shifts relative to external TMS. Poly-L-lysine hydrochloride MW = 17 000, 100 mg/ml, pD = 11.2.
[c] Poly-(γ-benzyl-L-glutamate), PBLG, spectra obtained on a Varian XL-100-15 spectrometer at 25.2 MHz with Fourier transform and decoupling. Chemical shifts relative to internal TMS. Polymer concentration 15% w/v, 30°. T_1 measurements taken on a Jeol PS-100 spectrometer at an ambient temperature on 30% w/v polymer solutions. PBLG (S 416), DP ~ 100.
[d] Allerhand and Oldfield (1973) reported T_1 measurements on a home built spectrometer operating at 15.18 MHz at 40 ± 2°. Helical samples were 0.8 M in 97% CDCl$_3$/3% TFA (v/v), random coil samples were 0.6 M in 71% CDCl$_3$/29% TFA (v/v).
[e] Poly-L-methionine, Poly-met, spectra obtained on a Jeol PS-100 spectrometer at 25.15 MHz. Polymer concentration was 20% (w/v).
[f] Poly-L-glutamic acid, PGA, spectra obtained using a Varian XL-100-15 spectrometer; PGA MW = 30 000.

5. LIPIDS

Lipids are one of the major structural components in living cells and encompass a myriad different types of compound. In most cases, they possess only one common property, which is their solubility in non-polar solvents such as chloroform, ether, hydrocarbons or alcohols and insolubility in water (Davenport and Johnson, 1971). Some of the types of compounds that are classified as lipids include fatty acids, glycerides, waxes, cholesterol esters, phosphoglycerides and sphingolipids.

Lipids are seldom found free in nature and they are invariably combined with carbohydrates or proteins, which then form macro-molecular bio-complexes (Gurr and James, 1971). Before achieving an understanding of such systems, the chemical and physical properties of the simpler lipid components need to be investigated in order to determine how lipids interact with other molecules.

Nuclear magnetic resonance has been shown to be a most effective method for the study of lipid chemistry (Chapman, 1965; 1972; Henrikson, 1971). With the advent of commercially available fast Fourier transform spectrometers, high resolution natural abundance ^{13}C spectra and relaxation times of lipids have become relatively commonplace. Utilization of these ^{13}C nmr techniques has yielded a considerable amount of information concerning the mobility and organization of lipids in liquid crystals and membranes (Oldfield and Chapman, 1971). ^{13}C Chemical shifts of lipids are given in Table 21. The rest of this discussion will be devoted to the interpretation of these results.

Phospholipids

The majority of ^{13}C nmr investigations of lipids have been concentrated on phospholipids. Phospholipids have been defined as any lipid that contains a radical derived from phosphoric acid. The reason why phospholipids have generated so much interest is that phospholipid bilayers are structural components of biological membranes (Engelman, 1970; Wilkins et al., 1971). Studies of model membrane systems using ^{13}C Fourier transform nmr have been made (Sears, 1971). Assignments of the ^{13}C chemical shifts of egg yolk lecithin (EYL) (Oldfield and Chapman, 1971; Keough et al., 1973), EYL-cholesterol (Keough et al., 1973), sphingomyelin (Keough et al., 1973) and dipalmitoyl-lecithin (DPL) (Oldfield and Chapman, 1971; Birdsall et al., 1972b; Metcalfe et al., 1973a) have been reported. ^{13}C Spectra for the phospholipids 1-stearoyl-2-linoleoyl-3-glycerophosphorylcholine, 1,2-distearoyl-3-glycerophosphorylcholine, enriched N-methyl phosphatidylcholine and sphingomyelin have also been recorded and the peaks have been fully assigned (Stoffel et al., 1972); data are given in Table 21. In addition, ^{13}C relaxation times of lecithins have been measured (Metcalfe et al., 1971, 1973a, b; Levine et al., 1972a, b; Metcalfe, 1972; Lee et al., 1973). Spin lattice-relaxation times, T_1, which are inversely proportional to correlation times for molecular motion, can be used as a measure of molecular motion in lipids without chemically altering these compounds (Levine et al., 1972a). Since chemically labeled groups are not necessary for ^{13}C spin lattice relaxation times, this technique is better suited to studies of molecular motion than ESR, which requires spin-labeled compounds.

The T_1-values for the side chains of DPL in deuterochloroform or sonicated DPL in deuterium oxide buffer were found to increase from the central glycerol moiety toward both the ends of the fatty acid chains and the trimethyl quaternary ammonium group of the choline (Metcalfe et al., 1971). This means that molecular motion becomes more restricted in moving from a methylene further from the glycerol skeleton to one that is closer to this backbone. Additionally, the ^{13}C nmr spectrum of human erythrocyte membranes (10% w/w) in D_2O buffer at 28° was also obtained (Metcalfe et al., 1971).

In a continuation of work on membranes, the ^{13}C nmr spectra of membranes from *Acholeplasma laidlawii* grown on ^{13}C-enriched palmitic acid were recorded (Metcalfe et al., 1972; 1973a; Metcalfe, 1972). The enhanced carboxyl resonance was the only peak observed under conditions where no natural abundance ^{13}C signals were measurable and the temperature exceeded the thermal transition temperature of the lipids in the membranes. The ability to reduce the ^{13}C nmr spectra of membranes to a few sharp resonances by incorporating ^{13}C-enriched lipids biosynthetically allows the T_1-values to be obtained.

The temperature dependence of the intensities and relaxation times of ^{13}C resonances in the spectra of sonicated vesicles of DPL has also been demonstrated (Fig. 13, Levine et al., 1972b; Metcalfe, 1972; Metcalfe et al., 1973a). Similar behavior has been observed in the ^{1}H nmr spectra of lipid-water systems and membrane systems

TABLE 21

^{13}C Chemical Shifts of Lipids

δ, ppm[a]

Compound	Solvent	C=O	Choline Me$_3$N$^+$	Choline CH$_2$N	Choline CH$_2$O	Glycerol 1CH$_2$O	Glycerol CHO	Glycerol 3CH$_2$O	2	3	–(CH$_2$)$_n$–	–CH$_2$CH$_2$CH$_3$	CH$_2$CH$_3$	CH$_3$	CH=CH	Other	Reference
Dipalmitoyl lecithin[b,c]	D$_2$O	174·8	55·3	67·3	60·7	64·4	72·0	64·4	35·4	26·0	31·0	33·1	23·7	14·9			Birdsall et al., 1972b
	CDCl$_3$	174·2	55·2	67·2	60·3	63·9	71·5	64·4	35·4	25·9	30·7	32·9	23·7	15·0			Birdsall et al., 1972b
		173·8							35·3								
	CHCl$_3$	173·6	54·6	66·6	59·9	63·3	71·3	63·3	34·6	25·2	30·0	32·2	22·9	14·0			Oldfield and Chapman, 1971
		173·2															
	CD$_3$OD	173·7	54·0	66·8	59·7	63·0	71·1	64·4	35·4	25·1	30·1	32·4	23·1	14·0			Birdsall et al., 1972b
		173·4							35·3								
sn-Glycero-3-phosphorylcholine[b]	D$_2$O		55·2	67·1	60·5	63·2	71·7	67·6									Birdsall et al., 1972b
1,2-sn-Dipalmitoyl glycerol[b]	CDCl$_3$	174·1				63·2	73·1	62·4	35·4	25·8	30·6	32·9	23·7	15·1			Birdsall et al., 1972b
		173·7							35·3								
Dipalmitoylphosphatidic acid[b]	CDCl$_3$	174·2				63·5	70·9	64·8	35·4	25·9	30·7	32·9	23·7	15·0			Birdsall et al., 1972b
		173·8							35·3								
Glycerol 3-phosphate[b]	D$_2$O, pD 8·0					63·6	72·4	66·0									Birdsall et al., 1972b
	D$_2$O, pD 4·3					63·4	72·0	67·2									Birdsall et al., 1972b
	D$_2$O, pD 0					63·2	71·6	68·7									Birdsall et al., 1972b
Methyl palmitoleate[d]	CDCl$_3$	174·0							33·8	24·8	28·9	31·6	22·5	13·8	129·7	OMe 51·2	Burlingame et al., 1972
											29·5					C-4 27·0	
											30·7						
Sphingomyelin[e]	D$_2$O	177·2	55·0	67·4	60·8						30·3		23·1	14·4	131·2		Keough et al., 1973
Egg yolk lecithin[e]	D$_2$O	175·5	54·4	66·9	59·3						30·6		23·1	14·3	128·0		Keough et al., 1973
Egg yolk lecithin-cholesterol[e]	D$_2$O	175·5	54·3	66·5	60·4								22·5	14·3	131·1		Keough et al., 1973
Mitochondrial membrane[e]	D$_2$O	177·0	54·7	66·2	61·4						29·9		23·0	14·4	128·5		Keough et al., 1973
Canine sciatic nerve[f]	solid	171·8	54·6			62·2	69·3	62·2	34·0	25·2	30·2	32·4	23·0	14·3	129·9	27·5	Williams et al., 1973
		171·5									29·7				128·3		
Stearic acid[g]	CCl$_4$/CDCl$_3$	180·7							34·4	25·1	29·7	32·3	23·1	14·2		C-15 29·5	Stoffel et al., 1972

Compound	Solvent									Reference
Oleic acid[g]	CCl₄/CDCl₃	180·0	33·9	24·5	29·0–29·6	31·8	22·6	14·1	129·4 129·7	C-8, 11 27·0	Stoffel et al., 1972
Linoleic acid[g]	CCl₄/CDCl₃	180·0	33·9	24·4	28·9–29·4	31·4	22·4	14·0	129·7 129·6 129·9 129·5	C-8, 14 27·0 C-11 25·5 C-15 29·1	Stoffel et al., 1972
α-Linolenic acid[g]	CCl₄/CDCl₃	180·0	33·2	24·4	28·9–29·4		20·4	14·1	127–131	C-8 27·0 C-11, 14 25·4	Stoffel et al., 1972
Arachidonic acid[g]	CCl₄/CDCl₃	180·0	33·2	24·3		31·4	22·4	14·1	127–130	C-4 26·3 C-7 25 C-10, 13 25 C-17 29·2	Stoffel et al., 1972
Triacetyl-4t-sphinganine[g]	CCl₄/CDCl₃		50·6	73·4	29·5–28·9	31·8	22·5	14·0	C-4 136·7 C-5 124·5	C-1 62·3 C-6 32·1	Stoffel et al., 1972
Diacetyl-3-dehydro-sphinganine[g]	CCl₄/CDCl₃		57·4	204·6	29·5–28·9	31·8	22·5	14·0	C-4 39·6 C-5 23·2	C-1 62·8	Stoffel et al., 1972
N-Acetyl-3-dehydro-sphinganine[g]			60·9	206·4	29·6–29·4	31·8	22·5	14·0		C-1 63 C-4 40·0 C-5 23·2	Stoffel et al., 1972
Sphinganine[b]			50–51	69·4	29·7–28·8	31·9	22·5	14·0		C-1 63 C-5 23–24	Birdsall et al., 1972b
Choline bromide[h]	D₂O		54·9	68·3	56·5						Behr and Lehn, 1972
Acetyl choline[h]	D₂O	175·1	56·2	66·7	60·6						Behr and Lehn, 1972
	CD₃OD		55·2	66·4	59·4						Behr and Lehn, 1972
Choline[h]	D₂O		56·3	69·8	58·0						Behr and Lehn, 1972

[a] Chemical shifts in parts per million relative to TMS; see cited references for additional details.
[b] Proton decoupled, natural abundance ¹³C FTnmr spectra obtained on a Varian XL-100-15 spectrometer at 25·2 MHz. Chemical shifts corrected to TMS scale using $\delta_{\text{dioxane}} = 67·4$ ppm.
[c] Data of Oldfield and Chapman (1971) obtained on a Bruker spectrometer. Chemical shifts corrected to TMS scale using 194 ppm.
[d] Instrumentation described elsewhere; see original paper for references. Chemical shifts corrected to TMS scale using $\delta\, CS_2 = 192·8$ ppm.
[e] Proton decoupled ¹³C FTnmr spectra obtained on a JEOL FFT-100 spectrometer at 25·152 MHz.
[f] Proton decoupled, natural abundance ¹³C FTnmr spectra obtained at 15-18 MHz. Chemical shifts corrected to TMS scale using $\delta\, CS_2 = 192·8$ ppm.
[g] Proton decoupled ¹³C FTnmr spectra obtained on a Bruker HFX-90 spectrometer at 22·628 MHz.
[h] Proton decoupled ¹³C FTnmr spectra obtained on a Varian XL-100-15 spectrometer.

(Chapman and Chen, 1972). The proposed explanation for the temperature dependence of peak intensities is that the fatty acid chains of the lipids undergo gradual crystallization or melting in the vesicles during the thermal transition (Levine et al., 1972b). Proton

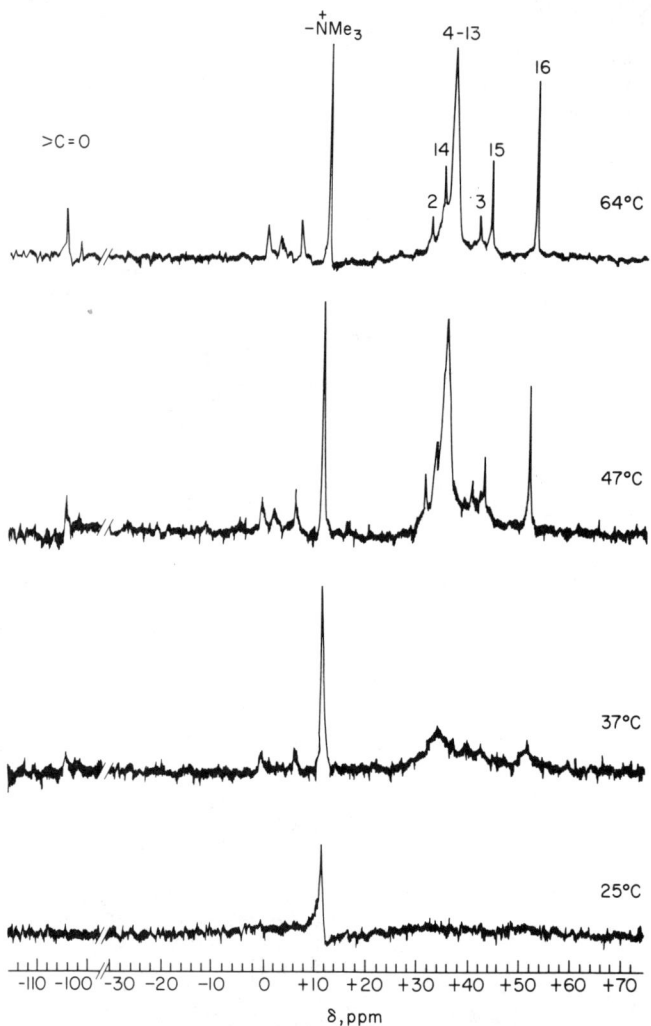

Figure 13. ^{13}C FTnmr spectra of sonicated DPL, 0·23M in D_2O buffer as a function of temperature (taken from Levine et al., 1972b).

nmr studies (Lee et al., 1972) and x-ray diffraction data on the thermal transition (Luzzati, 1968; Engelman, 1970) also support this reasoning. A change in conformation of the polar choline group occurs when the fatty acid portions of the lipids crystallize. This conformational change is indicated by the inflection point in the curve of temperature vs T_1 (Fig. 14; Levine et al., 1972b).

Additional applications of ^{13}C spectroscopy have been developed in an analysis of the conformation of lecithin in vesicles (Batchelor et al., 1972). It has been demonstrated that the ^{13}C resonances of alkanes will shift upfield 5 ppm when a γ-methyl is aligned with the methylene in a *gauche* configuration about the a–β bond (Cheney and Grant, 1967). This means that changes in chemical shift with temperature or other conditions can be used as a measure of the

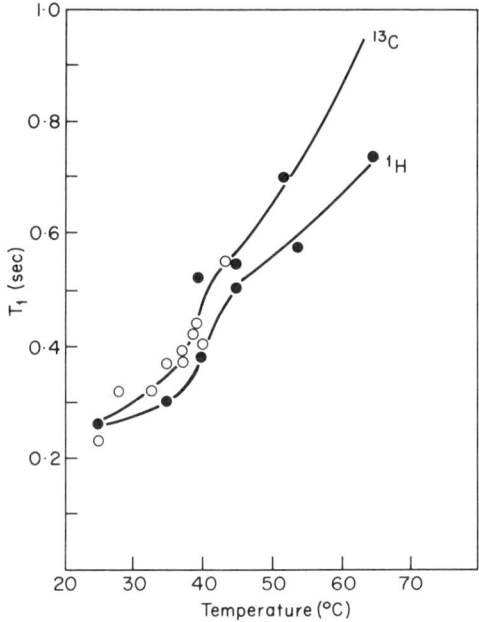

Figure 14. The temperature dependence of T_1 for the ^{13}C and ^{1}H nuclei of the $N^+(CH_3)_3$ group in sonicated DPL in D_2O (taken from Levine et al., 1972b).

populations of *gauche* and *trans* rotamers at any carbon along the chain. Further experimental work needs to be done before actual quantitative results of conformer ratios can be obtained.

Values of T_1 and activation energies of motion for the alkyl tail carbons of phospholipids in vesicles and multilayers have been determined (Gent and Prestegard, 1974). A change in configuration from *trans-trans-trans* to *gauche*$^+$-*trans-gauche*$^-$ agreed most reasonably with all the factors governing the motion that determined T_1.

The effect of nitroxide spin-labeled lipids on ^{13}C spin-lattice relaxation times of lecithin vesicles has been discussed (Levine et al.,

1972a; Godici and Landsberger, 1974). Information on the local environment about the stearic acid spin labels in lipid bilayers has been determined from T_1 measurements. This can easily be seen in plots of $1/T_{1N}$ vs chain position for EYL vesicles labeled with stearic acid spin labels (Fig. 15, Godici and Landsberger, 1974).

Miscellaneous Lipids

Some other classes of lipids in addition to phospholipids have also been studied by ^{13}C nmr spectroscopy. These lipids include fatty acids, both saturated and unsaturated (Stoffel et al., 1972; Batchelor et al., 1973), fatty acid esters (Burlingame et al., 1972; Batchelor et al., 1973), sphingolipids (Stoffel et al., 1972) and lipoproteins (Hamilton et al., 1973). ^{13}C Spectra of lipids in sarcoplasmic reticulum membranes (Robinson et al., 1972) and intact bovine retinal rod segments (Millet et al., 1973) as well as canine sciatic nerve (Williams et al., 1973a) have been recorded.

^{13}C chemical shifts of stearic, oleic, linoleic, α-linolenic and arachidonic acids have been assigned (Stoffel et al., 1972) and the data are presented in Table 21. Electric field effects associated with the dipolar head groups have been shown to be effective in influencing the ^{13}C chemical shifts of unsaturated fatty acids (Batchelor et al., 1973). This electric field effect (Horsley and Sternlicht, 1968, McFarlane, 1970), predicted to be linear and dependent on distance and orientation, can be applied to problems of conformational analysis. A nondestructive technique utilizing ^{13}C nmr to determine the relative concentrations of the unsaturated fatty acids present in soyabeans has recently been developed (Schaefer and Stejskal, 1974).

A number of sphingolipids have been studied by ^{13}C nmr (Stoffel et al., 1972) and all the carbon resonances have been well characterized. Attempts to record the ^{13}C spectra of gangliosides, which are glycosphingolipids containing sialic acids, ended in failure because of difficulties arising from line width problems (Behr and Lehn, 1973). However, the ^{13}C Fourier transform nmr spectrum of sialic acid (N-acetylneuraminic acid or NANA) in water has been recorded along with the changes in chemical shift accompanying the addition of excess calcium chloride (Behr and Lehn, 1973).

The applicability of ^{13}C nmr to human serum lipoproteins has been demonstrated (Hamilton et al., 1973, 1974). Spectra and partial peak assignments for high-density (HDL), low-density (LDL), and very-low-density lipoproteins (VDL) have been reported (Hamilton et al., 1973). Values of T_1 for several individual carbon resonances of the lipids of human plasma HDL, LDL and VDL have been reported (Hamilton et al., 1974). Relaxation times indicate that, in com-

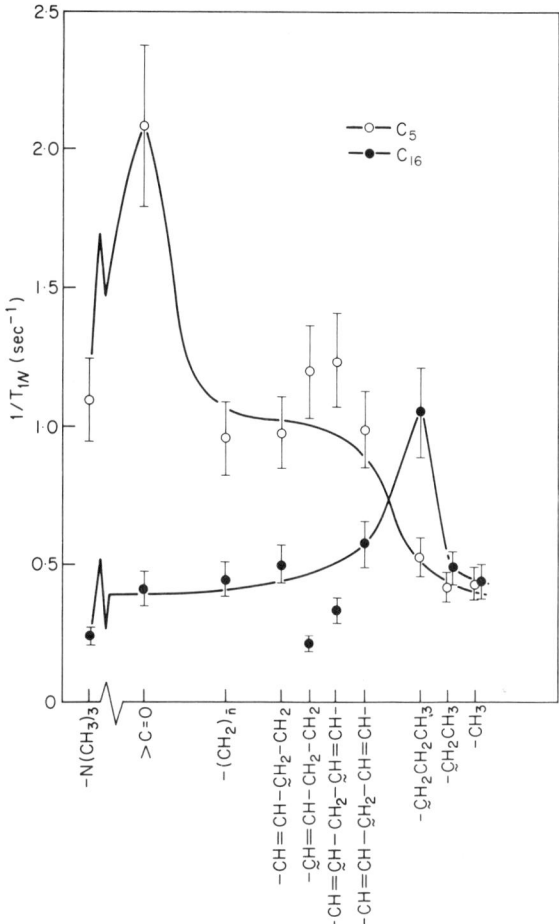

Figure 15. Plot of $1/T_{1N}$ vs approximate fatty acyl chain position for egg yolk phosphatidylcholine dispersions labelled within the C_5 and C_{16} spin labels. The $1/T_{1N}$ term is due to the relaxation mechanism in the presence of the unpaired electron of the nitroxide (taken from Godici and Landsberger, 1974).

parison to lipids in organic solvents, the lipids of human plasma lipoproteins are considerably more restricted in their rotational and segmental motions (Hamilton et al., 1974). In the recombination of HDL apoproteins with ^{13}C enriched lipids, it was found that the hydrophilic environment around the polar head groups of phosphatidylcholines and sphingomyelin in sonicated liposomes and reassembled lipoproteins was the same (Assman and Brewer, 1974; Assman et al., 1974).

Nmr studies of biological membranes have been summarized (Lee et al., 1973; Metcalfe et al., 1973a, b). Spin-lattice relaxation times for the resolved carbon resonances in sarcoplasmic reticulum membranes (Robinson et al., 1972) and intact bovine retinal rod outer segments (Millet et al., 1973) have been obtained. No changes in T_1-values and chemical shifts were detected upon bleaching the rod outer segments (Millet et al., 1973).

6. MACROMOLECULAR MODEL SYSTEMS

Micellar surfactants and macrocyclic compounds are being extensively investigated as possible models for the more complex enzyme-substrate interactions and catalyses. Substrate solubilization, either by dynamic association with micelles or by incorporation in the cavities of macrocyclic compounds, saturation kinetics, competitive inhibition and some degree of specificity observed in the different systems all support their utility as enzyme models (Cordes and Dunlap, 1969; Fendler and Fendler, 1970; Cordes and Gitler, 1973; Fendler and Fendler, 1974). Proton magnetic resonance investigations of model compounds in the absence and presence of interacting substrates considerably aided the understanding of the mechanisms involved (Fendler and Fendler, 1974). Complementary and additional ^{13}C nmr investigations are being increasingly carried out on these systems. Attention will be focused on the ^{13}C nmr data obtained, their interpretation and significance for macrocyclic compounds.

Micellar Surfactants

Long chain surfactants containing distinct hydrophobic and hydrophilic regions form large aggregates or micelles dynamically in solution. The concentration at which aggregates begin to appear (known as the critical micelle concentration or CMC), the number of monomers constituting the micelle (the aggregation number or n) and the fractional charge are the most important parameters determining the properties of micelles (Cordes and Dunlap, 1969; Fendler and Fendler, 1970; Cordes and Gitler, 1973). Proton and fluorine (Muller and Birkhahn, 1967, 1968; Muller and Johnson, 1969; Muller and Platko, 1971; Muller and Simsohn, 1971; Muller et al., 1972) magnetic resonance spectroscopy have been advantageously utilized for the elucidation of the CMC and aggregation number as well as the extent of water penetration of different micelles. The basic equations:

$$\delta = \delta_M + \frac{CMC}{C_D}(\delta_m - \delta_M)$$

$$\log(C_D - [S]) = \log nK + n \log [S]$$

(where δ, δ_M and δ_m are the observed, micellar and monomeric shift of the respective magnetic nuclei, C_D and $[S]$ are the stoichiometric and monomeric surfactant concentrations and K is the equilibrium constant for micelle formation) allowed the calculation of CMC, n, and K for several systems (Muller, 1973). The larger chemical shift range of ^{13}C magnetic resonance allows the separate observation of all the magnetically distinct carbon nuclei as a function of surfactant concentration. Such an investigation has been reported recently on sodium octanoate in water (Drakenberg and Lindman, 1973). The value for the CMC, obtained as the intercept of plots of δ vs $1/C_D$ (Fig. 16), 0·38–0·39 M, agreed well with those determined by other methods. More importantly ^{13}C nmr data indicated essential invariance of water penetration along the alkyl chain (Drakenberg and Lindman, 1973).

Spin-lattice relaxation times of the magnetically discrete carbons of monomeric and micellar n-alkyltrimethylammonium bromides in aqueous solution provided information on the rotational mobility along the hydrocarbon chain (Williams et al., 1973b). Table 22 gives the chemical shifts and relaxation times. Below their critical micelle concentration, the segmental motion only increases slightly along the hydrocarbon chain in going from the ionic head group toward the hydrocarbon tail. Above the critical micelle concentration, however, there are marked decreases in the mobility of the entire chain, the maximum restriction occurring at the polar end of the molecule. These results were rationalized in terms of a micellar interior whose properties are analogous to those of liquid hydrocarbon (Williams et al., 1973b).

TABLE 22

^{13}C Chemical Shifts and Spin Lattice Relaxation Times of Surfactants

Compound	Concentration, M	δ, ppm^a							Reference
		C-1	C-2	C-3	(CH$_2$)$_n$	CH$_2$CH$_2$CH$_3$	CH$_2$CH$_3$	CH$_3$	−N(CH$_3$)$_3$	
n-Hexyltrimethyl-ammonium bromide^b	1·0 (aq)	67·4 (4·4)	26·0 (5·0)	23·1 (5·2)		31·3 (6·3)	22·6 (8·6)	14·2 (14·3)	53·8 (6·0)	Williams et al., 1973b
n-Octyltrimethyl-ammonium bromide^b	0·2 (aq)	67·1 (4·7)	25·7 (4·7)	22·5	28·4 (4·7)	31·3 (7·8)	22·3	13·7 (12·9)	53·1 (6·3)	Williams et al., 1973b
n-Octyltrimethyl-ammonium bromide^b	2·0 (aq)	66·8 (0·9)	26·2 (1·0)	22·9 (1·0)	29·0 (1·6)	31·8 (2·4)	22·6 (·29)	14·0 (10·3)	53·4 (2·6)	Williams et al., 1973b
n-Hexadecyltri-methylammonium bromide^b	0·4 (aq)	66·8 (0·54)	26·6 (0·68)	23·0	30·4	32·3 (1·2)	23·0	14·2 (8·4)	53·5 (1·8)	Williams et al., 1973b
Sodium lauryl sulfate^c		69·86	26·16	30·52 (C$_{3-6}$)	30·15 (C$_{7,8}$) 29·67 (C$_9$)	32·63	23·50	14·44		Roberts and Chachaty, 1973

392

								Reference
Sodium lauryl sulfate[c]	0·86 (aq)	(0·35)	(0·35)	(0·38) (C$_{3-6}$)	(0·43) (C$_{7,8}$) (0·35) (C$_9$)	(0·55)	(1·20)	(2·20)	Roberts and Chachaty, 1973
Sodium lauryl sulfate[c]	0·43 (aq)	(0·35)	(0·37)	(0·39) (C$_{3-6}$)	(0·41) (C$_{7-9}$)	(0·66)	(0·75)	(2·10)	Roberts and Chachaty, 1973
Sodium lauryl sulfate and NaCl (0·15 M)[c]	0·43 (aq)	(0·53)	(0·37)	(0·35) (C$_{3-6}$)	(0·45) (C$_{7,8}$) (0·48) (C$_9$)	(0·60)	(0·90)	(2·00)	Roberts and Chachaty, 1973
Sodium lauryl sulfate and NaCl (0·30 M)[c]	0·43 (aq)	(0·70)	(0·40)	(0·40) (C$_{3-8}$)	(0·48) (C$_9$)	(0·50)	(0·85)	(2·20)	Roberts and Chachaty, 1973
Potassium caprate[d]	0·57 (aq)	187·2	388·4 (0·65)	27·16 (0·85)	30·0 (0·85) 30·2 (C$_7$) (0·92)	32·72 (1·55)	23·36 (2·2)	14·56 (4·2)	Alexandre et al., 1973

[a] Chemical shifts relative to TMS.
[b] Chemical Shifts corrected to TMS scale using δCS$_2$ = 193·7 ppm. Numbers in parentheses are NT_1-values in seconds where N is the number of directly attached hydrogens and T_1 is the spin lattice relaxation time. Measurements of T_1 made at 34°. Proton decoupled, 15·18 MHz, natural abundance ^{13}C FTnmr temp 43°.
[c] Numbers in parentheses are T_1-values in seconds.
[d] Chemical shifts in Hz relative to TMS converted to ppm using 25 MHz. Numbers in parentheses are T_1-values in seconds.

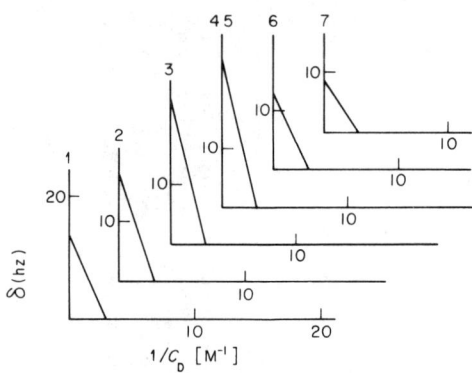

Figure 16. Change in the chemical shifts for the ^{13}C nmr signals of sodium octanoate vs the reciprocal octanoate concentration. (Taken from Drakenberg and Lindman, 1973.)

Molecular motion in micellar solutions of sodium lauryl sulfate with varying concentration of added sodium chloride has been studied by ^{13}C nmr techniques (Roberts and Chachaty, 1973). The effect of the salt was to increase the mobility of the methylene group adjacent to the polar head (Roberts and Chachaty, 1973). Values of T_1 for aqueous micellar potassium caprate ($C_9H_{19}COOK$) have been determined (Alexandre et al., 1973) and compared to the relaxation times of decanol (Doddrell and Allerhand, 1971a).

Macrocyclic Compounds

Synthetic and naturally occurring macrocyclic compounds complex with metal ions selectively both in aqueous and non-aqueous solvents. In non-aqueous solvents, such complex formation leads to anion desolvation and ion-pair separation. The novel chemistry of these macrocyclic compounds as well as their role in ion transport across cell membranes prompted the vigorous recent activity in this area of research (Christensen et al., 1971; Pedersen and Frensdorff, 1972; Lehn, 1973). ^{13}C spectroscopy is used to obtain structural and configurational information on these systems as well as to investigate their molecular dynamics.

Synthetic crown ethers

^{13}C Chemical shifts and spin lattice relaxation times of dicyclohexyl-18-crown-6 [127], dibenzo-18-crown-6 [128] and their pot-

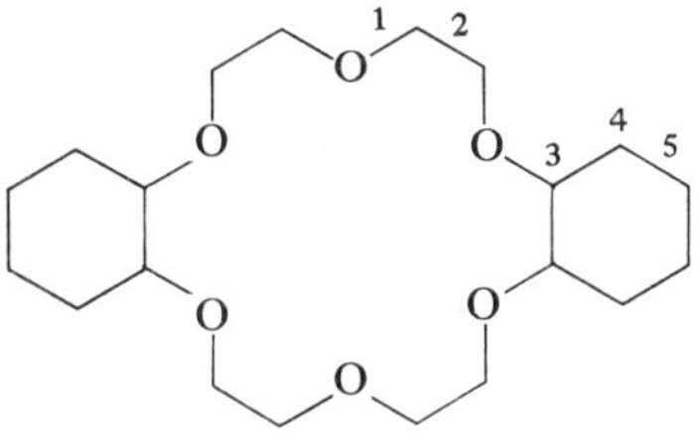

[127] 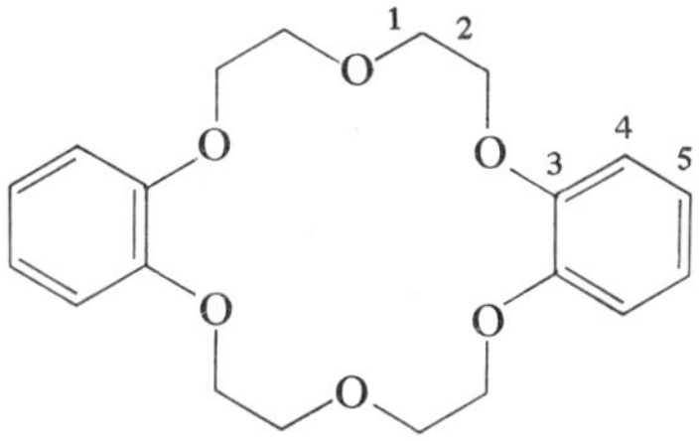   [128]

assium complexes in CD_2Cl_2 are given in Table 23. Both chemical shift and relaxation-time changes of complexes with respect to their parent ethers indicate interactions in which the potassium ion is trapped in the macrocyclic cavity and the counterion is an integral part of the complex (Ohnishi et al., 1972; Fedarko, 1973). Such interpretation substantiates data obtained by ^{1}H nmr and x-ray crystallography on macrocyclic polyethers and their complexes (Pedersen and Frensdorff, 1972).

Naturally occurring macrocyclic compounds

Naturally occurring valinomycin [129], nonactin [130], antamanide [131], beauvericin [132] and enterobactin [133] are analogous to synthetic crown ethers in that they possess polar cavities which can strongly and selectively bind cations and act as ion carriers in model and biological membranes (Fendler and Fendler, 1974). Inevitably their conformation is more complex than those for the crown ethers. ^{13}C spectroscopy has been profitably used to supplement ^{1}H nmr, ir and x-ray crystallography in the elucidation of conformations of naturally occurring macrocyclic compounds as functions of the complexing cations and of solvents. Initial work has centered on assignments of discrete ^{13}C resonances to appropriate carbon atoms (Pretsch et al., 1973; Fedarko, 1973; Ohnishi et al., 1972; Bystrov et al, 1972; Grell et al., 1972; Llinas et al., 1973; Patel 1973a, b, c; Tables 24–28). Subsequent and more detailed investi-

TABLE 23

^{13}C NMR Data for Crown Ethers and their Complexes[a]
(Numbered formulae are on p. 395)

Compound	δC-1	TC-1	δC-2	TC-2	δC-3	TC-3	δC	TC	δC-5	TC-5	Reference
Dicyclohexyl-18-crown-6 [127]	71·0 (70·7)[b] 71·2	1·26 1·30	68·2 (67·9) 68·2	1·22 1·40	77·5 (77·4) 77·6	1·70 1·64	27·8	0·89	22·3	0·82	Fedarko, 1973 Ohnishi et al., 1972
Dicyclohexyl-18-crown-6 K$^+$ complex	70·6 70·7	0·48	66·9 (66·4) 67·3	0·48 0·44	77·9 (78·4) 78·1	0·91 0·91	28·0 26·2 (25·8) 26·3	0·46 0·46	22·4 21·8 22·0	0·46	Fedarko, 1973 Ohnishi et al., 1972
Dibenzo-18-crown-6 [128]	70·2	1·19	68·2	1·24	149·0		121·3	1·73	113·5	1·72	Fedarko, 1973
Dibenzo-18-crown-6 K$^+$ complex	69·2	0·5	67·5	0·5	147·2		121·5	1·0	111·6	0·9	Fedarko, 1973

[a] Assignment for C-1 and C-2; C-4 and C-5 is not definite. Consult original paper; T in seconds.
[b] Parentheses denotes resonance of minor isomer in mixture. Identified as Pedersen's isomer B (Pedersen and Frensdorff, 1972).

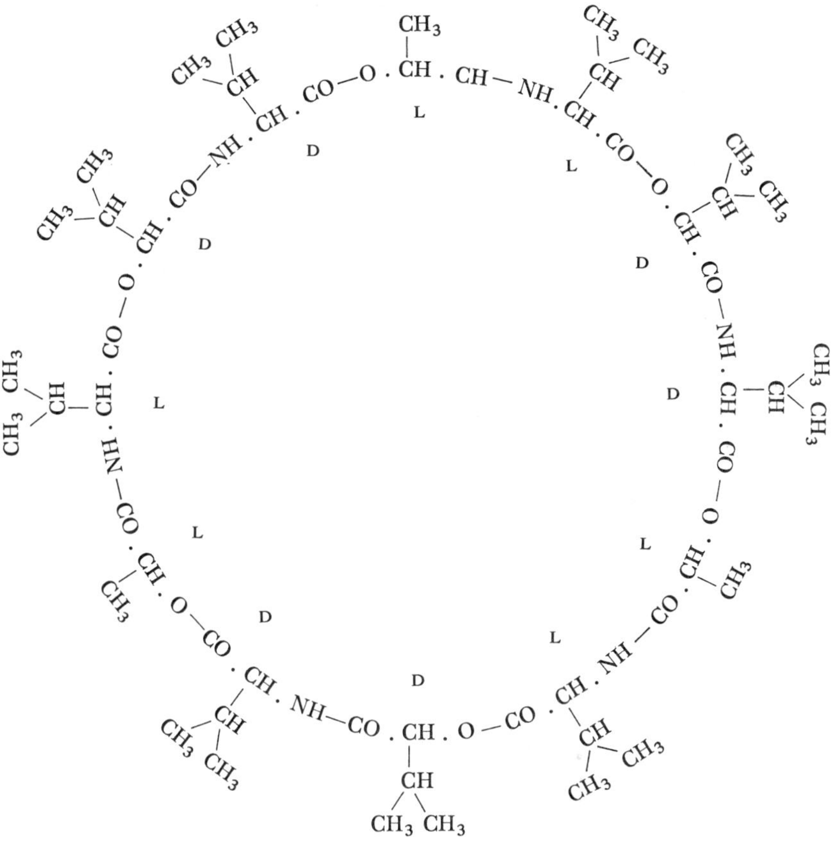

[129] Valinomycin

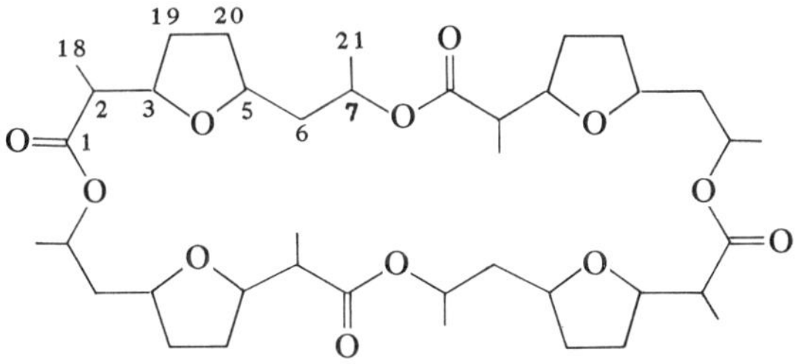

[130] Nonactin

8 9 10 1 2
Pro-Phe-Phe-Val Pro
| |
Pro-Phe-Phe-Ala-Pro
7 6 5 4 3

[131] Antamanide

TABLE 24

^{13}C Chemical Shifts for Valinomycin and Its Metal Complexes

δ, ppm^a

Compound	Solvent	Carbonyl esters	Amide	C_α C_α—O DHyIv	L-Lac	C_α—N L-Val	D-Val	C_β DHyIv	D,L-Val	Methyl	Reference
Valinomycin^b	CDCl$_3$/CD$_3$OD	171·9 170·5	172·4 171·1	78·9	70·7	59·8	58·8	30·6	29·7	19·2 19·0 18·8 17·0 19·7	Bystrov et al., 1972
Valinomycin^c	CD$_3$OD	172·5 171·5	173·1 171·9	80·0 (0·34)	71·7 (0·33)		59·8 (0·29)	31·7 (0·31)	31·1 30·8 (0·34, 0·30, 0·65, 0·60, 0·54)	19·7 19·6 19·3 19·2 18·8 17·7 17·5 (0·63, 0·58, 0·60)	Ohnishi et al., 1972 Fedarko, 1973 Grell et al., 1973
Valinomycin^d	CD$_3$OD	172·5 171·6	172·9 171·8	80·1	71·8	60·1	59·9				
Valinomycin^e	CD$_3$OD	171·68(4·0) 170·65(4·0)	172·16(4·0) 171·04(4·0)	79·19 (0·25)	70·88 (0·25)	59·15 (0·24)	58·95 (0·24)	30·73 (0·40)	30·15 30·05 (0·35)		Patel, 1973a
Valinomycin^d	Dioxane-d$_8$	171·7 170·6	172·2 170·6	79·3	71·2	60·1	59·9				Grell et al., 1973
Valinomycin^e	Dioxane	171·58 169·98	171·82 170·12	78·56	70·35	59·87	58·36	30·25	28·99 28·65		Patel, 1973a
Valinomycin^e	Octane-dioxane (3:1)	171·25 169·89	172·17 170·37	78·37	70·02	60·36	59·05	30·30	28·12 28·02		Patel, 1973a
Valinomycin^e	Dioxane-water (5:1)	171·53 170·27	171·63 160·61	78·76	70·50	58·90	58·31	30·40	29·67		Patel, 1973a
Valinomycin^e	THF ($-12°$)	171·48 168·66	172·40 189·97	77·92	69·52	59·62	56·37	30·30	30·59 28·74		Patel, 1973a
Valinomycin^e	THF (50°)	171·62 169·68	171·33 169·97	78·31	70·16	59·81	58·21	30·25	28·98 28·50		Patel, 1973a
Valinomycin^e	DMF	170·91 170·03	170·42 168·86	78·03	69·87	58·37	57·74	30·06	29·62 29·14		Patel, 1973a

Compound	Solvent	C=O (ester)	C=O (amide)	Cα (ester)	Cβ	Cα (amide)		Cβ (Val)	Cγ (methyls)	Methyls	Reference
Valinomycin[d]	Ethanol-d₆	172.1 / 170.8	172.5 / 171.2	79.4	71.2	60.0	59.3				Grell et al., 1973
Valinomycin[d]	Ethanol-d₆-cyclohexane d₁₂ (1·04:1)	171.1 / 170.5	172.6 / 171.2	79.3	71.0	60.6	59.3				Grell et al., 1973
Valinomycin[d]	Ethanol-d₆-cyclohexane d₁₂ (0·81:1)	171.9 / 170.4	172.6 / 171.2	79.0	70.8	60.8	59.4				Grell et al., 1973
Valinomycin[d]	Cyclohexane d₁₂	171.6 / 170.4	172.9 / 171.2	78.7	70.5	61.3	60.3				Grell et al., 1973
Valinomycin-K⁺ complex[b]	CDCl₃/CD₃OD	176.0 / 175.5	173.0 / 171.6	80.1	71.5	62.2	62.0	30.5	28.6	20.2, 19.6, 19.1, 18.9, 17.5, 16.9	Bystrov et al., 1972
Valinomycin-K⁺ complex[c]	CD₃OD	177.0 / 176.5	174.1 / 172.6	80.9 (0.31)	72.5 (0.32)	63.0 (0.31)	62.8 (0.31)	31.5 (0.35)	29.7, 29.6 (0.35, 0.61, 0.55)	20.6, 20.5, 19.9, 19.3, 17.8, 17.2 (0.60, 0.54)	Ohnishi et al., 1972; Fedarko, 1973
Valinomycin-Na⁺ complex[d]	CD₃OD	173.6 / 173.3	173.1 / 171.9	79.8	71.1		59.4				Grell et al., 1973
Valinomycin-K⁺ complex[d]	CD₃OD	176.8 / 176.2	174.1 / 172.5	81.0	72.6		63.0				Grell et al., 1973
Valinomycin-Rb⁺ complex[d]	CD₃OD	176.6 / 176.1	173.7 / 172.2	80.8	72.5		62.7				Grell et al., 1973
Valinomycin-Cs⁺ complex[d]	CD₃OD	176.1 / 175.5	173.7 / 172.2	80.7	72.4	62.7	62.5				Grell et al., 1973
Valinomycin-K⁺ complex[e]	CD₃OD	176.29(4.3) / 175.80(4.3)	173.47(4.3) / 171.97(4.3)	80.16 (0.32)	71.76 (0.32)	62.25 (0.29)	62.15 (0.40)	30.64 (0.40)	28.79 (0.35)		Patel, 1973a
Valinomycin-K⁺ complex[e]	DMF	173.97 / 173.38	171.88 / 170.42	78.76	80.6	60.12	59.92	29.87	28.51		Patel, 1973a

[a] All chemical shifts in parts per million relative to TMS; see cited reference for experimental details. Numbers in parentheses are spin-lattice relaxation times. T_1-values, in seconds.
[b] Proton noise-decoupled, natural abundance ¹³C FTnmr spectra recorded on a Bruker HX90 spectrometer at 22·63 MHz. Chemical shifts corrected to TMS using δ_{CS_2} = 192·8 ppm.
[c] Noise-decoupled, natural abundance ¹³C nmr spectra recorded on a Varian XL-100 FT spectrometer at 25·2 MHz.
[d] Proton noise-decoupled, natural abundance ¹³C FTnmr spectra recorded on a Varian XL-100 spectrometer at 25·2 MHz.
[e] Proton noise-decoupled, natural abundance ¹³C FTnmr spectra recorded on a Varian XL-100 spectrometer. Chemical shifts corrected to TMS scale using δ_{CS_2} = 192·8 ppm.

Abbreviations: DHylv, D-2-hydroxy-isovaleric acid; C_α—O, α-carbon atom of amino acid with ester linkage; C_α—N, α-carbon atom of amino acid with amine linkage.

TABLE 25

^{13}C Chemical Shifts of Nonactin and its Complexes
(Numbered formulae are on p. 397)

δ, ppm[a]

Compound	Solvent	1	2	3	5	6	7	18	19	20	21	Reference
Nonactin[b] [130]	CD$_2$Cl$_2$	174·2	45·7 (0·62)	76·8 (0·65)	69·4 (0·64)	42·7 (0·36)	40·7 (0·66)	12·9 (0·89)	28·5 (0·50)	31·9 (0·47)	20·7 (0·67)	Ohnishi et al., 1972 Fedarko, 1973
Nonactin-K+[b]	CD$_2$Cl$_2$	177·9	46·6 (0·76)	75·0 (0·83)	67·5 (0·78)	44·9 (0·42)	82·4 (0·78)	15·5 (1·03)	29·3 (0·43)	31·9 (0·43)	21·1 (0·70)	Ohnishi et al., 1972 Fedarko, 1973
Nonactin[c] [130]	CCl$_4$/ CDCl$_3$ (6:1)	173·2	44·8	79·9	75·9	42·4	68·5	12·6	28·0	31·4	20·6	Pretsch et al., 1972
Nonactin[c] [130]	CDCl$_3$	174·3	45·2	80·1	76·4	42·4	69·1	12·8	28·2	31·5	20·6	Pretsch et al., 1972
Nonactin[c] [130]	CDCl$_3$/ CD$_3$OD (1:1)	174·9	45·7	80·5	76·6	42·5	69·5	12·8	28·4	31·6	20·6	Pretsch et al., 1972
Nonactin[c] [130]	CD$_3$NO$_2$	175·7	46·7	81·8	77·5	43·6	70·2	13·5	29·2	32·5	21·0	Pretsch et al., 1972
Nonactin-Na+[c]	CDCl$_3$	177·4	45·9	80·9	74·5	44·3	67·9	15·1	29·0	31·3	20·9	Pretsch et al., 1972
Nonactin-K+[c]	CDCl$_3$	177·5	46·1	82·0	74·5	44·5	67·1	15·3	28·8	31·5	21·0	Pretsch et al., 1972
Nonactin-Ba^{2+}[c]	CDCl$_3$	179·5	45·2	80·8	75·0	43·8	69·9	14·8	29·4	31·3	21·6	Pretsch et al., 1972
Nonactin-NH$_4$+[c]	CDCl$_3$	176·3	46·5	82·8	74·8	43·6	66·9	15·0	28·9	31·0	20·8	Pretsch et al., 1972
Nonactin-Rb+[c]	CDCl$_3$	176·3	46·4	82·6	74·7	43·6	66·9	15·0	28·8	31·0	20·8	Pretsch et al., 1972
Nonactin-Cs+[c]	CDCl$_3$	176·6	46·1	81·6	74·7	43·4	68·1	14·5	28·8	31·2	20·9	Pretsch et al., 1972

[a] All chemical shifts relative to TMS. Numbers in parentheses are spin-lattice relaxation times, T_1, in seconds.
[b] Noise-decoupled, natural abundance ^{13}C nmr spectra recorded on a Varian XL100 FT spectrometer operating at 25·2 MHz.
[c] Decoupled, natural abundance ^{13}C nmr spectra recorded on a Bruker HFX-10 FT spectrometer operating at 22·6 MHz.

TABLE 26

^{13}C Chemical Shifts of Enterobactin, its Ga^{3+} Complex and its Monomer

Compound	Solvent	δ, ppm[a]									Reference
		—COO—	—CONH—	o-COH	m-COH	—C$_1$H	—C$_4$H	—C$_5$H	—C$_6$H	—C$_\beta$H$_2$O	—C$_\alpha$HNH—	
Monomer[b]	DMSO-d$_6$	172·4	169·5	149·4	147·0	119·8	119·5	119·2	117·2	62·1	56·2	Llinas et al., 1973
Enterobactin[b] [130]	DMSO-d$_6$	170·3	169·9	149·7	147·2	120·2	119·4	119·4	116·6	64·7	52·5	Llinas et al., 1973
Ga^{3+}-enterobactin[b]	DMSO-d$_6$	171·1	169·7	159·5	155·9	115·7	114·6	114·2	113·4	66·2	53·0	Llinas et al., 1973

[a] Chemical shifts corrected to TMS scale using $\delta_{\text{dioxane}} = 67·4$ ppm.
[b] Proton noise-decoupled, natural abundance ^{13}C FTnmr spectra recorded on a modified Varian HR-60.

TABLE 27

^{13}C Chemical Shifts of Antamanide, Val,^{6} Ala^9-Antamanide and their Sodium Complexes

		Valine			Alanine				Phenylalanine				Proline				Car-bonyls	Reference
Compound	Solvent	C_α	C_β	C_γ	C_α	C_β	C_α	C_β	C_1	C_{ortho}	C_{meta}	C_{para}	C_α	C_β	C_γ	C_δ		
Val,^{6} Ala^9-Antamanide^b	CDCl$_3$/CD$_3$OD	56·7 56·6	31·6	17·4	56·7 56·4	18·9	50·3	34·3	138·4	130·1	128·3	126·3	60·9 58·7	30·1 28·2 24·6 21·8	30·1 28·2 24·6 21·8	47·8 46·8	173·8 172·0 171·3 171·0	Bystrov et al., 1972
Na$^+$-Val,^{6} Ala^9-Antamanide^b	CDCl$_3$/CD$_3$OD	56·2 54·4	32·5	18·6 16·8	56·2 54·4	19·2	52·2	36·0	136·6	129·7	130·5	127·8	59·7 59·2	32·5 29·1 25·7 22·7	32·5 29·1 25·7 22·7	48·2 47·4	176·1 173·5 171·8 171·1 170·7	Bystrov et al., 1972
Na$^+$-Antamanide^c	CD$_3$CN	56·0 57·4	31·9 32·3	16·8 18·5 18·7	51·4 52·1	16·8 18·5 18·7	51·4 52·1 54·6 54·7 56·0 57·4	35·8 36·3 34·6 39·5					58·6 58·9 59·0 59·3	28·5 29·1 31·9 32·3	22·7 23·1 24·8 25·6	47·3 47·5 48·1		Patel, 1973c
Antamanide^c	CD$_3$CN	57·6 57·9	31·2 31·8 32·0	16·3 18·0 19·0	49·7 52·3	16·3 18·0 19·0	49·7 52·3 56·3 57·6 57·9	35·3 35·9 36·7 38·1					59·0 59·3 61·1 61·3	28·8 29·1 31·2 31·8 32·0	21·6 22·4 25·2	46·9 47·0 48·1 48·5		Patel, 1973b

^{a} ^{13}C chemical shifts corrected to TMS scale using δCS_2 = 192·8 ppm.
^{b} Proton noise-decoupled, natural abundance ^{13}C FTnmr spectra recorded on a Bruker HX-90 spectrometer operating at 22·63 MHz.
^{c} Proton noise-decoupled, natural abundance ^{13}C FTnmr recorded on a Varian XL-100 spectrometer.

TABLE 28

^{13}C Chemical Shifts of Beauvercin and its Complexes with Potassium and Sodium Ions

δ, ppm[a]

Compound	Solvent	Methyl phenylalanine							N-CH$_3$	Hydroxy isovaleric acid				Reference
		C$_\alpha$	C$_\beta$	C$_1$	C$_{ortho}$	C$_{meta}$	C$_{para}$			C$_\alpha$	C$_\beta$	CH$_3$	Carbonyls	
Beauvercin[b] [132]	CDCl$_3$/CD$_3$OD	56.6	34.6	136.5	128.7	128.7	127.0		30.0	76.0	31.5	18.4	171.2	Bystrov et al., 1972
												16.7	169.9	
K$^+$-Beauvercin[b]	CDCl$_3$/CD$_3$OD	57.5	33.9	136.4	128.9	128.5	127.4		30.1	77.0	31.5	17.4	172.6	Bystrov et al., 1972
												17.3	171.9	
Na$^+$-Beauvercin[b]	CDCl$_3$/CD$_3$OD	57.5	33.2	138.3	128.9	128.5	127.3		39.7		31.4	17.4	172.9	Bystrov et al., 1972
												17.2	172.2	

[a] ^{13}C chemical shifts corrected to TMS scale using δ CS$_2$ = 192.8 ppm.
[b] Proton noise-decoupled, natural abundance ^{13}C FTnmr spectra recorded on a Bruker HX-90 spectrometer operating at 22.63 MHz.

TABLE 29

^{13}C Chemical Shifts of Cyclodextrins

Compound	Solvent	C-1	C-2	C-3	C-4	C-5	C-6	CH$_3$	C=O	Reference
α-Cyclodextrin	D$_2$O	102.3	74.2	72.8	82.1	72.5	61.2			Takeo et al., 1973
	DMSO	103.1	74.5	73.3	83.3	73.3	61.3			
	1 N NaOH	103.7	74.8	74.0	82.7	72.8	61.4			
β-Cyclodextrin	D$_2$O	102.7	74.0	73.0	82.0	72.8	61.2			Takeo et al., 1973
	DMSO	103.1	74.3	73.6	82.7	73.2	61.2			
	1 N NaOH	103.7	74.8	74.0	82.7	72.8	61.4			
γ-Cyclodextrin	D$_2$	102.5	73.7	73.1	81.3	72.6	61.1			Takeo et al., 1973
	DMSO	192.9	74.5	73.9	82.6	73.4	61.3			
	1 N NaOH	103.9	75.0	74.4	82.5	72.8	61.4			
6-Deoxy-β-cyclo-dextrin	DMSO	103.4	74.3	73.7	89.4	67.7	18.5			Takeo et al., 1973
	1 N NaOH	103.7	75.0	74.3	88.9	68.4	17.9			
α-Cyclodextrin peracetate	CDCl$_3$	96.5	70.9	70.9	77.2	69.5	63.2	20.8	170.6 169.2 170.4 169.2	Takeo et al., 1973
	C$_6$D$_6$	96.5	71.5	71.1	77.3	70.1	63.6	20.6		
β-Cyclodextrin peracetate	CDCl$_3$	96.8	70.9	70.4	76.8	69.7	62.6	20.8	170.7, 170.4 169.4 170.4 169.3	Takeo et al., 1973
	C$_6$D$_6$	97.5	70.7	70.7	77.7	70.7	63.4	20.7		
γ-Cyclodextrin peracetate	CDCl$_3$	96.3	70.9	70.3	75.9	69.7	62.6	20.7	170.6, 170.3 169.3	Takeo et al., 1973
	C$_6$D$_6$	96.7	71.2	71.2	76.5	70.4	63.2	20.7, 20.6	170.6, 170.4	
6-Deoxy-β-cyclo-dextrin peracetate	CDCl$_3$	96.6	71.2	70.4	82.5	67.2	16.6	20.8	169.5 170.7 169.4	Takeo et al., 1973

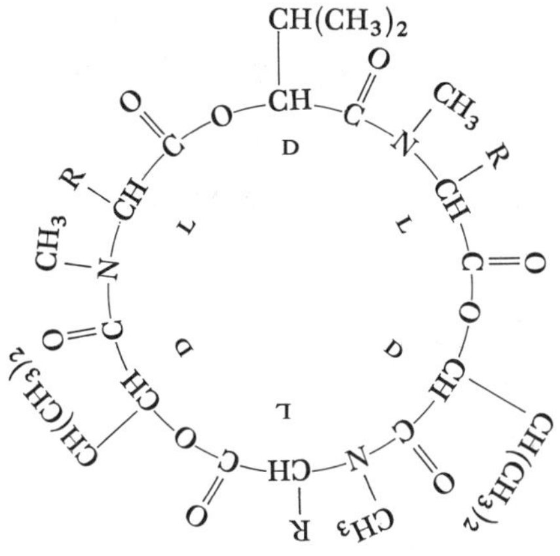

[132] Beauvericin; R = —CH$_2$C$_6$H$_5$

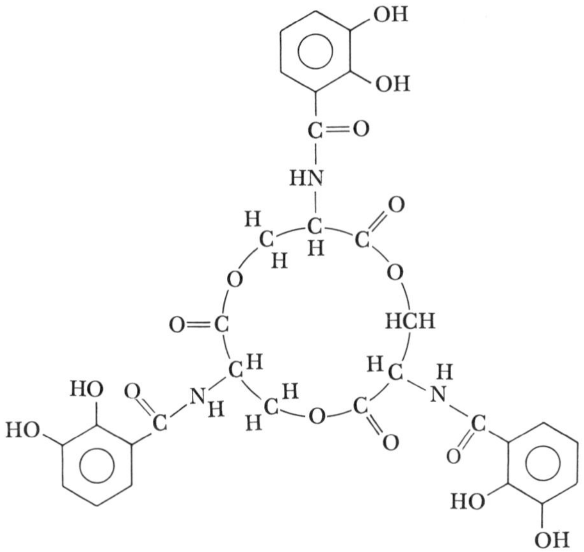

[133] Enterobactin

gations are aimed at the elucidation of the respective conformations. Thus, ^{13}C chemical shifts and spin-lattice relaxation times established valinomycin to be relatively rigid, moving isotropically both in the presence and absence of K$^+$. Conversely, nonactin [130] was found to move anisotropically while its K$^+$ complex is isotropic (Fedarko, 1973). In methanol, K$^+$, Rb$^+$, and Cs$^+$ coordinate with *all* the ester carbonyl groups of valinomycin [129] in addition to forming intramolecular hydrogen bonds with the amide groups; Na$^+$, however, utilizes reduced numbers of coordination and hydrogen bonding sites (Grell *et al.*, 1973). The role of the solvent in affecting conformation changes has not been unequivocally established (Ivanov *et al.*, 1971; Grell *et al.*, 1973). Combined ^{1}H and ^{13}C nmr have established that the cyclic decapeptide antaminide with all L-amino acids [131] exists in highly involved, solvent-dependent conformational equilibria, but it complexes with Na^+ to give a rigid structure (Patel, 1973b, c). Similarly, a structure has been proposed for the Ga^{3+} complex of enterobactin based on ^{13}C nmr (Llinas *et al.*, 1973).

ACKNOWLEDGMENTS

The support of this work by grants from the National Science Foundation and the United States Energy Research and Development Administration is gratefully acknowledged.

REFERENCES

Alexandre, M., Fouchet, C., and Rigny, P. (1973). *J. Chim. Phys.* 70, 1073.
Allerhand, A., and Doddrell, D. (1971). *J. Amer. Chem. Soc.* 93, 2777.
Allerhand, A., and Komoroski, R. A. (1973). *J. Amer. Chem. Soc.* 95, 8228.
Allerhand, A., and Oldfield, E. (1973). *Biochemistry* 12, 3428.
Allerhand, A., Cochran, D. W., and Doddrell, D. (1970). *Proc. Nat. Acad. Sci. USA* 67, 1093.
Allerhand, A., Doddrell, D., Glushko, V., Cochran, D. W., Wenkert, E., Lawson, P. J., and Gurd, F. R. N. (1971a). *J. Amer. Chem. Soc.* 93, 544.
Allerhand, A., Doddrell, D., and Komoroski, R. (1971b). *J. Chem. Phys.* 55, 189.
Allerhand, A., Childers, R. F., Goodman, R. A., Oldfield, E., and Ysern, S. (1972). *Amer. Lab.* 4, 19.
Allerhand, A., Childers, R. F., and Oldfield, E. (1973a). *Biochemistry* 12, 1335.
Allerhand, A., Childers, R. F., and Oldfield, E. (1973b). *J. Magn. Res.* 11, 272.
Allerhand, A., Childers, R. F., and Oldfield, E. (1973c). *Annals N.Y. Acad. Sci.* 222, 764.
Anderson, W. A., Freeman, R., and Hill, H. (1971). *Pure and Appl. Chem.* 32, 27.
Anet, F. A. L., and Levy, G. C. (1973). *Science* 180, 141.
Angyal, S. J. (1969). *Angew. Chem. I.E.* 8, 157.
Antonini, E., Brunori, M., Conti, F., and Geraci, G. (1973). *FEBS Lett.* 34, 69.
Arison, B. H. and Hoogsteen, K. (1970). *Biochemistry* 9, 3967.
Armitage, I. M., Huber, H., Pearson, H., and Roberts, J. D. (1974). *Proc. Nat. Acad. Sci. USA* 71, 2096.

Assmann, G., and Brewer, H. B., Jr. (1974). *Proc. Nat. Acad. Sci. USA* **71**, 1534.
Assmann, G., Highet, R. J., Sokoloski, E. A., and Brewer, H. B., Jr. (1974). *Proc. Nat. Acad. Sci. USA* **71**, 3701.
Batchelor, J. G., Prestegard, J. H., Cushley, R. J., and Lipsky, S. R. (1972). *Biochem. Biophys. Res. Comm.* **48**, 70.
Batchelor, J. G., Prestegard, J. H., Cushley, R. J., and Lipsky, S. R. (1973). *J. Amer. Chem. Soc.* **95**, 6358.
Becker, E. D., and Farrar, T. C. (1972). *Science* **178**, 361.
Becker, E. D., Shoup, R. R., and Farrar, T. C. (1971). *Pure and Appl. Chem.* **32**, 51.
Behr, J. P., and Lehn, J. M. (1973). *FEBS Lett.* **31**, 297.
Binkley, W. W., Horton, D., Bhacca, N. S., and Wander, J. D. (1972). *Carbohydrate Res.* **23**, 301.
Birdsall, B., and Feeney, J. (1972). *J.C.S. Perkin II*, 1643.
Birdsall, B., Birdsall, N. J. M., and Feeney, J. (1972a). *Chem. Comm.* 316.
Birdsall, N. J. M., Feeney, J., Lee, A. G., Levine, Y. K., and Metcalfe, J. C. (1972b). *J. C. S. Perkin II*, 1441.
Birdsall, B., Feeney, J., and Partington, P. (1973). *J.C.S. Perkin II* 2145.
Blumenstein, M., and Raftery, M. A. (1972). *Biochemistry* **11**, 1643.
Blumenstein, M., and Raftery, M. A. (1973). *Biochemistry* **12**, 3585.
Boccalon, G., Verdini, A. S., and Giacometti, G. (1972). *J. Amer. Chem. Soc.* **94**, 3639.
Bock, K., and Pedersen, C. (1974). *J.C.S. Perkin II*, 293.
Bock, K., Lundt, I., and Pedersen, C. (1973). *Tetrahedron Lett.* 1037.
Bock, K., Pedersen, C., and Heding, H. (1974). *J. Antibiotics* **27**, 139.
Bovey, F. A. (1972). "Chemistry and Biology of Peptides", Ann Arbor Science Publishers, Ann Arbor, Michigan.
Bradbury, J. H., and Norton, R. S. (1973). *Biochim. Biophys. Acta* **328**, 10.
Bradbury, E. M., Cary, P. D., Crane-Robinson, C., and Hartman, P. G. (1973). *Pure Appl. Chem.* **36**, 5.
Breitmaier, E. (1974). *Chimia* **28**, 120.
Breitmaier, E., Spohn, K. H., and Berger, S. (1975). *Angew Chem. I.E.* **14**, 145.
Breitmaier, E., and Voelter, W. (1972). *Eur. J. Biochem.* **31**, 234.
Breitmaier, E., and Voelter, W. (1973). *Tetrahedron* **29**, 227.
Breitmaier, E., and Voelter, W. (1974). "^{13}C NMR Spectroscopy, Methods and Applications", Verlag Chemie GmbH, Weinheim.
Breitmaier, E., Jung, G., and Voelter, W. (1971a). *Angew. Chem. I. E.* **10**, 673.
Breitmaier, E., Voelter, W., Jung, G., and Tanzer, C. (1971b). *Chem. Ber.* **104**, 1147.
Breitmaier, E., Jung, G., and Voelter, W. (1972). *Chimia* **26**, 139.
Breitmaier, E., Haas, G., and Voelter, W. (1974). In press. Heyden and Son.
Brewer, C. F., Sternlicht, H., Marcus, D. M., and Grollman, A. P. (1973a). *Proc. Nat. Acad. Sci. USA* **70**, 1007.
Brewer, C. F., Sternlicht, H., Marcus, D. M., and Grollman, A. P. (1973b). *Biochemistry* **12**, 4448.
Brewer, C. F., Marcus, D., and Grollman, A. P. (1973c). *Annals N.Y. Acad. Sci.* **222**, 978.
Brewster, A. I. R., Hruby, V. J., Spatola, A. F., and Bovey, F. A. (1973). *Biochemistry* **12**, 1643.
Browne, D. T., Kenyon, G. L., Packer, E. L., Wilson, D. M., and Sternlicht, H. (1973a). *Biochem. Biophys. Res. Commun.* **50**, 42.

Browne, D. T., Kenyon, G. L., Packer, E. L., Sternlicht, H., and Wilson, D. M. (1973b). *J. Amer. Chem. Soc.* **95**, 1316.
Bundle, D. R., Jennings, H. J., and Smith, I. C. P. (1973). *Can. J. Chem.* **51**, 3812.
Bundle, D. R., Smith, I. C. P., and Jennings, H. J. (1974). *J. Biol. Chem.*, **249**, 2275.
Burlingame, A. L., Balogh, B., Welch, J., Lewis, S., and Wilson, D. (1972). *Chem. Comm.* 318.
Bystrov, V. F., Ivanov, V. T., Koz'min, S. A., Mikhaleva, I. I., Khalilulina, K. Kh., Ovchinnikov, Yu. A., Fedin, E. I., and Petrovskii, P. V. (1972). *FEBS Lett.* **21**, 34.
Chaiken, I. M. (1974). *J. Biol. Chem.* **249**, 1247.
Chaiken, I. M., Freedman, M. H., Lyerla, J. R., Jr., and Cohen, J. S. (1973). *J. Biol. Chem.* **248**, 884.
Chapman, D. (1965). "The Structure of Lipids by Spectroscopy and X-Ray Techniques", John Wiley and Sons Inc., New York.
Chapman, D. (1972). *Annals N.Y. Acad. Sci.* **195**, 179.
Chapman, D., and Chen, S. (1972). *Chem. Phys. Lipids* **8**, 318.
Cheney, V. B., and Grant, D. M. (1967). *J. Amer. Chem. Soc.* **89**, 5319.
Chien, J. C. W., and Brandts, J. F. (1971). *Nature, New Biology* **230**, 209.
Chien, J. C. W., and Wise, W. B. (1973). *Biochemistry* **12**, 3418.
Christensen, J. J., Hill, J. O., and Izatt, R. M. (1971). *Science* **174**, 459.
Christl, M., and Roberts, J. D. (1972). *J. Amer. Chem. Soc.* **94**, 4565.
Conti, F., and Paci, M. (1971). *FEBS Lett.* **17**, 149.
Conway, E., Guthrie, R. D., Gero, S. D., Lukacs, G., Sepulchre, A.-M., Hagaman, E. W., and Wenkert, E. (1972). *Tetrahedron Lett.* 4879.
Cordes, E. H., and Dunlap, R. B. (1969). *Accounts Chem. Res.* **2**, 329.
Cordes, E. H., and Gitler, C. (1973). *Bioorganic Chemistry* **2**, 1.
Coxon, B. (1973). *Annals N.Y. Acad. Sci.* **222**, 952.
Danyluk, S. S., and Victor, T. A. (1970). *Jerusalem Sym. Quant. Chem. Biochem.* **2**, 395.
Davenport, J. B., and Johnson, A. R. (1971). "Biochemistry and Methodology of Lipids", Wiley-Interscience, New York.
Dea, P., Schweizer, M. P., and Kreishman, G. P. (1974). *Biochemistry* **13**, 1862.
Deber, C. M., Blout, E. R., Torchia, D. A., Dorman, D. E., and Bovey, F. A. (1972). In "Chemistry and Biology of Peptides", Ann Arbor Science Publishers, Ann Arbor, Michigan.
Deslauriers, R., Walter, R., and Smith, I. C. P. (1972). *Biochem. Biophys. Res. Comm.* **48**, 854.
Deslauriers, R., Garrigou-Lagrange, C., Bellocq, A. M., and Smith, I. C. P. (1973a). *FEBS Lett.* **31**, 59.
Deslauriers, R., Walter, R., and Smith, I. C. P. (1973b). *Biochem. Biophys. Res. Comm.* **53**, 244.
Deslauriers, R., Walter, R., and Smith, I. C. P. (1973c). *FEBS Lett.* **37**, 27.
Deslauriers, R., Smith, I. C. P., and Walter, R. (1974a). *J. Amer. Chem. Soc.* **96**, 2289.
Deslauriers, R., Walter, R., and Smith, I. C. P. (1974b). *J. Biol. Chem.* **249**, 7006.
Deslauriers, R., Walter, R., and Smith, I. C. P. (1974c). *Proc. Nat. Acad. Sci. USA* **71**, 265.
Doddrell, D., and Allerhand, A. (1971a). *J. Amer. Chem. Soc.* **93**, 1558.
Doddrell, D., and Allerhand, A. (1971b). *J. Amer. Chem. Soc.* **93**, 2781.
Dorman, D. E., and Bovey, F. A. (1973). *J. Org. Chem.* **38**, 2379.

Dorman, D. E., and Roberts, J. D. (1971). *J. Amer. Chem. Soc.* **93**, 4463.
Dorman, D. E., Torchia, D. A., and Bovey, F. A. (1973). *Macromolecules* **6**, 80.
Dorman, D. E., Brewster, A. I., Bovey, F. A., Deber, C. M., and Blout, E. R. (1974). In preparation. Cited in Dorman and Bovey, 1973.
Drakenberg, T., and Lindman, B. (1973). *J. Colloid and Interface Sci.* **44**, 184.
Ellis, P. D., Fisher, R. R., Dunlap, R. B., Zens, A. P., Bryson, T. A., Williams, T. J. (1973). *J. Biol. Chem.* **248**, 7677.
Elstner, E. F., Suhadolnik, R. J., and Allerhand, A. (1973). *J. Biol. Chem.* **248**, 5385.
Engelman, D. J. (1970). *J. Mol. Biol.* **47**, 115.
Farrar, T. C., and Becker, E. D. (1971). "Pulse and Fourier Transform NMR", Academic Press, New York.
Fedarko, M. D. (1973). *J. Magnetic Res.* **12**, 30.
Fendler, E. J., and Fendler, J. H. (1970). *Adv. Phys. Org. Chem.* **8**, 271.
Fendler, J. H., and Fendler, E. J. (1974). "Catalysis in Micellar and Macromolecular Systems", Academic Press, New York.
Feeney, J., Burgen, A. S. V., and Grell, E. (1973). *Eur. J. Biochem.* **34**, 107.
Flohe, L., Breitmaier, E., Gunzler, W. A., Voelter, W., and Jung, G. (1972). *Hoppe-Seyler's Z. Physiol. Chem.* **353**, 1159.
Freedman, M. H., Cohen, J. S., and Chaiken, I. M. (1971). *Biochem. Biophys. Res. Comm.* **42**, 1148.
Freedman, M. H., Lyerla, J. R., Chaiken, I. M., and Cohen, J. S. (1973). *Eur. J. Biochem.* **32**, 215.
Freeman, R., and Hill, H. D. W. (1971). *J. Chem. Phys.* **54**, 3367.
Frisch, L. (1966). "The Genetic Code", Cold Spring Harbor Symposia on Quantitative Biology, 31.
Fritzsche, H., Arnold, K., and Krusche, R. (1973). *Studia Biophysica* **40**, 103.
Fung, C. H., Mildvan, A. S., Allerhand, A., Komoroski, R., and Scrutton, M. C. (1973). *Biochemistry* **12**, 620.
Fung, C. H., Mildvan, A. S., and Leigh, J. S., Jr. (1974). *Biochemistry* **13**, 1160.
Gent, M. P. N., and Prestegard, J. H. (1974). *Biochem. Biophys. Res. Comm.* **58**, 549.
Gibbons, W. A., Sogn, J. A., Stern, A., Craig, L. C., and Johnson, L. F. (1970). *Nature* **227**, 840.
Glushko, V., Lawson, P. J., and Gurd, F. R. N. (1972). *J. Biol. Chem.* **247**, 2176.
Godici, P. E., and Landsberger, F. R. (1974). *Biochemistry* **13**, 362.
Gorin, P. A. J. (1973a). *Canad. J. Chem.* **51**, 2105.
Gorin, P. A. J. (1973b). *Canad. J. Chem.* **51**, 2375.
Gorin, P. A. J. (1974). *Canad. J. Chem.* **52**, 458.
Gorin, P. A. J., and Spencer, J. F. T. (1972). *Canad. J. Microbiol.* **18**, 1709.
Govil, G., and Smith, I. C. P. (1973). *Biopolymers* **12**, 2589.
Grathwohl, C., and Wüthrich, K. (1974). *J. Mag. Res.* **13**, 217.
Gray, G. A. (1973). CRC Critical Reviews in Biochemistry.
Grell, E., Funck, T. and Sauter, H. (1973). *Eur. J. Biochem.* **34**, 415.
Grutzner, J. B. (1972). *Lloydia* **35**, 375.
Gurd, F. R. N., and Keim, P. (1973). "Methods Enzymol", (C. H. W. Hirs, and S. N. Tunasheff, eds.), Vol. XXVII, Part D, p. 836, Academic Press, New York.
Gurd, F. R. N., Lawson, P. J., Cochran, D. W., and Wenkert, E. (1971). *J. Biol. Chem.* **246**, 3725.
Gurd, F. R. N., Keim, P., Glushko, V. G., Lawson, P. J., Marshall, R. C., Nigen, A. M., and Vigna, R. A. (1972). "Chemistry and Biology of Peptides"

(J. Meienhofer, ed.), pp. 45-49, Ann Arbor Science Publishers, Ann Arbor, Michigan.
Gurr, M. I., and James, A. T. (1971). "Lipid Biochemistry: An Introduction", Cornell University Press, Ithaca, N.Y.
Gurudata, N., and Rajabalee, F. J. M. (1973). *Canad. J. Chem.* 51, 1797.
Hall, L. D. (1964). *Adv. Carbohydrate Chem.* 19, 51.
Hamill, W. D., Jr., Pugmire, R. J., and Grant, D. M. (1974). *J. Amer. Chem. Soc.* 96, 2885.
Hamilton, J. A., Talkowski, C., Williams, E., Avila, E. M., Allerhand, A., Cordes, E. H., and Cameyo, G. (1973). *Science* 180, 193.
Hamilton, J. A., Talkowski, C., Childers, R. F., Williams, E., Allerhand, A., and Cordes, E. H. (1974). *J. Biol. Chem.* 249, 4872.
Haverkamp, J., van Dongen, J. P. C. M., and Vliegenthart, J. F. G. (1973). *Tetrahedron* 29, 3431.
Henrikson, K. P. (1971). In "Biochemistry and Methodology of Lipids" (A. R. Johnson and J. B. Davenport, eds.), Wiley-Interscience, New York.
Herve du Penhoat, P., and Perlin, A. S. (1974). *Carbohydrate Res.* 36, 111.
Honda, S., Yuki, H., and Takiura, K. (1973). *Carbohydrate Res.* 28, 150.
Horsley, W. J., and Sternlicht, H. (1968). *J. Amer. Chem. Soc.* 90, 3778.
Horsley, W., Sternlicht, H., and Cohen, J. S. (1969). *Biochem. Biophys. Res. Comm.* 37, 47.
Horsley, W., Sternlicht, H., and Cohen, J. S. (1970). *J. Amer. Chem. Soc.* 92, 680.
Hunkapiller, M. W., Smallcombe, S. H., Whitaker, D. R., and Richards, J. H. (1973a). *Biochemistry* 12, 4732.
Hunkapiller, M. W., Smallcombe, S. H., Whitaker, D. R., and Richards, J. H. (1973b). *J. Biol. Chem.* 248, 8306.
Ikehara, M., Tazawa, I., and Fukui, T. (1969). *Biochemistry* 8, 736.
Inch, T. D. (1969). In "Annual Reports on NMR Spectroscopy" (E. F. Mooney, ed.), Vol. 2, p. 305, Academic Press, London.
Inch, T. D. (1972). In "Annual Reports on NMR Spectroscopy" (E. F. Mooney, ed.), Vol. 5A, p. 305, Academic Press, London.
Ingram, V. (1965). "The Biosynthesis of Macromolecules", W. A. Benjamin, N.Y.
Ivanov, V. T., Miroshnikov, A. I., Abdullaev, N. D., Senyavina, L. B., Arkhipova, S. F., Uvarova, N. N., Khalilulina, K. Kh., Bystrov, V. F., and Ovchinnikov, Yu. A. (1971). *Biochem. Biophys. Res. Comm.* 42, 654.
Izatt, R. M., Christensen, J. J., and Rytteng, J. H. (1971). *Chem. Rev.* 71, 439.
Jennings, H. J., and Smith, I. C. P. (1973). *J. Amer. Chem. Soc.* 95, 606.
Johnson, L. F. (1971). *Anal. Chem.* 43, 28A.
Johnson, L. F. and Jankowski, W. C. (1972). "Carbon-13 Nuclear Magnetic Resonance Spectroscopy. A Collection of Assigned, Coded and Indexed Spectra", Wiley-Interscience, New York.
Jung, G., Breitmaier, E., Voelter, W., Keller, T., and Tanzer, C. (1970). *Angew. Chem. I. E.* 9, 894.
Jung, G., Ottnad, M., Voelter, W., and Breitmaier, E. (1972a). *Z. Anal. Chem.* 261, 328.
Jung, G., Breitmaier, E., and Voelter, W. (1972b). *Eur. J. Biochem.* 24, 438.
Jung, G., Breitmaier, E., Günzler, W. A., Ottnad, M., Voelter, W., and Flohe, L. (1974). In "Glutathione, Proceedings of the 16th Conference of the German Society of Biological Chemistry, Tübingen, March 1973" (L. Flohe, Ch. Benohr, H. Sies, H. D. Waller, and A. Wendel, eds.), pp. 1-15, Georg Thieme Publishers, Stuttgart.

Kapuler, A. M., and Reich, E. (1971). *Biochemistry* 10, 4050.
Kapuler, A. M., Monny, C., and Michelson, A. M. (1970). *Biochim. Biophys. Acta* 217, 18.
Kayushin, L. P., and Ajipa, Y. I. (1973). *Annals of N.Y. Acad. Sci.* 222, 255.
Keim, P., Vigna, R. A., Marshall, R. C., and Gurd, F. R. N. (1973a). *J. Biol. Chem.* 248, 6104.
Keim, P., Vigna, R. A., Morrow, J. S., Marshall, R. C., and Gurd, F. R. N. (1973b). *J. Biol. Chem.* 248, 7811.
Keough, K. M., Oldfield, E., Chapman, D., and Beynon, P. (1973). *Chem. Phys. Lipids* 10, 37.
Koch, H. J., and Perlin, A. S. (1970). *Carbohydrate Res.* 15, 403.
Koch, K. F., Rhoades, J. A., Hagaman, E. W., and Wenkert, E. (1974). *J. Amer. Chem. Soc.* 96, 3300.
Koenig, S. H., Brown, R. D., Needham, T. E., and Matwiyoff, N. A. (1973). *Biochem. Biophys. Res. Comm.* 53, 624.
Koerner, T. A. W., Jr., Cary, L. W., Bhacca, N. S., and Younathan, E. S. (1973). *Biochem. Biophys. Res. Comm.* 51, 543.
Komoroski, R. A., and Allerhand, A. (1972). *Proc. Nat. Acad. Sci. USA* 69, 1804.
Komoroski, R. A., and Allerhand, A. (1974). *Biochemistry* 13, 369.
Kotowycz, G. (1974). *Canad. J. Chem.* 52, 94.
Kotowycz, G., and Hayamizu, K. (1973). *Biochemistry* 12, 517.
Kotowycz, G., and Lemieux, R. U. (1973). *Chem. Rev.* 73, 669.
Kotowycz, G., and Suzuki, O. (1973a). *Biochemistry* 12, 3434.
Kotowycz, G., and Suzuki, O. (1973b). *Biochemistry* 12, 5325.
Kreishman, G. P., Witkowski, J. T., Robins, R. K., and Schweizer, M. P. (1972). *J. Amer. Chem. Soc.* 94, 3894.
Krugh, T. R. (1973). *J. Amer. Chem. Soc.* 95, 4761.
Krugh, T. R., and Neely, J. W. (1973). *Biochemistry* 12, 1775.
Lam, Y.-F., Kuntz, G. P. P., and Kotowycz, G. (1974). *J. Amer. Chem. Soc.* 96, 1834.
Lapper, R. D. and Smith, I. C. P. (1973). *J. Amer. Chem. Soc.* 95, 2880.
Lapper, R. D., Mantsch, H. H., and Smith, I. C. P. (1972). *J. Amer. Chem. Soc.* 94, 6243.
Lapper, R. D., Mantsch, H. H., and Smith, I. C. P. (1973). *J. Amer. Chem. Soc.* 95, 2878.
Lasker, S. E. and Chiu, M. L. (1973). *Annals N.Y. Acad. Sci.* 222, 971.
Lauterbur, P. C. (1970). *Applied Spectroscopy* 24, 460.
Lee, A. G., Birdsall, N. J. M., Levine, Y. K., and Metcalfe, J. C. (1972). *Biochim. Biophys. Acta* 255, 43.
Lee, A. G., Birdsall, N. J. M., and Metcalfe, J. C. (1973). *Chem. Brit.* 9, 116.
Lehn, J.-M. (1973). *Structure and Bonding*, 16, 1.
Lemieux, R. U., and Stevens, J. D. (1966). *Canad. J. Chem.* 44, 249.
Lemieux, R. U., Nagabhushan, T. L., and Paul, B. (1972). *Canad. J. Chem.* 50, 773.
Levine, Y. K., Partington, P., Roberts, G. C. K., Birdsall, N. J. M., Lee, A. G., and Metcalfe, J. C. (1972a). *FEBS Lett.* 23, 203.
Levine, Y. K., Birdsall, N. J. M., Lee, A. G., and Metcalfe, J. C. (1972b). *Biochemistry* 11, 1416.
Levy, G. C. (1973). *Accounts. Chem. Res.* 6, 161.
Levy, G. C., and Komoroski, R. A. (1974). *J. Amer. Chem. Soc.* 96, 678.
Levy, G. C., and Nelson, G. L. (1972). "Carbon-13 Nuclear Magnetic Resonance for Organic Chemists", John Wiley, New York.

Llinas, M., Wilson, D. M., and Neilands, J. B. (1973). *Biochemistry* **12**, 3836.
Lukacs, G., Sepulchre, A.-M., Gateau-Olesker, A., Vass, G., Gero, S. D., Guthrie, R. D., Voelter, W., and Breitmaier, E. (1972). *Tetrahedron Lett.* 5163.
Luzzati, V. (1968). In "Biological Membranes" (D. Chapman, ed.), Academic Press, New York.
Lyerla, J. R., Jr., and Freedman, M. H. (1972). *J. Biol. Chem.* **247**, 8183.
Lyerla, J. R., Jr., and Grant, D. M. (1972). MTP International Review of Science, Physical Chem. Series One, (C. A. McDowell, ed.), Vol. 4 Butterworths, London.
Lyerla, J. R., Jr., Barber, B. H., and Freedman, M. H. (1973). *Canad. J. Biochem.* **51**, 460.
Mantsch, H. H., and Smith, I. C. P. (1972). *Biochem. Biophys. Res. Comm.* **46**, 808.
Mantsch, H. H., and Smith, I. C. P. (1973). *Canad. J. Chem.* **51**, 1384.
Matwiyoff, N. A., and Needham, T. E. (1972). *Biochem. Biophys. Res. Comm.* **49**, 1158.
Matwiyoff, N. A., Vergamini, P. S., Needham, T. E., Gregg, C. T., Volpe, J. A., Caughey, W. S. (1973). *J. Amer. Chem. Soc.* **95**, 4429.
McFarlane, W. (1970). *Chem. Comm.* 2118.
Metcalfe, J. C. (1972). *Chem. Phys. Lipids* **8**, 333.
Metcalfe, J. C., Birdsall, N. J. M., Feeney, J., Lee, A. G., Levine, Y. K., and Partington, P. (1971). *Nature* **233**, 199.
Metcalfe, J. C., Birdsall, N. J. M., and Lee, A. G. (1972). *FEBS Lett.* **21**, 335.
Metcalfe, J. C., Birdsall, N. J. M., and Lee, A. G. (1973a). *FEBS Symp.* **28**, 197.
Metcalfe, J. C., Birdsall, N. J. M., and Lee, A. G. (1973b). *Annals N.Y. Acad. Sci.* **222**, 460.
Mildvan, A. S., Nowak, T., and Fung, C.-H. (1973). *Annals N.Y. Acad. Sci.* **222**, 192.
Millett, F., Hargrave, P. A., and Raftery, M. A. (1973). *Biochemistry* **12**, 3591.
Miziorko, H. M., and Mildvan, A. S. (1974). *J. Biol. Chem.* **249**, 2743.
Mochel, V. D. (1972). *J. Macromol. Sci.* **C8**, 289.
Moon, R. B., and Richards, J. H. (1972a). *Proc. Nat. Acad. Sci. USA* **69**, 2193.
Moon, R. B., and Richards, J. H. (1972b). *J. Amer. Chem. Soc.* **94**, 5093.
Moon, R. B., Nelson, M. J., Richards, J. H., and Powars, D. F. (1974). *Physiol. Chem. and Physics* **6**, 31.
Morrow, J. S., Keim, P., Visscher, R. B., Marshall, R. C., and Gurd, F. R. N. (1973). *Proc. Nat. Acad. Sci. USA* **70**, 1414.
Morton, B. J., Long, R. C., Daniels, P. J. L., Tkach, R. W., and Goldstein, J. H. (1973). *J. Amer. Chem. Soc.* **95**, 7464.
Muller, N. (1973). In "Reaction Kinetics in Micelles" (E. H. Cordes, ed.), p. 1, Plenum Press, New York.
Muller, N., and Birkhahn, R. H. (1967). *J. Phys. Chem.* **71**, 957.
Muller, N., and Birkhahn, R. H. (1968). *J. Phys. Chem.* **72**, 583.
Muller, N., and Johnson, T. W. (1969). *J. Phys. Chem.* **73**, 2042.
Muller, N., and Platko, F. E. (1971). *J. Phys. Chem.* **75**, 547.
Muller, N., and Simsohn, H. (1971). *J. Phys. Chem.* **75**, 942.
Muller, N., Pellerin, J. H., and Chen, W. W. (1972). *J. Phys. Chem.* **76**, 3012.
Neuss, N., Koch, K. F., Molloy, B. B., Day, W., Huckstep, L. L., Dorman, D. E., and Roberts, J. D. (1970). *Helv. Chim. Acta* **53**, 2314.
Nigen, A. M., Keim, P., Marshall, R. C., Morrow, J. S., and Gurd, F. R. N. (1972). *J. Biol. Chem.* **247**, 4100.
Nigen, A. M., Keim, P., Marshall, R. C., Glushko, V., Lawson, P. J., and Gurd, F. R. N. (1973a). *J. Biol. Chem.* **248**, 3616.

Nigen, A. M., Keim, P., Marshall, R. C., Morrow, J. S., Vigna, R. A., and Gurd, F. R. N. (1973b). *J. Biol. Chem.* 248, 3724.
Noggle, J. H., and Schirmer, R. E. (1971). "The Nuclear Overhauser Effect", Academic Press, New York.
Nouaille, F., Sepulchre, A.-M., Lukacs, G., Kornprobst, A., and Gero, S. D. (1974). *Bull. Soc. Chim. France*, 143.
Ohnishi, M., Fedarko, M. D., Baldeschwieler, J. D., and Johnson, L. F. (1972). *Biochim. Biophys. Res. Comm.* 46, 312.
Oldfield, E., and Allerhand, A. (1973). *Proc. Nat. Acad. Sci. USA* 70, 3531.
Oldfield, E., and Chapman, D. (1971). *Biochem. Biophys. Res. Comm.* 43, 949.
Olson, J. S., Andersen, M. E., and Gibson, Q. H. (1971). *J. Biol. Chem.* 246, 5919.
Omoto, S., Inouye, S., Kojima, M., and Niida, T. (1973). *J. Antibiotics* 26, 717.
Oster, O., Breitmaier, E., and Voelter, W. (1974). In "Nuclear Magnetic Resonance Spectroscopy of Nuclei Other Than Protons" (T. Axenrod, and G. A. Webb, eds.), John Wiley and Sons, New York.
Ovchinnikov, Yu. A., Ivanov, V. T., Bystrov, V. F., and Miroshinikov, A. I. (1972). In "Chemistry and Biology of Peptides" (J. Meienhofer, ed.), Ann Arbor Science Publishers, Ann Arbor, Michigan.
Packer, E. L., Sternlicht, H., and Raboniwitz, J. C. (1972). *Proc. Nat. Acad. Sci. USA* 69, 3278.
Paolillo, L., Tancredi, T., Temussi, P. A., Trivellone, E., Bradbury, E. M., and Crane-Robinson, C. (1972). *Chem. Comm.* 335.
Patel, D. J. (1973a) *Biochemistry* 12, 496.
Patel, D. J. (1973b). *Biochemistry* 12, 667.
Patel, D. J. (1973c). *Biochemistry* 12, 677.
Patel, D. J. (1974). *Biochemistry* 13, 1476.
Pease, L. G., Deber, C. M., and Blout, E. R. (1973). *J. Amer. Chem. Soc.* 95, 258.
Pedersen, C. J., and Frensdorff, H. K. (1972). *Angew. Chem. I. E.* 7, 16.
Perlin, A. S., and Casu, B. (1969). *Tetrahedron Lett.* 2921.
Perlin, A. S., Casu, B., Sanderson, G. R., and Tse, J. (1972a). *Carbohydrate Res.* 21, 123.
Perlin, A. S., Ng Ying Kin, N. M. K., Bhattacharjee, S. S., and Johnson, L. F. (1972b). *Canad. J. Chem.* 50, 2437.
Perlin, A. S., Herve du Penhoat, P., and Isbell, H. S. (1973). *Advan. in Chem.* 117, 39.
Phillips, R. (1966). *Chem. Rev.* 66, 501.
Pretsch, E., Vasak, M., and Simon, W. (1972). *Helv. Chim. Acta* 55, 1098.
Que, L., Willie, G. R., Cashel, M., Bodley, J. W., and Gray, G. R. (1973). *Proc. Nat. Acad. Sci. USA* 70, 2563.
Que, L., Jr., and Gray, G. R. (1974). *Biochemistry* 13, 146.
Quirt, A. R., Lyerla, J. R., Jr., Peat, I. R., Cohen, J. S., Reynolds, W. F., and Freedman, M. H. (1974). *J. Amer. Chem. Soc.* 96, 570.
Reynolds, W. F., Peat, I. R., Freedman, M. H., and Lyerla, J. R., Jr. (1973a). *Canad. J. Chem.* 51, 1857.
Reynolds, W. F., Peat, I. R., Freedman, M. H., and Lyerla, J. R., Jr. (1973b). *J. Amer. Chem. Soc.* 95, 328.
Rinehart, K. L., Jr., Malik, J. M., Mystrom, R. S., Stroshane, R. M., Truitt, S. T., Taniguchi, M., Rolls, J. P., Haak, W. J., and Ruff, B. A. (1974). *J. Amer. Chem. Soc.* 96, 2263.
Roberts, G. C. K., and Jardetzky, O. (1970). *Advan. Protein Chem.* 24, 447.
Roberts, J. D. (1971). 23rd International Congress of Pure and Applied Chemistry (Special Lectures) 7, 71.

Roberts, R. T., and Chachaty, C. (1973). *Chem. Phys. Letts.* 22, 348.
Robinson, J. D., Birdsall, N. J. M., Lee, A. G., and Metcalfe, J. C. (1972). *Biochemistry* 11, 2903.
Rodgers, P., and Roberts, G. C. K. (1973). *FEBS Lett.* 36, 330.
Saito, H., and Smith, I. C. P. (1973). *Arch. Biochem. Biophys.* 158, 154.
Saito, H., and Smith, I. C. P. (1974). *Arch. Biochem. Biophys.* 163, 699.
Schaefer, J., and Stejskel, E. O. (1974). *J. Amer. Oil Chem. Soc.* 51, 210.
Schwarcz, J. A., and Perlin, A. S. (1972). *Canad. J. Chem.* 50, 3667.
Schweizer, M. P., and Kreishman, G. P. (1973). *J. Mag. Resonance* 9, 334.
Schweizer, M. P., Broom, A. D., Ts'o, P. O. P., and Hollis, D. P. (1968). *J. Amer. Chem. Soc.* 90, 1042.
Schweizer, M. P., Witkowski, J. T., and Robins, R. K. (1971). *J. Amer. Chem. Soc.* 93, 277.
Schweizer, M. P., Banta, E. B., Witkowski, J. T., and Robins, R. K. (1973). *J. Amer. Chem. Soc.* 95, 3770.
Sears, B. D. (1971). Ph.D. Dissertation, Indiana University.
Sepulchre, A.-M., Septe, B., Lukacs, G., Gero, S. P., Voelter, W., and Breitmaier, E. (1974). *Tetrahedron* 30, 905.
Shulman, R. G., Ogawa, S., Mayer, A., and Castillo, C. L. (1973). *Annals N.Y. Acad. Sci.* 222, 9.
Smith, I. C. P., Deslauriers, R., and Walter, R. (1972). "Chemistry and Biology of Peptides", Ann Arbor Science Publishers, Ann Arbor, Michigan.
Smith, I. C. P., Mantsch, H. H., Lapper, R. D., Deslauriers, R., and Schleich, T. (1973a). In "Conformation of Biological Molecules and Polymer" (E. Bergman, and B. Pullman, eds.) Academic Press, New York, 381.
Smith, I. C. P., Deslauriers, R., Saito, H., Walter, R. W., Garrigou-Lagrange, C., McGregor, H., and Sarantakis, D. (1973b). *Annals N.Y. Acad. Sci.* 222, 597.
Sogn, J. A., Craig, L. C., and Gibbons, W. A. (1974). *J. Amer. Chem. Soc.* 96, 3306.
Sternlicht, H., Kenyon, G. L., Packer, E. L., and Sinclair, J. (1971). *J. Amer. Chem. Soc.* 93, 199.
Stoffel, W., Zierenberg, O., and Tunggal, B. D. (1972). *Hoppe-Seyler's Z. Physiol. Chem.* 354, 1962.
Stothers, J. B. (1972a). "Carbon-13 NMR Spectroscopy", Academic Press, New York.
Stothers, J. B. (1972b). *Applied Spectroscopy* 26, 1.
Tadokoro, S., Fujiwara, S., and Ichihara, Y. (1973). *Chem. Lett.* 849.
Takeo, K., Hirose, K., and Kuge, T. (1973). *Chem. Lett.* 1233.
Thomas, W. A., and Williams, M. K. (1972). *Chem. Comm.* 994.
Torchia, D. A. (1974). Unpublished results cited in Dorman and Bovey, 1973.
Torchia, D. A., and Lyerla, J. R., Jr. (1974). *Biopolymers* 13, 97.
Torchia, D. A., and Piez, K. A. (1973). *J. Mol. Biol.* 76, 419.
Tran-Dinh, S., Fermandjian, S., Sala, E., Mermet-Bouvier, R., Cohen, M., and Fromageot, P. (1974). *J. Amer. Chem. Soc.* 96, 1484.
Urry, D. W. (1974). *Research/Development* 25, 18.
Usui, T., Kobayashi, M., Yamaoka, N., Matsuda, K., Tuzimura, K., Sugiyama, H., and Seto, S. (1973a). *Tetrahedron Lett.* 3397.
Usui, T., Yamaoka, N., Matsuda, K., Tuzimura, K., Sugiyama, H., and Seto, S. (1973b). *J. C. S. Perkin I*, 2425.
Vergamini, P. S., Matwiyoff, N. A., Wohl, R. C., and Bradley, T. (1973). *Biochem. Biophys. Res. Comm.* 55, 453.
Viglino, P., Franconi, C., Lai, A., Brosio, E., and Conti, F. (1972). *Org. Mag. Res.* 4, 237.

Villafranca, J. J., and Viola, R. E. (1974). *Arch. Biochem. Biophys.* **160**, 465.
Voelter, W., and Breitmaier, E. (1973). *Org. Mag. Res.* **5**, 311.
Voelter, W., and Oster, O. (1972). *Chem.-Zeitung* **96**, 586.
Voelter, W., and Oster, O. (1973). *Org. Mag. Res.* **5**, 547.
Voelter, W., Jung, G., Breitmaier, E., and Bayer, E. (1971). *Z. Naturforsch* **26b**, 213.
Voelter, W., Bilik, U., and Breitmaier, E. (1973a). *Coll. Czech. Chem. Comm.* **38**, 2054.
Voelter, W., Oster, O., and Breitmaier, E. (1973b). *Z. Naturforsch.* **28b**, 370.
Walter, R., Prasad, K. U. M., Deslauriers, R., and Smith, I. C. P. (1973). *Proc. Nat. Acad. Sci. USA* **70**, 2086.
Walter, R., Smith, I. C. P., and Deslauriers, R. (1974). *Biochem. Biophys. Res. Comm.* **58**, 216.
Wenkert, Ernest, Hagaman, E. W., and Gutowski, Gerald E. (1973). *Biochem. Biophys. Res. Comm.* **51**, 318.
Wenkert, E., Bindra, J. S., Chang, C. J., Cochran, D. W., and Schell, F. M. (1974). *Accounts. Chem. Res.* **7**, 46.
Weser, U. (1968). *Struct. Bonding (Berlin)* **5**, 41.
Weser, U., Strobel, G.-J., and Voelter, W. (1974). *FEBS Lett.* **41**, 243.
Wilkins, M. H. F., Blaurock, A. E., and Engelman, D. M. (1971). *Nature* **230**, 72.
Williams, E., Hamilton, J. A., Jain, M. K., Allerhand, A., Cordes, E. H., and Ochs, S. (1973a). *Science* **181**, 869.
Williams, E., Sears, B., Allerhand, A., and Cordes, E. H. (1973b). *J. Amer. Chem. Soc.* **95**, 4871.
Wilson, N. K., and Stothers, J. B. (1974). *Topics in Stereochem.* **8**, 1.
Woo, P. W. K., and Westland, R. D. (1973). *Carbohydrate Res.* **31**, 27.
Wüthrich, K., Tun-Kyi, A., and Schwyzer, R. (1972). *FEBS Lett.* **25**, 104.
Yamaoka, N., Usui, T., Matsuda, K., Tuzimura, K., Sugiyama, H., and Seto, S. (1971). *Tetrahedron Lett.* 2047.
Zachau, H. G. (1969). *Angew. Chem. I. E.* **8**, 711.
Zimmer, S., Haar, W., Maurer, W., and Ruterjans, H. (1972). *Eur. J. Biochem.* **29**, 80.

ADDENDUM

^{13}C Nmr of macromolecular systems is an explosively growing area of research. Here we list relevant papers which came to our attention after submitting the manuscript.

I. *General*

Allerhand, A. (1975). *Pure and Applied Chemistry* **41**, 247. Natural-Abundance Carbon-13 Fourier Transform NMR Studies of Large Molecules.
Gray, G. A. (1975). *Anal. Chem.* **47**, 546A. Carbon-13 Nuclear Magnetic Resonance Spectroscopy.
Levy, G. C. (1974). "Topics in Carbon-13 NMR Spectroscopy", Vol. 1.
Levy, G. C. (1975). "Topics in Carbon-13 NMR Spectroscopy", Vol. 2.
Thermodynamics Research Center (1975). Serial Publication of American Petroleum Institute Research Project 44. *Selected* ^{13}C *Nuclear Magnetic Resonance Spectral Data.* Volume No. G-1, G-2.

II. *Carbohydrates*

Benkovic, S. J., Villafranca, J. J., and Kleinschuster, J. J. (1973). *Arch. Biochem. Biophys.* **155**, 458. ^{31}P and ^{13}C nmr Measurements of a Fructose-1-Phosphate, Mn(II) and Fructose 1,6-Diphosphatase Complex.

Bhattacharjee, A. K., Jennings, H. J., Kenny, C. P., Martin, A., and Smith, I. C. P. (1975). *J. Biol. Chem.* **250**, 1926. Structural Determination of the Sialic Acid Polysaccharide Antigens of *Neisseria Meningitidis* Serogroups B and C with Carbon 13 Nuclear Magnetic Resonance.

Binkley, R. W., Binkley, W. W., and Wickberg, B. (1974). *Carbohydrate Res.* **36**, 196. Application of cmr Spectroscopy to the Problem of the Anomeric Configuration of the di-D-fructose Dianhydrides.

Bock, K., and Hall, L. D. (1975). *Carbohydrate Res.* **40**, C3. Carbon-13 Spin-Lattice Relaxation-Times of Some Carbohydrate Derivatives.

Colson, P., Jennings, H. J., and Smith, I. C. P. (1974). *J. Amer. Chem. Soc.* **96**, 8081. Composition, Sequence, and Conformation of Polymers and Oligomers of Glucose as Revealed by Carbon-13 Nuclear Magnetic Resonance.

Colson, P., Slessor, K. N., Jennings, H. J., and Smith, I. C. P. (1975). *Can. J. Chem.* **53**, 1030. A Carbon-13 Nuclear Magnetic Resonance Study of Chlorinated and Polyol Analogs of Glucose and Related Oligomers.

Gorin, P. A. J. (1975). *Carbohydrate Res.* **39**, 3. Assignment of Signals of the Carbon-13 Magnetic Resonance Spectrum of a Selected Polysaccharide: Comments on Methodology.

Gorin, P. A. J., and Mazurek, M. (1974). *Can. J. Chem.* **52**, 3070. Structure of a Phosphonomannan, as Determined by the Effect of Lanthanide Ions on Its Carbon-13 Magnetic Resonance Spectrum.

Gorin, P. A. J., and Mazurek, M. (1975). *Can. J. Chem.* **53**, 1212. Further Studies on the Assignment of Signals in ^{13}C Magnetic Resonance Spectra of Aldoses and Derived Methyl Glycosides.

Haverkamp, J., de Bie, M. J. A., and Vliegenthart, J. F. G. (1974). *Carbohydrate Res.* **37**, 111. ^{13}C- and ^{1}H-Nuclear Magnetic Resonance Spectroscopy of Permethylated Disaccharides.

Haverkamp, J., van Dongen, J. P. C. M., and Vliegenthart, J. F. G. (1974). *Carbohydrate Res.* **33**, 319. ^{13}C- and ^{1}H-NMR Spectroscopy of Permethylated α- and β-D-Galactopyranoses.

Hutchinson, C. R. (1974). *J. Org. Chem.* **39**, 1854. A Synthesis of Tulipalin A and B and the Acylglucoside, Tuliposide A, Fungitoxic Agents from *Tulipa Gesneriana*. Carbon-13 Nuclear Magnetic Resonance Analysis of Anomeric Configuration in Acylglucosides.

Lemieux, R. U., and Driguez, H. (1975). *J. Amer. Chem. Soc.* **97**, 4063. The Chemical Synthesis of 2-Acetamido-2-deoxy-4-O-(α-L-fucopyranosyl)-3-O-(β-D-galactopyranosyl)-D-glucose. The Lewis a Blood-Group Antigenic Determinant.

Lemieux, R. U., and Driguez, H. (1975). *J. Amer. Chem. Soc.* **97**, 4069. The Chemical Synthesis of 2-O-(α-L-Fucopyranosyl)-3-O-(α-D)-galactopyranosyl-D-galactose. The Terminal Structure of the Blood-Group B Antigenic Determinant.

Lemieux, R. U., Bundle, D. R., and Baker, D. A. (1975). *J. Amer. Chem. Soc.* **97**, 4076. The Properties of a "Synthetic" Antigen Related to the Human Blood-Group Lewis a.

Lundt, I., and Pederson, C. (1974). *Carbohydrate Res.* **35**, 187. Preparation of Some Acylated D-arabino-Hexopyranosuloses and 4-Deoxy-D-Glycero-Hex-3-Enopyranosuloses.

Miljković, M., Gligorijević, M., Satoh, T., Glišin, D., and Pitcher, R. G. (1974). *J.*

Org. Chem. **39**, 3847. Carbon-13 Nuclear Magnetic Resonance Spectra of Branched-Chain Sugars, Configurational Assignment of the Branching Carbon Atom of Methyl Branched-Chain Sugars.

Munro, M. H. G., Taniguchi, M., Rinehart, K. L., Jr., Gottlieb, D., Stoudt, T. H., and Rogers, T. O. (1975). *J. Amer. Chem. Soc.* **97**, 4782. Carbon-13 Evidence for the Stereochemistry of Streptomycin Biosynthesis from Glucose.

Nagabhushan, T. L., and Daniels, P. J. L. (1975). *Tetrahedron Lett.* 747. Carbon-13 Magnetic Resonance Spectroscopy and Absolute Configuration of Anomeric Center in Axially Linked 4-O- and/or 6-O-Glycopyranosyl Derivatives of Deoxystreptamine.

Nair, V., and Walsh, R. H. (1974). *Carbohydrate Res.* **36**, 131. 2-Amino-2,5-Anhydro-2-Deoxy-D L-Ribitol: An Amino Sugar Derivative Having Nitrogen in the Ring.

Omura, S., Nakagawa, A., Neszmelyi, A., Gero, S. D., Sepulchre, A.-M., Piriou, F., and Lukacs, G. (1975). *J. Amer. Chem. Soc.* **97**, 4001. Carbon-13 Nuclear Magnetic Resonance Spectral Analysis of 16-Membered Macrolide Antibiotics.

Omura, S., Neszmelyi, A., Sangare, M., and Lukacs, G. (1975). *Tetrahedron Lett.* 2939. Conformational Homogeneity in Solution of 14-Membered Macrolide Antibiotics as Evidence by ^{13}C nmr Spectroscopy.

Perlin, A. S., Natsuko, C., Ritchie, R. G. S., and Parfondry, A. (1974). *Carbohydrate Res.* **37**, C1. $^{13}C-^{1}H$ Coupling in Natural Abundance ^{13}C Spectra. Applications to Branched Sugars, Disaccharides and Carbinol Groups.

Rao, G. V., Que, L., Jr., Hall, L. D., and Fondy, T. P. (1975). *Carbohydrate Res.* **40**, 311. Deoxyfluoroketohexoses: 4-Deoxy-4-fluoro-D-sorbose and -tagatose and 5-Deoxy-5-fluoro-L-sorbose.

Ritchie, R. G. S., Cyr, N., Korsch, B., Koch, H. J., and Perlin, A. S. (1975). *Can. J. Chem.* **53**, 1424. Carbon-13 Chemical Shifts of Furanosides and Cyclopentanols. Configurational and Conformational Influences.

Szarek, W. A., Vyas, D. M., Gero, S. D., and Lukacs, G. (1974). *Can. J. Chem.* **52**, 3394. Application of Carbon-13 Nuclear Magnetic Resonance Spectroscopy to the Structural Determination of Chlorodeoxy Sugars.

Taravel, F. R., and Vottero, Ph. J. A. (1975). *Tetrahedron Lett.* 2341. Application de la rmn du ^{13}C a L'Etude des Oligo et Polysaccharides. Configuration Anomerique.

Usui, T., Yamaoko, N., Matsuda, K., Tuzimura, K., Sugiyama, H., and Seto, S. (1975). *Agric, Biol. Chem.* **39**, 1071. Carbon-13 and Proton Nuclear Magnetic Resonance for Structural Investigation of Some Glucans.

Usui, T., Sugiyama, H., Seto, S., Araya, S., Nisizawa, T., Imai, S., and Kosaka, K. (1975). *J. Biochem. (Tokyo)* **78**, 225. Structural Determination of Glucans from Streptococcus Mutans JC-2(Dental Caries Bacterium) by Carbon-13 Nuclear Magnetic Resonance.

III. Nucleic acids and constituents

Birdsall, B., Birdsall, N. J. M., Feeney, J., and Thornton, J. (1975). *J. Amer. Chem. Soc.* **97**, 2845. A Nuclear Magnetic Resonance Investigation of the Conformation of Nicotinamide Mononucleotide in Aqueous Solution.

Chenon, M. T., Pugmire, R. J., Grant, D. M., Panzica, R. P., and Townsend, L. B. (1973). *J. Heterocyclic Chem.* **10**, 427. Carbon-13 nmr Spectra of C-Nucleosides. Showdomycin and β-Pseudouridine.

Chenon, M. T., Pugmire, R. J., Grant, D. M., Panzica, R. P., and Townsend, L. B. (1973). *J. Heterocyclic Chem.* **10**, 431. Carbon-13 nmr Spectra of C-Nucleo-

sides. II. A Study of the Tautomerism of Formycin and Formycin B by the use of cmr Spectroscopy.
Chenon, M.-T., Pugmire, R. J., Grant, D. M., Panzica, R. P., and Townsend, L. B. (1975). *J. Amer. Chem. Soc.* 97, 4627. Carbon-13 Magnetic Resonance. XXV. A Basic Set of Parameters for the Investigation of Tautomerism in Purines Established from Carbon-13 Magnetic Resonance Studies Using Certain Purines and Pyrrolo[2,3-d]-pyrimidines.
Chenon, M.-T., Pugmire, R. J., Grant, D. M., Panzica, R. P., and Townsend, L. B. (1975). *J. Amer. Chem. Soc.* 97, 4636. Carbon-13 Magnetic Resonance. XXVI. A Quantitative Determination of the Tautomeric Populations of Certain Purines.
Fritzsche, H., Arnold, K., and Krusche, R. (1974). *Stud. Biophys.* 45, 131. Interactions of DNA and Metal Ions. Carbon-13 nmr Study of Some Nucleosides and Nucleotides Adding Copper(II) and Manganese(II) Ions.
Imoto, T., Akasaka, K., and Hatano, H. (1975). *Chem. Phys. Lett.* 32, 86. Rotational Correlation Times of Adenosine 5'-Monophosphate in Solution as Deduced From Proton and Carbon-13 Spin-Lattice Relaxation Times.
Kuntz, G. P. P., and Kotowycz, G. (1975). *Biochemistry* 14, 4144. A Nuclear Magnetic Resonance Relaxation Time Study of the Manganese(II)-Inosine 5'-Triphosphate Complex in Solution.
Lüdemann, H.-D., and Röder, O. (1975). *J. Amer. Chem. Soc.* 97, 4402. Comment on the Carbon-13 Nuclear Relaxation Measurements in Adenosine Monophosphate.
Morr, M., Kula, M.-R., and Ernst, L. (1975). *Tetrahedron* 31, 1619. Synthesis and ^{13}C nmr Spectra of 3'-Amido Analogs of Adenosine 3',5'-Cyclic Monophosphate.
Röder, O., Lüdemann, H. D., and Von Goldammer, E. (1975). *Eur. J. Biochem.* 53, 517. Determination of the Activation Energy for Pseudorotation of the Furanose Ring in Nucleosides by ^{13}C Nuclear Magnetic Resonance Relaxation.
Roeder, S. B. W., Master, B. S., and Stewart, C. J. (1975). *Physiol. Chem. and Physics* 7, 115. ^{13}C nmr of Coenzyme A and Its Constituent Moieties.
Sugiyama, H., Yamaoka, N., Shimizu, B., Ishido, Y., and Seto, S. (1974). *Bull. Chem. Soc. Jap.* 47, 1815. Carbon-13 Nuclear Magnetic Resonance Spectra of α-Ribonucleosides.
Uchida, K., Breitmaier, E., and Koenig, W. A. (1975). *Tetrahedron* 31, 2315. ^{13}C Nmr Investigations of the Nucleoside Antibiotic Hikizimycin and its Constituents.
Way, J. L., Birdsall, B., Feeney, J., Roberts, G. C. K., and Burgen, A. S. V. (1975). *Biochemistry* 14, 3470. A Nuclear Magnetic Resonance Study of Nicotinamide Adenine Dinucleotide Phosphate Binding to *Lactobacillus casei* Dihydrofolate Reductase.
Yokono, T., Shimokawa, S., and Sohma, J. (1975). *J. Amer. Chem. Soc.* 97, 3827. Nuclear Magnetic Resonance Studies on Effects of Ions (Anions and Cations) to Nucleosides (Cytidine and Guanosine).

IV. *Proteins and their residues*

Bauer, D. R., Opella, S. J., Nelson, D. J., and Pecora, R. (1975). *J. Amer. Chem. Soc.* 97, 2580. Depolarized Light Scattering and Carbon Nuclear Resonance Measurements of the Isotropic Rotational Correlation Time of Muscle Calcium Binding Protein.
Baxter, C. S., and Byvoet, P. (1975). *Biochem. Biophys. Res. Comm.* 64, 514. Cmr Studies of Protein Modification. Progressive Decrease in Charge Density at the ϵ-Amino Function of Lysine with Increasing Methyl Substitution.

Blout, E. R., Bovey, F. A., Goodman, M., and Lotan, N. (1974). "Peptides Polypeptides and Proteins", John Wiley and Sons, New York.

Boettcher, B., and Martinez-Carrion, M. (1975). *Biochem. Biophys. Res. Comm.* 64, 28. Glutamate Aspartate Transaminase Modified at Cysteine 390 with Enriched Carbon-13 Cyanide.

Boni, R., DiBlasi, R., and Verdini, A. S. (1975). *Macromolecules* 8, 140. Conformational Properties of Poly(L-Azetidine-2-Carboxylic Acid) in Solution as Studied by Carbon-13 and Proton Nuclear Magnetic Resonance Spectroscopy.

Chaiken, I. M., Cohen, J. S., and Sokoloski, E. A. (1974). *J. Amer. Chem. Soc.* 96, 4703. Microenvironment of Histidine 12 in Ribonuclease-S as Detected by ^{13}C Nuclear Magnetic Resonance.

Chaiken, I. M., Randolph, R. E., and Taylor, H. C. (1975). *Ann. N.Y. Acad. Sci.* 248, 442. Conformational Effects Associated with the Interaction of Polypeptide Ligands with Neurophysins.

Chein, J. C. W., and Wise, W. B. (1975). *Biochemistry* 14, 2786. A ^{13}C Nuclear Magnetic Resonance and Circular Dichroism Study of the Collagen-Gelatin Transformation in Enzyme Solubilized Collagen.

Climie, I. J. G., Evans, D. A., and Akhtar, M. (1975). *Chem. Comm.* 160. Labelling of Amino-Acid Side-Chains with ^{13}C-Labelled Electrophiles; Potential Application to the Probing of Active Sites of Enzymes.

Cozzone, P. J., Opella, S. J., Jardetzky, O., Berthou, J., and Jolles, P. (1975). *Proc. Nat. Acad. Sci. U.S.A.* 72, 2095. Detection of New Temperature-Dependent Conformational Transition in Lysozyme by Carbon-13 Nuclear Magnetic Resonance Spectroscopy.

Cutnell, J. D., Glasel, J. A., and Hruby, V. J. (1975). *Ann. N.Y. Acad. Sci.* 248, 458. Nonpolar Contributions to ^{13}C Relaxation in Molecules.

Cutnell, J. D., Glasel, J. A., and Hruby, V. J. (1975). *Org. Mag. Res.* 7, 256. An Investigation of Contributions to Carbon-13 Spin-Lattice Relaxation in Amino Acids and Peptide Hormones.

Darke, A., and Finer, E. G. (1975). *Biopolymers* 14, 441. Nmr Studies of Mixtures of Poly-L-Lysine Hydrobromide with Water.

Deber, C. M., and Blout, E. R. (1974). *J. Amer. Chem. Soc.* 96, 7566. Amino Acid-Cyclic Peptide Complexes.

Deber, C. M., Fossel, E. T., and Blout, E. R. (1974). *J. Amer. Chem. Soc.* 96, 4015. Cyclic Peptides. VIII. ^{13}C and ^{1}H Nuclear Magnetic Resonance Evidence for Slow $Cis'-Trans'$ Rotation in a Cyclic Tetrapeptide.

Deslauriers, R., McGregor, W. H., Sarantakis, D., and Smith, I. C. P. (1974). *Biochemistry* 13, 3443. Carbon-13 Nuclear Magnetic Resonance Studies of Structure and Function in Thyrotropin-Releasing Factor. Determination of the Tautomeric Form of Histidine and Relationship to Biology Activity.

Deslauriers, R., Paiva, A. C. M., Schaumburg, K., and Smith, I. C. P. (1975). *Biochemistry* 14, 878. Conformational Flexibility of Angiotensin II. A Carbon-13 Spin-Lattice Relaxation Study.

Deslauriers, R., Grzonka, Z., Schaumburg, K., Shiba, T., and Walter, R. (1975). *J. Amer. Chem. Soc.* 97, 5093. Carbon-13 Nuclear Magnetic Resonance Studies of the Conformations of Cyclic Dipeptides.

Deslauriers, R., Levy, G. C., McGregor, W. H., Sarantakis, D., and Smith, I. C. P. (1975). *Biochemistry* 14, 4335. Conformational Flexibility of Luteinzing Hormone-Releasing Hormone in Aqueous Solution. A Carbon-13 Spin-Lattice Relaxation Time Study.

DiBlasi, R., and Verdini, A. S. (1974). *Biopolymers* 13, 765. ^{13}C Nmr Study of the Helix-Coil Transition of Poly-N^{5}-(30Hydroxypropyl)-L-Glutamine.

DiBlasi, R., and Verdini, A. S. (1974). *Biopolymers* 13, 2209. Conformational

Behavior of Poly-N^5-(3-Hydroxypropyl)-L-Glutamine in Water–Methanol Mixtures Studied by ^{13}C nmr and CD Spectroscopy.

DisBlasi, R., and Kopple, K. D. (1975). *Chem. Comm.* **33**. Vicinal $^{13}C-^{15}N$ Coupling Constants in Peptides.

Dwek, R. A. (1973). "Monographs on Physical Biochemistry, Nuclear Magnetic Resonance in Biochemistry: Applications to Enzyme Systems", Clarendon Press, Oxford.

Eakin, R. T., Morgan, L. O., and Matwiyoff, N. A. (1975). *Biochemistry* **14**, 4538. Carbon-13 Nuclear Magnetic Resonance Spectroscopy of [2-^{13}C]Carboxymethylcytochrome c.

Evans, C. A., and Rabenstein, D. L. (1974). *J. Amer. Chem. Soc.* **96**, 7312. Nuclear Magnetic Resonance Studies of the Acid-Base Chemistry of Amino Acids and Peptides. II. Dependence of the Acidity of the C-Terminal Carboxyl Group on the Conformation of the C-Terminal Peptide Bond.

Feeney, J., Hansen, P. E., and Roberts, G. C. K. (1974). *Chem. Comm.* **465**. Use of $^{13}C-^1H$ Spin-Coupling Constants in the Determination of Side-Chain Conformations of Amino-Acids.

Fossel, E. T., Easwaran, K. R. K., and Blout, E. R. (1975). *Biopolymers* **14**, 927. A ^{13}C Spin-Lattice Relaxation Study of Dipeptides Containing Glycine and Proline: Mobility of the Cyclic Proline Side Chain.

Fossel, E. T., Veatch, W. R., Ovchinnikov, Y. A., and Blout, E. R. (1974). *Biochemistry* **13**, 5264. A ^{13}C Nuclear Magnetic Resonance Study of Gramicidin A in Monomer and Dimer Forms.

Grathwohl, C., Tun-Kyi, A., Bundi, A., Schwyzer, R., and Wüthrich, K. (1975). *Helv. Chim. Acta* **58**, 415. ^1H- and ^{13}C-Nmr Studies of the Molecular Conformations of Cyclo-Tetraglycyl.

Griffin, J. H., Alazard, R., Dibello, C., Sala, E., Mermet-Bouvier, R., and Cohen, P. (1975). *FEBS Letters* **50**, 168. Carbon-13 Nuclear Magnetic Resonances Studies on (85% ^{13}C-Enriched Gly^9) Oxytocin.

Hansen, P. E., Feeney, J., and Roberts, G. C. K. (1975). *J. Mag. Res.* **17**, 249. Long Range $^{13}C-^1H$ Spin-Spin Coupling Constants in Amino Acids. Conformational Applications.

Harris, D. C., Gray, G. A., and Aisen, P. (1974). *J. Biol. Chem.* **249**, 5261. ^{13}C Nuclear Magnetic Resonance Study of the Spatial Relation of the Metal- and Anion-binding Sites of Human Transferrin.

Hollstein, U., Breitmaier, E., and Jung, G. (1974). *J. Amer. Chem. Soc.* **96**, 8036. ^{13}C Nuclear Magnetic Resonance Study of Actinomycin D.

Hunkapiller, M. W., Smallcombe, S. H., and Richards, J. H. (1975). *Org. Mag. Res.* **7**, 262. Mechanism of Serine Protease Action. Ionization Behaviour of Tetrahedral Adduct Between α-Lytic Protease and Tripeptide Aldehyde Studied by Carbon-13 Magnetic Resonance.

Jones, W. C., Jr., Rothgeb, T. M., and Gurd, F. R. N. (1975). *J. Amer. Chem. Soc.* **97**, 3875. Specific Enrichment with ^{13}C of the Methionine Methyl Groups of Sperm Whale Myoglobin.

Jung, G., Dubischar, N., and Leibfritz, D. (1975). *Eur. J. Biochem.* **54**, 395. Conformational Changes of Alamethicin Induced by Solvent and Temperature. A ^{13}C nmr and Circular-Dichroism Study.

Keim, P., Vigna, R. A., Nigen, A. M., Morrow, J. S., and Gurd, F. R. N. (1974). *J. Biol. Chem.* **249**, 4149. Carbon-13 Nuclear Magnetic Resonance of Pentapeptides of Glycine Containing Central Residues of Methionine, Proline, Arginine and Lysine.

Kessler, H., and Molter, M. (1974). *Angew. Chem. I. E.* **13**, 538. Use of the $Eu(fod)_3$-Induced Shift for Determination of the Complexation Site and for Signal Assignment in the ^{13}C-nmr Spectra of Protected Dipeptides.

Koenig, S. H., Brown, R. D., London, R. E., Needham, T. E., and Matwiyoff, N. A. (1974). *Pure Appl. Chem.* 40, 103. Kinetic Parameters of Carbonic Anhydrase by Carbon-13 nmr.

Komoroski, R. A., Peat, I. R., and Levy, G. C. (1975). *Biochem. Biophys. Res. Comm.* 65, 272. High Field Carbon-13 nmr Spectroscopy. Conformational Mobility in Gramicidin S and Frequency Dependence of ^{13}C Spin-Lattice Relaxation Times.

Lackner, H. (1975). *Angew. Chem. I. E.* 14, 375. Three-Dimensional Structure of the Actinomycins.

Led, J. J., Grant, D. M., Horton, W. J., Sunby, F., and Vilhelmsen, K. (1975). *J. Amer. Chem. Soc.* 97, 5997. Carbon-13 Magnetic Resonance Study of Structural and Dynamical Features in Carbamylated Insulins.

Maggio, E. T., Kenyon, G. L., Mildvan, A. S., and Hegeman, G. D. (1975). *Biochemistry* 14, 1131. Mandelate Racemase from *Pseudomonas Putida*. Magnetic Resonance and Kinetic Studies of the Mechanism of Catalysis.

Matwiyoff, N. A., and Burnham, B. F. (1973). *Ann. N. Y. Acad. Sci.* 206, 365. Carbon-13 nmr Spectroscopy of Tetrapyrroles.

Moon, R. B., and Richards, J. H. (1974). *Biochemistry* 13, 3437. ^{13}C Magnetic Resonance Studies of the Binding of Carbon Monoxide to Various Hemoglobins.

Morrow, J. S., Keim, P., and Gurd, F. R. N. (1974). *J. Biol. Chem.* 249, 7484. CO_2 Adducts of Certain Amino Acids, Peptides, and Sperm Whale Myoglobin Studied by Carbon-13 and Proton Nuclear Magnetic Resonance.

Neville, G. A., and Drakenberg, T. (1974). *Can. J. Chem.* 52, 616. Mercury(II) Complexation of Cysteine, Methyl Cysteineate, and S-Methylcysteine in Acidic Media.

Norton, R. S., and Bradbury, J. H. (1974). *Chem. Comm.* 870. Carbon-13 Nuclear Magnetic Resonance Study of Tyrosine Titrations.

Oldfield, E., and Allerhand, A. (1975). *J. Amer. Chem. Soc.* 97, 221. Identification of Tryptophan Resonances in Natural Abundance Carbon-13 Nuclear Magnetic Resonance Spectra of Proteins. Application of Partially Relaxed Fourier Transform Spectroscopy.

Oldfield, E., Norton, R. S., and Allerhand, A. (1975). *J. Biol. Chem.* 250, 6368. Studies of Individual Carbon Sites of Proteins in Solution by Natural Abundance Carbon-13 Nuclear Magnetic Resonance Spectroscopy. Relaxation Behavior.

Oldfield, E., Norton, R. S., and Allerhand, A. (1975). *J. Biol. Chem.* 250, 6381. Studies of Individual Carbon Sites of Proteins in Solution by Natural Abundance Carbon-13 Nuclear Magnetic Resonance Spectroscopy. Strategies for Assignments.

Oldfield, E., and Allerhand, A. (1975). *J. Biol. Chem.* 250, 6403. Studies of Individual Carbon Sites of Hemoglobins in Solution by Natural Abundance Carbon-13 Nuclear Magnetic Resonance Spectroscopy.

Packer, E., Sternlicht, H., Lode, E. T., and Rabinowitz, J. C. (1975). *J. Biol. Chem.* 250, 2062. The use of ^{13}C Nuclear Magnetic Resonance of Aromatic Amino Acid Residues to Determine the Midpoint Oxidation-Reduction Potential of Each Iron–Sulfur Cluster of *Clostridium Acidic-Urici* and *Clostridium Pasteurianum* Ferredoxins.

Pearson, H., Gust, D., Armitage, I. M., Huber, H., Roberts, J. D., Stark, R. E., Vold, R. R., and Vold, R. L. (1975). *Proc. Nat. Acad. Sci. U.S.A.* 72, 1599. Nuclear Magnetic Resonance Spectroscopy: Reinvestigation of Carbon-13 Spin-Lattice Relaxation Time Measurements of Amino Acids.

Robillard, G., Shaw, E., and Shulman, R. G. (1974). *Proc. Nat. Acad. Sci. U.S.A.* 71, 2623. ^{13}C High-Resolution Nuclear Magnetic Resonance Studies of

Enzyme-Substrate Reactions at Equilibrium. Substrate Strain Studies of Chymotrypsin-N-Acetyltyrosine Semicarbazide Complexes.

Servis, K. L., and Patel, D. J. (1975). *Tetrahedron* 31, 1359. Aspects of the Valinomycin Backbone Conformation in Chloroform Solution Using Lanthanide Shift Reagents, Evaluation of ϕ(D-Val) and ψ(D-HyIv) Rotation Angles.

Smith, I. C. P., Jennings, H. J., and Deslauriers, R. (1975). *Accounts Chem. Res.* 8, 306. Carbon-13 Nuclear Magnetic Resonance and the Conformations of Biological Molecules.

Sogn, J. A., Craig, L. C., and Gibbons, W. A. (1974). *J. Amer. Chem. Soc.* 96, 4696. $^{13}C-^{13}C$ Coupling Constants in a Series of ^{13}C-Enriched Amino Acids.

Suzuki, Y., Inoue, Y., and Chujo, R. (1975). *Biopolymers* 14, 1223. Carbon-13 and Proton nmr Studies of Helix-Coil Transition of Poly(γ-Benzyl-L-Glytamate).

Torchia, D. A., Lyerla, J. R., Jr., and Quatrone, A. J. (1975). *Biochemistry* 14, 887. Molecular Dynamics and Structure of the Random Coil and Helical States of the Collagen Peptide, α1-CB2, as Determined by ^{13}C Magnetic Resonance.

Tran-Dinh, S., Fermandjian, S., Sala, E., Mermet-Bouvier, R., and Fromageot, P. (1975). *J. Amer. Chem. Soc.* 97, 1267. Germinal and Vicinal $^{13}C-^{13}C$ Coupling Constants of 85% ^{13}C-Enriched Amino Acids.

Urry, D. W., Mitchell, L. W., and Ohnishi, T. (1974). *Biochemistry* 13, 4083. Carbon-13 Magnetic Resonance Assignments of the Repeat Peptides of Elastin and Solvent Delineation of Carbonyls.

Urry, D. W., Mitchell, L. W., and Ohnishi, T. (1974). *Biochem. Biophys. Res. Comm.* 59, 62. Solvent Dependence of Peptide Carbonyl Carbon Chemical Shifts and Polypeptide Secondary Structure: The Repeat Tetrapeptide of Elastin.

Urry, D. W., Mitchell, L. W., and Ohnishi, T. (1974). *Proc. Nat. Acad. Sci. U.S.A.* 71, 3265. Carbon-13 Magnetic Resonance Evaluation of Polypeptide Secondary Structure and Correlation with Proton Magnetic Resonance Studies.

Van-Binst, G., Biesemans, M., and Barel, A. O. (1975). *Bull. Soc. Chim. Belg.* 84, 1. Comparative Carbon-13 Nuclear Magnetic Resonance Study at 67·9 MHz on Lysozyme (Human and Egg-White) and α-Lactalbumin (Human and Bovine) in Their Native and Denatured State.

Visscher, R. B., and Gurd, F. R. N. (1975). *J. Biol. Chem.* 250, 2238. Rotational Motions in Myoglobin Assessed by Carbon-13 Relaxation Measurements at Two Magnetic Field Strengths.

Voelter, W., Fuchs, St., Seuffer, R. H., and Zech, K. (1974). *Monatshefte für Chemie* 105, 1110. ^{13}C-Nmr Studies of Protected Amino Acids.

Wüthrich, K. (1974). *Pure Appl. Chem.* 40, 127. Carbon-13 in Hemes and Hemo Proteins.

Wüthrich, K. (1974). *Pure Appl. Chem.* 37, 235. Studies of the Molecular Conformations in Proteins by ^1H and ^{13}C nmr Spectroscopy.

Yeagle, P. L., Lochmüller, C. H., and Henkens, R. W. (1975). *Proc. Nat. Acad. Sci. U.S.A.* 72, 454. ^{13}C Nuclear Magnetic Resonance Studies on the Mechanism of Action of Carbonic Anhydrase.

V. *Lipids*

Bermejo, J., Barbadillo, A., Tato, F., and Chapman, D. (1975). *FEBS Lett.* 52, 69. Magnetic Resonance Studies on the Interaction of Antidepressants with Lipid Model Membranes.

Brûlet, P., and McConnell, H. (1975). *Proc. Nat. Acad. Sci. U.S.A.* 72, 1451. Magnetic Resonance Spectra of Membranes.

Godici, P. E., and Landsberger, F. R. (1975). *Biochemistry* 14, 3927. ^{13}C Nuclear Magnetic Resonance Study of the Dynamic Structure of Lecithin-Cholesterol Membranes and the Position of Stearic Acid Spin-Labels.

Nicolau, Cl., Dietrich, W., Steiner, M. R., Steiner, S., and Melnick, J. L. (1975). *Biochim. Biophys. Acta* 382, 311. ^{1}H and ^{13}C Nuclear Magnetic Resonance Spectra of the Lipids in Normal and SV 40 Virus-Transformed Hamster Embryo Fibroblast Membranes.

Nicolau, Cl., Dreeskamp, H., and Schulte-Frohlinde, D. (1974). *FEBS Lett.* 43, 148. ^{13}C Nuclear Magnetic Resonance Relaxation Measurements of α-Lecithin-Peptide Interaction in Model Membranes.

Sears, B. (1975). *J. Membrane Biol.* 20, 59. ^{13}C Nuclear Magnetic Resonance Studies of Egg Phosphatidylcholine.

Shapiro, Y. E., Viktorov, A. V., Volkova, V. I., Barsukov, L. I., Bystrov, V. F., and Nergel'son, L. D. (1975). *Chem. Phys. Lipids* 14, 227. Carbon-13 Nmr Investigation of Phospholipid Membranes with the Aid of Shift Reagents.

Stoffel, W., and Bister, K. (1975). *Biochemistry* 14, 2841. ^{13}C Nuclear Magnetic Resonance Studies on the Lipid Organization in Enveloped Virions (Vesicular Stomatitis Virus).

Stoffel, W., Tunggal, B. D., Zierenberg, O., Schreiber, E., and Binczek, E. (1974). *Hoppe-Seyler's Z. Physiol. Chem.* 355, 1367. ^{13}C Nuclear Magnetic Resonance Studies of Lipid Interactions in Single- and Multi-Component Lipid Vesicles.

Stoffel, W., Zierenberg, O., Tunggal, B. D., and Schreiber, E. (1974). *Hoppe-Seyler's Z. Physiol. Chem.* 355, 1381. ^{13}C Nuclear Magnetic Resonance Spectroscopic Studies on Lipid-Protein Interactions in Human High-Density Lipoprotein (HDL) A Model of the HDL Particle.

Stoffel, W., Zierenberg, O., Tunggal, B., and Schreiber, E. (1974). *Proc. Nat. Acad. Sci. U.S.A.* 71, 3696. ^{13}C Nuclear Magnetic Resonance Spectroscopic Evidence for Hydrophobic Lipid-Protein Interactions in Human High Density Lipoproteins.

Urbina, J., and Waugh, J. S. (1974). *Proc. Nat. Acad. Sci. U.S.A.* 71, 5062. Proton-Enhanced ^{13}C Nuclear Magnetic Resonance of Lipids and Biomembranes.

VI. Macromolecular model systems

Brown, J. M., and Schofield, J. D. (1975). *Chem. Comm.* 434. Localized Regions of Reduced Mobility in Micelles. Carbon-13 Nmr Spin-Lattice Relaxation Times of Functional Surfactants in Aqueous Solution.

Davis, D. G., and Tosteson, D. C. (1975). *Biochemistry* 14, 3962. Nuclear Magnetic Resonance Studies of the Interactions of Anions and Solvent with Cation Complexes of Valinomycin.

Kalyanasundaram, K., Grätzel, M., and Thomas, J. K. (1975). *J. Amer. Chem. Soc.* 97, 3915. Electrolyte-Induced Phase Transitions in Micellar Systems. A Proton and Carbon-13 Nuclear Magnetic Resonances Relaxation and Photochemical Study.

Lackner, H. (1975). *Tetrahedron Lett.* 1921. ^{13}C-Nmr-Spektren der Pentapeptidlactonringe von Actinomycinen.

Levy, G. C., Komoroski, R. A., and Halstead, J. A. (1974). *J. Amer. Chem. Soc.* 96, 5456. Solvation and Segmental Motions of n-Alkylammonium Ions. A ^{13}C Spin-Lattice Relaxation Study.

Nourse, J. G., and Roberts, J. D. (1975). *J. Amer. Chem. Soc.* 97, 4584. Nuclear Magnetic Resonance Spectroscopy. Carbon-13 Spectra of Some Macrolide Antibiotics and Derivatives. Substituent and Conformational Effects.

Pines, A., and Chang, J. J. (1974). *J. Amer. Chem. Soc.* 96, 5590. Effect of Phase Transitions on Carbon-13 Nuclear Magnetic Resonance Spectra in p-Azoxydianisole, a Nematic Liquid Crystal.

Author Index

Numbers in italics refer to the pages on which references are listed at the end of each article. The authors in the Appendum in pp. 415ff. are not included in the index.

A

Aalbersberg, W. L., 158, 159, 163, 166, 167, 229, *265*
Abdullaev, N. D., 406, *410*
Abe, M., 62, *76*
Abraham, R. J., 59, 71, 72, *76*
Achiba, Y., 185, *272*
Adams, J. Q., 166, *265*
Adams, R. N., 159, 195, 198, 203, 207, 228, 233, *265, 267, 271, 273, 274, 276*
Addison, C. C., 85, *151*
Adler, A. D., 196, *269*
Ajipa, Y. I., 341, *411*
Akamatu, H., 165, 169, 195, 196, *275*
Akbulut, U., 251, *265*
Akhavein, A. A., 254, *265*
Akkerman, O. S., 207, *267*
Albert, A., 114, *148*
Albery, W. J., 202, *265*
Albrecht, A. C., 180, 181, *265, 267, 276*
Alcacer, L., 178, *265*
Alcais, P., 235, *265*
Alcock, N. W., 169, *265*
Alexandre, M., 393, 394, *406*
Allen, T. L., 40, *80*
Allerhand, A., 282, 286, 300, 301, 304, 306, 308, 309, 313, 318, 326, 327, 328, 329, 346, 347, 349, 350, 363, 364, 371, 372, 377, 381, 385, 388, 389, 391, 392, 394, *406, 408, 410, 411, 413, 415*
Allinger, J., 34, 35, 72, *76*
Allinger, N. L., 15, 16, 18, 21, 22, 24, 29, 30, 32, 33, 34, 35, 38, 39, 44, 47, 48, 49, 50, 53, 54, 55, 56, 57, 58, 59, 60, 61, 62, 63, 64, 65, 66, 67, 68, 69, 70, 72, 74, 75, 76, 77, *79, 82*
Almenningen, A., 34, 35, *77*
Altona, C., 29, 38, 59, 77, *79*
Alyev, I. Y., 234, *275*
Ambrose, J. F., 208, *265*
Anand, S. P., 235, *265*
Andersen, B., 34, 35, *77*
Andersen, M. E., 378, *413*
Anderson, J. N., 98, *148*
Anderson, W. A., 283, *406*
Andose, J. D., 16, 21, 23, 33, 38, 41, 44, 58, *79*
Andreades, S., 232, *265*
Andrews, D. H., 10, *77*
Andrieux, C. P., 201, 205, *265, 266, 276*
Andrulis, P. J., Jr. 170, 171, 175, *266*
Anet, F. A. L., 49, 56, 63, 64, 77, 280, 281, 282, *406*
Angyal, S. J., 22, 61, 63, 64, *79*, 300, *406*
Antonini, E., 374, 378, 379, *406*
Aoyama, Y., 166, *275*
Arison, B. H., 344, *406*
Arkhipova, S. F., 406, *410*
Armitage, I. M., 352, *406*
Armstrong, V. C., 92, *148*
Arnett, E. M., 84, 85, 86, 87, 88, 89, 90, 91, 92, 95, 96, 98, 99, 100, 104, 105, 108, 109, 114, 117, 119, 121, 122, 123, 124, 126, 127, 128, 132, 134, 135, 136, 137, 138, 139, 140, 142, 143, 144, 146, *148*
Arnold, D. R., 252, *274*, 276
Arnold, K., 339, *409*
Arrhenius, S., 86, *148*
Asai, S., 212, *272*
Asanuma, T., 252, *266, 278*
Asmus, K-D., 187, *274*
Assman, G., 390, *407*
Atkinson, T. V., 198, *266*
Auborn, J. J., 215, *266*
Aue, D. H., 135, *148*
Au-Young, Y. K., 231, *266*
Avila, E. M., 388, 389, *410*

B

Badger, B., 211, *266*
Bagner, J. E., 255, *278*
Baird. S. L., 177, *271*
Baizer, M. M., 198, *266*
Bak, B., 48, *77*

Baldeschwieler, J. D., 87, *148*, 395, 396, 398, 399, 400, *413*
Baldt, J. H., 51, *79*
Balogh, B., 385, 388, *408*
Bandlish, B. K., 233, 240, 243, 253, *266*
Banks, R. E., 166, *266*
Bansal, S. R., 236, *266*
Banta, E. B., 330, *414*
Barber, B. H., 372, 373, 381, *412*
Barbey, G., 227, 230, *266, 268, 273*
Bard, A. J., 160, 198, 202, 203, 205, 208, 216, 220, 221, 222, 223, 224, 225, 226, *266, 268, 269, 272, 273, 275, 277, 278*
Bardet, L., 36, *79*
Barmby, D. S., 188, 190, *271*
Barnes, K. K., 198, *273*
Barnett, J. R., 261, *266*
Baron, D., 59, *81*
Baron, P. A., 56, *77*
Barry, C., 230, *266*
Bartell, L. S., 12, 16, 18, 24, 30, 33, 47, 49, 58, *77, 78, 80*
Barter, C., 192, *275*
Bartlett, P. D., 89, *148*
Barton, D. H. R., 2, 74, *78*, 194, *266*
Bastiansen, O., 33, *78*
Batchelor, J. G., 387, 388, *407*
Bates, R. G., 113, 116, *148, 150*
Bauer, D., 167, *266*
Bauer, S. H., 49, *81*
Baughan, E. C., 167, *275*
Baumgärtel, H., 208, *276*
Bawn, C. E. H., 173, 251, *266*
Bayer, E., 306, 350, *415*
Beagley, B., 33, *78*
Beauchamp, J. L., 87, 108, 135, 136, 139, 146, *148, 151, 152*
Bechgaard, K., 217, 230, *277*
Beck, B. H., 52, 68, *80*
Beck, J. P., 167, *266*
Becker, E. D., 281, 283, *407, 409*
Behr, J. P., 386, 388, *407*
Bell, F. A., 165, 169, 194, 251, *266*
Bell, R. P., 106, 114, 133, *148*
Belleau, B., 231, *266, 277*
Bellobono, I. R., 113, *148*
Bellocq, A. M., 360, 361, *408*
Belozerov, A. I., 231, *266*
Beltrame, P., 113, *148*
Bennema, P., 162, *266*
Bennion, B. C., 215, *266*
Benson, S. W., 50, 51, *78*
Bentrude, W. G., 119, 121, *148*
Beresford, P., 194, 251, *266*
Berger, S. 283, *407*
Bergstrom, R. G., 215, *266*
Bernasconi, C. F., 215, *266*

Bernath, T., 172, 173, 248, *272*
Bertini, F., 247, *274*
Beveridge, D. L., 5, 6, *81*
Beynon, P., 382, 385, *411*
Bez, W., 239, *278*
Bezman, R., 224, *266*
Bhacca, N. S., 288, 300, 310, 306, 313, 318, *407, 411*
Bhattacharjee, S. S., 320, 322, *413*
Biebl, G., 196, 203, *276*
Bierbaum, V. M., 87, *151*
Bigeleisen, J., 179, *273*
Biggs, A. I., 114, *149*
Bilik, U., 305, 306, 307, 308, *415*
Billings, W. E., 247, *268*
Billon, J-P., 195, *267*
Bindra, J·S., 280, *415*
Bingham, R. C., 7, *78*
Binkley, W. W., 306, 313, 318, *407*
Bird, R. B., 14, 17, *80*
Birdsall, B., 332, 333, 334, 335, 336, 337, 338, 341, 342, *407*
Birdsall, N. J. M., 333, 383, 385, 386, 387, 388, 390, *411, 412, 414*
Birke, R. L., 251, *265*
Birkhahn, R. H., 391, *412*
Blackburn, G. M., 239, *267*
Blackburne, I. D., 59, *78*
Blanchard, K. R., 45, *81*
Blanchi, J-P., 184, *267*
Blatter, H. M., 59, 63, *77*
Blaurock, A. E., 382, *415*
Bloch, A. N., 177, *267, 269*
Bloemhoff, W., 166, *267*
Blois, M. S., 158, *272*
Blomgren, G. E., 165, *266*
Blout, E. R., 358, 360, 361, *409, 413*
Blount, H. N., 196, 228, 233, 239, 242, *266, 276*
Blumenstein, M., 334, 335, 336, 337, 341, 342, *407*
Boccalon, G., 372, *407*
Bock, K., 302, 311, *407*
Bodley, J. W., 341, 342, *413*
Bodot, H., 59, *79*
Bohme, D. K., 87, *149*
Boleijn, P. T., 263, *276*
Bolhuis, P. A., 207, *267*
Bolton, J. R., 161, 193, *267*
Bolton, P. D., 109, 114, *117, 118, 149*
Bondi, A., 15, *78*
Bonifaviec, M., 187, *274*
Bonvicini, P., 92, 101, 102, 103, 104, 143, *149*
Bordner, J., 66, *82*
Borg, D. C., 196, 203, 213, 218, 219, 235, *268, 269*

AUTHOR INDEX

Borror, A. L., 257, *278*
Bothner-By, A. A., 92, *148*
Bourn, A. J. R., 56, 77
Bovey, F. A., 354, 355, 357, 358, 360, 361, 364, *407*, *409*
Bowden, K., 84, *149*
Bowers, M. T., 135, *148*
Boyd, R. H., 4, 11, 16, 20, 24, 53, 73, *78*, *81*, *82*, 84, 85, 98, 101, 102, *149*
Bracke, W., 251, *267*
Bradbury, E. M., 372, 373, 380, *407*
Bradbury, J. H., 371, 377, *407*
Bradley, T., 374, 378, 379, *414*
Brandts, J. F., 371, *408*
Brass, K., 211, *267*
Bregman, J., 57, *78*
Breitmaier, E., 280, 282, 283, 285, 288, 289, 290, 293, 295, 303, 305, 306, 307, 308, 327, 328, 335, 337, 344, 350, 353, 360, *407*, *409*, *410*, *413*, *414*, *415*
Brett, T. J., 65, *79*
Brewer, C. F., 374, *407*
Brewer, H. B., Jr. 390, *407*
Brewster, A. I. R., 360, 364, *407*, *409*
Briegleb, G., 177, *273*
Briggs, A. G., 118, 127, *152*
Briggs, J. P., 108, *149*
Bright, D., 59, *78*
Britt, A. D., 221, *267*
Britt, W. J., 219, *270*
Brivati, J. A., 161, *267*
Brixius, D. W., 174, *277*
Brockelhurst, B., 211, *266*
Bromberg., A., 59, *78*
Brönsted, J. N., 86, 149
Broom, A. D., 344, *414*
Brosio, E., 361, *414*
Brouwer, D. M., 189, 190, 219, *267*
Brown, D. P., 33, *78*
Brown, E. A., 231, *277*
Brown, H. C., 114, *149*, 164, 166, *267*
Brown, N. M. D., 167, 257, *267*
Brown, R. D., 56, 77, 378, *411*
Brown, W. A. C., 36, *78*
Browne, D. T., 371, *407*, *408*
Brownlee, R. T. C., 108, *152*
Broyde, S. B., 59, *82*
Brubaker, G. R., 59, *78*
Bruce, C. R., 221, *267*
Bruck, D., 256, *267*
Bruni, P., 196, *267*
Bruning, W. H., 195, 220, 221, 222, *267*, *271*, *275*, *276*
Brunori, M., 374, 378, 379, *406*
Bryce-Smith, D., 183, *267*
Bryson, A., 114, *149*
Bryson, T. A., 335, 336, 342, *409*

Buck, H. M., 166, 216, *267*, *268*
Buckingham, D. A., 59, *78*
Bucourt, R., 49, 59, *78*
Buemi, G., 47, 49, 50, *78*, *79*, *82*
Bukhtiarov, A. V., 234, *272*, *275*
Bundle, D. R., 292, 293, 320, 321, 323, *408*
Bunnett, J. F., 96, 101, 106, *149*
Burden, F. R., 56, *77*
Burdon, J., 235, *267*
Burgen, A. S. V., 378, *409*
Burgi, H. B., 33, *78*
Burke, J. J., 90, 99, 100, 108, 119, 121, 122, 123, 124, 125, 134, 135, 136, 138, 142, 146, *148*, *149*
Burlingame, A. L., 385, 388, *408*
Burrows, H. D., 187, *267*
Bushick, R. D., 98, 117, *148*
Bushweller, C. H., 52, 60, 68, *80*
Bussey, R. J., 162, 169, 195, 234, 235, 240, *276*
Butler, M. A., 177, *267*
Buys, H. R., 33, *78*, *79*
Bystrov, V. F., 360, 395, 398, 399, 402, 403, 406, *407*, *408*, *410*, *413*

C

Cabani, S., 113, *149*
Cadogan, K. D., 181, *267*
Calvin, M., 163, 177, 178, 185, 239, *268*, *269*, *271*
Camaioni, D. M., 237, *277*
Cameron, R., 89, 92, 104, *151*
Cameyo, G., 388, 389, *410*
Canfield, N. D., 204, *268*
Cargioli, J. D., 92, *151*
Carreira, L., 55, *78*
Carrington, A., 158, 159, 161, 193, *267*
Carter, J. V., 108, 122, 123, 124, 127, 138, 142, 146, *148*
Carter, M. K., 161, 162, 211, *267*
Cary, L. W., 288, 300, 301, *411*
Cary, P. D., 372, *407*
Casalino, A., 86, *152*
Cashel, M., 341, 342, *413*
Castillo, C. L., 374, *414*
Casu, B., 300, 302, 320, *413*
Caughey, W. S., 374, 378, *412*
Caullet, C., 204, 227, 230, *266*, *267*, *268*, *273*
Cauquis, G., 196, 198, 209, 213, 227, 229, 230, *266*, *267*
Chachaty, C., 392, 393, 394, *414*
Chaiken, I. M., 371, 372, *408*, *409*
Chakrabarty, M. R., 113, *149*
Chambers, J. Q., 204, 217, *268*, *270*

Chandross, E. A., 219, 223, *267*, 277
Chang, C. J., 280, *415*
Chang, C. K., 62, *81*
Chang, S., 11, 16, 20, 24, 53, *78*
Chang, T. C., 196, *275*
Chapman, N. B., 101, *149*
Chapman, D., 382, 383, 385, 386, *408*, *411*, *413*
Chekhecheva, I. P., 178, *276*
Chen, H. J., 96, *151*
Chen, S., 386, *408*
Chen, W. W., 391, *412*
Cheney, V. B., 387, *408*
Cheng, A. K., 64, *77*
Cheng, W. J., 251, *267*, *269*
Chester, A. W., 174, 175, *268*, *276*
Chiang, J. F., 35, *78*
Chiang, T. C., 211, *268*
Chiang, Y., 96, *151*
Chien, J. C. W., 371, 377, *408*
Childers, R. F., 286, 346, 349, 371, 377, 388, 389, *406*, *410*
Chiu, M. L., 322, *411*
Chmurny, G. N., 63, *77*
Chow, Y. L., 245, 247, *268*
Christensen, J. J., 109, *149*, 338, 394, *408*, *410*
Christl, M., 350, 352, 357, *408*
Chung, A. L. H., 40, *80*
Chung, D. 56, *79*
Ciampolini, M., 113, *152*
Clark, D. B., 157, 234, *268*, *272*
Clark, R. G., 6, *78*
Clarke, M. T., 183, *267*
Clerici, A., 245, *268*
Cochran, D. W., 280, 350, 352, 371, *406*, *409*, *415*
Coffen, D. L., 204, *268*
Cohen, J. S., 350, 352, 371, *408*, *409*, *410*, *413*
Cohen, M., 352, *414*
Cohen, M. I., 257, *278*
Cohen, N. C., 59, *78*
Cohen, S. G., 184, *268*
Coleman, J., 163, *275*
Collenbrander, D. P., 260, *274*
Collins, L. J., 70, *78*
Collumeau, A., 85, *149*
Colonna, M., 196, *267*
Conradi, J. J., 158, 164, *277*
Conti, F., 361, 371, 374, 378, 379, *406*, *408*, *414*
Conti, G., 113, *149*
Conway, E., 291, 303, *408*
Cook, R. D., 92, *150*
Cooksey, D., 256, *268*
Cooper, A., 37, *78*
Cooper, J. R., 211, *277*

Cooper, J. T., 211, *268*
Cooper, M. A., 72, *76*
Cordes, E. H., 385, 388, 389, 390, 391, 392, *408*, *410*, *415*
Corio, P. L., 189, *268*
Corkill, J. M., 138, *149*
Cotter, J. L., 255, *272*
Cowan, D. O., 177, *267*, *269*, *275*
Cowell, G. W., 166, *268*
Cowley, D. J., 167, 257, *267*
Cox, D. E., 56, *80*
Cox, J. D., 4, 43, 50, 51, 68, *78*
Cox, M. C., 116, *149*
Cox, M. T., 252, *268*
Coxon, B., 293, 294, *408*
Crabbé, P., 60, *78*
Craig, L. C., 358, 360, 363, 364, 371, *409*, *413*
Crane-Robinson, C., 372, 373, 380, *407*, *413*
Creason, S. C., 207, *268*
Crellin, R. A., 252, *268*
Cros, J-L., 209, *267*
Cross, P. C., 10, *82*
Cruickshank, F. R., 50, 51, *78*
Cruser, S. A., 226, *266*, *268*
Csizmadia, I. G., 69, *81*
Cukman, D., 217, *268*
Curci, R., 92, 94, *149*
Curtiss, C. F., 14, 17, *80*
Cushley, R. J., 387, 388, *407*

D

Dallinga, G., 35, 49, *78*, 159, *268*
Dam, H. T., van, 263, *276*
Danen, W. C., 245, *268*
Daniels, P. J. L., 309, 311, 314, 318, *412*
Dannenberg, J. J., 173, *268*
d'Annibale, A., 62, *78*
Danyluk, S. S., 168, *268*, 344, *408*
Dashevskii, V. G., 49, 59, *78*, *80*
Davenport, J. B., 382, *408*
Davidson, R. S., 183, *268*
Davis, C. T., 91, *149*
Davis, D. G., 196, *270*
Davis, M. M., 85, *149*
Davy, H., 86, *149*
Day, W., 309, 311, 313, *412*
Dea, P., 324, 329, 330, *408*
Deber, C. M., 358, 360, 361, *409*, *413*
Decius, J. C., 10, *82*
de Boer, E., 158, 164, 167, 211, *268*, *277*
de Boer, J., 59, *78*
de Courville, A., 114, *149*
de Graaff, R. A. G., 59, *77*
Delahaye, D., 227, *266*, *268*
de la Mare, P. B. D., 73, *78*
Del Re, G., 72, *78*

Deno, N. C., 84, 91, 95, 104, *149*, 247, *268*
Dennis, E. A., 59, *82*
Deslauriers, R., 334, 335, 341, 342, 344, 345, 346, 354, 359, 360, 361, 363, 364, 367, 370, *408*, *414*, *415*
Dessau, R. M., 170, 171, 172, 173, 211, *268*, *270*, *276*
DeTar, D. F., 74, *78*
Dewar, M. J. S., 7, 53, 57, *78*, 170, 171, 173, 175, *266*, *268*
Deyrup, A. J., 86, 88, 94, *150*
Diani, E., 113, *148*
Dietz, R., 171, 175, 206, *266*, *268*
Dingwall, A., 91, *150*
Dirlan, J. P., 231, *274*
Diserens, L., 168, 195, *272*
Djerassi, C., 66, *82*
Dobkin, J., 185, *277*
Dobler, M., 36, *79*
Doddrell, D., 282, 286, 300, 301, 304, 306, 308, 309, 313, 318, 371, 394, *406*, *408*
Dodge, A. D., 255, 264, *268*
Dodziuk, H., 59, *79*
Doering, W. von E., 50, 51, *82*
Dosen-Micovic, L., 89, *79*
Dollish, F. R., 190, 191, *268*
Dolphin, D., 196, 203, 213, 218, 219, *268*, *269*
Domaille, P. J., 56, *77*
Dongen, J. P. C. M., van, 292, *410*
Doorn, J. A., van, 263, *267*
Doorn, R. A. van, 219, *276*
Dorfman, L. M., 220, *276*
Dorman, D. E., 304, 306, 308, 309, 311, 313, 320, 322, 354, 355, 357, 358, 360, 361, *408*, *409*, *412*
dos Santas Viega, J., 193, *267*
Douty, C. F., 108, 122, 123, 124, 138, 142, 146, *148*, *150*
Drakenberg, T., 391, 394, *409*
Dralants, A., 56, *79*
Dravnieks, F., 158, 159, *267*
Dreisbach, R. R., 136, *150*
Duffey, W., Jr. 193, *268*
Duggleby, P. McC., 119, 121, *148*
Duke, R. P., 59, *78*
Dunitz, J. D., 36, 48, *79*, *82*
Dunlap, R. B., 335, 336, 342, 390, 391, *409*
Dwyer, M., 59, *79*

E

Eastman, J. W., 177, 178, *268*
Eberson, L., 156, 198, 205, 226, 230, 231, 233, 240, 247, 248, 250, *268*, *274*
Ebert, M., 264, *269*
Edlund, O., 187, 212, *268*
Edsberg, R. L., 255, 263, *268*

Edward, J. T., 85, 92, 101, 102, *150*
Effenberger, F., 251, *268*
Eliel, E. L., 22, 61, 63, 64, 65, *79*
Ellis, P. D., 335, 336, 342, *409*
Elofson, R. M., 255, 263, *268*
Elson, I. H., 173, *268*
Elstner, E. F., 329, *409*
Engelman, D. J., 382, 386, *409*
Engelman, D. M., 382, *415*
Engelsma, G., 177, 178, *268*
Engler, E. M., 16, 21, 23, 33, 36, 38, 41, 44, 58, 74, 75, *79*, *81*, 178, *268*
Ermer, O., 39, 47, 48, 52, *79*
Essery, J. M., 113, *150*
Euler, R. A., 59, *78*
Evans, J., 242, *267*
Evans, M. G., 143, *150*
Evans, M. W., 107, *150*
Evans, T. R., 252, *269*
Everett, D. H., 116, *149*, *150*
Exner, O., 106, *150*
Eyler, J. R., 136, *151*
Eyring, E. M., 215, *266*
Eyring, H., 71, *81*

F

Faber, D. H., 59, *79*
Fahey, M. R., 194, *274*
Fajer, J., 196, 203, 213, 218, 219, 235, *268*, *269*
Fanta, K., 211, *267*
Farcasiu, D., 36, *79*
Farid, S., 252, *269*
Farmell, L. F., 166, *266*
Farrar, T. C., 281, 283, *407*, *409*
Farrington, J. A., 255, 264, *269*
Faulkner, L. R., 212, 223, 224, 225, 226, *266*, *269*
Fauvelot, G., 230, *267*
Fava, A., 163, *269*
Favini, G., 47, 49, 50, 75, *78*, *79*, *81*, *82*
Fedarko, M. D., 395, 396, 398, 399, 400, 405, *409*, *413*
Fedin, E. I., 395, 398, 399, 402, 403, *408*
Feeney, J., 332, 333, 334, 335, 336, 337, 338, 341, 342, 378, 383, 385, 386, *407*, *409*, *412*
Feitelson, J., 180, 181, *269*
Feldberg, S., 217, *269*
Feldberg, S. W., 208, 224, 233, *269*, *274*
Felton, R. H., 196, 203, 213, 218, 219, 235, *268*, *269*
Fenoglio, D. J., 61, *80*
Fenandez, J. E., 251, *265*
Fendler, E. J., 390, 391, 395, *409*
Fendler, J. H., 390, 391, 395, *409*

Ferguson, J. A., 204, 258, *270*
Fermandjian, S., 352, *414*
Fernholt, L., 33, *78*
Ferraris, J. P., 177, *267, 269, 275*
Ferre, Y., 59, *79*
Ferro, D. R., 15, *79*
Fickling, M. M., 114, *150*
Figeys, H. P., 56, *79*
Filinovsky, V. Yu., 202, *275*
Filipescu, N., 257, *270, 274*
Filler, R., 235, *265, 276*
Finder, C. J., 56, *77, 79*
Finocchiaro, P., 59, *81*
Fioshin, M. Ya, 230, *269*
Fischer, A., 114, *150*
Fischer-Hjalmars, I., 59, *79*
Fishbein, R., 247, *268*
Fisher, R. R., 335, 336, 342, *409*
Fitzgerald, E. A., Jr. 168, 182, *269*
Fleischer, E. B., 34, *79*
Fleischfresser, B. E., 251, *269*
Fleischmann, M., 157, 234, *268, 272*
Fleming, K. A., 109, 117, *149*
Fletcher, K., 264, *269*
Flexser, L. A., 91, *150*
Flockhart, B. D., 188, 189, 191, *269, 276*
Flohe, L., 360, *409, 410*
Flory, P. J., 59, *80*
Flurry, R. L., 54, *79*
Forbes, W. F., 164, 166, 211, *268, 269, 277*
Ford, R. A., 70, *79*
Forman, A., 196, 203, 213, 218, 219, 235, *269*
Foster, R., 175, 177, 179, 185, *269*
Fouchet, C., 393, 394, *406*
Fournier, J., 59, *79, 82*
Fowden, L., 73, *78*
Fraenkel, G. K., 158, 211, *271*
Franconi, C., 361, *414*
Frank, H. F., 143, *150*
Frank, S. N., 208, *269*
Franklin, J. L., 136, *150*
Freed, D. J., 212, 225, 226, *269*
Freedman, M. H., 349, 352, 360, 367, 370, 371, 372, 373, 381, *408, 409, 412, 413*
Freeman, R., 281, 283, 364, *409*
Freiberg, L. A., 59, 63, *77*
Frensdorff, H. K., 394, 395, 396, *413*
Fried, J., 228, *269*
Frisch, L., 323, *409*
Fritsch, J. M., 195, 203, 204, 205, 207, *269, 273, 276*
Fritzsche, H., 339, *409*
Froimowitz, M., 59, *79*
Fromageot, P., 352, *414*
Fry, A. J., 198, *269*
Fry, J. L., 74, 75, *79*
Fueno, T., 187, 232, *278*

Fuhrhop, J-H., 196, 213, 214, *270*
Fujiwara, S., 372, 380, 381, *414*
Fukui, T., 330, *410*
Fukuyama, T., 29, *80*
Funck, T., 395, 398, 399, 406, *409*
Fung, C. H., 373, 374, *409, 412*
Fung, K. W., 217, *270*
Funt, B. L., 251, *277*

G

Gaaf, J., 159, *265*
Galli, R., 244, *274*
Gal'pern, E. G., 234, *275*
Gans, P. J., 59, *79*
Garbisch, E. W., Jr. 55, *80*
Gardini, G. P., 247, *274*
Gardner, D. M., 158, *271*
Garrett, P. E., 204, *268*
Garrigou-Lagrange, C., 360, 361, 364, *408, 414*
Gassner, S., 168, *275*
Gateau-Olesker, A., 293, 295, 303, *412*
Gaugler, R. W., 91, *149*
Gavezzotti, A., *79*
Geiger, F. E., 257, *270*
Geise, H. J., 33, 63, *78, 79*
Geissman, T. A., 91, *149*
Gel'bschtein, A. I., 117, *150*
Geue, R. J., 59, *79*
Genies, M. 209, 227, *267*
Gent, M. P. N., 387, *409*
Geraci, G., 374, 378, 379, *406*
Gerlach, O., 251, *268*
Gero, S. D., 291, 293, 295, 303, 324, 330, *408, 412, 413*
Gero, S. P., 303, *414*
Giacometti, G., 372, *407*
Gibbons, W. A., 180, *270*, 358, 360, 363, 364, 371, *409, 413*
Gibson, Q. H., 378, *413*
Gilbert, A., 183, *267*
Gillespie, R. J., 91, 125, *150*
Girard, J. P., 36, *79*
Girault-Vexlearschi, G., 116, *150*
Gitler, C., 390, 381, *408*
Glasel, A., 251, *277*
Glass, R. S., 219, *270*
Gleason, J. G., 66, *82*
Gleicher, G. J., 74, 75, *79*
Gleicher, G. L., 59, *79*
Glushkov, V., 371, 377, *406, 409, 412*
Glushko, V. G., 352, *409*

Godfrey, T. S., 181, *270*
Godici, P. E., 388, 389, *409*
Gold, V., 91, 95, *150*
Golden, D. M., 50, 51, *78*
Goldin, M. M., 234, *272*
Goldstein, J. H., 309, 311, 314, 318, *412*
Golebiewski, A., 54, *79*
Goodin, R. D., 178, *276*
Goodman, R. A., 286, 346, *406*
Goodman, J. F., 138, *149*
Gordon, M., 184, *270*
Gorin, P. A. J., 283, 289, 323, *409*
Gough, T. A., 203, *270*
Goursot-Leray, A., 59, *79*
Govil, G., 346, *409*
Grace, J. A., 160, *270*
Gradowski, M., 182, *270*
Graham, J. C., 54, 55, 76, *79*
Gramaccioni, P., 75, *81*
Grandmougin, E., 160, *272*
Granger, R., 36, *79*
Granick, S., 160, 168, 193, *273*
Grant, D. M., 283, 342, 344, 387, *408*, *410*, *412*
Grathwohl, C., 351, 352, *409*
Gray, G. A., 280, 283, 324, 350, 352, *409*, *415*
Gray, G. R., 288, 294, 295, 300, 301, 341, 342, *413*
Grayson, M., 164, 166, *267*
Greatorex, D., 187, *267*
Greci, L., 196, *267*
Gregg, C. T., 374, 378, *412*
Greig, C. C., 97, *150*
Grell, E., 378, 395, 398, 399, 406, *409*
Grollman, A. P., 374, *407*
Gross, J., 263, *271*
Grossweiner, L. I., 180, *271*
Grunwald, E., 97, 106, 107, *150*, *151*
Grutzner, J. B., 280, 283, *409*
Gruver, G. A., 216, *270*
Guglielmetti, R., 59, *79*
Guilbault, G. G., 256, *272*
Gund, P., 74, *82*
Gund, T. M., 36, *79*
Günzler, W. A., 360, *409*, *410*
Gupta, A., 260, *270*
Gurd, F. R. N., 280, 348, 349, 350, 352, 371, 378, *406*, *409*, *411*, *413*
Gurr, M. I., 383, *410*
Gurudata, N., 292, 303, *410*
Gust, D., 56, *79*, *81*
Guthrie, J. P., 143, *150*, 291, 293, 295, 303, *408*, *412*
Gutowski, Gerald E., 324, 325, 326, 327, 328, 329, *415*
Gutowsky, H. S., 255, *271*
Guttman, D. W., 256, *267*
Gwinn, W. D., 42, 63, *80*, *81*

H

Haak, W. J., 311, 315, *413*
Haake, P., 92, *150*
Haar, W., 360, 361, 363, *415*
Haas, G., 282, *407*
Haase, J., 52, *79*
Haddon, R. C., 57, *79*
Häfliger, O., 114, *149*
Hagaman, E. W., 291, 303, 311, 313, 324, 325, 326, 327, 328, 329, *408*, *411*, *415*
Hainaut, D., 49, *78*
Haines, W. J., 69, *81*
Haldna, U. L., 85, 91 *150*, *152*
Hall, F. M., 109, 114, 117, *149*
Hall, F. R., 168, *272*
Hall, H. K., 51, *79*
Hall, L. D., 287, *410*
Hall, N. F., 114, 116, *150*
Hall, R. E., 74, 75, *80*
Hall, W. K., 190, 191, 192, *268*, *278*
Hamill, W. D., Jr. 342, 344, *410*
Hamilton, E. J., 260, *270*
Hamilton, J. A., 385, 388, 389, *410*, *415*
Hamilton, R. G., 201, *274*
Hammerich, O., 163, 172, 196, 204, 207, 217, 219, 228, 230, 235, 250, *270*, *275*, *277*
Hammett, L. P., 86, 88, 91, 94, 95, 96, 97, 104, 106, 107, 108, *150*
Hammond, G. S., 260, *270*
Hamori, E., 59, *80*
Handloser, C. S., 113, *149*
Haney, M. A., 136, *150*
Hanotier, J., 173, *270*
Hanotier-Bridoux, M., 173, *270*
Hansen, L. D., 109, *149*
Hanson, P., 162, *270*
Hantzsch, A., 86, 91, *150*
Happ, J. W., 204, 258, *270*
Haq, M. Z., 49, *77*
Hargrave, P. A., 388, 390, *412*
Hariharan, P. C., 6, *79*
Harmony, M. D., 56, *82*
Harpp, D. N., 66, *82*
Hartman, P. G., 372, *407*
Hartman, S. E., 252, *269*
Hasselbalch, K. A., 95, *150*
Haszeldine, R. N., 166, *266*
Haugen, G. R., 50, 51, *78*
Hauptman, H., 37, *78*
Hausser, K. H., 164, 165, 193, 195, 213, *270*, *271*
Haverkamp, J., 292, *410*
Hawes, B. W. V., 91, 95, *150*
Hayamizu, K., 335, 338, 339, 340, 341, *411*
Hayashi, M., 62, *81*
Hayon, E., 180, 181, *269*

Head, A. J., 74, 78
Heding, H., 311, *407*
Hefter, H. J., 260, *270*
Heiba, E. I., 170, 171, 172, 173, *268*, *270*
Henderson, L. J., 95, *150*
Henderson, W. G., 87, 108, 135, 139, 146, *148*, *152*
Hendrickson, J. B., 19, 22, 33, 34, *79*, *80*
Henrichs, P. M., 64, *71*
Henrikson, K. P., 382, *410*
Hepler, L. G., 106, 107, 108, 109, 114, *150*, *151*
Hercules, D. M., 223, *270*
Hermans, J., 15, *79*
Herve du Penhoat, P., 288, 295, 300, 301, *410*, *413*
Hetzer, H. B., 113, 116, *148*, *150*
Hickey, M. J., 11, 16, 20, 24, 53, 62, 66, 67, 76, 77, *78*, *80*
Higginbotham, H. K., 33, *78*
Highet, R. J., 390, *407*
High, D. F., 36, *80*
Hilderbrandt, R. L., 33, 38, 39, *80*
Hill, H., 283, *406*
Hill, H. D. W., 281, *409*
Hill, J. O., 394, *408*
Hill, T. L., 13, 14, 17, *80*
Hilpert, S., 164, *270*
Hine, J., 143, *150*
Hingerty, B., 59, *82*
Hinman, R. L., 95, 104, *150*
Hintz, P. J., 194, *274*
Hirao, K-I., 187, *270*
Hirose, K., 404, *414*
Hirsch, J. A., 15, 16, 21, 30, 32, 47, 55, 65, 68, 69, 75, 77, *80*
Hirschfeld, F. L., 57, *78*
Hirschfelder, J. O., 14, 17, *80*
Hirschler, A. E., 188, 190, 192, *271*
Hirschmann, H., 38, *77*
Hirschon, J. M., 158, *271*
Hitchman, M. L., 202, *265*
Hodgson, R. L., 189, 190, *271*
Hoefs, E. V., 220, 221, *271*
Hoerr, C. W., 116, *151*
Hoffman, M. Z., 217, *278*
Hoffmann, R., 7, *80*
Hoijtink, G. J., 158, 162, 163, 166, 167, 198, 203, 229, *265*, *266*, *271*
Holdroyd, R. A., 182, *271*
Hollis, D. P., 344, *414*
Holtz, D., 87, 108, 135, 136, 139, 146, *148*, *151*, *152*
Holtz, H. D., 175, *271*
Holz, J. B., 220, *271*
Honda, A., 213, *271*
Honda, S., 293, *410*
Hoogsteen, K., 344, *406*

Hopkins, A. S., 255, 257, 259, 261, 263, *266*, *271*
Hopkins, H. P., 114, *151*
Horsley, W. J., 350, 352, 388, *410*
Horton, D., 306, 313, 318, *407*
Howard, C. J., 87, *150*
Howard, P. B., 136, *151*
Howarth, O. W., 211, *271*
Howarth, T. T., 252, *268*
Hruby, V. J., 360, 364, *407*
Hsi, N., 61, *80*
Huber, H., 352, *406*
Huckstep, L. L., 309, 311, 313, *412*
Hudson, J. O., 192, *270*
Hughes, E. D., 73, *78*
Hughes, F., 187, *271*
Huguet, J., 204, *273*
Huizer, A. H., 216, *268*
Hull, V. J., 244, 249, *272*
Hulme, R., 161, 162, *267*, *271*
Hünig, S., 204, 215, 216, 263, *266*, *271*
Hunkapiller, M. W., 371, *410*
Hunt, C. R., 257, *274*
Hunt, R. L., 171, 175, *266*
Hurst, G. H., 92, *150*
Hutton, R. S., 219, *277*
Hyde, P., 186, 251, 259, 261, 264, *271*
Hyman, H. H., 235, *265*, *276*

I

Ichihara, Y., 372, 380, 381, *414*
Ichikawa, T., 212, *271*
Iida, Y., 195, *271*
Ijima, T., 33, *80*
Ikehara, M., 330, *410*
Iles, D. H., 194, *266*
Ilten, D. B., 185, *271*
Imamura, M., 212, *272*
Imura, T., 182, *271*
Inch, T. D., 287, *410*
Ingold, C. K., 73, *78*, *80*
Ingold, K. U., 247, *276*
Ingram, V., 323, *410*
Inouye, S., 311, 313, 315, *413*
Isaacs, N. S., 179, *271*
Isbell, H. S., 288, 295, *413*
Isenberg, I., 177, *271*
Itoh, M., 187, *271*
Ivanov, V. T., 360, 395, 398, 399, 402, 403, 406, *408*, *410*, *413*
Izatt, R. M., 109, *149*, 338, 394, *408*, *410*

J

Jackson, A. H., 252, *268*
Jacob, E. J., 12, 18, 24, 33, 47, 58, *80*
Jaenicke, O., 252, *269*

AUTHOR INDEX

Jain, M. K., 385, 388, *415*
James, A. T., 382, *410*
James, R. L., 188, 190, *271*
Janata, J., 99, *150*, 204, *271*
Jander, G., 85, *151*
Jankowski, W. C., 282, 364, *410*
Jansen, G., 99, *151*
Jardetzky, O., 350, *413*
Jaruzelski, J. J., 95, 104, *149*
Jeftic, L. J., 228, *271*
Jennings, H. J., 292, 293, 320, 321, 322, 323, *408*, *410*
Jenson, B. S., 205, 206, *271*
Jensen, F. R., 52, 60, 68, *80*
Johnson, A. R., 382, *408*
Johnson, A. W., 254, *271*
Johnson, C. D., 97, 118, 127, 134, *149, 150, 151*
Johnson, C. S., Jr. 220, 221, 255, *271*
Johnson, D. H., 76, *80*
Johnson, L. F., 282, 320, 322, 358, 360, 363, 364, 395, 396, 398, 399, 400, *409, 410, 413*
Johnson, M. D., 239, 256, *268, 271*
Johnson, P. M., 162, *277*
Johnson, P. V., 167, *271*
Johnson, R. A., 237, *277*
Johnson, T. W., 391, *412*
Jones, F. M., III, 85, 87, 108, 114, 121, 126, 135, 137, 139, 146, *148, 151*
Jones, J. R., 85, *151*
Jones, R. A. Y., 59, *78*
Joo, N., 213, *271*
Joschek, J-I., 180, *271*
Joslin, T., 234, *272*
Jung, G., 280, 285, 288, 289, 290, 306, 327, 328, 350, 353, 360, *407, 409, 410, 415*

K

Kadish, K. M., 196, *270*
Kainer, H., 164, 165, 177, 195, *271*
Kalb, A. J., 40, *80*
Kambara, H., 33, *78*
Kamiya, Y., 170, *275*
Kapuler, A. M., 330, *411*
Karabatsos, G. J., 61, *80*
Karkowski, F. M., 59, 63, *77*
Karplus, M., 54, 55, *82*
Katritzky, A. R., 59, *78*, 91, 95, 104, 118, 127, 134, *149, 151, 153*
Katsumata, S., 185, 213, 214, *272*
Katz, T. J., 222, *271*
Kaufman, F., 87, *151*
Kaufman, J. J., 69, *80*
Kawabe, K., 182, *271*
Kawai, N., 178, *275*
Kawamori, A., 213, *271*
Kawasaki, K., 249, *277*

Kay, M. I., 56, *80*
Kayushin, L. P., 341, *411*
Kebarle, P., 87, 108, *149, 151*
Kehrmann, F., 160, 168, 195, *272*
Keii, T., 191, *277*
Keim, P., 280, 348, 349, 352, 371, 377, 378, *409, 411, 413*
Keller, T., 360, *410*
Kemball, C., 190, *276*
Kemp, T. J., 187, *267*
Kenner, G. W., 252, *268*
Kent, F. J., 36, *79*
Kent, G. E., 56, *77*
Kenyon, G. L., 350, 371, *408, 413*
Keough, K. M., 382, 385, *411*
Keszathelyi, C. P., 203, 225, 226, *266, 272, 277*
Kevan, L., 180, 181, 187, *272, 274*
Khalilulina, K. Kh., 395, 398, 399, 402, 403, 406, *408, 410*
Khan, Z. H., 162, *272*
Khanna, B. N., 162, *272*
Kharasch, M. S., 95, *153*
Kiesslich, G., 204, 263, *271*
Kim, H., 63, *80*
Kim, K., 219, 232, 233, 240, 244, 249, 253, *272, 276*
Kimura, K., 185, 211, 213, 214, *272, 278*
Kinell, P-O., 187, 212, *268, 272*
King, D. L., 116, *152*
King, D. M., 198, *272*
Kinoshita, M., 165, 169, 195, 196, *275*
Kira, A., 212, *272*
Kirk, D. N., 70, *78*
Kirk, R. D., 187, *271*
Kistenmacher, T. J., 177, *275*
Kitaigorodskii, A. I., 22, 59, *78, 80*
Klofutar, C., 116, *151*
Klunklin, G., 183, *267*
Klyne, W., 64, *80*
Knunyants, I. L., 234, *272, 275*
Knutson, R. S., 36, *81*
Ko, H. C., 109, *151*
Kobayashi, M., 320, 322, *414*
Koch, H. J., 283, *411*
Koch, K. F., 309, 311, 313, *410, 412*
Koch, V. R., 234, *272*
Kochi, J. K., 172, 173, 237, 248, *268, 272*
Koehl, W. J., Jr. 170, 171, 172, 173, *270*
Koenig, S. H., 378, *411*
Koerner, T. A. W., Jr. 288, 300, 301, *411*
Kohl, D. A., 30, 49, *78*
Koizumi, M., 216, *274*
Kojima, M., 311, 313, 315, *413*
Komatsu, T., 187, *272*
Kommandeur, J., 165, 168, *267, 272, 276*
Komorski, R., 282, 286, 300, 373, *406, 409*
Komoroski, R. A., 283, 326, 327, 328, 329, 346, 347, 350, 363, 364, 371, *406, 411*

Kon, H., 158, *272*
Konicek, J., 136, *151*
Kooyman, E. C., 172, *275*
Kornprobst, A., 324, 330, *413*
Korte, R. W., de, 172, *275*
Kosower, E. M., 176, 255, *272*
Kotowycz, G., 287, 309, 311, 255, 338, 339, 340, 341, *411*
Kowert, B. A., 220, 221, 222, *272*
Koyama, K., 232, *272*, *276*
Koz'min, S. A., 395, 398, 399, 402, 403, *408*
Kramer, D. N., 256, *272*
Krane, J., 63, 64, 77
Krasnovsky, A. A., 262, *272*
Kraut, J., 36, *80*
Krebs, A., 52, *79*
Kreishman, G. P., 324, 329, 330, *408*, *411*, *414*
Kremser, D., 116, *151*
Kresge, A. J., 96, *151*
Kricka, L. J., 194, 251, 252, *266*, *272*
Krugh, T. R., 324, 326, 327, 328, 329, 330, 344, *411*
Krusche, E., 339, *409*
Kuchitsu, K., 29, 33, 35, 62, *76*, *78*, *80*, *82*
Kuck, V. J., 219, *277*
Kudirka, P. J., 201, *272*
Kudryavtsev, R. V., 234, *272*
Kuge, T., 404, *414*
Kunieda, N., 230, *274*
Kuntz, G. P. P., 338, 339, *411*
Kurita, Y., 189, *272*
Kuszkat, K. A., 185, *277*
Kuura, H. J., 91, *150*
Kuwana, T., 196, 216, 223, 228, 233, 262, *267*, *270*, *272*, *276*

L

Laane, J., 49, *80*
Laaneste, H. E., 91, *150*
Lagowski, J. J., 85, *151*
Lai, A., 361, *414*
Laidler, K. J., 113, *151*
Lakshminarayanan, A. U., *81*
Lam, Y. F., 338, 339, *411*
Lambert, J. B., 67, *80*
Lambert, M. C., 194, *266*
Lamola, A. A., 185, *275*
Lampman, G. M., 21, *82*
Land, E. J., 180, 264, *269*, *272*
Landini, G., 92, *151*
Landsberg, A., 206, *272*
Landsberger, F. R., 388, 389, *409*
Landsman, D. A., 116, *149*
Lane, G. A., 65, 74, 77

Langridge, R., 59, *82*
Lapper, R. D., 334, 335, 341, 342, 344, 345, 346, *411*, *414*
Larcombe, B. E., 206, *268*
Larsen, J. W., 90, 106, 108, 114, 126, 128, 134, 135, 136, *148*, *151*, 215, *275*
Lasker, S. E., 322, *411*
Latowski, T., 182, *270*
Lau, M. P., 247, *268*
Laurie, V. W., 62, *81*
Lauterbur, P. C., 371, *411*
Lawson, P. J., 350, 352, 371, 377, *406*, *409*, *412*
Leane, J. B., 92, *150*
Leclerc, G., 194, *266*
Led, J. J., 48, 77
Ledwith, A., 165, 166, 169, 174, 185, 186, 187, 194, 208, 236, 237, 240, 251, 252, 255, 257, 259, 261, 264, *266*, *268*, *269*, *271*, *273*
Lee, A. G., 383, 385, 386, 387, 388, 390, *407*, *411*, *412*, *414*
Lee, D. G., 89, 92, 104, 143, *151*
Lee, Y. N., 59, *80*
Leedy, D. W., 195, 207, *276*
Lee-Ruff, E., 87, *149*
Leeuwestein, C. H., 59, 77
Leffler, J. E., 97, 106, 107, *151*
Lehn, J-M., 70, *80*, 386, 388, 394, *407*, *411*
Leigh, J. S., Jr. 374, *409*
Lemieux, R. U., 287, 300, 309, 311, 331, *411*
Lennard-Jones, J. E., 14, *80*
Lenoir, D., 74, 75, *80*
Levi, A., 92, 94, 101, 102, 103, 104, 105, 143, *149*, *151*
Levine, Y. K., 383, 385, 386, 387, *407*, *411*, *412*
Levy, G. C., 92, *151*, 280, 281, 282, 283, 285, 286, 324, 350, 352, *406*, *411*, *415*
Lewis, G. N., 179, *273*
Lewis, I. C., 161, 164, 211, *273*, *276*
Lewis, S., 385, 388, *408*
Lexa, D., 203, *273*
Libert, M., 204, 230, *273*
Lichtin, N. N., 217, *278*
Lide, D. R., Jr. *80*
Liebig, J. von 86, *151*
Lifson, S., 11, 15, 16, 24, 30, 33, 39, 47, 48, 52, *79*, *80*, *82*
Liler, M., 84, 85, 92, *151*
Lin, Y. S., 56, 77
Lindenmeyer, P. H., 16, *81*
Lindman, B., 391, 394, *409*
Lindner, H. J., 55, *80*
Lindsay Smith, J. R., 245, 247, *273*
Line, L. L., 208, *275*
Linnel, R. H., 104, *152*
Linscott, D. L., 254, *265*

Liotta, C. L., 114, *151*
Lipkin, D., 179, *273*
Lipnick, R. L., 55, *80*
Lipsky, S. R., 387, 388, *407*
Liptay, W., 177, *273*
Littler, J. S., 174, *273*
Llinas, M., 395, 401, 406, *412*
Lo, D. H., 55, *80*
Loewenstein, A., *150*
Loftus, P., 59, *76*
Long, F. A., 84, 85, 95, 98, 99, *151*, *152*
Long, J., 95, 104, 136, *150*, *151*
Long, R. C., 309, 311, 314, 318, *412*
Lord, R. C., 49, *80*
Los, J. M., 207, *267*
Lowen, A. M., 95, *151*
Lucchini, V., 92, 94, 101, 102, 103, 104, 143, *149*
Lucken, E. A. C., 195, *273*
Ludman, C. J., 234, *273*
Lugovskoi, A. A., 49, 59, *78*, *80*
Lukacs, G., 291, 293, 295, 303, 324, 330, *408*, *412*, *413*, *414*
Lumry, R., 107, *151*
Lunazzi, L., 62, *78*
Lund, A., 187, 212, *268*, *272*
Lundt, I., 302, *407*
Lupinski, J. H., 162, *266*
Luzzati, V., 386, *412*
Lyerla, J. R., Jr. 283, 349, 352, 357, 358, 360, 364, 367, 370, 371, 372, 373, 381, *408*, *409*, *412*, *413*, *414*
Lynden-Bell, R. M., 193, *273*

M

McAlister, J., 59, *81*
McCall, M. T., 258, *270*
McCarron, E. M., 234, *273*
McClelland, R. A., 85, 96, 98, 99, 100, 101, 102, 142, 146, *153*, 170, *273*
McCollum, J. D., 89, *148*
McConnell, H. M., 193, *273*
McCorkle, M. R., 116, *151*
McCullough, R. L., 16, *81*
McDaniel, D. H., 114, *149*
McDevit, W. F., 98, 99, *151*
McFarlane, W., 388, *412*
McGregor, H., 361, 364, *414*
McGuire, R. F., 71, *81*
Mach, G. W., 88, 89, 95. 96, 99, 104, *148*
McIntyre, D., 98, *151*
Mackie, J. D. H., 73, *78*
Mackor, E. L., 158, 159, 163, 166, 167, 229, *265*, *268*
MacLean, C., 159, *273*

McManus, S. P., 215, *275*
McNally, D., 4, 11, 16, 20, 24, 53, *78*, *81*
Magee, J. L., 71, *81*
Magnus, P. D., 194, *266*
Maki, A. H., 178, *265*
Malachesky, P. A., 159, *273*
Malik, J. M., 311, 315, *413*
Mallon, B. J., 50, 51, *82*
Maloy, J. T., 226, *273*
Mamantov, G., 217, *270*
Manassen, J., 203, *278*
Mancilla, J. M., 236, *266*
Mani, S. R., 219, 232, 233, 240, 243, 253, *266*, *273*, *276*
Mann, B. R., 114, *150*
Mann, C. K., 198, *273*
Manning, C., 183, *267*
Mansson, M., 4, *80*
Mantsch, H. H., 326, 331, 332, 334, 335, 341, 342, 344, 345, 346, 347, *412*, *414*
Marchessault, R. H., 59, *82*
Marcoux, L. S., 159, 195, 203, 207, 220, 221, 222, 239, *267*, *272*, *273*, *276*
Marcus, D. M., 374, *407*
Marcus, R. A., 220, *273*
Mariani, C., 75, *81*
Mark, J. E., 59, 72, *80*
Marshall, R. C., 348, 349, 352, 371, 377, 378, *409*, *411*, *413*
Martin, J., 36, *78*
Marvell, E. N., 36, *81*
Masuhara, H., 185, *276*
Masui, M., 240, *273*
Mataga, N., 185, 186, *273*, *276*
Matsuda, K., 288, 289, 306, 307, 313, 320, 322, *414*, *415*
Matsui, T., 109, *151*
Matsunaga, Y., 164, 195, *273*
Matsuura, T., 160, *274*
Matsuyama, Y., 249, *277*
Matwiyoff, N. A., 374, 378, 379, *411*, *412*, *414*
Maurer, W., 360, 361, 363, *415*
Maurey, M., 230, *266*
Maurey-Mey, M., 196, 213, 229, 230, *267*
Mauzerall, D., 196, 213, 214, *269*
Maxwell, I. E., 59, *78*
May, P. J., 74, *78*
Mayeda, E. A., 156, 157, 203, 225, *273*
Mayer, A., 374, *414*
Medvdev, S. S., 178, *276*
Medzhikov, A. A., 195, *273*
Meiboom, S., *150*
Meites, I., 203, *273*
Melby, E. G., 136, *151*
Menzies, I. D., 194, *266*
Mermet-Bouvier, R., 352, *414*
Metcalfe, J. C., 383, 385, 386, 387, 388, 390, *407*, *411*, *412*, *414*

Metzger, J., 52, 59, *79*
Meyer, A. Y., 20, *81*
Meyer, K., 164, *273*
Meyer, W. C., 182, *273*
Michaelis, L., 160, 168, 193, *273*
Michejda, C. J., 221, *275*
Michelson, A. M., 330, *411*
Migliorese, K. G., 235, *265*
Mijlhoff, F. C., 33, 63, *79*
Mikhaleva, I. L., 395, 398, 399, 402, 403, *408*
Mildvan, A. S., 373, 374, *409*, *412*
Miller, F. M., 256, *272*
Miller, L. L., 156, 157, 203, 234, *272*, *273*
Miller, M. A., 15, 16, 21, 24, 29, 30, 32, 47, 48, 55, 60, 61, 62, 63, 64, 65, 66, 68, 69, 75, 77
Miller, W. N., 219, *270*
Millett, F., 388, 390, *412*
Millington, J. P., 236, *273*
Minn, F. L., 257, *270*, *274*
Minisci, F., 244, 245, 247, *268*, *273*, *274*
Mirkind, L. A., 230, *269*
Miroshnikov, A. I., 360, 406, *410*, *413*
Misina, V. P., 178, *276*
Mislow, K., 56, *79*, *81*
Mitchell, E. J., 86, 91, 100, 108, 125, 127, 128, 132, 136, *148*, *151*
Mitsky, J., 139, *152*
Miura, M., 189, *277*
Mixan, C. E., 67, *80*
Miyashita, I., 158, *278*
Miziorko, H. M., 374, *412*
Mizushima, S., 70, *81*
Mochel, V. D., 280, *412*
Möckel, H., 181, 187, *274*
Modena, G., 92, 93, 101, 102, 103, 104, 105, 119, 120, 143, *149*, *151*
Moe, N. S., 163, 172, 196, *270*
Molloy, B. B., 309, 311, 313, *412*
Momany, F. A., 71, *81*
Monaghan, J. J., 33, *78*
Monetti, M. A., 113, *148*
Monny, C., 330, *411*
Montaudo, G., 59, *81*
Montgomery, L. K., 33, 38, 39, *80*
Moodie, R. B., 92, *148*
Moon, R. B., 371, 374, 378, 379, *412*
Moriarity, T. C., 109, *148*
Morino, Y., 29, 35, *80*, *82*
Moriyama, M., 230, *274*
Morrison, G. A., 22, 61, 63, 64, *79*
Morrow, J. S., 348, 349, 352, 371, 378, *411*, *413*
Mortensen, E. M., 71, *81*
Mortimer, C. T., 113, *151*
Morton, B. J., 309, 311, 314, 318, *412*
Mosher, M. W., 113, *149*

Mugnoli, A., *79*
Muller, N., 391, *412*
Muljiani, Z., 196, 219, *268*
Müller, S., 206, *272*
Mulliken, R. S., 175, 176, *274*
Munn, R. J., 116, *149*
Munson, B., 136, *151*
Murata, Y., 164, 169, 195, 217, 229, 234, *274*, *276*
Murphy, W. J., 167, 257, *267*
Murray, M. A., 95, 127, *152*
Murrell, J. N., 213, *270*
Murty, T. S. S. R., 86, 91, 100, 132, *148*
Muszkat, K. A., 59, *78*
Myles, D., 255, *278*
Mystrom, R. S., 311, 315, *413*

N

Nadjo, L., 201, *266*, *276*
Nagabhushan, T. L., 331, *411*
Nagai, S., 212, *274*
Nagakura, S., 193, *275*
Nakamaru, K., 216, *274*
Nakaya, T., 173, *268*
Nakayama, S., 213, *277*
Nava, A., 47, 49, 50, *79*
Neale, R. S., 245, *274*
Neckars, D. C., 260, *274*
Needham, T. E., 374, 378, *411*, *412*
Neely, J. W., 344, *411*
Neikam, W. C., 188, 190, *271*
Neilands, J. B., 395, 401, 406, *412*
Newman, M. B., 195, *273*
Nelson, G. L., 280, 281, 282, 283, 285, 324, 350, 352, *411*
Nelson, M. J., 378, *412*
Nelson, R., 62, *81*
Nelson, R. F., 195, 207, 208, *265*, *267*, *268*, *274*, *275*, *276*
Nelson, S. F., 194, *274*
Nepgodina, O. I., 178, *276*
Neunteufel, R. A., 252, *274*
Neuss, N., 309, 311, 313, *412*
Nicholson, R. S., 201, *272*, *274*
Nicksie, S. W., 166, *265*
Nigen, A. M., 352, 371, 377, *409*, *412*, *413*
Niida, T., 311, 313, 315, *413*
Niizuma, S., 216, *274*
Nilsson, S., 232, 233, *274*
Nishijima, Y., 252, *266*, *278*
Nishinaga, A., 160, *274*
Nitta, I., 212, *274*
Noggle, J. H., 281, *413*
Nonhebel, D. C., 236, *266*
Norberg, R. E., 221, *267*
Nordbloom, G. D., 156, 157, 203, *273*
Norman, L. J., 166, *276*

Norman, R. O. C., 162, 170, 172, 245, 247, 270, 273, 274
Northcott, D., 116, 152
Norton, D. A., 37, 78
Norton, R. S., 371, 377, 407
Nouaille, F., 324, 330, 413
Nowak, T., 373, 412
Numata, T., 230, 274
Nyberg, K., 156, 198, 205, 247, 248, 250, 268, 274
Nyburg, S. C., 30, 81
Nygaard, L., 48, 77
Nyhus, B. A., 34, 35, 77

O

Oae, S., 230, 274
Oancea, D. J., 127, 148
Oberhammer, H., 49, 81
O'Brien, D. H., 91, 125, 152
Ochs, S., 385, 388, 415
Oddo, G., 86, 152
Ogawa, S., 374, 414
O'Hara, W. F., 114, 152
Ohashi, M., 187, 278
Ohmori, H., 240, 273
Ohnesorge, W. E., 209, 276
Ohnishi, M., 395, 396, 398, 399, 400, 413
Ohnishi, S-I., 212, 274
Ohta, N., 170, 212, 271, 275
Okaya, Y., 56, 80
Okuyama, T., 162, 169, 195, 234, 235, 240, 276
Olah, G. A., 91, 125, 152
Oldenburg, S. J., 216, 268
Oldfield, E., 286, 346, 349, 371, 372, 377, 381, 382, 383, 385, 386, 406, 411, 413
O'Leary, M. H., 256, 274
Ollis, W. D., 59, 81
Olmstead, M. L., 201, 274
Olp, D., 194, 274
Olsen, F. P., 96, 101, 106, 149
Olson, J. S., 378, 413
O'Malley, R. F., 234, 273
Omori, A., 232, 276
Omoto, S., 311, 313, 315, 413
O'Neal, H. E., 50, 51, 78
Ono, Y., 191, 277
Onodera, A., 178, 275
Onopchenko, A., 175, 274
Oohashi, Y., 193, 274
Ooshika, Y., 213, 271, 277
Oosterhoff, L. J., 166, 267
Osawa, E., 36, 79
Oster, G. K., 162, 274
Oster, O., 280, 355, 356, 357, 359, 360, 413, 415

Ostwald, W., 86, 152
Ottnad, M., 353, 360, 410
Ottolenghi, M., 186, 274
Ouellette, R. J., 59, 60, 70, 81
Ourisson, G., 70, 80
Ovchinnikov, Yu. A., 360, 395, 398, 399, 402, 403, 406, 408, 410, 413

P

Pac, C., 182, 274
Paci, M., 371, 374, 408
Packer, E. L., 350, 371, 408, 413
Packer, J., 114, 150
Padilla, A. G., 233, 240, 243, 253, 266
Palik, S., 116, 151
Pal'm, V. A., 85, 91, 150, 152
Paoletti, P., 113, 152
Paolillo, L., 372, 373, 380, 413
Papouchado, L., 233, 274
Parezewski, A., 54, 79
Park, S-M., 224, 225, 266
Parker, V. D., 163, 164, 172, 196, 198, 204, 205, 206, 207, 208, 217, 219, 228, 230, 231, 233, 235, 239, 240, 248, 250, 267, 270, 271, 274, 275, 277
Parola, A., 184, 268
Parry, K., 59, 76
Parsons, G. H., 184, 268
Parsons, I. W., 235, 267
Parsons, R., 238, 274
Partington, P., 341, 383, 387, 407, 411, 412
Paskovich, D. H., 212, 274
Pasternak, R., 20, 81
Patel, D. J., 344, 358, 359, 360, 395, 398, 399, 402, 406, 413
Patel, V. V., 178, 268
Patten, F. W., 187, 271
Paul, B., 331, 411
Paul, M. A., 84, 85, 95, 152
Pauli, G. H., 33, 81
Pauling, L., 15, 81, 152
Pearson, H., 352, 406
Pearson, J. M., 251, 267, 269
Pease, L. G., 360, 361, 413
Peat, I. R., 352, 413
Pedersen, C., 302, 311, 407
Pedersen, C. J., 394, 395, 396, 413
Pellerin, J. H., 391, 412
Peltier, D., 114, 149
Peover, M. E., 198, 203, 270, 274, 275
Perchinunno, M., 245, 268
Perdonein, G., 93, 119, 120, 151
Perdue, E. M., 114, 151
Perkampus, H. H., 113, 152
Perlin, A. S., 283, 288, 295, 300, 301, 302, 320, 322, 411, 414

Perrin, D. D., 114, *152*
Perry, R. A., 247, *268*
Person, W. B., 175, *274*
Petrovskii, P. V., 395, 398, 399, 402, 403, *408*
Phelps, J., 203, 205, 216, *266, 275*
Phillips, R., 338, *413*
Phillips, T. E., 177, *275*
Pierce, L., 62, *81*
Pierson, C., 247, *268*
Piette, L., 163, 219, *276*
Piez, K. A., 371, *414*
Pilcher, G., 4, 43, 50, 51, 68, *78*
Pinching, G. D., 116, *148*
Pink, R. C., 166, 188, 189, 190, 191, *269, 275, 276*
Pittman, C. U., Jr. 215, *275*
Pitzer, K. S., 42, *81*
Platko, F. E., 391, *412*
Pleskov, Yu. V., 202, *275*
Pletcher, D., 157, 234, *268, 272*
Pocchler, T. O., 177, *269*
Pohland, A. E., 163, *278*
Polanyi, M., 107, *150*
Ponjee, J. J., 263, *276*
Pople, J. A., 5, 6, *79, 81*
Popp, G., 210, *275*
Porta, O., 245, *268*
Porter, G., 180, 181, *270, 272*
Porter, G. B., 167, *275*
Powars, D. F., 378, *412*
Praat, A. P., 167, *268*
Prasad, K. U. M., 360, 364, 367, 370, *415*
Prater, K. B., 202, 226, *273, 275*
Pravdic, V., 217, *268*
Prescher, G., 113, *152*
Prestegard, J. H., 387, 388, *407, 409*
Preston, P. N., 166, *266*
Pretsch, E., 395, 400, *413*
Prueckner, H., 136, *152*
Puglisi, V. J., 202, *275*
Pugmire, R. J., 342, 344, *410*
Pummerer, R., 168, *275*
Purbrick, M. B., 185, *272*
Puss, R. K., 91, *150*

Q

Quarterman, L. A., 235, *256*
Quast, H., 263, *271*
Que, L., Jr. 288, 294, 295, 300, 301, 341, 342, *413*
Quirk, R. P., 90, 108, 126, 128, 134, 135, 136, *148*
Quirt, A. R., 352, *413*

R

Rabinovich, D., 57, *78*
Raboniwitz, J. C., 371, *413*
Racela, W., 92, *151*
Radzitsky, P., de, 172, *270*
Raftery, M. A., 334, 335, 336, 337, 341, 342, 388, 390, *407, 412*
Rahman, M., 160, *276*
Raimondi, M., 47, 49, 50, *79*
Rajabalee, F. J. M., 292, 303, *410*
Rajender, S., 107, *151*
Raley, J. H., 189, 190, *271*
Ralston, A. W., 116, *151*
Ramachandran, G. N., 10, *81*
Ramakrishnan, V., 183, 185, *275*
Rao, V. R., 183, 185, *275*
Rao, V. S. R., 59, *82*
Rapport, N., 4, *80*
Rastrup-Andersen, J., 48, *77*
Rau, M-C., 235, *265*
Rawlinson, D. J., 245, *276*
Reagan, M. T., 104, *152*
Record, K. A. F., 59, *78*
Reddoch, A. H., 196, 211, 212, *268, 274, 275*
Reddy, T. B., 231, *277*
Redmond, W., 116, *152*
Ree, T., 71, *81*
Reich, E., 330, *411*
Reix, M., 203, *273*
Reuss, R. H., 257, *275*
Reynolds, R., 208, *275*
Reynolds, W. F., 352, *413*
Rhoades, J. A., 311, 313, *411*
Riccobono, P. X., 56, *81*
Richards, J. H., 371, 374, 378, 379, *410, 412*
Richrol, H. H., 168, 182, *269*
Rickard, R. C., 245, *268*
Riesner, D., 213, 214, *269*
Rigaudy, J., 230, *267*
Rigny, P., 393, 394, *406*
Rinehart, K. L., Jr. 311, 315, *413*
Ristagno, C. V., 169, 196, 219, 228, 233, 234, 249, *275, 276*
Robb, M. A., 69, *81*
Roberts, G. C. K., 350, 374, 383, 387, *411, 413*
Roberts, J. D., 280, 304, 306, 308, 309, 311, 313, 320, 322, 350, 352, 357, *409, 412, 414*
Roberts, J. L., Jr. 198, *276*
Roberts, R. M., 192, *275*
Roberts, R. T., 392, 393, 394, *414*
Robertson, R. E., 116, *152*
Robiette, A. G., 28, 29, *81*
Robins, R. K., 324, 330, *411, 414*

Robinson, J. D., 388, 390, *414*
Robinson, R. A., 116, *150*
Rochester, C. H., 84, 85, *152*
Rochlitz, J., 239, *275*, *278*
Rodgers, A. S., 50, 51, *78*
Rodgers, P., 374, *414*
Rolls, J. P., 311, 315, *413*
Romans, D., 221, *275*
Romers, C., 59, *77*
Ronlan, A., 163, 164, 208, 217, 230, 248, *275*, *277*
Rooney, J. J., 166, 188, 189, 190, 191, *275*, *276*
Rosdahl, A., 56, *81*
Rosenblum, A., 59, *81*
Rossetti, Z. L., 71, *76*
Roth, H. D., 185, *275*
Roth, W. R., 50, 51, *82*
Rousseau, K., 196, 219, *268*
Rowley, A. G., 245, 247, *273*
Rozantsev, E. G., 195, *273*
Rozhkov, I. N., 234, *272*, *275*
Rudolph, J. P., 109, *148*
Ruff, B. A., 311, 315, *413*
Rundel, W., 195, *275*
Rundle, H. W., 87, *151*
Russell, C. D., 227, *275*
Russell, P. J., 174, 236, 237, 240, *272*
Russell, R. L., 182, 271
Ruterjans, H., 360, 361, 363, *415*
Rytteng, J. H., 338, *410*

S

Sablayrolles, C., 36, *79*
Sacconi, L., 113, *152*
Sadar, M., 143, *151*
Saeki, T., 187, 232, *278*
Saito, H., 361, 364, 372, 373, 380, 381, *414*
Sakata, K., 170, *274*
Sakata, T., 178, 193, *274*, *275*
Sakurai, H., 182, *274*, *277*
Sala, E., 352, *414*
Salem, L., 20, *82*
Samat, A., 59, *79*
Sanderson, G. R., 302, 320, *413*
Sandoz, M., 160, *271*
Sandstrom, J., 56, *81*
Sano, M., 165, 169, 195, 196, *275*
Santhanam, K. S. V., 160, 203, *266*, *275*, *277*
Sanwal, S. N., 4, 11, 16, 20, 24, 53, *78*
Sarantakis, D., 361, 364, *414*
Sargerson, A. M., 59, *78*
Sasisekharan, V., 10, *81*

Sata, M., 189, *272*
Sato, H., 166, *275*
Sato, T., 59, *82*
Sato, Y., 165, 169, 195, 196, *275*
Sauter, H., 395, 398, 399, 406, *409*
Savadatti, M. I., 180, *270*
Saveant, J. M., 201, 205, *265*, *266*, *275*, *276*
Sawyer, D. T., 198, *276*
Scandola, E., 86, *152*
Schachtschneider, J. H., 10, 13, *82*
Schaefer, J., 388, *412*, *414*
Schaefer, L., 33, *81*
Schaefer, M. F., 6, *81*
Schäfer, H., 251, *276*
Scharpen, L. H., 62, *81*
Scheffler, K., 195, *275*
Schell, F. M., 280, *415*
Schenk, W., 263, *271*
Scheraga, H. A., 10, 71, *81*
Scheutzow, D., 204, 216, 263, *271*
Schindler, K., 177, *273*
Schipper, P., 216, *268*
Schirmer, R. E., 281, *413*
Schlaf, H., 204, 216, *271*
Schleich, T., 334, 335, 341, 342, 344, 345, 346, *414*
Schleyer, P. von. R., 7, 16, 19, 21, 23, 33, 36, 38, 41, 44, 45, 58, 74, 75, *79*, *80*, *81*, *82*
Schmidt, G. M. J., 57, *78*
Schneider, W. G., 168, *268*
Schofield, K., 113, *150*
Schoot, C. J., 263, *276*
Schor, R., *82*
Schriesheim, A., 95, 104, *149*
Schroder, G., 50, 51, *82*
Schubert, M. P., 160, 168, 193, *273*
Schubert, W., 33, *81*
Schulz, J. G. D., 175, *274*
Schulze, T., 91, *149*
Schumm, D. E., 228, *269*
Schwarz, F. P., 181, *276*
Schwarz, J. A., 302, *414*
Schwechten, H. W., 169, 194, *278*
Schweizer, M. P., 324, 329, 330, 344, *408*, *411*, *414*
Schwyzer, R., 360, 361, *415*
Scorrano, G., 88, 90, 92, 93, 94, 101, 102, 103, 104, 105, 119, 120, 143, *149*, *151*, *152*
Scott, E. J. Y., 175, *276*
Scott, J. A. N., 188, 189, 191, *269*, *276*
Scrutton, M. C., 373, *409*
Searle, G. H., 59, *79*
Sears, B. D., 382, 391, 392, *414*, *415*
Seeger, H., 160, *276*
Seekircher, R., 175, *274*
Seip, H. M., 33, *78*

Sekigawa, K., 59, *81*
Sendtner, R., 193, *278*
Senkler, G. H., 56, *79*, *81*
Senyavina, L. B., 406, *410*
Seo, E. T., 195, 207, *276*
Septe, B., 303, *414*
Sepulchre, A.-M., 291, 293, 295, 303, 324, 330, *408*, *412*, *413*, *414*
Serjeant, E. P., 114, *148*
Seto, S., 288, 289, 306, 307, 313, 320, 322, *414*, *415*
Shade, L. R., 196, 229, *276*
Shain, I., 201, *274*
Shang, D. T., 242, *276*
Shank, N. E., 220, *276*
Shapiro, S. A., 118, 127, 134, *149*, *151*
Sharp, J. A., 173, *266*
Shary-Tehrany, S., 4, 11, 16, 20, 24, 53, *78*
Shaw, R., 50, 51, *78*
Shcheglova, G., 117, *150*
Shealer, S. E., 252, *269*
Shegal, I. L., 231, *266*
Shepard, F. E., 190, *276*
Sherrington, D. C., 165, 169, 194, 255, *266*, *273*
Sherzhaknova, L. M., 231, *266*
Shieh, C. F., 11, 16, 20, 24, 53, *81*
Shigemitsu, Y., 252, *276*
Shih, S., 170, 189, 211, *268*, *276*
Shimada, M., 185, *276*
Shimanouchi, T., 62, *76*
Shimizu, A., 187, 212, *268*
Shine, H. J., 160, 162, 163, 164, 169, 178, 195, 196, 217, 219, 228, 232, 233, 234, 235, 240, 243, 244, 249, 253, *272*, *273*, *274*, *275*, *276*
Shorter, J., 101, *149*
Shoup, R. R., 283, *407*
Shulman, R. G., 374, *414*
Sidgwick, N. V., 193, *276*
Silber, J. J., 162, 169, 195, 234, 235, 240, 249, *276*
Sim, G. A., 36, *78*, *82*
Simon, W., 395, 400, *413*
Simonetta, M., 75, *79*, *81*
Simpson, J., 167, *275*
Simsohn, H., 391, *412*
Sinclair, J., 350, *413*
Singer, L. S., 161, 164, 168, 211, *273*, *276*
Sioda, R. E., 207, 228, 234, 239, *276*
Sinke, G. C., 50, 51, *82*
Slootmaekers, P. J., 114, *152*
Slutsky, J., 36, *79*, *81*
Small, L. E., 109, *148*
Smallcombe, S. H., 371, *410*
Smith, C. D., 51, *79*
Smith, I. C. P., 292, 293, 320, 321, 322, 323, 326, 331, 332, 334, 335, 341, 342, 344, 345, 346, 347, 354, 359, 360, 361, 363, 364, 367, 370, 372, 373, 380, 381, *408*, *409*, *410*, *412*, *414*, *415*
Smith, N. G., 257, *278*
Smith, R. P., 71, *81*
Snow, M. R., 59, *79*
Snyder, R. G., 10, 13, *82*
Sogn, J. A., 358, 360, 363, 364, 371, *409*, *414*
Sogo, P. B., 163, *269*
Sokoloski, E. A., 390, *407*
Sonoda, T., 189, *272*
Sorensen, G. O., 48, *77*
Sørensen, S. P., 220, 221, 222, *276*
Sorgo, M., de, 213, *268*
Sosnovsky, G., 245, *276*
Southern, J. F., 33, *81*
Southwick, E. W., 177, *278*
Spandau, H., 85, *151*
Spanswick, J., 247, *276*
Spatola, A. F., 360, 364, *407*
Speitel, J., 160, *272*
Spencer, J. F. T., 323, *409*
Spohn, K. H., 283, *407*
Sprague, J. T., 47, 48, 49, 50, 53, 54, 55, 56, 57, 58, *77*
Sprinkle, M. R., 114, 116, *150*
Srinivasan, R., 35, *77*
Stach, R. W., 256, *274*
Staley, R. H., 136, *152*
Stam, M. F., 255, 257, *269*, *271*
Stamires, D. M., 191, *276*
Stang, P. J., 7, 16, 19, *82*
Stanienda, A., 196, 203, *276*
Starkova, S. D., 178, *276*
Steckham, E., 251, 262, *276*
Stejskel, E. O., 388, *414*
Stellman, S. D., 59, *82*
Stern, A., 358, *409*
Sternlicht, H., 350, 352, 371, 374, 388, *408*, *410*, *413*, *414*
Stevens, J. B., 95, 104, *153*
Stevens, J. D., 300, *411*
Stewart, E. T., 6, *78*
St. Jacques, M., 64, *77*
Stoddart, J. F., 59, *81*
Stoffel, W., 383, 385, 386, 388, *414*
Stolfo, J., 59, *81*
Stone, H., 192, *275*
Stothers, J. B., 280, 282, 283, 287, 288, 289, 290, 291, 296, 300, 308, 324, 326, 327, 328, 329, 334, 348, 349, 350, 352, *414*, *415*
Strobel, G-J., 339, *415*
Stroshane, R. M., 311, 315, *413*
Stuart, J. D., 209, *276*
Stuart, T. W., 15, 30, 55, 65, 75, *77*
Student, P. J., 57, *79*

Stull, D. R., 50, 51, *82*
Sturtevant, J. M., 119, *152*
Stüwe, A., 208, *276*
Subramanian, E., 59, *82*
Sugiyama, H., 288, 289, 306, 307, 313, 320, 322, *414, 415*
Suhadolnik, R. J., 329, *409*
Sullivan, P. D., 164, 166, 218, *269, 276,*
Summerhays, K. D., 139, *152*
Sundararajan, P. R., 59, *82*
Sundaralingam, M., 59, 77, *81, 82*
Suneram, R. D., 56, *82*
Susuki, T., 232, *272, 277*
Sutcliffe, L. H., 166, *266*
Sutherland, I. O., 59, *81*
Sutton, C., 72, *80*
Suzuki, K., 213, *271, 276*
Suzuki, O., 338, 339, *411*
Suzuki, Y., 195, *273*
Svanholm, U., 163, 217, 219, 230, 239, 250, *277*
Sweeting, L. M., 98, *152*
Symons, M. C. R., 158, 159, 160, 161, 162, *267, 270, 271*
Szwarc, M., 213, 218, 251, *268, 269, 277*
Szymanski, J. T., 30, *81*
Szkrybalo, W., 30, *77*

T

Taagepera, M., 87, 108, 135, 139, 146, *148, 152*
Tachikawa, H., 223, 225, 226, *266, 272, 277*
Taddei, F., 92, *151*
Tadokoro, S., 372, 380, 381, *414*
Taft, R. W., Jr. 87, 106, 108, 135, 139, 146, *148, 152*
Tai, J. C., 15, 30, 55, 65, 75, *77*
Takeo, K., 404, *414*
Takimoto, K., 189, 213, *277*
Takiura, K., 293, *410*
Talkowski, C., 388, 389, *410*
Talvik, A. J., 85, *152*
Tam, J. N. S., 247, *268*
Tanaseichuk, B. S., 231, *266*
Tancredi, T., 372, 373, 380, *413*
Tanei, T., 181, 189, *277*
Tang, R. T., 172, 173, 248, *272*
Taniguchi, M., 311, 315, *413*
Tanzer, C., 288, 289, 290, 360, *407, 410*
Tate, J. R., 138, *149*
Tazawa, I., 330, *410*

Temkin, M. I., 117, *150*
Temussi, P. A., 372, 373, 380, *413*
Tengler, E., 211, *267*
Texier, P., 167, *266*
Thielecke, W., 36, *79*
Thomas, C. B., 170, 172, *273, 274*
Thomas, W. A., 360, *414*
Thompson, H. B., 12, 18, 24, 33, 47, 58, *80*
Tichy, M., 50, 51, *82*
Tickle, P., 118, 127, *152*
Timmermans, J., 143, *152*
Tissier, C., 104, *152*
Tissier, M., 104, *152*
Tkach, R. W., 309, 314, 318, *412*
Tobolsky, A. V., 260, *277*
Tokel, N. E., 203, 225, 226, *266, 277*
Tokunaga, H., 191, *277*
Toneman, L. H., 35, 49, *78*
Torchia, D. A., 355, 357, 358, 360, 361, 371, *409, 414*
Torii, S., 249, *277*
Tosa, T., 182, *274, 277*
Traetteberg, M., 56, *82*
Trahanovsky, W. S., 174, *277*
Tran-Dinh, S., 352, *414*
Treinin, A., 181, *269*
Tribble, M. T., 24, 29, 48, 59, 60, 61, 62, 63, 64, 66, 77, *82*
Trichilo, C. L., 257, *270, 274*
Trivellone, E., 372, 373, 380, *413*
Trotman-Dickenson, A. F., 105, *152*
Truitt, S. T., 311, 315, *413*
Tse, J., 302, 320, *413*
Ts'o, P. O. P., 344, *414*
Tsubomura, H., 178, 182, 213, *271, 272, 275*
Tsujino, Y., 162, *277*
Tsutsumi, S., 232, *272, 276*
Tunggal, B. D., 383, 385, 386, 388, *414*
Tun-Kyi, A., 360, 361, *415*
Turcot. L., 251, *277*
Turkevich, J., 191, *276*
Turner, R. B., 50, 51, *82*
Tuzimura, K., 288, 289, 306, 307, 313, 320, 322, *414, 415*
Tyminski, I. J., 16, 21, 32, 47, 68, 69, 75, *77*

U

Uneyama, K., 249, *277*
Urry, D. W., 364, *414*
Usher, D. A., 59, *82*
Usui, T., 288, 289, 306, 307, 313, 320, 322, *414, 415*
Uvarova, N. N., 406, *410*

V

Van-Catledge, F. A., 15, 21, 30, 32, 55, 57, 65, 75, 77, *82*
Van de Poel, W., 114, *152*
Van der Linde, W., 116, *152*
van der Ploeg, R. E., 172, *277*
van der Waals, J. H., 159, *273*
Vanderzee, C. E., 116, *152*
Van Hook, J. P., 260, *277*
Van Horn, A. R., 66, *82*
van't Hoff, J. H., 2, *82*
van Willigen, H., 211, *277*
Varadi, V., 196, *269*
Vasak, M., 395, 400, *413*
Vass, G., 293, 295, 303, *412*
Vaughan, J., 114, *150*
Vegh, L., 196, 218, 219, *269*
Verdini, A. S., 372, *407*
Vergamini, P. S., 374, 378, 379, *412*, *414*
Verrijn Stuart, A. A., 159, *268*
Vianello, E., 201, *275*
Victor, T. A., 344, *408*
Viglino, P., 361, *414*
Vigna, R. A., 348, 349, 352, 371, *409*, *411*, *413*
Villafranca, J. J., 374, *415*
Vincent, E-J., 59, *79*
Vincow, G., 158, 161, 162, 211, *267*, *277*
Vinogradov, S. N., 104, *152*
Viola, R. E., 374, *415*
Vischer, R. B., 378, *412*
Viskocil, J., 56, 72, *82*
Vliegenthart, J. F. G., 292, *410*
Voelter, W., 280, 282, 285, 288, 289, 290, 293, 295, 303, 305, 306, 307, 308, 327, 328, 335, 337, 339, 344, 350, 353, 355, 356, 357, 359, 360, *407*, *409*, *410*, *413*, *414*, *415*
Volke, J., 265, *277*
Volpe, J. A., 374, 378, *412*

W

Waddington, T. C., 85, *153*, 169, *265*
Wadso, I., 116, 136, *151*, *153*
Waegell, B., 59, *79*, *82*
Wahl, G. H., 66, *82*
Wai, H., *153*
Wake, R. W., 252, *269*
Walling, C., 237, 238, *277*
Walsh, R., 50, 51, *78*
Walter, R., 354, 359, 360, 361, 363, 364, 367, 370, *408*, *415*
Walter, R. I., 194, *277*
Walter, R. W., 361, 364, *414*
Wander, J. D., 306, 313, 318, *407*
Wang, G. L., 65, *77*
Wang, I. C., 92, *150*
Ward, P. J., 172, *274*
Ward, R. L., 185, *277*
Ware, W. R., 184, *270*
Waring, A. J., 91, *151*
Warshel, A., 11, 15, 16, 24, 30, 33, 48, 54, 55, *82*
Wasser, P., 213, 214, *269*
Wasserman, B., 213, *268*
Wasserman, E., 219, *277*
Watkins, A. R., 184, *267*
Weast, R. C., 136, *153*
Webb, H. M., 135, *148*
Weber, P., 59, *81*
Weber-Schäfer, M., 208, *276*
Wecker, E., 168, 194, *278*
Weijland, W. P., 158, 163, 166, 167, 229, 265, *271*
Weimar, R. D., 143, *150*
Weinberg, N. L., 231, *277*
Weiner, S. A., 260, *270*
Weingarten, H., 204, 205, *269*
Weinstein, M., 185, *277*
Weisman, G. R., 194, *274*
Weiss, J., 161, *277*
Weissman, S. I., 158, 164, 221, *267*, *277*
Weitz, E., 169, 194, *277*
Welbourn, M. J., 187, *267*
Welch, J., 385, 388, *408*
Weller, A., 183, 223, 224, *277*
Welch, G., *153*
Wells, P. R., 106, *153*
Wenkert, E., 280, 291, 303, 311, 313, 324, 325, 326, 327, 328, 329, 350, 352, 371, *406*, *408*, *409*, *411*, *415*
Wertz, D. H., 16, 18, 21, 24, 25, 29, 33, 34, 35, 38, 39, 41, 44, 65, 67, 77, *82*
Weser, U., 338, 339, *415*
Westheimer, F. H., 22, 59, *82*, 95, *153*
Westland, R. D., 311, 313, 315, *415*
Westrum, E. F., Jr., 4, 50, 51, *80*, *82*
Wettermark, G., *82*
Whalen, R., 247, *268*
Wheeler, J., 207, *268*
Wheeler, L. O., 160, *277*
Wherle, A., 177, *271*
Whetsel, K. B., 114, *153*
Whitaker, D. R., 371, *410*
White, A. C., 166, *268*
White, A. M., 91, 125, *152*
White, B. S., 203, *275*
White, D. N. J., *82*
Whitehead, M. A., 55, *80*
Whitten, D. G., 204, 258, *270*
Wiberg, K. B., 21, 22, 73, *82*
Wieland, H., 168, 194, *277*
Wieser, J. D., 33, 38, 39, *80*
Wildes, P. D., 217, *278*

Wilhoit, R. C., 121, *153*
Wilk, M., 239, *278*
Wilkins, M. H. F., 382, *415*
Wilkinson, R. G., 248, *268*
Will, J. P., 239, *267*
Williams, D. E., 15, 29, *82*
Williams, D. F., 196, *275*
Williams, D. R., 204, *268*
Williams, E., 385, 388, 389, 391, 392, *410*, *415*
Williams, G., 95, *152*
Williams, J. E., 7, 16, 19, 45, *81*, *82*
Williams, M. B., 204, *271*
Williams, M. K., 360, *414*
Williams, S. H., 70, *81*
Williams, T. J., 335, 336, 342, *409*
Willie, G. R., 341, 342, *413*
Wilson, D., 385, 388, *408*
Wilson, D. M., 371, 395, 401, 406, *408*, *412*
Wilson, E. B., 10, *82*
Wilson, G. E., *82*
Wilson, G. S., 219, *270*
Wilson, J. D., 204, 205, *269*
Wilson, J. M., 118, 127, *152*
Wilson, N. K., 280, *415*
Wilson, T. S., 170, *274*
Winkler, F. K., 48, *82*
Winograd, N., 233, *267*
Winstein, S., 97, *150*
Winters, L. J., 257, *275*, *278*
Wipke, W. T., 74, *82*
Wise, W. B., 371, 377, *408*
Wiskott, E., 36, *79*
Witkowski, J. T., 324, 330, *411*, *414*
Wohl, R. C., 374, 378, 379, *414*
Wolberg, A., 203, *278*
Wolf, J. F., 126, 128, 135, 136, 137, 140, *148*
Wolf, L., 164, *270*
Wolff, M. E., 246, *278*
Wong, L., 64, *77*
Woo, E. P., 71, *82*
Woo, P. W. K., 311, 313, 315, *415*
Wood, G., 71, *82*
Woodgate, S. S., 87, *148*
Woods, H. J., 166, *268*
Woorst, J. D. W., van, 162, *266*
Wrathall, D. P., 116, *149*
Wright, J. S., 20, *82*
Wu, C. Y., 91, 98, *148*, 192, *278*
Wu, F. Z., 38, *77*
Wu, G. S., 160, *276*
Wudl, F., 177, *278*
Wuelfing, P., Jr. 168, 182, *269*
Wurster, C., 193, *278*
Wüthrich, K., 351, 352, 360, 361, *409*, *415*
Wyckoff, J. C., 247, *268*
Wynne-Jones, W. F. K., *150*

Y

Yamada, H., 211, 213, *272*, *278*
Yamamoto, M., 252, *266*, *278*
Yamamoto, N., 182, *271*
Yamaoka, N., 288, 289, 306, 307, 313, 320, 322, *414*, *415*
Yamazaki, T., 213, 214, *272*, *278*
Yamdagni, R., 108, *149*
Yang, G. C., 165, *278*
Yang, N-L., 162, *274*
Yates, K., 85, 91, 95, 96, 98, 99, 100, 101, 102, 104, 142, 146, *151*, *152*, *153*
Yathindra, N., 59, *81*, *82*
Ying Kin, N. M. K. Ng., 320, 322, *413*
Yokosawa, Y., 158, *278*
Yokozeki, A., 35, *82*
Yonemitsu, O., 187, *270*
Yonezawa, T., 187, *278*
Yoshida, K., 187, 232, *278*
Yoshino, A., 187, *278*
Younathan, E. S., 288, 300, 301, *411*
Young, L. B., 87, *149*
Ysern, S., 286, 346, *406*
Yuen, J., 181, 187, *274*
Yuen, S. H., 255, *278*
Yuki, H., 293, *410*

Z

Zachariasse, K., 223, 224, 225, *277*, *278*
Zachau, H. G., 346, *415*
Zahnow, E. W., 232, *265*
Zens, A. P., 335, 336, 342, *409*
Zhurinov, M. Zh., 230, *269*
Ziemeck, P., 160, *274*
Zierenberg, O., 383, 385, 386, 388, *414*
Zimmer, S., 360, 361, 363, *415*
Zuccarello, F., 47, 49, 50, 78, *82*
Zweig, A., 223, *278*

Cumulative Index to Authors

Allinger, N. L., **13**, 1
Anbar, M., **7**, 115
Arnett, E. M., **13**, 83
Bard, A. J., **13**, 155
Bell, R. P., **4**, 1
Bennett, J. E., **8**, 1
Bentley, T. W., **8**, 151
Bethell, D., **7**, 153; **10**, 53
Brand, J. C. D., **1**, 365
Brinkman, M. R., **10**, 53
Brown, H. C., **1**, 35
Cabell-Whiting, P. W., **10**, 129
Cacace, F., **8**, 79
Carter, R. E., **10**, 1
Collins, C. J., **2**, 1
Cornelisse, J., **11**, 225
Crampton, M. R., **7**, 211
de Gunst, G. P., **11**, 225
Eberson, L., **12**, 1
Farnum, D. G., **11**, 123
Fendler, E. J., **8**, 271
Fendler, J. H., **8**, 271; **13**, 279
Ferguson, G., **1**, 203
Fields, E. K., **6**, 1
Fife, T. H., **11**, 1
Fleischmann, M., **10**, 155
Frey, H. M., **4**, 147
Gilbert, B. C., **5**, 53
Gillespie, R. J., **9**, 1
Gold, V., **7**, 259
Greenwood, H. H., **4**, 73
Havinga, E., **11**, 225
Hogeveen, H., **10**, 29, 129
Ireland, J. F., **12**, 131
Johnson, S. L., **5**, 237
Johnstone, R. A. W., **8**, 151
Kohnstam, G., **5**, 121
Kramer, G. M., **11**, 177
Kreevoy, M. M., **6**, 63
Liler, M., **11**, 267
Ledwith, A., **13**, 155

Long, F. A., **1**, 1
Maccoll, A., **3**, 91
McWeeny, R., **4**, 73
Melander, L., **10**, 1
Mile, B., **8**, 1
Miller, S. I., **6**, 185
Modena, G., **9**, 185
More O'Ferrall, R. A., **5**, 331
Neta, P., **12**, 223
Norman, R. O. C., **5**, 53
Nyberg, K., **12**, 1
Olah, G. A., **4**, 305
Parker, A. J., **5**, 173
Peel, T. E., **9**, 1
Perkampus, H. H., **4**, 195
Pittmann, C. U., Jr., **4**, 305
Pletcher, D., **10**, 155
Ramirez, F., **9**, 25
Rappoport, Z., **7**, 1
Reeves, L. W., **3**, 187
Robertson, J. M., **1**, 203
Rosenthal, S. N., **13**, 279
Samuel, D., **3**, 123
Schaleger, L. L., **1**, 1
Scheraga, H. A., **6**, 103
Scorrano, G., **13**, 83
Shatenshtein, A. I., **1**, 156
Shine, H. J., **13**, 155
Silver, B. L., **3**, 123
Simonyi, M., **9**, 127
Stock, L. M., **1**, 35
Symons, M. C. R., **1**, 284
Thomas, A., **8**, 1
Tonellato, U., **9**, 185
Tüdös, F., **9**, 127
Turner, D. W., **4**, 31
Ugi, I., **9**, 25
Ward, B., **8**, 1
Whalley, E., **2**, 93
Williams, J. M., Jr., **6**, 63
Williamson, D. G., **1**, 365
Wolf, A. P., **2**, 201
Wyatt, P. A. H., **12**, 131
Zollinger, H., **2**, 163
Zuman, P., **5**, 1

Cumulative Index of Titles

Abstraction, hydrogen atom, from O—H bonds, 9, 127
Acid solutions, strong, spectroscopic observation of alkylcarbonium ions in, 4, 305
Acid-base properties of electronically excited states of organic molecules, 12, 131
Acids, reactions of aliphatic diazo compounds with, 5, 331
Acids, strong aqueous, protonation and solvation in, 13, 83
Activation, entropies of, and mechanisms of reactions in solution, 1, 1
Activation, heat capacities of, and their uses in mechanistic studies, 5, 121
Activation, volumes of, use for determining reaction mechanisms, 2, 93
Aliphatic diazo compounds, reactions with acids, 5, 331
Alkylcarbonium ions, spectroscopic observation in strong acid solutions, 4, 305
Ambident conjugated systems, alternative protonation sites in, 11, 267
Ammonia, liquid, isotope exchange reactions of organic ions in, 1, 156
Aromatic photosubstitution, nucleophilic, 11, 225
Aromatic substitution, a quantitative treatment of directive effects in, 1, 35
Aromatic substitution reactions, hydrogen isotope effects in, 2, 163
Aromatic systems, planar and non-planar, 1, 203
Arynes, mechanisms of formation and reactions at high temperatures, 6, 1
A-S_E2 reactions, developments in the study of, 6, 63

Base catalysis, general, of ester hydrolysis and related reactions, 5, 237
Basicity of unsaturated compounds, 4, 195
Bimolecular substitution reactions in protic and dipolar aprotic solvents, 5, 173

^{13}C N.M.R. spectroscopy in macromolecular systems of biochemical interest, 13, 279
Carbene chemistry, structure and mechanism in, 7, 163
Carbon atoms, energetic, reactions with organic compounds, 3, 201
Carbon monoxide, reactivity of carbonium ions towards, 10, 29
Carbonium ions (alkyl), spectroscopic observation in strong acid solutions, 4, 305
Carbonium ions, gaseous, from the decay of tritiated molecules, 8, 79
Carbonium ions, photochemistry of 10, 129.
Carbonium ions, reactivity towards carbon monoxide, 10, 29
Carbonyl compounds, reversible hydration of, 4, 1
Catalysis, enzymatic, physical organic model systems and the problem of, 11, 1
Catalysis, general base and nucleophilic, of ester hydrolysis and related reactions, 5, 237
Catalysis, micellar, in organic reactions; kinetic and mechanistic implications, 8, 271
Cation radicals in solution, formation, properties and reactions of, 13, 155
Cations, vinyl, 9, 185
Charge density—N.M.R. chemical shift correlations in organic ions, 11, 125
Chemically induced dynamic nuclear spin polarization and its applications, 10, 53
CIDNP and its applications, 10, 53

Conformations of polypeptides, calculations of, 6, 103
Conjugated molecules, reactivity indices in, 4, 73

D_2O-H_2O Mixtures, protolytic processes in, 7, 259
Diazo compounds, aliphatic, reactions with acids, 5, 331
Dipolar aprotic and protic solvents, rates of bimolecular substitution reactions in, 5, 173
Directive effects in aromatic substitution, a quantitative treatment of, 1, 35

Electrochemistry, organic, structure and mechanism in, 12, 1
Electrode processes, physical parameters for the control of, 10, 155
Electron spin resonance, identification of organic free radicals by, 1, 284
Electron spin resonance studies of short-lived organic radicals, 5, 23
Electronically excited molecules, structure of, 1, 365
Electronically excited states of organic molecules, acid-base properties of, 12, 131
Energetic tritium and carbon atoms, reactions of, with organic compounds, 2, 201
Entropies of activation and mechanisms of reactions in solution, 1, 1
Enzymatic catalysis, physical organic model systems and the problem of, 11, 1
Equilibrium constants, N.M.R. measurements of, as a function of temperature, 3, 187
Ester hydrolysis, general base and nucleophilic catalysis, 5, 237
Exchange reactions, hydrogen isotope, of organic compounds in liquid ammonia, 1, 156
Exchange reactions, oxygen isotope, of organic compounds, 3, 123
Excited molecules, structure of electronically, 1, 365

Force-field methods, calculation of molecular structure and energy by, 13, 1
Free radicals, identification by electron spin resonance, 1, 284
Free radicals and their reactions at low temperature using a rotating cryostat, study of, 8, 1

Gaseous carbonium ions from the decay of tritiated molecules, 8, 79
Gas-phase heterolysis, 3, 91
Gas-phase pyrolysis of small-ring hydrocarbons, 4, 147
General base and nucleophilic catalysis of ester hydrolysis and related reactions, 5, 237

H_2O-D_2O Mixtures, protolytic processes in, 7, 259
Heat capacities of activation and their uses in mechanistic studies, 5, 121
Heterolysis, gas-phase, 3, 91
Hydrated electrons, reactions of, with organic compounds, 7, 115
Hydration, reversible, of carbonyl compounds, 4, 1
Hydrocarbons, small-ring, gas-phase pyrolysis of, 4, 147
Hydrogen atom abstraction from O—H bonds, 9, 127
Hydrogen isotope effects in aromatic substitution reactions, 2, 163
Hydrogen isotope exchange reactions of organic compounds in liquid ammonia, 1, 156
Hydrolysis, ester, and related reactions, general base and nucleophilic catalysis of, 5, 237

Ionization potentials, 4, 31
Ions, organic, charge density—N.M.R. chemical shift correlations, 11, 125
Isomerization, permutational, of pentavalent phosphorus compounds, 9, 25
Isotope effects, hydrogen, in aromatic substitution reactions, 2, 163
Isotope effects, steric, experiments on the nature of, 10, 1
Isotope exchange reactions, hydrogen, of organic compounds in liquid ammonia, 1, 150
Isotope exchange reactions, oxygen, of organic compounds, 3, 123
Isotopes and organic reaction mechanisms, 2, 1

Kinetics, reaction, polarography and, 5, 1

Macromolecular systems of biochemical interest, ^{13}C N.M.R. spectroscopy in, 13, 279
Mass spectrometry, mechanism and structure in: a comparison with other chemical processes, 8, 152
Mechanism and structure in carbene chemistry, 7, 153

Mechanism and structure in mass spectrometry: a comparison with other chemical processes, 8, 152
Mechanism and structure in organic electrochemistry, 12, 1
Mechanisms, organic reaction, isotopes and, 2, 1
Mechanisms, reaction, use of volumes of activation for determining, 2, 93
Mechanisms of formation and reactions of arynes at high temperatures, 6, 1
Mechanisms of reactions in solution, entropies of activation and, 1, 1
Mechanistic studies, heat capacities of activation and their use in, 5, 121
Meisenheimer complexes, 7, 211
Micellar catalysis in organic reactions: kinetic and mechanistic implications, 8, 271
Molecular structure and energy, calculation of, by force-field methods, 13, 1

N.M.R. chemical shift—charge density correlations, 11, 125
N.M.R. measurements of reaction velocities and equilibrium constants as a function of temperature, 3, 187
N.M.R. spectroscopy, ^{13}C, in macromolecular systems of biochemical interest, 13, 279
Non-planar and planar aromatic systems, 1, 203
Norbornyl cation: reappraisal of structure, 11, 179
Nuclear magnetic resonance, see N.M.R.
Nucleophilic aromatic photosubstitution, 11, 225
Nucleophilic catalysis of hydrolysis and related reactions, 4, 237
Nucleophilic vinylic substitution, 7, 1

O—H bonds, hydrogen atom abstraction from, 9, 127
Oxygen isotope exchange reactions of organic compounds, 3, 123

Permutational isomerization of pentavalent phosphorus compounds, 9, 25
Phosphorus compounds, pentavalent, turnstile rearrangement and pseudorotation in permutational isomerization, 9, 25
Photochemistry of carbonium ions, 10, 129
Photosubstitution, nucleophilic aromatic, 11, 225
Planar and non-planar aromatic systems, 1, 203
Polarizability, molecular refractivity and, 3, 1
Polarography and reaction kinetics, 5, 1
Polypeptides, calculations of conformations of, 6, 103
Protic and dipolar aprotic solvents, rates of bimolecular substitution reactions in, 5, 173
Protolytic processes in $H_2O–D_2O$ mixtures, 7, 259
Protonation and solvation in strong aqueous acids, 13, 83
Protonation sites in ambident conjugated systems, 11, 267
Pseudorotation in isomerization of pentavalent phosphorus compounds, 9, 25
Pyrolysis, gas-phase, of small-ring hydrocarbons, 4, 147

Radiation techniques, application to the study of organic radicals, 12, 223
Radicals, cation, in solution, formation, properties and reactions of, 13, 155
Radicals, organic, application of radiation techniques, 12, 223
Radicals, organic free, identification by electron spin resonance, 1, 284
Radicals, short-lived organic, electron spin resonance studies of, 5, 53
Reaction kinetics, polarography and, 5, 1
Reactions mechanisms, use of volumes of activation for determining, 2, 93
Reaction mechanisms in solution, entropies of activation and, 1, 1
Reaction velocities and equilibrium constants, N.M.R. measurements of, as a function of temperature, 3, 187
Reactions of hydrated electrons with organic compounds, 7, 115
Reactivity indices in conjugated molecules, 4, 73
Refractivity, molecular, and polarizability, 3, 1
Resonance, electron-spin, identification of organic free radicals, by, 1, 284
Resonance, electron-spin, studies of short-lived organic radicals, 5, 63

Short-lived organic radicals, electron-spin resonance studies of, 5, 53

Small-ring hydrocarbons, gas-phase pyrolysis of, 4, 147
Solution, reactions in, entropies of activation and mechanisms, 1, 1
Solvation, and protonation, in strong aqueous acids, 13, 83
Solvents, protic and dipolar aprotic, rates of bimolecular substitution reactions in, 5, 173
Spectroscopic observation of alkylcarbonium ions in strong acid solutions, 4, 305
Spectroscopy, ^{13}C N.M.R., in macromolecular systems of biochemical interest, 13, 279
Stereoselection in elementary steps of organic reactions, 6, 185
Steric isotope effects, experiments on the nature of, 10, 1
Structure and mechanism in carbene chemistry, 7, 153
Structure and mechanism in organic electrochemistry, 12, 1
Structure of electronically excited molecules, 1, 365
Substitution, aromatic, a quantitative treatment of directive effects in, 1, 35
Substitution reactions, aromatic, hydrogen isotope effects in, 2, 163
Substitution reactions, bimolecular, in protic and dipolar aprotic solvents, 5, 173
Superacid systems, 9, 1

Temperature, N.M.R. measurements of reaction velocities and equilibrium constants as a function of, 3, 187
Tritiated molecules, gaseous carbonium ions from the decay of, 8, 79
Tritium atoms, energetic, reactions with organic compounds, 2, 201
Turnstile rearrangement in isomerization of pentavalent phosphorus compounds, 9, 25

Unsaturated compounds, basicity of, 4, 195

Vinyl cations, 9, 185
Volumes of activation, use of, for determining reaction mechanisms, 2, 93